Lecture Notes in Computer Science　13366

More information about this series at https://link.springer.com/bookseries/558

François Boulier · Matthew England ·
Timur M. Sadykov · Evgenii V. Vorozhtsov (Eds.)

Computer Algebra in Scientific Computing

24th International Workshop, CASC 2022
Gebze, Turkey, August 22–26, 2022
Proceedings

 Springer

Editors
François Boulier ⓘ
Université de Lille
Villeneuve d'Ascq, France

Matthew England ⓘ
Coventry University
Coventry, UK

Timur M. Sadykov ⓘ
Plekhanov Russian University of Economics
Moscow, Russia

Evgenii V. Vorozhtsov ⓘ
Institute of Theoretical and Applied
Mechanics
Novosibirsk, Russia

ISSN 0302-9743   ISSN 1611-3349 (electronic)
Lecture Notes in Computer Science
ISBN 978-3-031-14787-6   ISBN 978-3-031-14788-3 (eBook)
https://doi.org/10.1007/978-3-031-14788-3

This Springer imprint is published by the registered company Springer Nature Switzerland AG
The registered company address is: Gewerbestrasse 11, 6330 Cham, Switzerland

Preface

The International Workshop on Computer Algebra in Scientific Computing (CASC) is an annual forum which aims to bring together the leading scientists, scholars, and engineers from the various disciplines including computer algebra and to attract original research papers of high quality. This workshop provides a platform for the delegates to exchange new ideas and application experiences, share research results, and discuss existing issues and challenges.

This year, CASC was hosted near Istanbul, the largest city in Turkey and the financial and cultural center of the country. Gebze Technical University (GTU) is situated close to the border of the provinces of Istanbul and Kocaeli. GTU (previously the Gebze Institute of Advanced Technologies – GYTE) is one of the leading research institutes and one of 10 research universities in Turkey.

During the two years of the COVID-19 pandemic, many conferences were either canceled or took place fully online. This impacted our community and especially the early career researchers. A decision was made to gather many symbolic computation/computer algebra events together in the same place, driven by the need to restore ties in the community. SCALE (Symbolic Computation: Algorithms, Learning, and Engineering), a three-week long event, was the result of this effort. CASC 2022 was the concluding event of SCALE.

The choice of Istanbul/Gebze for CASC 2022 was made because Turkey has an increasing number of symbolic computation researchers. Some years ago, symbolic computation and computer algebra were almost nonexistent in Turkey, but today many researchers who obtained their Ph.D. or spent time as post-doctoral researchers abroad are based in Turkey, together with a number of foreign researchers.

The ALCYON lab (Institute of Information Technologies, GTU) was started by Zafeirakis Zafeirakopoulos six years ago. Its main focus is symbolic computation. During this period, three nationally and internationally funded research projects were carried out, more than 10 researchers were employed, and more than 30 students supervised. In the last three years, a further three research projects related to computer algebra were granted to researchers based around Istanbul and a joint effort between computer algebra researchers from different universities was initiated. This led to the Symbolic Computation Istanbul Meetings (organized by Tülay Ayyıldız Akoğlu, Türkü Özlüm Çelik, and Zafeirakis Zafeirakopoulos): a biweekly seminar and social event series that has successfully ran for a year with local and international speakers and participants. In addition, the International Mathematics Union approved a grant for computation in Turkey. In this context there are monthly workshops (organized by Tülay Ayyıldız Akoğlu, Can Ozan Oğuz, and Zafeirakis Zafeirakopoulos) related to computational mathematics (focused on computer algebra). The first event was the Sagedays Workshop, held as part of SCALE (two weeks before CASC 2022).

This year, the CASC International Workshop had two categories of participation: (1) talks with accompanying papers to appear in these proceedings and (2) talks with accompanying extended abstracts for distribution at the conference only. The latter was for work either already published or not yet ready for publication, but in either case still new and of interest to the CASC audience. The former was strictly for new and original research results or review articles, ready for publication.

All papers submitted to the LNCS proceedings received a minimum of three reviews. In addition, the whole Program Committee (PC) was invited to comment on and debate all papers. In total this volume contains 20 contributed papers. Along with the contributed talks, CASC 2022 had two invited speakers.

The invited talk of Michael Nikitas Vrahatis was devoted to the generalizations of the intermediate value theorem in several variables. This theorem is very useful in various approaches including the existence of solutions of systems of nonlinear algebraic and/or transcendental equations, the existence of fixed points of continuous functions, the localization of extrema of objective functions, and the localization of periodic orbits of nonlinear mappings and periodic orbits (fixed points) on Poincaré's surface of section (the Poincaré map).

Methods emanating from the theorem are of major importance for studying and tackling problems with imprecise (not exactly known) information because, in a large variety of applications, precise function values are either impossible or time-consuming and computationally expensive to obtain. Furthermore, these methods are particularly useful for investigating various problems where the corresponding functions assume very large and/or very small values.

Applications related to systems of nonlinear algebraic and/or transcendental equations, as well as fixed points of continuous functions, were presented. Furthermore, an application was presented which concerns the computation of all the periodic orbits (stable and unstable) of any period and accuracy which occur, among others, in the study of beam dynamics in circular particle accelerators, such as the Large Hadron Collider machine at the European Organization for Nuclear Research.

The invited talk of Marc Moreno Maza was on the topic of implementation techniques for power, Laurent, and Puiseux series in several variables. This theme was motivated by the fact that limits of multivariate functions and more advanced notions of limits, like topological closures, are almost absent from such general-purpose computer algebra systems as Maple and Mathematica.

The discussion of the application of above-mentioned implementation techniques started with the implementation of arithmetic operations which are sometimes easier than one may think (for instance, the substitution of unit formal power series into formal power series) and sometimes harder (for instance, the inversion of multivariate Laurent series). The impact of the implementation environment on the implementation techniques, considering both interpreted and compiled code, was also discussed. The main points of the talk were illustrated with Maple's MultivariatePowerSeries package and the Basic Polynomial Algebra Subroutines Library. In that latter environment, it was shown how different parallel programming patterns can be used to obtain efficient multi-threaded implementation of arithmetic operations on power series and factorization of univariate polynomials over such series.

The CASC 2022 program covered a wide range of topics. Polynomial algebra, which is at the computer algebra core, was represented by contributions devoted to the development of a new accelerated subdivision algorithm for finding the complex roots of univariate polynomials, the use of Gröbner bases and invariant manifolds for finding the equilibrium positions in the problem of the motion of a system of two bodies in a uniform gravity field, the application of the quaternion Fourier transform for locating the nearest singularity in a polynomial homotopy, a new interpolation algorithm for the solution of polynomial equations with parameters by using the Dixon resultant, and the computations in the computer algebra system (CAS) Maple and in the C programming language. Polynomial computer algebra is also the foundation of the contributions to the present proceedings that expose a new speculative algorithm for computing sub-resultant chains over rings of multivariate polynomials with the aid of the Bézout matrix, new chordality-preserving top-down algorithms for triangular decomposition of polynomial sets, new heuristics for choosing a cylindrical algebraic decomposition variable ordering motivated by complexity analysis, a comparison of several algorithms implemented in the CASs SageMath and Mathematica for proving positivity of linearly recurrent sequences with polynomial coefficients, and the derivation with the aid of Gröbner bases and Sylvester resultants of new optimal symplectic fourth-order partitioned Runge–Kutta methods for the numerical solution of Hamiltonian mechanics problems.

Two papers deal with the application of symbolic manipulations for obtaining the solutions of both ordinary and partial differential equations. These are the contributions devoted to finding the solutions to linear ordinary differential equations under the condition of incomplete information about the coefficients and the use of the parametrization of boundary conditions at the solution of boundary-value problems for partial differential equations to ensure well-posedness.

Two papers are devoted to the applications of symbolic-numerical algorithms developed in the language of the CAS Mathematica for the calculation of energy spectrum and eigenfunctions in the geometric collective model of atomic nucleus and the analysis of gyroscopic stabilization of equilibriums of a gyrostat, respectively. A further application of CASs in mechanics is included, specifically for the stability analysis of periodic motion of the swinging Atwood machine.

The remaining topics include a new algorithm for finding the Frobenius distance from a given matrix to the set of matrices possessing multiple eigenvalues, the parallel implementation of the fast Fourier transform in a ring and a finite field that can be used for homomorphic encryption and polynomial multiplication, a new symbolic computation method for constructing a small neighborhood around a known local optimal point of a given multivariate function that contains radical or rational expressions, new algorithms implemented in Maple and in the C/C++ programming language for computing the integer hull of a convex polyhedral set, a new algorithm implemented in the CAS SageMath for computing the equivalent Hilbert series of automorphisms acting on canonical rings of projective curves with the application to Fermat curves, and a survey of the state of the art and of the future of computer science for continuous data as a bridge between pure and applied mathematics and as an expansion of computer algebra to analytic data types.

We want to thank all the members of the CASC 2022 Program Committee for their thorough work in selecting and preparing the technical program. We also thank the external referees who provided reviews as part of this process.

We are grateful to the members of the group headed by Timur Sadykov for their technical help in the preparation of the camera-ready manuscript for this volume. We are grateful to the CASC publicity chair Dmitry Lyakhov for the management of the conference web page (http://www.casc-conference.org) and for the design of the conference poster.

The local organization of the CASC 2022 at Gebze Technical University was conducted as part of the larger SCALE event. Our particular thanks are due to the members of the CASC 2022 local organizing committee at the GTU, in particular Zafeirakis Zafeirakopoulos (chair), Tülay Ayyıldız Akoğlu, Hadi Alizadeh, Hülya Öztürk, and Ali Kemal Uncu, who ably handled the local arrangements. In addition, Zafeirakis Zafeirakopoulos kindly provided us with the information above about computer algebra activities at GTU.

SCALE would not have been possible without the enormous help of a large local team. This includes members of the Institute of Information Technologies and the Department of Mathematics of Gebze Technical University, members of other universities in Istanbul, and the (student) Mathematics Club of Turkey (TMK). Among them, Hadi Alizadeh and Başak Karakaş especially worked tirelessly to deal with the many problems a post-pandemic conference had to face. In total there were more than 20 people volunteering work to ensure the success of the event.

Finally, we acknowledge that the success of CASC and SCALE has been financially supported by the sponsors: the Scientific and Technological Research Institution of Turkey (TÜBITAK), the Turkish Mathematics Society (TMD), the European Mathematics Society (EMS), the International Mathematics Union (IMU), the ACM Special Interest Group on Symbolic and Algebraic Manipulation (SIGSAM), the MATRIS lab of SBA Research (Austria), the CARGO lab of Wilfrid Laurier University (Canada), and Maplesoft.

July 2022

François Boulier
Matthew England
Timur M. Sadykov
Evgenii V. Vorozhtsov

Organization

CASC 2022 was hosted by Gebze Technical University, Gebze, Turkey.

General Chairs

François Boulier   Université de Lille, France
Timur M. Sadykov   Plekhanov Russian University of Economics,
Russia

Program Committee Chairs

Matthew England   Coventry University, UK
Evgenii V. Vorozhtsov   Khristianovich Institute of Theoretical and
Applied Mechanics, Russia

Program Committee

François Boulier	University of Lille, France
Changbo Chen	Chinese Academy of Sciences, China
Jin-San Cheng	Academy of Mathematics and Systems Science, China
Türkü Özlüm Çelik	Boğaziçi University, Turkey
Victor F. Edneral	Lomonosov Moscow State University, Russia
Jaime Gutierrez	University of Cantabria, Spain
Sergey Gutnik	Moscow State Institute of International Relations, Russia
Amir Hashemi	Isfahan University of Technology, Iran
Gabriela Jeronimo	Universidad de Buenos Aires, Argentina
Rui-Juan Jing	Jinagsu University, China
Wen-Shin Lee	University of Stirling, UK
François Lemaire	University of Lille, France
Viktor Levandovskyy	University of Kassel, Germany
Marc Moreno Maza	University of Western Ontario, Canada
Dominik L. Michels	KAUST, Saudi Arabia
Chenqi Mou	Beihang University, China
Sonia Perez-Diaz	Universidad de Alcalá, Spain
Veronika Pillwein	JKU Linz, Austria
Alexander Prokopenya	Warsaw University of Life Sciences, Poland

Hamid Rahkooy	Max Planck Institute for Informatics, Germany
Timur M. Sadykov	Plekhanov Russian University of Economics, Russia
Svetlana Selivanova	KAIST, South Korea
Ekaterina Shemyakova	University of Toledo, USA
Thomas Sturm	CNRS, France
Akira Terui	University of Tsukuba, Japan
Elias Tsigaridas	Inria, France
Ali Kemal Uncu	University of Bath, UK, and RICAM, Austrian Academy of Sciences, Austria
Jan Verschelde	University of Illinois, USA
Evgenii V. Vorozhtsov	Khristianovich Institute of Theoretical and Applied Mechanics, Russia
Zafeirakis Zafeirakopoulos	Gebze Technical University, Turkey

Local Organization

Zafeirakis Zafeirakopoulos (Chair)	Gebze Technical University, Turkey
Tülay Ayyıldız Akoğlu	Istanbul Technical University, Turkey
Hadi Alizadeh	Gebze Technical University, Turkey
Hülya Öztürk	Gebze Technical University, Turkey
Ali Kemal Uncu	Gebze Technical University, Turkey

Publicity Chair

| Dmitry Lyakhov | KAUST, Saudi Arabia |

Advisory Board

Wolfram Koepf	Universität Kassel, Germany
Ernst W. Mayr	Technische Universität München, Germany
Werner M. Seiler	Universität Kassel, Germany

Implementation Techniques for Power, Laurent, and Puiseux Series in Several Variables (Abstract of Invited Talk)

Marc Moreno Maza

University of Western Ontario, London, ON, CA
mmorenom@uwo.ca

While computer algebra systems can perform highly sophisticated algebraic tasks, they are much less equipped for solving problems from mathematical analysis in a symbolic manner. Elementary problems in analysis, such as the manipulation of Taylor series and the calculation of limits of univariate functions, are supported, with some limitations, in general-purpose computer algebra systems such as Maple and Mathematica. However, limits of multivariate functions and more advanced notions of limits, like topological closures, are almost absent from such systems. For instance, Maple is not always capable of computing finite limits of a multivariate rational function at a zero of the denominator that is not an isolated pole. Many fundamental concepts in mathematics are defined in terms of limits and it is highly desirable for computer algebra systems to implement these concepts. However, limits are, by their essence, hard to compute, in the sense of performing finitely many rational operations on polynomials or matrices. A first helper tool is the famous Weierstrass preparation theorem (and its extensions) which essentially reduces the local study of analytic functions (and more general functions) to the local study of polynomials via the manipulation of power series. A second helper tool is the famous Newton–Puiseux algorithm (and its extensions) which essentially allows for the local study of curves (separating their branches about a point) via the manipulation of Laurent and Puiseux series.

The Newton–Puiseux algorithm, and more generally the factorization of univariate polynomials over power, Laurent, and Puiseux series, has been a very active research area in the computer algebra community in the past 50 years. This research effort, however, has been mainly focusing on the development of algorithms and the analysis of their algebraic complexity. Relatively little has been done in terms of implementation, except for the case of univariate series.

In this talk, we will discuss recent findings on (1) the implementation of power, Laurent, and Puiseux series in several variables and, (2) its application to the factorization of univariate polynomials over such series. We will start with the implementation of arithmetic operations which can be sometimes easier than one may think (for instance, the substitution of unit formal power series into formal power series) and sometimes harder (for instance, the inversion of multivariate Laurent series). We will also discuss the impact of the implementation environment on the implementation techniques, considering both interpreted and compiled code. We will illustrate our points with Maple's MultivariatePowerSeries package and the Basic Polynomial Algebra Subroutines. In that latter environment, we will show how different parallel programming patterns can be used to obtain efficient multithreaded implementation of arithmetic operations on power series and factorization of univariate polynomials over such series.

Contents

Survey on Generalizations
of the Intermediate Value Theorem
and Applications

Michael N. Vrahatis$^{(\boxtimes)}$ ⓘ

Department of Mathematics, University of Patras, 26110 Patras, Greece
`vrahatis@math.upatras.gr`

Abstract. Generalizations of the intermediate value theorem in several variables are presented. These theorems are very useful in various approaches including the existence of solutions of systems of nonlinear equations, the existence of fixed points of continuous functions as well as the existence of periodic orbits of nonlinear mappings and similarly, fixed points of the Poincaré map on a surface of section. Based on the corresponding criteria for the existence of a solution or a fixed point emanated by the intermediate value theorems, generalized bisection methods for approximating zeros or fixed points of continuous functions are given. These bisection methods require only the algebraic signs of the function values and are of major importance for studying and tackling problems with imprecise information.

Keywords: Generalizations of the intermediate value theorem · Existence theorems · Zeros · Fixed points · Systems of nonlinear algebraic and/or transcendental equations · Periodic orbits · Poincaré map

1 Introduction

Assume that $F_n = (f_1, f_2, \ldots, f_n) \colon \mathcal{D} \subset \mathbb{R}^n \to \mathbb{R}^n$ is a nonlinear mapping and $\theta^n = (0, 0, \ldots, 0)$ is the origin of $\mathbb{R}^n$. The problem of solving the equation:

$$F_n(x) = \theta^n, \tag{1}$$

is to find a *zero* $x^* = (x_1^*, x_2^*, \ldots, x_n^*) \in \mathcal{D}$ for which $F_n(x^*) = \theta^n$. The problem (1) may be represented as follows:

$$
\begin{aligned}
f_1(x_1, x_2, \ldots, x_n) &= 0, \\
f_2(x_1, x_2, \ldots, x_n) &= 0, \\
&\vdots \\
f_n(x_1, x_2, \ldots, x_n) &= 0.
\end{aligned}
\tag{2}
$$

© The Author(s), under exclusive license to Springer Nature Switzerland AG 2022
F. Boulier et al. (Eds.): CASC 2022, LNCS 13366, pp. 1–17, 2022.
https://doi.org/10.1007/978-3-031-14788-3_1

The problem of computing the *extrema of an objective function* $f \colon \mathcal{D} \subset \mathbb{R}^n \to \mathbb{R}$ can be studied and tackled by solving the following equation:

$$\nabla f(x) = \theta^n, \tag{3}$$

where $\nabla f(x) = \left(\frac{\partial f(x)}{\partial x_1}, \frac{\partial f(x)}{\partial x_2}, \ldots, \frac{\partial f(x)}{\partial x_n} \right)$, denotes the gradient of f at $x \in \mathcal{D}$.

Furthermore, the problem of finding a *fixed point* of F_n in $\mathcal{D} \subset \mathbb{R}^n$ is to find a point $x^\star \in \mathcal{D}$ which satisfies the equation:

$$F_n(x^\star) = x^\star. \tag{4}$$

Obviously, the problem of finding a fixed point is equivalent to the problem of solving Eq. (1) by using the mapping $G_n = I_n - F_n$ (where I_n indicates the identity mapping) instead of F_n and solving the equation:

$$G_n(x) = \theta^n. \tag{5}$$

The problem of computing *periodic orbits of nonlinear mappings* or *fixed points of the Poincaré map on a surface of section* can be studied and tackled by using fixed points [31]. More specifically the problem of finding periodic orbits of nonlinear mappings: $\Phi_n = (\varphi_1, \varphi_2, \ldots, \varphi_n) \colon \mathcal{D} \subset \mathbb{R}^n \to \mathbb{R}^n$, of *period p* amounts to finding *fixed points* $x^\star = (x_1^\star, x_2^\star, \ldots, x_n^\star) \in \mathcal{D}$ of *period p* which satisfy the following equation:

$$\Phi_n^p(x^\star) = \underbrace{\Phi_n \left(\Phi_n \left(\cdots \Phi_n \left(\Phi_n(x^\star) \right) \cdots \right) \right)}_{p \text{ times}} = x^\star. \tag{6}$$

The problem of finding periodic orbits of period p of *dynamical systems* in $\mathbb{R}^{n+1}$ amounts to fixing one of the variables, say $x_{n+1} = \text{const}$, and locating points $x^\star = (x_1^\star, x_2^\star, \ldots, x_n^\star)$ on an n-dimensional *surface of section* Σ_{t_0} which satisfy Eq. (6). where $\Phi_n^p = P_{t_0} \colon \Sigma_{t_0} \to \Sigma_{t_0}$ is the *Poincaré map* of the system. For example, let us consider a conservative dynamical system of the form:

$$, \qquad \dot{\mathbf{x}} = \mathbf{f}(\mathbf{x}, t), \tag{7}$$

with $\mathbf{x} = (x, \dot{x}) \in \mathbb{R}^2$ and $\mathbf{f} = (f_1, f_2)$ periodic in t with frequency ω. We obtain periodic orbits of period p of System (7) by taking as initial conditions of these orbits the points which the orbits intersect the surface of section:

$$\Sigma_{t_0} = \left\{ (x(t_k), \dot{x}(t_k)), \quad \text{with} \quad t_k = t_0 + k\frac{2\pi}{\omega}, \quad k \in \mathbb{N} \right\}, \tag{8}$$

at a finite number of points p. Thus the dynamics is studied in connection with a Poincaré map $\Phi_n^p = P_{t_0} \colon \Sigma_{t_0} \to \Sigma_{t_0}$, constructed by following the solutions of (7) in continuous time.

In the paper at hand, *generalizations of the intermediate value theorem* in several variables are presented. These theorems are very useful in various approaches including, among others, those mentioned previously. Specifically, using these theorems we can study and analyze (a) the existence of solutions of systems of nonlinear algebraic and/or transcendental equations, (b) the localization of extrema of objective functions, (c) the existence of fixed points of continuous functions, as well as (d) the existence of periodic orbits of nonlinear mappings and similarly, fixed points of the Poincaré map on a surface of section. We notice that, these theorems are of major importance for tackling problems with imprecise (not exactly known) information.

Based on the corresponding existence criteria emanated by the above theorems, methods, named *generalized bisection methods*, are given. The only computable information required by the generalized bisection methods is the algebraic sign of the function value which is the minimum possible information (one bit of information) necessary for the purpose needed, and not any additional information. Thus, these methods are of major importance for studying and tackling problems with imprecise (not exactly known) information. These problems appear in various fields of science and technology, because, in a large variety of applications, precise function values are either impossible or time consuming and computationally expensive to obtain. In other cases, it may be necessary to integrate numerically a system of differential equations in order to obtain a function value, so that the precision of the computed value is limited. Furthermore, these methods are particularly useful for studying and tackling problems where the corresponding functions obtain very large and/or very small values.

It is worthy to mention that regarding the case of *algebraic equations*, it is well known that these equations are very important in studying and solving problems on geometric, kinematic, and other constraints in various fields of science and technology including, among others, robotics, vision, modeling and graphics, molecular biology, signal processing, and computational economics. In addition, regarding the algebraic signs of algebraic expressions there are various efficient approaches in obtaining this information, see [4,8,9] and the references thereof.

Applications of the presented generalizations of the intermediate value theorem for obtaining methods related to systems of nonlinear algebraic and/or transcendental equations, as well as fixed points of continuous functions are presented. Furthermore, an application is presented which concerns the computation of all the periodic orbits (stable and unstable) of any period and accuracy which occur, among others, in the study of beam dynamics in circular particle accelerators like the Large Hadron Collider (LHC) machine at the European Organization for Nuclear Research (CERN).

2 Generalizations of the Intermediate Value Theorem

2.1 Definitions and Notations

Let us give some necessary definitions and notations.

Notation 1. We denote by ϑA the boundary of a set A, by cl$A$ its closure, by int$A$ its interior, by card$\{A\}$ its cardinality (i.e., the number of elements in the set A) and by co$A$ its convex hull (i.e., the set of all finite convex combinations of elements of A).

Notation 2. We shall use the index sets $N^n = \{0, 1, \ldots, n\}$, $N^n_{\neg 0} = \{1, 2, \ldots, n\}$ and $N^n_{\neg i} = \{0, 1, \ldots, i-1, i+1, \ldots, n\}$. Also, for a given set $I = \{i, j, \ldots, \ell\} \subset N^n$ we denote by $N^n_{\neg I}$ or equivalently by $N^n_{\neg ij \ldots \ell}$ the set $\{k \in N^n \mid k \notin I\}$.

Definition 1. For any positive integer n, and for any set of points $V = \{v^0, v^1, \ldots, v^n\}$ in some linear space which are affinely independent (i.e., the vectors $\{v^1 - v^0, v^2 - v^0, \ldots, v^n - v^0\}$ are linearly independent) the convex hull $\mathrm{co}\{v^0, v^1, \ldots, v^n\} = [v^0, v^1, \ldots, v^n]$ is called the n-*simplex with vertices* $v^0, v^1, \ldots, v^n$. For each subset of $(m+1)$ elements $\{\omega^0, \omega^1, \ldots, \omega^m\} \subset \{v^0, v^1, \ldots, v^n\}$, the m-simplex $[\omega^0, \omega^1, \ldots, \omega^m]$ is called an m-*face* of $[v^0, v^1, \ldots, v^n]$. In particular, 0-faces are vertices and 1-faces are edges. The m-faces are also called *facets* of the n-simplex. An m-face of the n-simplex is called the *carrier* of a point p if p lies on this m-face and not on any sub-face of this m-face.

Notation 3. We denote the n-simplex with set of vertices $V = \{v^0, v^1, \ldots, v^n\}$ by $\sigma^n = [v^0, v^1, \ldots, v^n]$. Also, we denote the $(n-1)$-simplex that determines the i-th $(n-1)$-face of σ^n by $\sigma^n_{\neg i} = [v^0, v^1, \ldots, v^{i-1}, v^{i+1}, \ldots, v^n]$. Furthermore, for a given index set $I = \{i, j, \ldots, \ell\} \subset N^n$ with cardinality $\mathrm{card}\{I\} = \kappa$, we denote by $\sigma^n_{\neg I}$ or equivalently by $\sigma^n_{\neg ij \ldots \ell}$ the $(n-\kappa)$-face of σ^n with vertices $v^m, m \in N^n_{\neg I}$.

Definition 2 [26,29]. The *diameter* of an m-simplex σ^m in $\mathbb{R}^n$, $m \leqslant n$, denoted by $\mathrm{diam}(\sigma^m)$, is defined to be the length of the longest edge (1-face) of σ^m while the *microdiameter*, $\mu\mathrm{diam}(\sigma^m)$, of σ^m is defined to be the length of the shortest edge of σ^m.

Definition 3. Let $\sigma^m = [v^0, v^1, \ldots, v^m]$ be an m-simplex in $\mathbb{R}^n$, $m \leqslant n$. Then the *barycenter* of σ^m denoted by K is the point $K = (m+1)^{-1} \sum_{i=0}^{m} v^i$ in $\mathbb{R}^n$.

Remark 1. By convexity it is obvious that the barycenter of any m-simplex σ^m in $\mathbb{R}^n$ is a point in the relative interior of σ^m.

Definition 4. An n-simplex is *oriented* if an order has been assigned to its vertices. If $\langle v^0, v^1, \ldots, v^n \rangle$ is an orientation of $\{v^0, v^1, \ldots, v^n\}$ this is regarded as being the same as any orientation obtained from it by an even permutation of the vertices and as the opposite of any orientation obtained by an odd permutation of the vertices. We shall denote oriented n-simplices by $\sigma^n = \langle v^0, v^1, \ldots, v^n \rangle$, and we shall write, for example, $\langle v^0, v^1, v^2, \ldots, v^n \rangle = -\langle v^1, v^0, v^2, \ldots, v^n \rangle = \langle v^2, v^0, v^1, \ldots, v^n \rangle$. The *boundary* $\vartheta\sigma^n$ of an oriented n-simplex $\sigma^n = \langle v^0, v^1, \ldots, v^n \rangle$ is given by $\vartheta\sigma^n = \sum_{i=0}^{n} (-1)^i \langle v^0, v^1, \ldots, v^{i-1}, v^{i+1}, \ldots, v^n \rangle$. The oriented $(n-1)$-simplex $\langle v^0, v^1, \ldots, v^{i-1}, v^{i+1}, \ldots, v^n \rangle$ will be called the ith *face* of σ^n.

Definition 5. An n-dimensional *polyhedron* Π^n is a union of a finite number of oriented n-simplices σ^n_i, $i = 1, 2, \ldots, k$ such that the σ^n_i have pairwise-disjoint interiors. We write $\Pi^n = \sum_{i=1}^{k} \sigma^n_i$ and $\vartheta\Pi^n = \sum_{i=1}^{k} \vartheta\sigma^n_i$.

Definition 6. Let $\psi \in \mathbb{R}$, then the *sign (or signum) function*, denoted by sgn, maps ψ to the set $\{-1, 0, 1\}$ as follows:

$$\operatorname{sgn}\psi = \begin{cases} -1, & \text{if } \psi < 0, \\ 0, & \text{if } \psi = 0, \\ 1, & \text{if } \psi > 0. \end{cases} \tag{9}$$

Furthermore, for any $a = (a_1, a_2, \ldots, a_n) \in \mathbb{R}^n$ the *sign* of a, denoted $\operatorname{sgn} a$, is defined as $\operatorname{sgn} a = (\operatorname{sgn} a_1, \operatorname{sgn} a_2, \ldots, \operatorname{sgn} a_n)$.

2.2 Bolzano's Intermediate Value Theorem

The fundamental and pioneering well-known and widely applied Bolzano's theorem states the following [3,12]:

Theorem 1 (Bolzano's theorem). *If $f: [a, b] \subset \mathbb{R} \to \mathbb{R}$ is a continuous function and if it holds that $f(a)f(b) < 0$, then there is at least one $x \in (a, b)$ such that $f(x) = 0$.*

The above theorem is also called *intermediate value theorem* since it can be easily given as follows:

Theorem 2 (Bolzano's intermediate value theorem). *If $f: [a, b] \subset \mathbb{R} \to \mathbb{R}$ is a continuous function and if y_0 is a real number such that:*

$$\min\{f(a), f(b)\} < y_0 < \max\{f(a), f(b)\},$$

then there is at least one $x_0 \in (a, b)$ such that $f(x_0) = y_0$.

Remark 2. Obviously, Theorem 2 can be deduced from Theorem 1 by considering the function $g(x) = f(x) - y_0$.

Remark 3. The above theorem has been independently proved by Bolzano in 1817 [3] and Cauchy in 1821 [6]. These proofs were crucial in the procedure of *arithmetization of analysis*, which was a research program in the foundations of mathematics during the second half of the 19th century.

2.3 Bolzano-Poincaré-Miranda Intermediate Value Theorem

A straightforward generalization of Bolzano's intermediate value theorem to continuous mappings in several variables was proposed (without proof) by Poincaré in 1883 and 1884 in his work on the *three body problem* [20,21]. This generalization, known as Bolzano-Poincaré-Miranda theorem, states that [17,25,30]:

Theorem 3 (Bolzano - Poincaré - Miranda theorem). *Suppose that $P = \{x \in \mathbb{R}^n \mid |x_i| < L, \text{ for } 1 \leqslant i \leqslant n\}$ and let the mapping $F_n = (f_1, f_2, \ldots, f_n): P \to \mathbb{R}^n$ be continuous on $\operatorname{cl}P$ such that $\theta^n \notin F_n(\vartheta P)$, and*

(a) $f_i(x_1, x_2, \ldots, x_{i-1}, -L, x_{i+1}, \ldots, x_n) \geqslant 0,$ for $1 \leqslant i \leqslant n,$
(b) $f_i(x_1, x_2, \ldots, x_{i-1}, +L, x_{i+1}, \ldots, x_n) \leqslant 0,$ for $1 \leqslant i \leqslant n.$

Then, there is at least one $x \in P$ such that $F_n(x) = \theta^n$.

Remark 4. The Bolzano-Poincaré-Miranda theorem is closely related to important theorems in analysis and topology and constitutes an invaluable tool for verified solutions of numerical problems by means of interval arithmetic. For various interesting relations between the theorems of Bolzano-Poincaré-Miranda, Borsuk, Kantorovich and Smale with respect to the existence of a solution of a system of nonlinear equations, we refer the interested reader to [1].

Remark 5. Theorem 3 it has come to be known as Miranda's theorem since in 1940 Miranda [17] proved that it is equivalent to the traditional Brouwer fixed point theorem [5]. Also, this theorem has been named Miranda-Vrahatis theorem [2]. For a short proof and a generalization of the Bolzano-Poincaré-Miranda theorem using topological degree theory we refer the interested reader to [30]. Following the proof of [30] it is easy to see that Theorem 3 is also true, if L is dependent of i. That is, P can also be an n-dimensional rectangle and need not to be necessarily an n-dimensional cube. In addition, for generalizations with respect to an arbitrary basis of $\mathbb{R}^n$ that eliminate the dependence of the Bolzano-Poincaré-Miranda theorem on the standard basis of $\mathbb{R}^n$ see [11,30].

2.4   Intermediate Value Theorem for Simplices

The intermediate value theorem for simplices (cf. Theorem 4 below) is proposed in [33]. The obtained proof is based on the following *Knaster-Kuratowski-Mazurkiewicz covering principle* [15]:

Lemma 1 (Knaster-Kuratowski-Mazurkiewicz). *Let C_i, $i \in N^n$ be a family of $(n+1)$ closed subsets of an n-simplex $\sigma^n = [v^0, v^1, \ldots, v^n]$ in $\mathbb{R}^n$ satisfying the following hypotheses:*

(a) $\sigma^n = \bigcup_{i \in N^n} C_i$ and
(b) For each $\emptyset \neq I \subset N^n$ it holds that $\bigcap_{i \in I} \sigma^n_{\neg i} \subset \bigcup_{j \in N^n_{\neg I}} C_j$.

Then, it holds that $\bigcap_{i \in N^n} C_i \neq \emptyset$.

Remark 6. Lemma 1 is often referred in the literature as *KKM Lemma*.

Remark 7. The three well known and widely applied fundamental and pioneering classical results, namely, the Brouwer fixed point theorem [5], the Sperner lemma [24], and the KKM lemma [15] are mutually equivalent in the sense that each one can be deduced from another.

Similar to KKM covering principle, the following covering principles have been proposed by Sperner [24]:

Lemma 2 (Sperner covering principle). *Let C_i, $i \in N^n$ be a family of $(n+1)$ closed subsets of an n-simplex $\sigma^n = [v^0, v^1, \ldots, v^n]$ in $\mathbb{R}^n$ satisfying the following hypotheses:*

(a) $\sigma^n = \bigcup_{i \in N^n} C_i$ and
(b) $\sigma^n_{\neg i} \cap C_i = \emptyset$, $\forall i \in N^n$.

Then, it holds that $\bigcap_{i \in N^n} C_i \neq \emptyset$.

A similar covering principle is the following:

Lemma 3 (Sperner covering principle). *Let C_i, $i \in N^n$ be a family of $(n+1)$ closed subsets of an n-simplex $\sigma^n = [v^0, v^1, \ldots, v^n]$ in $\mathbb{R}^n$ satisfying the following hypotheses:*

(a) $\sigma^n = \bigcup_{i \in N^n} C_i$ and
(b) $\sigma^n_{\neg i} \subset C_i$, $\forall i \in N^n$.

Then, it holds that $\bigcap_{i \in N^n} C_i \neq \emptyset$.

Remark 8. Based on the above *Sperner covering principles* two short proofs of the intermediate value theorem for simplices (cf. Theorem 4 below) are given in [34].

Next, we give the intermediate value theorem for simplices [33, 34]:

Theorem 4 (Intermediate value theorem for simplices). *Assume that $\sigma^n = [v^0, v^1, \ldots, v^n]$ is an n-simplex in $\mathbb{R}^n$. Let $F_n = (f_1, f_2, \ldots, f_n) \colon \sigma^n \to \mathbb{R}^n$ be a continuous function such that $f_j(v^i) \neq 0$, $\forall j \in N^n_{\neg 0} = \{1, 2, \ldots, n\}$, $i \in N^n = \{0, 1, \ldots, n\}$ and $\theta^n \notin F_n(\partial \sigma^n)$. Assume that the vertices v^i, $i \in N^n$ are reordered such that the following hypotheses are fulfilled:*

$$(a) \quad \mathrm{sgn} f_j(v^j)\, \mathrm{sgn} f_j(x) = -1, \quad \forall x \in \sigma^n_{\neg j}, \quad j \in N^n_{\neg 0}, \tag{10}$$

$$(b) \quad \mathrm{sgn} F_n(v^0) \neq \mathrm{sgn} F_n(x), \quad \forall x \in \sigma^n_{\neg 0}, \tag{11}$$

where $\mathrm{sgn} F_n(x) = \big(\mathrm{sgn} f_1(x), \mathrm{sgn} f_2(x), \ldots, \mathrm{sgn} f_n(x)\big)$ and $\sigma^n_{\neg i}$ denotes the face opposite to vertex v^i. Then, there is at least one point $x \in \mathrm{int}\,\sigma^n$ such that $F_n(x) = \theta^n$.

Remark 9. The only computable information required by the hypotheses (10) and (11) of Theorem 4 is the algebraic sign of the function values on the boundary of the n-simplex σ^n. Thus, Theorem 4 is applicable whenever the signs of the function values are computed correctly. Theorem 4 has been applied for the localization and approximation of fixed points and zeros of continuous mappings using a simplicial subdivision of a simplex [34]. For an interesting application of this theorem see [16].

3   Applications of the Intermediate Value Theorems

Applications of the corresponding existence criteria emanated by the above intermediate value theorems are given below.

3.1   Bisection Method

Based on the hypotheses of Bolzano's theorem (Theorem 1), a very useful criterion for the existence of a zero of a continuous mapping $f : [a, b] \subset \mathbb{R} \rightarrow \mathbb{R}$ within an interval (a, b) is the following *Bolzano's existence criterion*:

$$f(a) f(b) < 0, \tag{12}$$

or equivalently:

$$\operatorname{sgn} f(a) \operatorname{sgn} f(b) = -1, \tag{13}$$

where sgn denotes the sign function (9).

Remark 10. The Bolzano existence criterion is well-known and widely used and it can be generalized to higher dimensions, see [30,33] (cf. Sect. 2.3 and Sect. 2.4). Note that when the condition (12) (or the condition (13)) is not fulfilled, then in the interval (a, b) either no zero exists or there are zeros for which the sum of their multiplicities is an even number (e.g., two simple zeros, one double and two simple zeros, one triple and one simple zeros etc.).

The well-know and widely applied *bisection method* is based on the Bolzano existence criterion in order to approximate a zero of a continuous function $f : [a, b] \subset \mathbb{R} \rightarrow \mathbb{R}$ in a given interval (a, b). A simplified version described in [27] is the following:

$$x^{p+1} = x^p + c \operatorname{sgn} f(x^p) / 2^{p+1}, \quad p = 0, 1, \ldots, \tag{14}$$

where $x^0 = a$ and $c = \operatorname{sgn} f(a) (b - a)$. Instead of the iterative formula (14) we can also use the following [27]:

$$x^{p+1} = x^p - \hat{c} \operatorname{sgn} f(x^p) / 2^{p+1}, \quad p = 0, 1, \ldots, \tag{15}$$

where $x^0 = b$ and $\hat{c} = \operatorname{sgn} f(b) (b - a)$.

The sequences (14) and (15) converge with certainty to a zero $r \in (a, b)$ if for some x^p it holds that:

$$\operatorname{sgn} f(x^0) \operatorname{sgn} f(x^p) = -1, \quad \text{for } p = 1, 2, \ldots.$$

Furthermore, the number of iterations ν required to obtain an approximate zero r^* such that $|r - r^*| \leqslant \varepsilon$ for some $\varepsilon \in (0, 1)$ is given by:

$$\nu = \left\lceil \log_2(b - a) \varepsilon^{-1} \right\rceil, \tag{16}$$

where $\lceil x \rceil = \operatorname{ceil}(x)$ denotes the ceiling function that maps a real number x to the least integer greater than or equal to x.

Remark 11. The main characteristics of the iterative schemes (14) and (15) are the following:

(a) They *converge with certainty* within the given interval (a, b).
(b) They are *globally convergent methods* in the sense that they converge to a zero from remote initial guesses.
(c) Using relation (16), the *number of iterations* that are required for the attainment of an approximate zero to a given accuracy is known a priori.
(d) They are *worst-case optimal.* That is, they possess asymptotically the best possible rate of convergence in the worst case [23]. This means that they are guaranteed to converge within the predefined number of iterations, and, moreover, no other method has this important property.
(e) They require only the *algebraic signs* of the function values to be computed, as is evident from (14) and (15); thus they can be applied to problems with imprecise function values.

For applications of the iterative schemes (14) and (15) we refer the interested reader, among others, to [27, 28].

3.2   Generalized Bisection Methods

The conditions of the Bolzano-Poincaré-Miranda theorem give an invaluable existence criterion for a solution of Eq. (1). Similarly to Bolzano's criterion, the Bolzano - Poincaré - Miranda criterion requires only the algebraic sings of the function values to be computed on the boundary of the n-cube P. On the other hand, for general continuous functions, in contrary to Bolzano's criterion, the hypotheses (a) and (b) of Theorem 3 are not always fulfilled or it is impossible to be verified for a given n-cube P.

Next, the characteristic polyhedron criterion and the characteristic bisection method are briefly presented. These approaches, in contrary to Bolzano - Poincaré - Miranda criterion require only the algebraic sings of the function values to be computed on the vertices of the considered polyhedron.

There are various generalized bisection methods that require the computation of the topological degree [19] in order to localize a solution of Eq. (1) (see, e.g., [14, 26]). The important Kronecker's theorem [19] states that if the value of topological degree is not zero Eq. (1) has at least one zero within $\mathcal{D}$. To this end, several methods for the computation of the topological degree have been proposed in the past few years (see, e.g., [14, 25]). One such method is the fundamental and pioneering Stenger's method [25] that in some classes of functions is an almost optimal complexity algorithm (see, e.g., [18, 23, 25]).

Once we have obtained a domain for which the value of the topological degree relative to this domain is nonzero, we are able to obtain upper and lower bounds for solution values. To this end, by computing a sequence of bounded domains with nonzero values of topological degree and decreasing diameters, we are able to obtain a region with arbitrarily small diameter that contains at least one solution of Eq. (1). However, although the nonzero value of topological degree plays an important role in the existence of a solution of Eq. (1), the computation of this value is a time-consuming procedure.

The bisection method which is briefly described below, avoids all calculations concerning the topological degree by implementing the concept of the *characteristic n-polyhedron criterion* for the existence of a solution of Eq. (1) within a given bounded domain. This criterion is based on the construction of a *characteristic n-polyhedron (CP)* [27,28,35,37]. This can be done as follows. Let $\mathcal{M}_n$ be the $2^n \times n$ matrix whose rows are formed by all possible combinations of -1 and 1. Consider now an oriented n-polyhedron Π^n, with vertices V_k, $k = 1, 2, \ldots, 2^n$. If the $2^n \times n$ *matrix of signs associated with F_n and Π^n*, $\mathcal{S}(F_n; \Pi^n)$, whose entries are the vectors $\operatorname{sgn} F_n(V_k) = \bigl(\operatorname{sgn} f_1(V_k), \operatorname{sgn} f_2(V_k), \ldots, \operatorname{sgn} f_n(V_k)\bigr)$, is identical to $\mathcal{M}_n$, possibly after some permutations of these rows, then Π_n is called *characteristic polyhedron relative to F_n*. Furthermore, if F_n is continuous, then, under some suitable assumptions on the boundary of Π^n, the topological degree of F_n relative to Π^n is not zero (see [37] for a proof), which implies the existence of a solution within Π^n. For more details on how to construct a CP and locate a desired solution see [27,31].

Next, we describe a generalized bisection method. This method combined with the above mentioned CP criterion, produces a sequence of characteristic polyhedra of decreasing size always containing the desired solution. We call it *characteristic bisection method*. This version of bisection does not require the computation of the topological degree at each step, as others do [14,26]. It can be applied to problems with imprecise function values, since it depends only on their signs.

The method simply amounts to constructing another refined characteristic polyhedron, by bisecting a known one, say Π^n. To do this, we compute the midpoint M of the longest edge $\langle V_i, V_j \rangle$, of Π^n (where the distances are measured in Euclidean norms). Then we obtain another characteristic polyhedron, Π^n_*, by comparing the sign, $\operatorname{sgn} F_n(M)$, of $F_n(M)$ with that of $F_n(V_i)$ and $F_n(V_j)$ and substituting M for that vertex for which the signs are identical [27,28,31]. Then we select the longest edge of Π^n_* and continue the above process. If one of the $\operatorname{sgn} F_n(V_i)$, $\operatorname{sgn} F_n(V_j)$ does not coincide with $\operatorname{sgn} F_n(M)$, we either continue with another edge or perform a relaxation process (for details see [27,28,31]).

The minimum number ζ of bisections of the edges of Π^n required to obtain a characteristic polyhedron Π^n_* whose longest edge length satisfies $\Delta(\Pi^n_*) \leqslant \varepsilon$, for some accuracy $\varepsilon \in (0, 1)$, is given by [37]:

$$\zeta = \left\lceil \log_2 \left(\Delta(\Pi^n) \, \varepsilon^{-1} \right) \right\rceil. \tag{17}$$

Remark 12. Notice that ζ is independent of n and that the bisection algorithm has the same number of iterations as the bisection in one-dimension which is optimal and possesses asymptotically the best rate of convergence [22].

3.3 Generalized Method of Bisection for Simplices

Definition 7 [13]. Let $\sigma^m_0 = \langle v^0, v^1, \ldots, v^m \rangle$ be an oriented m-simplex in $\mathbb{R}^n$, $m \leqslant n$, suppose that $\langle v^i, v^j \rangle$ is the longest edge of σ^m_0 and let $\Upsilon = (v^i + v^j)/2$ be the midpoint of $\langle v^i, v^j \rangle$. Then the *bisection* of σ^m_0 is the order pair of m-simplices $\langle \sigma^m_{10}, \sigma^m_{11} \rangle$ where:

$$\sigma_{10}^m = \langle v^0, v^1, \ldots, v^{i-1}, \Upsilon, v^{i+1}, \ldots, v^j, \ldots, v^m \rangle,$$
$$\sigma_{11}^m = \langle v^0, v^1, \ldots, v^i, \ldots, v^{j-1}, \Upsilon, v^{j+1}, \ldots, v^m \rangle.$$

The m-simplices σ_{10}^m and σ_{11}^m will be called *lower simplex* and *upper simplex* respectively corresponding to σ_0^m while both σ_{10}^m and σ_{11}^m will be called *elements of the bisection* of σ_0^m. Suppose that $\sigma_0^n = \langle v^0, v^1, \ldots, v^n \rangle$ is an oriented n-simplex in $\mathbb{R}^n$ which includes at least one solution of Eq. (1). Suppose further that $\langle \sigma_{10}^n, \sigma_{11}^n \rangle$ is the bisection of σ_0^n and that there is at least one solution of the system (1) in some of its elements. Then this element will be called *selected n-simplex produced after one bisection of σ_0^n* and it will be denoted by σ_1^n. Moreover if there is at least one solution of the system (1) in both elements, then the selected n-simplex will be the lower simplex corresponding to σ_0^n. Suppose now that the bisection is applied with σ_1^n replacing σ_0^n giving thus the σ_2^n. Suppose further that this process continues for p iterations. Then we call σ_p^n *the selected n-simplex produced after p iterations of the bisection of σ_0^n* .

Definition 8 [29]. The *barycentric radius* $\beta(\sigma^m)$ of an m-simplex σ^m in $\mathbb{R}^n$ is the radius of the smallest ball centered at the barycenter of σ^m and containing the simplex. The barycentric radius $\beta(A)$ of a subset A of $\mathbb{R}^n$ is the supremum of the barycentric radii of simplices with vertices in A.

Theorem 5 [29]. *Any m-simplex $\sigma^m = [v^0, v^1, \ldots, v^m]$ in $\mathbb{R}^n$, $m \leqslant n$ is enclosable by the spherical surface S_β^{m-1} with radius $\beta(\sigma^m)$ given by:*

$$\beta(\sigma^m) = \frac{1}{m+1} \max_i \left(m \sum_{\substack{j=0 \\ j \neq i}}^m \|v^i - v^j\|_2^2 - \sum_{\substack{p=0 \\ p \neq i}}^{m-1} \sum_{\substack{q=p+1 \\ q \neq i}}^m \|v^p - v^q\|_2^2 \right)^{1/2}.$$

Remark 13. The barycentric radius $\beta(\sigma^n)$ of a n-simplex σ^n in $\mathbb{R}^n$ can be used to estimate error bounds for approximate fixed points or approximate roots of mappings in $\mathbb{R}^n$, by approximating a fixed point or a root by the barycenter of σ^n. Note that the computation of $\beta(\sigma^n)$ requires only the lengths of the edges of σ^n, which are also required in order to compute the diameter $\operatorname{diam}(\sigma^n)$ of σ^n. Furthermore, since the distance of the barycenter K of an n-simplex $\sigma^n = [v^0, v^1, \ldots, v^n]$ in $\mathbb{R}^n$ from the barycenter K_i of the i-th face $\sigma_{-i}^n = [v^0, v^1, \ldots, v^{i-1}, v^{i+1}, \ldots, v^n]$ of σ^n is equal to $\|K - v^i\|_2/n$ [26,29], then using Theorem 5 we can easily compute the value of $\gamma(\sigma^n) = \min_i \|K - K_i\|_2/\operatorname{diam}(\sigma^n)$. The value $\gamma(\sigma^n)$ can be used to estimate the thickness $\theta(\sigma^n)$ of σ^n, that is:

$$\theta(\sigma^n) = \min_i \left\{ \min_{x \in \sigma_{-i}^n} \|K - x\|_2 \right\} /\operatorname{diam}(\sigma^n).$$

In general, the thickness $\theta(\sigma^n)$ is important to piecewise linear approximations of smooth mappings and, in general, to simplicial and continuation methods for approximating fixed points or roots of systems of nonlinear equations.

Theorem 6 [13]. *Suppose that σ_0^m is an m-simplex in $\mathbb{R}^n$ and let σ_p^m be any m-simplex produced after p bisections of σ_0^m. Then*

$$\mathrm{diam}(\sigma_p^m) \leqslant \left(\sqrt{3}/2\right)^{\lfloor p/m \rfloor} \mathrm{diam}(\sigma_0^m), \tag{18}$$

where $\mathrm{diam}(\sigma_p^m)$ and $\mathrm{diam}(\sigma_0^m)$ are the diameters of σ_p^m and σ_0^m respectively and $\lfloor p/m \rfloor$ is the largest integer less than or equal to p/m.

Theorem 7 [26,32]. *Suppose that σ_0^m, σ_p^m, $\mathrm{diam}(\sigma_0^m)$ and $\mathrm{diam}(\sigma_p^m)$ are as in Theorem 6 and let K_p^m be the barycenter of σ_p^m. Then for any point T in σ_p^m the following relationship is valid*

$$\|T - K_p^m\|_2 \leqslant \frac{m}{m+1} \left(\sqrt{3}/2\right)^{\lfloor p/m \rfloor} \mathrm{diam}(\sigma_0^m). \tag{19}$$

Definition 9. Let σ^n be an n-simplex in $\mathbb{R}^n$ and let $\mathrm{diam}(\sigma^n)$ and $\mu\mathrm{diam}(\sigma^n)$ be the diameter and the microdiameter of σ^n respectively. Suppose that r is a solution of Eq. (1) in σ^n. Then we define the barycenter K^n of σ^n to be an *approximation* of r and the quantity

$$\varepsilon(\sigma^n) = \frac{n}{n+1} \left(\left(\mathrm{diam}(\sigma^n)\right)^2 - \frac{n-1}{2n} \left(\mu\mathrm{diam}(\sigma^n)\right)^2 \right)^{1/2}, \tag{20}$$

to be an *error estimate* for K^n.

Theorem 8 [26,32]. *Suppose that σ_p^n is the selected n-simplex produced after p bisections of an n-simplex σ_0^n in $\mathbb{R}^n$. Let r be a solution of Eq. (1) which is included in σ_p^n and that K_p^n and $\varepsilon(\sigma_p^n)$ are the approximation of r and the error estimate for K_p^n respectively. Then the following hold:*

(a) $\quad \varepsilon(\sigma_p^n) \leqslant \dfrac{n}{n+1} \left(\sqrt{3}/2\right)^{\lfloor p/n \rfloor} \mathrm{diam}(\sigma_0^n),$

(b) $\quad \varepsilon(\sigma_p^n) \leqslant \left(\sqrt{3}/2\right)^{\lfloor p/n \rfloor} \varepsilon(\sigma_0^n),$

(c) $\quad \lim\limits_{p \to \infty} \varepsilon_p = 0,$

(d) $\quad \lim\limits_{p \to \infty} K_p^n = r.$

3.4   Locating and Computing Periodic Orbits

Our approaches are illustrated here for methods for *locating and computing periodic orbits of nonlinear mappings* as well as *fixed points of the Poincaré map on a surface of section*. In general, analytic expressions for locating and computing these periodic orbits on fixed points are not available.

Many problems in a variety of areas of science and technology can be studied and tackled using periodic orbits of nonlinear mappings or dynamical systems. For example, such problems appear in Quantum Mechanics where a weighted

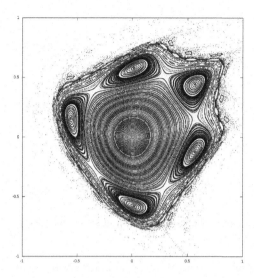

Fig. 1. Hénon mapping for $\cos\omega = 0.24$ and $g(x_1) = -x_1^2$

sum over unstable periodic orbits yields quantum mechanical energy level spacings as well as in Statistical Mechanics where a weighted, according to the values of their Liapunov exponents, sum over unstable periodic orbits can be used to calculate thermodynamic averages (see, e.g., [10]). Furthermore, periodic orbits play a major role in assigning the vibrational levels of highly excited polyatomic molecules. as well as in Celestial Mechanics and Galactic Dynamics.

Let us illustrated our approaches for the following *quadratic area-preserving two-dimensional Hénon's mapping* [31]:

$$\Phi_2 : \begin{pmatrix} \widehat{x}_1 \\ \widehat{x}_2 \end{pmatrix} = \begin{pmatrix} \cos\omega & -\sin\omega \\ \sin\omega & \cos\omega \end{pmatrix} \begin{pmatrix} x_1 \\ x_2 + g(x_1) \end{pmatrix}, \tag{21}$$

where $(x_1, x_2) \in \mathbb{R}^2$ and $\omega \in [0, \pi]$ is the *rotation angle*. By choosing $\cos\omega = 0.24$ and $g(x_1) = -x_1^2$, we observe in the corresponding Hénon's mapping phase plot, illustrated in Fig. 1, that there is a chain of five "islands" around the center of the rectangle. The center points of each island contain a *stable elliptic periodic orbit* of *period five* ($p = 5$). Additionally, the five points where the islands connect consist an *unstable hyperbolic periodic orbit* of period five [31]. These points can be computed by applying the aforementioned methods. When one of these points is computed we can either subsequently apply the same method with different starting conditions and find another point of the periodic orbit or we can iterate the mapping using one of the computed points as starting point. For example, to produce the stable periodic orbit we can iterate the mapping using the following starting point: $(x_1, x_2) = (0.5672405470221847, -0.1223202134278941)$. The rotation number of this orbit is $\sigma = m_1/m_2 = 1/5$. It produces $m_2 = 5$ points by rotating around the origin $m_1 = 1$ times. Additionally, to compute the

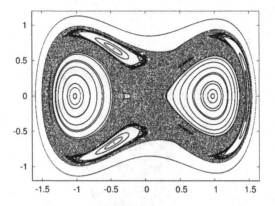

Fig. 2. A Poincaré surface of section of Duffing's oscillator for $\alpha = 0.05$ and $\beta = 2$

unstable periodic orbit, one can iterate the mapping using at starting point the $(x_1, x_2) = (0.2942106885737921, -0.4274862418615337)$ (for details see [31]).

Also, periodic orbits can be used in the study of the structure and breakdown properties of invariant tori in the case of symplectic mappings of direct relevance of the beam stability problem in circular accelerators like the Large Hadron Collider (LHC) machine at the European Organization for Nuclear Research (CERN). Such a *4-D symplectic mapping* can be defined as follows [31,36,38]:

$$
\Phi_4 : \begin{pmatrix} \widehat{x}_1 \\ \widehat{x}_2 \\ \widehat{x}_3 \\ \widehat{x}_4 \end{pmatrix} = \begin{pmatrix} \cos\omega_1 & -\sin\omega_1 & 0 & 0 \\ \sin\omega_1 & \cos\omega_1 & 0 & 0 \\ 0 & 0 & \cos\omega_2 & -\sin\omega_2 \\ 0 & 0 & \sin\omega_2 & \cos\omega_2 \end{pmatrix} \begin{pmatrix} x_1 \\ x_2 + x_1^2 - x_3^2 \\ x_3 \\ x_4 - 2x_1x_3 \end{pmatrix}. \quad (22)
$$

This mapping describes the (instantaneous) effect experienced by a hadronic particle as it passes through a magnetic focusing element of the FODO cell type, where x_1 and x_3 are the particle's deflections from the ideal (circular) orbit, in the horizontal and vertical directions respectively, and x_2, x_4 are the associated "momenta", while ω_1, ω_2 are related to the accelerator's betatron frequencies (or "tunes") q_x, q_y by $\omega_1 = 2\pi q_x$ and $\omega_2 = 2\pi q_y$ and constitute the main parameters that can be varied by an experimentalist see, e.g., [31,36,38] and the references thereof.

Next we consider a Poincaré surface of section for the conservative Duffing's oscillator. More specifically, the conservative Duffing's oscillator [7] can be described by the following equation:

$$
\ddot{x} = x - x^3 + \alpha \cos \beta t, \quad (23)
$$

which can be written as:

$$
\begin{cases} \dot{x}_1 = x_2, \\ \dot{x}_2 = x_1 - x_1^3 + \alpha \cos \beta t. \end{cases} \quad (24)
$$

For the aforementioned dynamical system, we consider the Poincaré surface of section for the parameter values of $\alpha = 0.05$ and $\beta = 2$. Figure 2 illustrates the phase plot of this surface, in the $[-1.6, 1.6] \times [-1.2, 1.2]$ rectangle. For this example, we can observe two distinct islands along the $x_2 = 0$ axis. The center points of each island correspond to fixed points of period one ($p = 1$). Once again we can easily compute these two points by applying the aforementioned methods. The two center points correspond to $(x_1, x_2) = (-1.024572461190486, 0.0)$, and $(x_1, x_2) = (0.9746253482044169, 0.0)$.

In conclusion, our experience is that the generalized methods of bisection are very efficient and effective applied on the problems (21), (22) and (23). These is so, because, we have succeeded to compute rapidly and accurately periodic orbits (stable and unstable) for periods which reach up to the thousands. For detailed results we refer the interested reader to [7, 31, 36, 38].

4 Synopsis

Generalizations the intermediate value theorems in several variables are presented. These theorems are very useful for the existence of solutions of systems of nonlinear equations, the existence of fixed points of continuous functions as well as the existence of periodic orbits of nonlinear mappings and similarly, fixed points of the Poincaré map on a surface of section. Based on the corresponding criteria for the existence of a solution or a fixed point emanated by the intermediate value theorems, generalized bisection methods for approximating zeros or fixed points of continuous functions are given. These bisection methods require only the algebraic signs of the function values and are of major importance for studying and tackling problems with imprecise information.

Acknowledgments. The author would like to thank the editors for their kind invitation.

References

1. Alefeld, G., Frommer, A., Heindl, G., Mayer, J.: On the existence theorems of Kantorovich, Miranda and Borsuk. Electron. Trans. Numer. Anal. **17**, 102–111 (2004)
2. Bánhelyi, B., Csendes, T., Hatvan, L.: On the existence and stabilization of an upper unstable limit cycle of the damped forced pendulum. J. Comput. Appl. Math. **371**, 112702 (2020)
3. Bolzano, B.: Rein analytischer Beweis des Lehrsatzes, dass zwischen je zwei Werten, die ein entgegengesetztes Resultat gewähren, wenigstens eine reelle Wurzel der Gleichung liege. Prague (1817)
4. Brönnimann, H., Emiris, I.Z., Pan, V., Pion, S.: Sign determination in residue number systems. Theor. Comput. Sci. **210**, 173–197 (1999)
5. Brouwer, L.E.J.: Über Abbildungen von Mannigfaltigkeiten. Math. Ann. **71**, 97–115 (1912)

6. Cauchy, A.-L.: Cours d'Analyse de l'École Royale Polytechnique, Paris (1821). (Reprinted in Oeuvres Completes, Series 2, vol. 3)
7. Drossos, L., Ragos, O., Vrahatis, M.N., Bountis, T.C.: Method for computing long periodic orbits of dynamical systems. Phys. Rev. E **53**(1), 1206–1211 (1996)
8. Emiris I.Z., Mourrain B., Vrahatis M.N.: Sign methods for counting and computing real roots of algebraic systems. RR-3669, Inria (1999). inria-00073003
9. Emiris I.Z., Mourrain B., Vrahatis M.N.: Sign methods for enumerating solutions of nonlinear algebraic systems. In: Proceedings of the Fifth Hellenic European Conference on Computer Mathematics and Its Applications, vol. 2, pp. 469–473, Athens, Greece (2002)
10. Gutzwiller, M.C.: Chaos in Classical and Quantum Mechanics. Springer, New York (1990). https://doi.org/10.1007/978-1-4612-0983-6
11. Heindl, G.: Generalizations of theorems of Rohn and Vrahatis. Reliable Comput. **21**, 109–116 (2016)
12. Jarník, V.: Bernard Bolzano and the foundations of mathematical analysis. In: Bolzano and the Foundations of Mathematical Analysis, pp. 33–42. Society of Czechoslovak Mathematicians and Physicists, Prague (1981)
13. Kearfott, R.B.: A proof of convergence and an error bound for the method of bisection in $\mathbb{R}^n$. Math. Comp. **32**(144), 1147–1153 (1978)
14. Kearfott, R.B.: An efficient degree-computation method for a generalized method of bisection. Numer. Math. **32**, 109–127 (1979). https://doi.org/10.1007/BF01404868
15. Knaster, B., Kuratowski, K., Mazurkiewicz, S.: Ein Beweis des Fixpunkt-satzes für n-dimensionale Simplexe. Fund. Math. **14**, 132–137 (1929)
16. Milgrom, P., Mollner, J.: Equilibrium selection in auctions and high stakes games. Econometrica **86**(1), 219–261 (2018)
17. Miranda, C.: Un' osservatione su un teorema di Brouwer. Bollettino dell'U.M.I. **3**, 5–7 (1940)
18. Mourrain, B., Vrahatis, M.N., Yakoubsohn, J.C.: On the complexity of isolating real roots and computing with certainty the topological degree. J. Complex. **18**(2), 612–640 (2002)
19. Ortega, J.M., Rheinboldt, W.C.: Iterative Solution of Nonlinear Equations in Several Variables. Classics in Applied Mathematics vol. 30. Society for Industrial and Applied Mathematics, Philadelphia, PA, USA (2000)
20. Poincaré, H.: Sur certaines solutions particulières du problème des trois corps. Comptes rendus de l'Académie des Sciences Paris **91**, 251–252 (1883)
21. Poincaré, H.: Sur certaines solutions particulières du problème des trois corps. Bull. Astronomique **1**, 63–74 (1884)
22. Sikorski, K.: Bisection is optimal. Numer. Math. **40**, 111–117 (1982)
23. Sikorski, K.: Optimal Solution of Nonlinear Equations. Oxford University Press, New York (2001)
24. Sperner, E.: Neuer Beweis für die Invarianz der Dimensionszahl und des Gebietes. Abh. Math. Sem. Hamburg **6**, 265–272 (1928)
25. Stenger, F.: Computing the topological degree of a mapping in $\mathbb{R}^n$. Numer. Math. **25**, 23–38 (1975)
26. Vrahatis, M.N.: An error estimation for the method of bisection in $\mathbb{R}^n$. Bull. Greek Math. Soc. **27**, 161–174 (1986)
27. Vrahatis, M.N.: Solving systems of nonlinear equations using the nonzero value of the topological degree. ACM Trans. Math. Softw. **14**, 312–329 (1988)
28. Vrahatis, M.N.: CHABIS: a mathematical software package for locating and evaluating roots of systems of nonlinear equations. ACM Trans. Math. Softw. **14**, 330–336 (1988)

29. Vrahatis, M.N.: A variant of Jung's theorem. Bull. Greek Math. Soc. **29**, 1–6 (1988)
30. Vrahatis, M.N.: A short proof and a generalization of Miranda's existence theorem. Proc. Amer. Math. Soc. **107**, 701–703 (1989)
31. Vrahatis, M.N.: An efficient method for locating and computing periodic orbits of nonlinear mappings. J. Comput. Phys. **119**, 105–119 (1995)
32. Vrahatis, M.N.: Simplex bisection and Sperner simplices. Bull. Greek Math. Soc. **44**, 171–180 (2000)
33. Vrahatis, M.N.: Generalization of the Bolzano theorem for simplices. Topol. Appl. **202**, 40–46 (2016)
34. Vrahatis, M.N.: Intermediate value theorem for simplices for simplicial approximation of fixed points and zeros. Topol. Appl. **275**, 107036 (2020)
35. Vrahatis, M.N.: Generalizations of the intermediate value theorem for approximating fixed points and zeros of continuous functions. In: Sergeyev, Y.D., Kvasov, D.E. (eds.) NUMTA 2019, Part II. LNCS, vol. 11974, pp. 223–238. Springer, Cham (2020). https://doi.org/10.1007/978-3-030-40616-5_17
36. Vrahatis, M.N., Bountis, T.C., Kollmann, M.: Periodic orbits and invariant surfaces of 4D nonlinear mappings. Int. J. Bifurcat. Chaos **6**, 1425–1437 (1996)
37. Vrahatis, M.N., Iordanidis, K.I.: A rapid generalized method of bisection for solving systems of non-linear equations. Numer. Math. **49**, 123–138 (1986). https://doi.org/10.1007/BF01389620
38. Vrahatis, M.N., Isliker, H., Bountis, T.C.: Structure and breakdown of invariant tori in a 4D mapping model of accelerator dynamics. Int. J. Bifurcat. Chaos **7**, 2707–2722 (1997)

On Truncated Series Involved in Exponential-Logarithmic Solutions of Truncated LODEs

S. A. Abramov$^{(\boxtimes)}$![ORCID], D. E. Khmelnov ![ORCID], and A. A. Ryabenko ![ORCID]

Federal Research Center "Computer Science and Control" of the Russian Academy of Sciences, Vavilova, 40, Moscow 119333, Russia
sergeyabramov@mail.ru

Abstract. Previously, the authors proposed algorithms for finding exponential-logarithmic solutions of linear ordinary differential equations with coefficients in the form of series, for which only a finite number of initial terms is known. Each solution involves a finite set of power series, for which the maximum possible number of terms is calculated. Below, these algorithms are supplemented with the option to confirm the impossibility of obtaining a larger number of terms in the series without using additional information about the given equation. Such a confirmation has the form of a counterexample to the assumption that it is possible to obtain additional terms of the series involved in the solution that are invariant under all prolongations of the given equation.

Keywords: Differential equations · Truncated power series · Computer algebra systems

1 Introduction

The representation of solutions of linear ordinary differential equations requires the use of power and Laurent series. This is the subject of many theoretical studies (see, e.g., [19–23,26,27]) and found numerous application in computer algebra (see, e.g., [1–5,8,11,17,28]).

The proposed paper is a continuation of the series of works by the authors on LODE with coefficients, having the form of such power series, with respect to which only their first terms are known. Thus, about the considered equations, there is only some incomplete information. In our previous papers, we proposed algorithms for finding solutions of such equations in the form of Laurent series, as well as the search for regular and exponential-logarithmic solutions. It has been proven that these algorithms allow one to find the maximum possible number of terms of those series that are included in the solutions. The algorithms are implemented by the authors as a package of procedures. The user of these procedures may find it is desirable to obtain some visual arguments in favor of the maximum number of found terms of the series. Below, the authors proposed

© The Author(s), under exclusive license to Springer Nature Switzerland AG 2022
F. Boulier et al. (Eds.): CASC 2022, LNCS 13366, pp. 18–28, 2022.
https://doi.org/10.1007/978-3-031-14788-3_2

such visual arguments: for an arbitrary equation with truncated coefficients, a new algorithm presents two prolonged versions of the original equation whose solutions differ from each other in subsequent (not included in the number of previously found) terms of the series included in the solutions.

2   Truncated Equations

Suppose that K is an algebraically closed field of characteristics 0. The standard notation $K[x]$ is used below for a ring of polynomials in x over K. A ring of formal power series in x over K is denoted by $K[[x]]$, a field of formal Laurent series is denoted by $K((x))$. It is clear that $K[x] \subset K[[x]] \subset K((x))$. For any nonzero element $a(x) = \sum a_i x^i$ in $K((x))$, its *valuation* $\mathrm{val}\, a(x)$ is defined by the equality $\mathrm{val}\, a(x) = \min \{i \mid a_i \neq 0\}$, while $\mathrm{val}\, 0 = \infty$.

The differential equations in the paper are represented with $\theta = x\frac{d}{dx}$ instead of $\frac{d}{dx}$. It is convenient for the algorithms to solve linear ordinary differential equations with coefficients in the form of truncated series (see [6,7,13,14,16]). We consider such equations in the form

$$a_r(x)\theta^r y(x) + a_{r-1}(x)\theta^{r-1}y(x) + \cdots + a_0(x)y(x) = 0, \tag{1}$$

where $y(x)$ is an unknown function of x. *The equation coefficients $a_0(x), a_1(x),$ $\ldots, a_r(x)$ are truncated series*, i.e., for each $i = 0, 1, \ldots, r$ we have

$$a_i(x) = \sum_{j=0}^{t_i} a_{ij}x^j + O(x^{t_i+1}), \tag{2}$$

where $a_{ij} \in K$; t_i is an integer such that $t_i \geqslant -1$ (if $t_i = -1$ then the sum in (2) is 0). Hereinafter, the symbol $O(x^t)$ involved in the formal expressions denotes some series, whose valuation is not less than t. For a series

$$\sum_{k=l}^{t} a_k x^k + O(x^{t+1}),$$

$a_k \in K$, l, t are integer, $t \geqslant l$, we call t *the truncation degree*. Note that a coefficient in (1) can be in the form $O(x^m)$, $m \geqslant 0$.

We refer as *a prolongation* of equation (1) to any equation

$$\tilde{a}_r(x)\theta^r y(x) + \tilde{a}_{r-1}(x)\theta^{r-1}y(x) + \cdots + \tilde{a}_0(x)y(x) = 0,$$

such that $\tilde{a}_i(x) - a_i(x) = O(x^{t_i+1})$, i.e., $\mathrm{val}\,(\tilde{a}_i(x) - a_i(x)) > t_i$, $i = 0, 1, \ldots, r$. We consider as prolongations both equations with truncated coefficients, and equations with completely specified series coefficients, i.e., equations

$$\left(\sum_{j=0}^{\infty} \tilde{a}_{rj}x^j \right) \theta^r y(x) + \left(\sum_{j=0}^{\infty} \tilde{a}_{r-1,j}x^j \right) \theta^{r-1}y(x) + \cdots$$

$$\cdots + \left(\sum_{j=0}^{\infty} \tilde{a}_{0j}x^j \right) y(x) = 0. \tag{3}$$

3   Truncated Solutions

Formal *exponential-logarithmic* solutions of equation (3) are solutions in the form

$$e^{Q(x^{-1/q})} x^\lambda w(x^{1/q}),$$ (4)

where Q is a polynomial with coefficients in K, $q \in \mathbb{Z}_{>0}$, $\lambda \in K$,

$$w(x) = \sum_{s=0}^{m} w_s(x) \ln^s x,$$

$m \in \mathbb{Z}_{>0}$, $w_s(x) \in K((x))$, $s = 0, \ldots, m$, and $w_m(x) \neq 0$. In (4), the factor $x^\lambda w(x^{1/q})$ is *the regular part*, $Q(x^{-1/q})$ is *the exponent of irregular part*, and q is *the ramification index*.

When $q = 1$ and $Q \in K$, solution (4) is called *formal regular* solution, otherwise it is called *irregular*. When $q = 1$, $Q \in K$, $\lambda \in \mathbb{Z}$ and $w(x) \in K((x))$, formal regular solution (4) is called *Laurent* one. In the further references of solutions in the paper we skip the word "formal", but it is assumed.

Suppose that the leading coefficient $\tilde{a}_r(x)$ is nonzero in Eq. (3) with completely specified coefficients. It is known (see e.g. [20, Ch. V], [17,26,29]) that for Eq. (3), there exist r solutions in form (4), which are linearly independent over K. Algorithms are proposed in [17,18,26,29] for finding the ramification index q and the exponent of irregular part $Q(x^{-1/q})$ for r linearly independent solutions of the form (4). Suppose that the valuation of at least one of the coefficients in (3) is equal to 0. Then, to construct the ramification index q and the exponent of irregular part $Q(x^{-1/q})$ for all solutions, it is sufficient to know r val $\tilde{a}_r(x)$ initial coefficients of all $\tilde{a}_i(x)$, $i = 0, 1, \ldots, r$ (see e.g. [25]). To construct the regular part of the solution with any given truncation degree of the series in $w(x)$, the algorithms proposed in [20, ch. IV], [21], [22, ch. II, VIII] may be used. For this construction, it is also sufficient to know some finite number of initial coefficients of all $\tilde{a}_i(x)$ ([3, Proposition 1]).

Let $Q(x^{-1/q}) \in K[x^{-1/q}]$, $q \in \mathbb{Z}_{>0}$, $\lambda \in K$ and

$$w_s^{\langle k_s \rangle}(x) = \sum_{j=j_s}^{k_s} w_{s,j} x^j + O(x^{k_s+1}),$$

$j_s, k_s \in \mathbb{Z}$, $k_s \geq j_s$, $s = 0, \ldots, m$, and $w_{m,j_m} \neq 0$. For equation (1) with truncated coefficients, the expression

$$e^{Q(x^{-1/q})} x^\lambda \sum_{s=0}^{m} w_s^{\langle k_s \rangle}(x^{1/q}) \ln^s x$$ (5)

is referred to as *a solution with a truncated regular part* if any equation that is a prolongation of (1) has the solution $e^{Q(x^{-1/q})} x^\lambda \tilde{w}(x^{1/q})$ that is *a prolongation of solution* (5), i.e., $\tilde{w}(x)$ has a form

$$\tilde{w}(x) = \sum_{s=0}^{m} \tilde{w}_s(x) \ln^s x$$

and it is satisfied that $\tilde{w}_s(x) - w_s^{\langle k_s \rangle}(x) = O(x^{k_s+1})$, i.e., val $(\tilde{w}_s(x) - w_s^{\langle k_s \rangle}(x)) > k_s$, $s = 0, 1, \ldots, m$. Such truncated solution is described as *invariant* to the prolongations of equation (1).

In [6, 7, 13, 14, 16] it is shown that for an equation of the form (1), it is possible to construct all truncated solutions with the maximum possible truncation degree of the series involved in the solution. The maximum possible truncation degree in the invariant solution s_{max} means that there is no invariant solution s that is a prolongation of s_{max} such that the truncation degree of at least one series in s is greater than the truncation degree of the corresponding series in s_{max}. We describe this case as the exhaustive use of information on a given equation in constructing truncated solutions. The above articles present algorithms for solving this problem and their implementation in Maple.

In [12, 24], we have considered the question of automatic confirmation of such an exhaustive use of information about a given equation for the construction of Laurent and regular truncated solutions. Confirmation is presented as a counterexample with two different prolongations of the given equation, which lead to the appearance of different additional terms in the solutions.

Algorithms for constructing both the truncated solutions themselves and counterexamples of the described type are based on finding solutions with *literals*, i.e., symbols used to represent unspecified coefficients of a series involved in the equation (see [7]). Literals denote the coefficients of the terms of the series, the degrees of which are greater than the truncation degree of the series. Finding solutions using literals means representing subsequent (non-invariant for all possible prolongations of the equation) terms of the series by formulas containing literals, i.e., unspecified coefficients. This allows us to clarify the influence of unspecified coefficients on the subsequent terms of the series in the solution.

Remark 1. *Thus, literals are something close to undetermined coefficients. But for literals, it is not supposed to find specific values that allow one to find out all the solutions to the original differential equation. Here the goal is to find out whether the unknown coefficients of the series included in the equation have an effect on the initial terms of those series that are included in the solutions.*

In this article, we extend the results obtained in [12, 24] to the case of exponential-logarithmic solutions with a truncated regular part. The problem of presenting two different prolongations of the original equation, which form a counterexample to the assumption about the possibility of adding invariant terms to the series involved in the truncated exponential-logarithmic solutions of the given truncated equation, is solved.

4   The Case of Exponential-Logarithmic Solutions

Prolongations of equation (1) which contain literals $U_{[i,j]}$ look like the following:

$$\left(\sum_{j=0}^{t_r} a_{rj} x^j + \sum_{j=t_r+1}^{\infty} U_{[r,j]} x^j \right) \theta^r y(x)$$

$$+ \left(\sum_{j=0}^{t_{r-1}} a_{r-1,j} x^j + \sum_{j=t_{r-1}+1}^{\infty} U_{[r-1,j]} x^j \right) \theta^{r-1} y(x) + \cdots$$

$$\cdots + \left(\sum_{j=0}^{t_0} a_{0j} x^j + \sum_{j=t_0+1}^{\infty} U_{[0,j]} x^j \right) y(x) = 0 \quad (6)$$

(we use the notation $U_{[i,j]}$ rather than, say $U_{i,j}$ to emphasize the special status of these unknowns).

The algorithms from [17, 18] allow computing exponential parts $e^{Q(x^{-1/q})}$ of all solutions in form (4) for Eq. (6). We are only interested in the exponential parts that have ramification indices q and coefficients of polynomials Q that do not depend on literals. For each of such pairs q, Q the substitution

$$x = t^q, \quad y(x) = e^{Q(1/t)} z(t) \quad (7)$$

is made in Eq. (6), where t is a new independent variable, and $z(t)$ is a new unknown function. As a result of the substitution with further multiplication of the equation by $e^{-Q(1/t)}$, we obtain a new equation, whose coefficients are Laurent series in t. The coefficients of the series are polynomials in literals over K. The regular solutions $t^\lambda w(t)$ of the new equation are then constructed using the version of the algorithm ([14, Sect. 4.2]). For each series involved in the regular solutions, the version of the algorithm computes the maximum number of terms which are invariant under the prolongations of the equation, and one more term which depends on literals. Such a coefficient will be a polynomial over K in a finite number of literals.

In such a way we get a finite set of polynomials in literals for the exponential-logarithmic solution with regular part (5). The set may be used to construct a counterexample.

In [12], we proved the following theorem for the case of truncated Laurent and regular solutions.

Theorem 1 ([12], Theorem 1). *Suppose that solutions of equation (6) involve m truncated power series*

$$c_{i0} + c_{i1} x + \cdots + c_{ik_i} x^{k_i} + p_i(u_1, \ldots, u_l) x^{k_i+1} + O(x^{k_i+2}), \quad (8)$$

where $u_1, \ldots, u_l$ are literals, the coefficients c_{ij} are independent from the literals, while the coefficient $p_i(u_1, \ldots, u_l)$ is a non-constant polynomial in the literals, $i = 1, \ldots, m$. Then, there are $\alpha_1, \ldots, \alpha_l, \beta_1, \ldots, \beta_l \in K$ such that two prolongations of the equation that correspond to $u_j = \alpha_j$, $u_j = \beta_j$, $j = 1, \ldots, l$, lead to the occurrence of different very first additional terms in the truncated series involved in the solutions.

Now we show that a similar statement is valid for exponential-logarithmic solutions with a truncated regular part.

Theorem 2. *Let $\mathcal{E}$ be an equation of the form (1) and s be its truncated solution of the form (5), computed using the algorithm from [16]. Then there exist $\mathcal{E}_1$ and $\mathcal{E}_2$, which are two different prolongations of the equation $\mathcal{E}$ such that $\mathcal{E}_1$ has a truncated solution s_1, $\mathcal{E}_2$ has a truncated solution s_2, both solutions s_1 and s_2 are prolongations of s, and any truncated series involved in s has a prolongation both in s_1, and in s_2, while the very first additional terms of those prolongations are different.*

Proof. The algorithm from [16] is based on the construction of the truncated solutions in form (5), each series in the solutions being constructed up to the first term that contains literals and that is not included in the resulting truncated solutions. Before dropping the terms with literals each series in the truncated solutions is in form (8). Theorem 1 can be applied to all these truncated series together. Thus, there are two different sets of values $\alpha_1, \ldots, \alpha_l, \beta_1, \ldots, \beta_l \in K$ for the literals $u_1, \ldots, u_l$, which are used to construct the prolongations $\mathcal{E}_1$ and $\mathcal{E}_2$ that have truncated solutions s_1 and s_2 with different additional terms $p_i(\alpha_1, \ldots, \alpha_l)x^{k_i+1}$ and $p_i(\beta_1, \ldots, \beta_l)x^{k_i+1}$ not containing literals. $\square$

An algorithm to compute two different sets $\alpha_1, \ldots, \alpha_l, \beta_1, \ldots, \beta_l \in K$ may be based on the approach used in [12] to prove Theorem 1.

5   Automatic Confirmation of the Solutions Truncation Degree Maximality

The counterexample computation is implemented by us as an extension of *FormalSolution* procedure from *TruncatedSeries* package. The package contains our implementation of the algorithms presented in [6,7,10,12–14,16,24] in Maple. The Maple library with the *TruncatedSeries* package and Maple worksheets with examples of using its commands are available from [30].

The first argument of *FormalSolution* procedure is a differential equation in the form (1). The application of θ^k to the unknown function $y(x)$ is written as $\theta(y(x), x, k)$. The truncated coefficients $a_i(x)$ of the equation, i.e., the coefficients in the form (2) are written as $b_i(x) + O(x^{t_i+1})$, where $b_i(x)$ is a polynomial of the degree not higher than t_i over the field of algebraic numbers. An unknown function of the equation is specified as the second argument of the procedure.

A row of optional arguments are also supported in the procedure (see [7,9,15] for details). We introduce a new optional argument $'counterexample' = 'Eqs'$, which allows obtaining the automatically constructed counterexample assigned to the variable Eqs in addition to the computed solution itself. The use of some optional parameters are demonstrated below.

In order to use the package download TruncatedSeries2021.zip from [30]. This archive includes two files: maple.ind and maple.lib. Put these files to some directory, for example to ''/usr/userlib''. Assign

> $libname := "/usr/userlib", libname:$

in the Maple session. Make the short form name of *FormalSolution* procedure available:

> $with(TruncatedSeries):$

Consider the third-order equation with coefficients truncated to different degrees:

> $eq := (x^4 + O(x^7))\theta(y(x), x, 3) + (3x + O(x^5))\theta(y(x), x, 2) +$
  $(1 + 3x^3 + 2x^2 + x + O(x^4))\theta(y(x), x, 1) + O(x^5)y(x) = 0:$

Using the *FormalSolution* command we obtain exponential-logarithmic solutions whose regular parts are calculated to the maximum possible degrees:

> $FormalSolution(eq, y(x))$

$$\left[\_c_1 + O(x^5) + e^{\frac{1}{3x}} x^{\frac{2}{3}} \left(\_c_2 + \frac{35 \_c_2 x}{27} + \frac{8947 \_c_2 x^2}{1458} + O(x^3) \right) + \right.$$
$$\left. e^{\frac{1}{x^3} - \frac{1}{3x}} y_{reg}(x) \right] \tag{9}$$

The first two terms of the result, i.e., $\_c_1 + O(x^5)$, mean that all prolongations of *eq* have Laurent solutions with valuation 0, and their initial segment till the degree 4 is equal to $\_c_1$ where $\_c_1$ is an arbitrary constant $\_c_1$.

The third term means that all prolongations of the equation *eq* have irregular solutions with the exponential part $e^{1/(3x)}$ and the regular part, which is the same up to an arbitrary constant $\_c_2$ for all prolongations of the original equation.

The fourth term means that all prolongations of the equation *eq* have irregular solutions with the exponential part $e^{1/(x^3)-1/(3x)}$. Moreover, there are such prolongations that their regular parts differ by λ.

If, when calling the *FormalSolution* command, the optional argument $'output' = 'literal'$ is used, then the regular parts of the solution are calculated to the maximum degree and, furthermore, terms are added with coefficients depending on literals. In some cases, it is possible to obtain the expression for λ which also depends on literals.

> $FormalSolution(eq, y(x), 'output' = 'literal')$

$$\_c_1 - \frac{U_{[0,5]} \_c_1 x^5}{5} + O(x^6) + e^{\frac{1}{3x}} x^{\frac{2}{3}} \left(\_c_2 + \frac{35 \_c_2 x}{27} + \frac{8947 \_c_2 x^2}{1458} \right.$$
$$\left. + \left(\frac{5832431}{118098} \_c_2 - \frac{1}{9} \_c_2 U_{[1,4]} + \frac{1}{27} \_c_2 U_{[2,5]} \right) x^3 + O(x^4) \right) \tag{10}$$
$$+ e^{\frac{1}{x^3} - \frac{1}{3x}} x^{\frac{19}{3} + 3U_{[3,7]}} (\_c_3 + O(x))$$

Here the literal $U_{[i,k]}$ denotes the coefficient of $x^k \theta^i$. There are two sets of values from $\bar{\mathbb{Q}}$ for these literals such that the expressions

$$\frac{U_{[0,5]\_}c_1}{5}, \qquad \frac{5832431}{118098}\_c_2 - \frac{1}{9}\_c_2 U_{[1,4]} + \frac{1}{27}\_c_2 U_{[2,5]}, \qquad \frac{19}{3} + 3U_{[3,7]}$$

take different values. These two sets correspond to two prolongations of the equation eq. Their solutions are different prolongations of solution (9) and all regular parts of the solution are prolonged. We call such prolongations a counterexample. Obviously, there are an infinite number of counterexamples. As a result of running the *FormalSolution* command with the new optional argument $'counterexample' = 'Eqs'$, the variable Eqs will be assigned a pair of the equations which forms one of the possible counterexamples:

> $FormalSolution(eq, y(x), 'counterexample' = 'Eqs')$:

For the first counterexample equation

> $Eqs[1]$

$$\left(x^5 + O(x^6)\right) y(x) + \left(3x^3 + 2x^2 + x + 1 + 4x^4 + O(x^5)\right) \theta(y(x), x, 1)$$
$$+ \left(3x + O(x^6)\right) \theta(y(x), x, 2) + \left(x^4 - 4x^7 + O(x^8)\right) \theta(y(x), x, 3) = 0 \tag{11}$$

using *FormalSolution* we obtain a truncated solution

> $FormalSolution(Eqs[1], y(x))$

$$\left[\_c_1 - \frac{\_c_1 x^5}{5} + O(x^6) \right.$$

$$+ e^{\frac{1}{3x}} x^{\frac{2}{3}} \left(\_c_2 + \frac{35\_c_2 x}{27} + \frac{8947\_c_2 x^2}{1458} + \frac{5779943\_c_2 x^3}{118098} + O(x^4) \right) \tag{12}$$

$$\left. + \frac{e^{\frac{1}{x^3} - \frac{1}{3x}} (\_c_3 + O(x))}{x^{\frac{17}{3}}} \right]$$

For the second counterexample equation

> $Eqs[2]$

$$\left(5x^5 + O(x^6)\right) y(x) + \left(3x^3 + 2x^2 + x + 1 - 2x^4 + O(x^5)\right) \theta(y(x), x, 1)$$
$$+ \left(3x + O(x^6)\right) \theta(y(x), x, 2) + \left(x^4 - x^7 + O(x^8)\right) \theta(y(x), x, 3) \tag{13}$$

we obtain

> *FormalSolution* $(Eqs\,[2], y(x))$

$$\left[-\_c_1 x^5 + \_c_1 + \mathrm{O}\!\left(x^6\right)\right.$$

$$+ e^{\frac{1}{3x}} x^{\frac{2}{3}} \left(\_c_2 + \frac{35\_c_2 x}{27} + \frac{8947\_c_2 x^2}{1458} + \frac{5858675\_c_2 x^3}{118098} + \mathrm{O}\!\left(x^4\right)\right) \tag{14}$$

$$\left. + e^{\frac{1}{x^3} - \frac{1}{3x}} x^{\frac{10}{3}} \left(\_c_3 + \mathrm{O}(x)\right)\right]$$

It can be seen that (12) and (14) are prolongations of (9), they differ in all regular parts. The exponents λ of the third regular part are also different: $\lambda = -\frac{17}{3}$ for (12) and $\lambda = \frac{10}{3}$ for (14).

6 Conclusion

In this paper, we have described an algorithm which confirms the exhaustive use of the information contained in a truncated LODE in the process of finding truncated exponential-logarithmic solutions by our algorithms which were published earlier.

The mathematical techniques we employ in this paper use the algebras of differential operators and polynomials, and we give the explicit counterexample for the supposition that additional terms of solutions of a given LODE can be obtained.

From our work, new questions arise. For example, can our results be extended to systems of LODEs? We will continue to investigate this line of enquiry.

Acknowledgments. The authors are grateful to anonymous referees for their helpful comments, as well as Maplesoft (Waterloo, Canada) for consultations and discussions.

References

1. Abramov, S.A., Barkatou, M.A.: Computable infinite power series in the role of coefficients of linear differential systems. In: Gerdt, V.P., Koepf, W., Seiler, W.M., Vorozhtsov, E.V. (eds.) CASC 2014. LNCS, vol. 8660, pp. 1–12. Springer, Cham (2014). https://doi.org/10.1007/978-3-319-10515-4_1
2. Abramov, S., Barkatou, M., Khmelnov, D.: On full rank differential systems with power series coefficients. J. Symb. Comput. **68**, 120–137 (2015)
3. Abramov, S.A., Barkatou, M.A., Pflügel, E.: Higher-order linear differential systems with truncated coefficients. In: Gerdt, V.P., Koepf, W., Mayr, E.W., Vorozhtsov, E.V. (eds.) CASC 2011. LNCS, vol. 6885, pp. 10–24. Springer, Heidelberg (2011). https://doi.org/10.1007/978-3-642-23568-9_2
4. Abramov, S., Bronstein, M., Petkovšek, M.: On polynomial solutions of linear operator equations. In: ISSAC 1995: Proceedings of the 1995 International Symposium on Symbolic and Algebraic Computation, pp. 290–296 (1995)

5. Abramov, S.A., Khmelnov, D.E.: Regular solutions of linear differential systems with power series coefficients. Program. Comput. Softw. **40**(2), 98–106 (2014). https://doi.org/10.1134/S0361768814020029
6. Abramov, S., Khmelnov, D., Ryabenko, A.: Laurent solutions of linear ordinary differential equations with coefficients in the form of truncated power series. In: Computer Algebra: 3rd International Conference Materials, Moscow, 17–21 June 2019, International Conference Materials, pp. 75–82 (2019)
7. Abramov, S., Khmelnov, D., Ryabenko, A.: Procedures for searching Laurent and regular solutions of linear differential equations with the coefficients in the form of truncated power series. Program. Comput. Softw. **46**, 67–75 (2020)
8. Abramov, S.A., Ryabenko, A.A., Khmelnov, D.E.: Procedures for searching local solutions of linear differential systems with infinite power series in the role of coefficients. Program. Comput. Softw. **42**(2), 55–64 (2016). https://doi.org/10.1134/S036176881602002X
9. Abramov, S.A., Khmelnov, D.E., Ryabenko, A.A.: The `TruncatedSeries` package for solving linear ordinary differential equations having truncated series coefficients. In: Corless, R.M., Gerhard, J., Kotsireas, I.S. (eds.) MC 2020. CCIS, vol. 1414, pp. 19–33. Springer, Cham (2021). https://doi.org/10.1007/978-3-030-81698-8_2
10. Abramov, S.A., Khmelnov, D.E., Ryabenko, A.A.: Truncated and infinite power series in the role of coefficients of linear ordinary differential equations. In: Boulier, F., England, M., Sadykov, T.M., Vorozhtsov, E.V. (eds.) CASC 2020. LNCS, vol. 12291, pp. 63–76. Springer, Cham (2020). https://doi.org/10.1007/978-3-030-60026-6_4
11. Abramov, S., Petkovšek, M.: Special power series solutions of linear differential equations. In: Proceedings FPSAC 1996, pp. 1–8 (1996)
12. Abramov, S., Ryabenko, A., Khmelnov, D.: Exhaustive use of information on an equation with truncated coefficients. Program. Comput. Softw. **48**, 116–124 (2022). https://doi.org/10.1134/S0361768822020025
13. Abramov, S., Ryabenko, A., Khmelnov, D.: Linear ordinary differential equations and truncated series. Comput. Math. Math. Phys. **59**, 1649–1659 (2019). https://doi.org/10.1134/S0965542519100026
14. Abramov, S., Ryabenko, A., Khmelnov, D.: Regular solutions of linear ordinary differential equations and truncated series. Comput. Math. Math. Phys. **60**, 1–14 (2020). https://doi.org/10.1134/S0965542520010029
15. Abramov, S., Ryabenko, A., Khmelnov, D.: Procedures for constructing truncated solutions of linear differential equations with infinite and truncated power series in the role of coefficients. Program. Comput. Softw. **47**, 144–152 (2021). https://doi.org/10.1134/S036176882102002X
16. Abramov, S., Ryabenko, A., Khmelnov, D.: Truncated series and formal exponential-logarithmic solutions of linear ordinary differential equations. Comput. Math. Math. Phys. **60**, 1609–1620 (2020). https://doi.org/10.1134/S0965542520100024
17. Barkatou, M.A.: Rational Newton algorithm for computing formal solutions of linear differential equations. In: Gianni, P. (ed.) ISSAC 1988. LNCS, vol. 358, pp. 183–195. Springer, Heidelberg (1989). https://doi.org/10.1007/3-540-51084-2_17
18. Barkatou, M., Richard-Jung, F.: Formal solutions of linear differential and difference equations. Program. Comput. Software **23**(1), 17–30 (1997)
19. Bruno, A.D.: Asymptotic behavior and expansions of solutions of an ordinary differential equation. Russ. Math. Surv. **59**(3), 31–80 (2004)
20. Coddington, E.A., Levinson, N.: Theory of Ordinary Differential Equations. Krieger, Malabar (1984)

21. Frobenius, G.: Über die Integration der linearen Differentialgleichungen durch Reihen. J. für die reine und angewandte Mathematik **76**, 214–235 (1873)
22. Heffter, L.: Einleitung in die Theorie der linearen Differentialgleichungen. Teubner, Leipzig (1894)
23. Ince, E.: Ordinary Differential Equations. Longmans, London, New York, Bombay (1926)
24. Khmelnov, D., Ryabenko, A., Abramov, S.: Automatic confirmation of exhaustive use of information on a given equation. In: Computer Algebra: 4th International Conference Materials, pp. 69–72. MAKS Press, Moscow (2021) (2021)
25. Lutz, D.A., Schäfke, R.: On the identification and stability of formal invariants for singular differential equations. Linear Algebra Appl. **72**, 1–46 (1985)
26. Malgrange, B.: Sur la réduction formelle des équations différentielles a singularités irrégulières. Université Scientifique et Médicale de Grenoble (1979)
27. Schlesinger, L.: Handbuch der Theorie der linearen Differentialgleichungen, vol. 1. Teubner, Leipzig (1895)
28. Singer, M.F.: Formal solutions of differential equations. J. Symb. Comput. **10**(1), 59–94 (1990)
29. Tournier, E.: Solutions formelles d'équations différentielles. Le logiciel de calcul formel DESIR. Étude théorique et réalisation. Thèse d'Etat, Université de Grenoble (1987)
30. TruncatedSeries website. http://www.ccas.ru/ca/TruncatedSeries. Accessed 11 May 2022

Subresultant Chains Using Bézout Matrices

Mohammadali Asadi, Alexander Brandt⬤, David J. Jeffrey⬤,
and Marc Moreno Maza$^{(\boxtimes)}$

ORCCA, The University of Western Ontario, London, Canada
{masadi4,abrandt5,djeffrey}@uwo.ca, moreno@csd.uwo.ca

Abstract. Subresultant chains over rings of multivariate polynomials
are calculated using a speculative approach based on the Bézout matrix.
Our experimental results yield significant speedup factors for the proposed approach against comparable methods. The determinant computations are based on fraction-free Gaussian elimination using various pivoting strategies.

Keywords: Subresultant chain · Speculative algorithm ·
Multithreaded algorithm · Bézout matrix

1 Introduction

Subresultants are one of the most fundamental tools in computer algebra. They
are at the core of numerous algorithms including, but not limited to, polynomial
GCD computations, polynomial system solving, and symbolic integration. When
the subresultant chain of two polynomials is required in a procedure, not all
polynomials of the chain, or not all coefficients of a given subresultant, may be
needed. Based on that observation, the authors of [5] studied different practical
schemes, and their implementation, for efficiently computing subresultants.

The main objective of [5] is, given two univariate polynomials $a, b \in \mathcal{A}[y]$
over some commutative ring $\mathcal{A}$, to compute the subresultant chain of $a, b \in \mathcal{A}[y]$
speculatively. To be precise, the objective is to compute the subresultants of
index 0 and 1, delaying the computation of subresultants of higher index until it
is proven necessary. The practical importance of this objective, as well as related
works, are discussed extensively in [5].

Taking advantage of the Half-GCD algorithm and evaluation-interpolation
methods, the authors of [5] consider the cases in which the coefficient ring $\mathcal{A}$
is a polynomial ring with one or two variables, and with coefficients in a field,
$\mathbb{Q}$ or $\mathbb{Z}/p\mathbb{Z}$, for a prime number p. The reported experimentation demonstrates
the benefits of computing subresultant chains speculatively in the context of
polynomial system solving.

That strategy, however, based on the Half-GCD algorithm, cannot scale to
situations in which the coefficient ring $\mathcal{A}$ is a polynomial ring in many variables,
say 5 or more. The reason is that, for the Half-GCD algorithm to bring benefits,

F. Boulier et al. (Eds.): CASC 2022, LNCS 13366, pp. 29–50, 2022.
https://doi.org/10.1007/978-3-031-14788-3_3

the degree in y of the polynomials a, b must be in the 100's, implying that the resultant of a, b is likely to have very large degrees in the variables of $\mathcal{A}$, thus making computations not feasible in practice when $\mathcal{A}$ has many variables.

Therefore, for this latter situation, one should consider an alternative approach in order to compute subresultant chains speculatively, which is the objective of the present paper. To this end, we consider subresultant chain computations using Bézout matrices. Most notably, [1] introduced an algorithm to compute the nominal coefficients of subresultants by calculating the determinants of sub-matrices of a modified version of the Bézout matrix. Later, [15] generalized this approach to compute all subresultants instead of only the nominal coefficients. Although the approach is theoretically slower than Ducos' subresultant chain algorithm [10], early experimental results in MAPLE, collected during the development of the SubresultantChain method in the *RegularChains* library [16], indicate that approaches based on the Bézout matrix are particularly well-suited for sparse polynomials with many variables.

In this paper, we report on further work following this approach. In Sect. 2, we discuss how to compute the necessary determinants of the sub-matrices of the Bézout matrix. We modify and optimize the fraction-free LU decomposition (FFLU) of a matrix over a polynomial ring presented in [14]. We demonstrate the efficacy of the proposed methods using implementations in MAPLE and the Basic Polynomial Algebra Subprograms (BPAS) library [3]. Our optimization techniques include smart-pivoting and using the BPAS multithreaded interface to parallelize the row elimination step. All of our code, is open source and part of the BPAS library available at www.bpaslib.org.

In Sect. 3, we focus on the computation of subresultants using the Bézout matrix. In Sect. 3.1, we review the definitions of the Bézout matrix and a modified version of it, known as the Hybrid Bézout matrix. Then, we introduce a speculative approach for computing subresultants by modifying the fraction-free LU factorization and utilizing the Hybrid Bézout matrices in Sect. 3.2. We have implemented these computational schemes for subresultant chains and our experimental results, presented in Sect. 3.3, illustrate the benefits of the proposed methods.

2   Fraction-Free LU Decomposition

A standard way to compute the determinant of a matrix A is to reduce it to a triangular form and then take the product of the resulting diagonal elements [18]. One such triangular form is given by an LU matrix decomposition. When the input matrix A has elements in a polynomial ring, standard LU decomposition algorithms lead to matrices with rational functions as elements. In order to keep the elements in the ring of polynomials, while controlling expression swell, one can use a *fraction-free* LU decomposition (FFLU), taking advantage of Bareiss' algorithm [6], which was originally developed for integer matrices. Although in an FFLU decomposition the matrices contain only elements from the ring of polynomials, the intermediate computations do require exact divisions. Reducing the cost of these divisions is a practical challenge, one which we discuss in this section. The main algorithm on which we rely has been described in [12, Ch. 9] and [14]. The main theorem is the following.

Theorem 1. *A rectangular matrix A with elements from an integral domain $\mathbb{B}$, having dimensions $m \times n$ and rank r, may be factored into matrices containing only elements from $\mathbb{B}$ in the form,*

$$A = P_r L D^{-1} U P_c = P_r \begin{pmatrix} \mathcal{L} \\ \mathcal{M} \end{pmatrix} D^{-1} \left(\mathcal{U} \; \mathcal{V} \right) P_c,$$

where the permutation matrix P_r is $m \times m$; the permutation matrix P_c is $n \times n$; $\mathcal{L}$ is $r \times r$, lower triangular and has full rank:

$$\mathcal{L} = \begin{pmatrix} p_1 & 0 & \dots & 0 \\ l_{21} & p_2 & \ddots & \vdots \\ \vdots & \vdots & \ddots & 0 \\ l_{r1} & l_{r2} & \dots & p_r \end{pmatrix},$$

where the $p_i \neq 0$ are the pivots in a Gaussian elimination; $\mathcal{M}$ is $(m-r) \times r$ and is null when $m = n$ holds; D is $r \times r$ and diagonal:

$$D = diag(p_1, p_1 p_2, p_2 p_3, \cdots, p_{r-2} p_{r-1}, p_{r-1} p_r),$$

$\mathcal{U}$ is $r \times r$ and upper triangular, while $\mathcal{V}$ is $r \times (n-r)$ and is null when $m = n$ holds:

$$\mathcal{U} = \begin{pmatrix} p_1 & u_{12} & \dots & u_{1r} \\ 0 & p_2 & \dots & u_{2r} \\ \vdots & \ddots & \ddots & \vdots \\ 0 & \dots & 0 & p_r \end{pmatrix}.$$

PROOF [14, Theorem 2]. Note that the elements of the matrix D belong to $\mathbb{B}$, but the matrix D^{-1}, if explicitly calculated, lies in the quotient field.

Algorithm 3 implements Theorem 1 while Algorithm 2 utilizes Theorem 1 for computing the determinant of A, when A is square. Both Algorithm 3 and Algorithm 2 rely on Algorithm 1, which is a helper-function. This latter algorithm updates the input matrix A in-place, to record the upper triangular matrix U; it also computes the "denominator" d, the rank r of the matrix A and the column permutation of the input matrix. This is sufficient information to calculate the determinant of a square matrix.

In Algorithm 2, the routine CHECK-PARITY calculates the parity of the given permutation modulo 2. Note that in both Algorithms 1 and 3, we only consider row-operations to find the pivot and store the row permutation patterns in the list P_r of size m. Column-permutations, and the corresponding list P_c, are used in Sect. 2.1.

To optimize the FFLU algorithm, we use a *smart-pivoting* strategy, discussed in Sect. 2.1. The idea is to find a "best" pivot by searching through the matrix to pick a non-zero coefficient (actually a polynomial) with the minimum number of terms in each iteration. The goal of this technique is to reduce the cost of the exact divisions in Bareiss' algorithm; see Sect. 2.1 for the details.

In addition, we discuss the parallel opportunities of this algorithm in Sect. 2.2, taking advantage of the BPAS multithreaded interface. Finally, Sect. 2.3 highlights the performance of these algorithms in the BPAS library, utilizing *sparse multivariate polynomial arithmetic*.

Algorithm 1. FFLU-HELPER(A)

Input: an $m \times n$ matrix $A = (a_{i,j})_{0 \leq i < m,\ 0 \leq j < n}$ over $\mathbb{B}$ $(a_{i,j} \in \mathbb{B})$.
Output: r, d, P_r where r is the rank of A, d is the "denominator", so that, $d = s\ det(S)$ where S is an appropriate sub-matrix of A ($S = A$ if A is square and non-singular). $s \in (-1, 1)$ is decided by the parity of row permutations (encoded by P_r).

1: $k := 0;\ d := 1;\ k := 0;\ c := 0;\ P_r := [0, 1, \ldots, m-1]$
2: **while** $k < m$ **and** $c < n$ **do**
3:     **if** $a_{k,c} = 0$ **then**
4:         $i := k + 1$
5:         **while** $i < m$ **do**
6:             **if** $a_{i,c} \neq 0$ **then**
7:                 SWAP i-th and k-th rows of A
8:                 $P_r[i], P_r[k] := P_r[k], P_r[i]$
9:                 **break**
10:            $i := i + 1$
11:         **if** $m \leq i$ **then**
12:             $c := c + 1$
13:             **continue**
14:     $r := r + 1$
15:     **for** $i = k + 1, \ldots, m - 1$ **do**
16:         **for** $j = c + 1, \ldots, n - 1$ **do**
17:             $a_{i,j} := a_{i,c}\, a_{k,j} - a_{i,j}\, a_{k,c}$
18:             **if** $k = 0$ **then** $a_{i,j} := -a_{i,j}$
19:             **else** $a_{i,j} := $ EXACTQUOTIENT$(a_{i,j}, d)$
20:     $d := -a_{k,c};\ k := k + 1;\ c := c + 1$
21: **return** $r, -d, P_r$

Algorithm 2. DET(A)

Input: a $n \times n$ matrix A over $\mathbb{B}$
Output: $det(A)$, the determinant of A
1: $r, d, P_r := $ FFLU-HELPER(A)
2: **if** $r < n$ **then return** 0
3: $p := $ CHECK-PARITY(P_r)
4: **if** $p \neq 0$ **then** $d := -d$
5: **return** d

Algorithm 3. FFLU(A)

Input: an $m \times n$ matrix $A = (a_{i,j})_{0 \le i < m,\ 0 \le j < n}$ over $\mathbb{B}$ ($a_{i,j} \in \mathbb{B}$).
Output: r, d, P, L, U where r is the rank of A, d is the "denominator", so that, $d =$
  $s \det(S)$ where S is an appropriate sub-matrix of A ($S = A$ if A is square and
  non-singular) and $s \in (-1, 1)$ is decided by the parity of the row permutations
  (encoded by the matrix P) performed on A, L is the lower triangular matrix, and
  U is the upper triangular matrix s.t. $PA = LDU$.
1: $U := A$; $i = 0$; $j = 0$; $k = 0$
2: $r, d, P_r := $ FFLU-HELPER(U)
3: INITIALIZE P to the null square matrix of order m
4: LET $P[i, j] := 1$ iff $Pr[i] = j$ for all $0 \le i, j, \le m - 1$
5: **while** $i < m$ **and** $j < n$ **do**
6:   **if** $U[i, j] \ne 0$ **then**
7:     **for** $l = 0, \ldots, i - 1$ **do** $L_{l,k} := 0$
8:     $L_{i,k} := U_{i,j}$
9:     **for** $l = 0, \ldots, m - 1$ **do** $L_{l,k} := U_{l,j}$; $U_{l,j} := 0$
10:     $i := i + 1$; $k := k + 1$
11:   $j := j + 1$
12: **while** $k < m$ **do**
13:   **for** $l = 0, \ldots, k - 1$ **do** $L_{l,k} := 0$
14:   $L_{k,k} := 1$
15:   **for** $l = k + 1, \ldots, m$ **do** $L_{l,k} := 0$
16:   $k := k + 1$
17: **return** r, d, P_r, L, U

Example 1. *Consider matrix $A \in \mathbb{B}^{4 \times 4}$ where $\mathbb{B} = \mathbb{Z}[x]$. $A =$*

$$\begin{pmatrix} 11x^2 - 11x + 3 & -3(x-1)(2x-3) & 0 & 0 \\ 0 & 11x^2 - 11x + 3 & -3(x-1)(2x-3) & 0 \\ 0 & 0 & 11x^2 - 11x + 3 & -3(x-1)(2x-3) \\ -2x + 3 & 0 & 0 & -x \end{pmatrix}.$$

To compute the determinant of this matrix, Algorithm 1 starts with $d = 1$, $k = 0$, $c = 0$, $P_r = [0, 1, 2, 3]$, $A_{0,0} = 11x^2 - 11x + 3 \ne 0$, and $r = 1$. After the first iteration, the nested for-loops update the (bottom-right) sub-matrix from the second row and column; we have $A^{(1)} =$

$$\left(\begin{array}{c|ccc} A_{0,0} & -3(x-1)(2x-3) & 0 & 0 \\ 0 & (A_{0,0})^2 & -3(x-1)(2x-3)A_{0,0} & 0 \\ 0 & 0 & (A_{0,0})^2 & -3(x-1)(2x-3)A_{0,0} \\ -2x + 3 & -3(x-1)(2x-3)^2 & 0 & -xA_{0,0} \end{array} \right),$$

where $A_{1,2}^{(1)} = A_{2,3}^{(1)} = -3(x-1)(2x-3)(11x^2 - 11x + 3)$. *In the second iteration of the while-loop, we have* $d = -11x^2 + 11x - 3$, $k = 1$, $c = 1$, $A_{1,1}^{(1)} = (11x^2 - 11x + 3)^2 \neq 0$, *and* $r = 2$. *Then,* $A^{(2)} =$

$$
\left(
\begin{array}{cc|cc}
A_{0,0} & A_{0,1} & 0 & 0 \\
0 & (A_{0,0})^2 & -3(x-1)(2x-3)A_{0,0} & 0 \\
0 & 0 & (A_{0,0})^3 & -3(x-1)(2x-3)(A_{0,0})^2 \\
-2x+3 & (2x-3)A_{0,1} & -9(x-1)^2(2x-3)^3 & -x(A_{0,0})^2
\end{array}
\right).
$$

In the third iteration of the while-loop, we have $d = -(11x^2 - 11x + 3)^2$, $k = 2$, $c = 2$, $A_{2,2}^{(2)} = -(11x^2 - 11x + 3)^3 \neq 0$, *and* $r = 3$. *And so,* $A^{(3)} =$

$$
\left(
\begin{array}{ccc|c}
A_{0,0} & A_{0,1} & 0 & 0 \\
0 & (A_{0,0})^2 & -3(x-1)(2x-3)A_{0,0} & 0 \\
0 & 0 & (A_{0,0})^3 & -3(x-1)(2x-3)(A_{0,0})^2 \\
-2x+3 & (2x-3)A_{0,1} & -9(x-1)^2(2x-3)^3 & A_{3,3}^{(3)}
\end{array}
\right),
$$

where $A_{3,3}^{(3)} = -1763x^7 + 7881x^6 - 19986x^5 + 35045x^4 - 41157x^3 + 30186x^2 - 12420x + 2187$. *In fact, one can check that* $A_{3,3}^{(3)}$ *is the determinant of the full-rank* $(r = 4)$ *matrix* $A \in \mathbb{Z}[x]^{4\times4}$.

In [6], Bareiss introduced an alternative version of this algorithm, known as *multi-step* Bareiss' algorithm to compute fraction-free LU decomposition. This method reduces the computation of row eliminations by adding three cheaper divisions to compute each row in the while-loop and removing one multiplication in each iteration of the nested for-loops; see the results in Table 1 and [12, Chapter 9] for more details.

In the next sections, we investigate optimizations of Algorithm 1 to compute the determinant of matrices over multivariate polynomials. These optimizations are achieved by reducing the cost of exact divisions by finding better pivots and utilizing the BPAS multithreaded interface to parallelize this algorithm.

2.1 Smart-Pivoting in FFLU Algorithm

Returning to Example 1, we performed exact divisions for the following divisors in the second and third iterations,

$$
d^{(1)} = -11x^2 + 11x - 3,
$$
$$
d^{(2)} = -121x^4 + 242x^3 - 187x^2 + 66x - 9.
$$

However, we could pick a polynomial with fewer terms as our pivot in every iteration to reduce the cost of these exact divisions. Such a method, which finds a polynomial with the minimum number of terms in each column as the pivot of each iteration, is referred to as column-wise smart-pivoting. For matrix A of

Example 1, one can pick $A_{3,0} = -2x+3$ as the first pivot. Applying this method yields, after the first iteration, $A^{(1)} =$

$$
\left(
\begin{array}{c|ccc}
-2x+3 & 0 & 0 & -x \\
\hline
0 & -(2x-3)A_{0,0} & 3(x-1)(2x-3)^2 & 0 \\
0 & 0 & -(2x-3)A_{0,0} & 3(x-1)(2x-3)^2 \\
A_{0,0} & 3(x-1)(2x-3)^2 & 0 & xA_{0,0}
\end{array}
\right),
$$

where $d = 2x-3$. Continuing this method from Algorithm 1, we get the following matrix for $r = 4$, $A^{(4)} =$

$$
\left(
\begin{array}{ccc|c}
-2x+3 & 0 & 0 & -x \\
0 & -(2x-3)A_{0,0} & 3(x-1)(2x-3)^2 & 0 \\
0 & 0 & -(2x-3)A_{0,0}^2 & 3(x-1)(2x-3)^2 A_{0,0} \\
\hline
A_{0,0} & 3(x-1)(2x-3)^2 & 9(x-1)^2(2x-3)^3 & A_{3,3}^{(4)}
\end{array}
\right),
$$

where $A_{3,3}^{(4)} = 1763x^7 - 7881x^6 + 19986x^5 - 35045x^4 + 41157x^3 - 30186x^2 + 12420x - 2187$, $P_r = [3,1,2,0]$, and we have $\mathrm{DET}(A) = -A_{3,3}^{(4)}$ from Algorithm 2.

In *column-wise smart-pivoting*, we limited our search for the best pivot to the corresponding column of the current row. To extend this method, one can try searching for the best pivot in the sub-matrix starting from the next current row and column. To perform this method, referred to as *(fully) smart pivoting*, we need to use column-operations and a column-wise permutation matrix P_c. The column operations along with row operations are not cache-friendly. This is certainly an issue for matrices with (large) multivariate polynomial entries while this may not be an issue with (relatively small) matrices with numerical entries. Therefore, we avoid column swapping within the decomposition, and instead we keep track of column permutations in the list of column-wise permutation patterns P_c to calculate the parity check later in Algorithm 2.

Algorithm 4 presents the pseudo-code of the *smart pivoting* fraction-free LU decomposition utilizing both row-wise and column-wise permutation patterns P_r, P_c. This algorithm updates A in-place, to become the upper triangular matrix U, and returns the rank and denominator of the given matrix $A \in \mathbb{B}^{m \times n}$.

2.2   Parallel FFLU Algorithm

For further practical performance, we now investigate opportunities for parallelism alongside our schemes for cache-efficiency. In particular, notice that during the row reduction step (the `for` loops on lines 24–28 of Algorithm 4) the update of each element is independent. Implementing this step as a `parallel_for` loop is easily achieved with the multithreading support provided in the BPAS library; see further details in [4].

Algorithm 4. SPFFLU-HELPER(A)

Input: an $m \times n$ matrix $A = (a_{i,j})_{0 \le i < m, \, 0 \le j < n}$ over $\mathbb{B}$ ($a_{i,j} \in \mathbb{B}$).

Output: r, d, P_r, P_c where r is the rank, d is the denominator, so that, $d = s \, det(S)$ where S is an appropriate sub-matrix of A ($S = A$ if A is square and non-singular) $s \in (-1, 1)$ is decided by the parity of row and column permutations, P_r, P_c.

1: $k := 0; \; d := 1; \; \ell := 0$
2: $P_r := [0, 1, \ldots, m-1]; \; P_c := [0, 1, \ldots, n-1]$
3: **while** $k < m$ **and** $\ell < n$ **do**
4:    **if** $a_{k,\ell} = 0$ **then**
5:       $i := k + 1$
6:       **while** $i < m$ **do**
7:          **if** $a_{i,\ell} \neq 0$ **then**
8:             $(i, j) := $ FINDBESTPIVOT(A, i, ℓ)
9:             SWAP i-th and k-th rows of A
10:             $P_r[i], P_r[k] := P_r[k], P_r[i]$
11:             $P_c[j], P_c[\ell] := P_c[\ell], P_c[j]$
12:             **break**
13:          $i := i + 1$
14:       **if** $m \le i$ **then**
15:          $\ell := \ell + 1$
16:          **continue**
17:    **else**
18:       $(i, j) := $ FINDBESTPIVOT(A, k, ℓ)
19:       SWAP i-th and k-th rows of A
20:       $P_r[i], P_r[k] := P_r[k], P_r[i]$
21:       $P_c[j], P_c[\ell] := P_c[\ell], P_c[j]$
22:    $r := r + 1$
23:    **for** $i = k+1, \ldots, m-1$ **do**
24:       **for** $j = \ell+1, \ldots, n-1$ **do**
25:          $a_{i,P_c[j]} := a_{i,P_c[\ell]} \, a_{k,P_c[j]} - a_{i,P_c[j]} \, a_{k,P_c[\ell]}$
26:          **if** $k = 0$ **then** $a_{i,P_c[j]} := -a_{i,P_c[j]}$
27:          **else** $a_{i,P_c[j]} := $ EXACTQUOTIENT$(a_{i,P_c[j]}, d)$
28:    $d := -a_{k,P_c[\ell]}; \; k := k + 1; \; \ell := \ell + 1$
29: **return** $r, -d, P_r, P_c$

Algorithm 5. PARALLEL-SPFFLU-HELPER(A)

   // -snip-
1: **parallel_for** $i = k+1, \ldots, m-1$
2:    **parallel_for** $j = \ell+1, \ldots, n-1$
3:       $a_{i,P_c[j]} := a_{i,P_c[\ell]} \, a_{k,P_c[j]} - a_{i,P_c[j]} \, a_{k,P_c[\ell]}$
4:       **if** $k = 0$ **then** $a_{i,P_c[j]} := -a_{i,P_c[j]}$
5:       **else** $a_{i,P_c[j]} := $ EXACTQUOTIENT$(a_{i,P_c[j]}, d)$
6:    **end for**
7: **end for**
   // -snip-

Algorithm 5 shows a naïve implementation of this parallel algorithm. Note that in a `parallel_for` loop, each iteration is (potentially) executed in parallel. In Algorithm 5, this means lines 3–5 are executed independently and in parallel for each possible value of (i, j). If that number of such possible values exceeds a pre-determined limit (e.g. the number of hardware threads supported), then the number of iterations will be divided as evenly as possible among the available threads.

A difficulty to this parallelization scheme is that the size of the sub-matrices decreases with each iteration. Therefore, the amount of work executed by each thread also decreases. In practice, to address this *load-balancing* and to maximize parallelism, we only parallelize the outer loop (line 1 of Algorithm 5).

2.3 Experimentation

In this section, we compare the fraction-free LU decomposition algorithms for Bézout matrix (Definition 2) of randomly generated, non-zero and sparse polynomials in $\mathbb{Z}[x_1, \ldots, x_v]$ for $v \geq 5$ in the BPAS library. We recall that methods based on the Bézout matrix have been observed (during development of the *RegularChains* library [16], and later in Sect. 3.3) to be well-suited for sparse polynomials with many variables. Throughout this paper, our benchmarks were collected on a machine running Ubuntu 18.04.4 and GMP 6.1.2, with an Intel Xeon X5650 processor running at 2.67 GHz, with 12×4 GB DDR3 memory at 1.33 GHz.

Table 1 shows the comparison between the standard implementation of the fraction-free LU decomposition (Algorithm 1; denoted `plain`), the *column-wise* smart pivoting (denoted `col-wise SP`), the *fully* smart-pivoting method (Algorithm 4; denoted `fully SP`), and Bareiss' multi-step technique added to Algorithm 4 (denoted `multi-step`). Here, $v = 5$ and the generated polynomials have a *sparsity ratio* (the fraction of zero terms to the total possible number of terms in a fully dense polynomial of the same partial degrees) of 0.98.

This table indicates that using smart-pivoting yields up to a factor of 3 speed-up. Comparing `col-wise SP` and `fully SP` shows that calculating P_c (column-wise permutation patterns) along with P_r (row-wise permutation patterns) does not cause any slow-down in the calculation of d.

Moreover, using both multi-step technique and smart-pivoting does not bring any additional speed-up. The smart-pivoting technique is already minimized the cost of exact divisions in each iteration. Table 2 shows $^{\text{plain}}/_{\text{fully SP}}$, $^{\text{plain}}/_{\text{multi-step}}$, and $^{\text{fully SP}}/_{\text{multi-step}}$ ratios from Table 1.

To analyze the performance of parallel FFLU algorithm, we compare Algorithm 5 and Algorithm 4 for $n \times n$ matrices of randomly generated non-zero univariate polynomials with integer coefficients and degree 1. Table 3 summarizes these results. For $n = 75$, $2.14\times$ parallel speed-up is achived, and speed-up continues to increase with increasing n.

Table 1. Compare the execution time (in seconds) of fraction-free LU decomposition algorithms for Bézout matrix of randomly generated, non-zero and sparse polynomials $a, b \in \mathbb{Z}[x_1, x_2, \ldots, x_5]$ with $x_5 < \cdots < x_2 < x_1$, $\deg(a, x_1) = \deg(b, x_1) + 1 = d$, $\deg(a, x_2) = \deg(b, x_2) + 1 = 5$, $\deg(a, x_3) = \deg(b, x_3) = 1$, $\deg(a, x_4) = \deg(b, x_4) = 1$, $\deg(a, x_5) = \deg(b, x_5) = 1$

d	plain	col-wise SP	fully SP	multi-step
6	0.048346	0.018623	0.021154	0.021257
7	2.379480	0.941655	0.954981	0.953532
8	3.997310	0.444759	0.426654	0.475043
9	73.860600	32.531600	31.764200	30.882500
10	2726.690000	1431.430000	1408.140000	1398.370000
11	9059.290000	5113.530000	4768.950000	5348.520000
12	5953.150000	3937.250000	3521.140000	3711.790000
13	81411.900000	42858.500000	42043.600000	41850.800000

Table 2. Ratios of FFLU algorithms for polynomials in Table 1

d	plain/fully SP	plain/multi-step	fully SP/multi-step
6	2.285431	2.274357	0.995155
7	2.491652	2.495438	1.001520
8	9.368973	8.414628	0.898138
9	2.325278	2.391665	1.028550
10	1.936377	1.949906	1.006987
11	1.899640	1.693794	0.891639
12	1.690688	1.603849	0.948637
13	1.936368	1.945289	1.004607

Table 3. Comparing the execution time (in seconds) of Algorithm 4 and Algorithm 5 for $n \times n$ matrices of random non-zero degree 1 univariate integer polynomials

n	serial FFLU	parallel FFLU	serial/parallel
10	00.11976	0.012765	0.938109
15	0.118972	0.076118	1.562994
20	0.628613	0.339738	1.850288
25	2.299270	1.126620	2.040857
30	6.241600	3.109840	2.007049
35	15.305100	7.552200	2.026575
40	33.831800	16.387200	2.064526
45	67.702600	32.307100	2.095595
50	127.438000	60.420000	2.109202
55	224.681000	106.043000	2.118773
60	392.795000	177.456000	2.213478
65	607.089000	284.659000	2.132689
70	947.805000	444.181000	2.133826
75	1432.180000	668.991000	2.140806

3   Bézout Subresultant Algorithms

In this section, we continue exploring the subresultant algorithms for multivariate polynomials based on calculating the determinant of (Hybrid) Bézout matrices.

3.1   Bézout Matrix and Subresultants

A traditional way to define subresultants is via computing determinants of submatrices of the Sylvester matrix (see, e.g. [5] or [11, Ch. 6]). Li [17] presented an elegant way to calculate subresultants directly from the following matrices. This method follows the same idea as subresultants based on Sylvester matrix.

Theorem 1. *The k-th subresultant $S_k(a, b)$ of $a = \sum_{i=0}^{m} a_i y^i, b = \sum_{i=0}^{n} b_i y^i \in \mathbb{B}[y]$ is calculated by the determinant of the following $(m + n - k) \times (m + n - k)$ matrix:*

$$
E_k := \begin{array}{c}
\left. \begin{array}{c} \\ n-k \\ \\ \end{array} \right\{ \\
\left. \begin{array}{c} \\ k \\ \\ \end{array} \right\{ \\
\left. \begin{array}{c} \\ m-k \\ \\ \end{array} \right\{
\end{array}
\begin{bmatrix}
a_m & a_{m-1} & \cdots & a_2 & a_1 & a_0 & & & \\
 & \ddots & & & & & \ddots & & \\
 & & a_m & a_{m-1} & \cdots & a_2 & a_1 & a_0 & \\
 & & & & & 1 & -y & & \\
 & & & & & & \ddots & \ddots & \\
 & & & & & & & 1 & -y \\
b_n & b_{n-1} & \cdots & b_2 & b_1 & b_0 & & & \\
 & \ddots & & & & & \ddots & & \\
 & & b_n & b_{n-1} & \cdots & b_2 & b_1 & b_0 &
\end{bmatrix},
\tag{1}
$$

so that,

$$
S_k(a, b) = (-1)^{k(m-k+1)} \det(E_k).
$$

PROOF. [17, Section 2]

Another practical division-free approach is through utilizing the Bézout matrix to compute the subresultant chain of multivariate polynomials by calculating the determinant of the Bézout matrix of the input polynomials [13]. From [7], we define the *symmetric* Bézout matrix as follows.

Definition 1. *The Bézout matrix associated with $a, b \in \mathbb{B}[y]$, where $m := \deg(a) \geq n := \deg(b)$ is the symmetric matrix:*

$$
Bez(a, b) := \begin{pmatrix}
c_{0,0} & \cdots & c_{0,m-1} \\
\vdots & \ddots & \vdots \\
c_{m-1,0} & \cdots & c_{m-1,m-1}
\end{pmatrix},
$$

where the coefficients $c_{i,j}$, for $0 \le i, j < m$, are defined by the so-called Cayley expression as follows,

$$\frac{a(x)b(y) - a(y)b(x)}{x - y} = \sum_{i,j=0}^{m-1} c_{i,j} y^i x^j.$$

The relations between the *Sylvester* and Bézout matrices have been studied for decades yielding an efficient algorithm to construct the Bézout matrix [2] using a so-called Hybrid Bézout matrix.

Definition 2. *The Hybrid Bézout matrix of* $a = \sum_{i=0}^{m} a_i y^i$ *and* $b = \sum_{i=0}^{n} b_i y^i$ *is defined as the* $m \times m$ *matrix*

$$HBez(a, b) := \begin{pmatrix} h_{0,0} & \cdots & h_{0,m-1} \\ \vdots & \ddots & \vdots \\ h_{m-1,0} & \cdots & h_{m-1,m-1} \end{pmatrix},$$

where the coefficients $h_{i,j}$, *for* $0 \le i, j < m$, *are defined as:*

$$h_{i,j} = \text{coeff}(H_{m-i+1}, m - j) \text{ for } 1 \le i \le n,$$
$$h_{i,j} = \text{coeff}(x^{m-i} b, m - j) \text{ for } m + 1 \le i \le n,$$

with,

$$H_i = (a_m y^{i-1} + \cdots + a_{m-i+1})(b_{n-i} y^{m-i} + \cdots + b_0 y^{m-n})$$
$$- (a_{m-i} y^{m-i} + \cdots + a_0)(b_n y^{i-1} + \cdots + b_{n-i+1}).$$

Example 2. *Consider the polynomials* $a = 5y^5 + y^3 + 2y + 1$ *and* $b = 3y^3 + y + 3$ *in* $\mathbb{Z}[y]$. *The Sylvester matrix of* a, b *is:*

$$Sylv(a, b) = \begin{pmatrix} 5 & 0 & 1 & 0 & 2 & 1 & 0 & 0 \\ 0 & 5 & 0 & 1 & 0 & 2 & 1 & 0 \\ 0 & 0 & 5 & 0 & 1 & 0 & 2 & 1 \\ 3 & 0 & 1 & 3 & 0 & 0 & 0 & 0 \\ 0 & 3 & 0 & 1 & 3 & 0 & 0 & 0 \\ 0 & 0 & 3 & 0 & 1 & 3 & 0 & 0 \\ 0 & 0 & 0 & 3 & 0 & 1 & 3 & 0 \\ 0 & 0 & 0 & 0 & 3 & 0 & 1 & 3 \end{pmatrix},$$

and the Bézout matrix of a, b *is:*

$$Bez(a, b) = \begin{pmatrix} 0 & -15 & 0 & -5 & -15 \\ -15 & 0 & -5 & -15 & 0 \\ 0 & -5 & -15 & 5 & 0 \\ -5 & -15 & 5 & 0 & 0 \\ -15 & 0 & 0 & 0 & -5 \end{pmatrix},$$

while the Hybrid Bézout matrix of a, b *is:*

$$HBez(a, b) = \begin{pmatrix} 15 & -6 & 0 & -2 & -1 \\ 2 & 15 & -6 & -3 & 0 \\ 0 & 2 & 15 & -6 & -3 \\ 3 & 0 & 1 & 3 & 0 \\ 0 & 3 & 0 & 1 & 3 \end{pmatrix}.$$

Diaz-Toca and Gonzalez-Vega examined the relations between Bézout matrices and subresultants in [9]. Hou and Wang studied to apply the Hybrid Bézout matrix for the calculation of subresultants in [13].

Notation 1. *Let* J_m *denote the backward identity matrix of order* m *and let* B *and* H *be defined as follows:*

$$B := J_m \ Bez(a, b) \ J_m = \begin{pmatrix} c_{m-1,m-1} & \cdots & c_{m-1,0} \\ \vdots & \ddots & \vdots \\ c_{0,m-1} & \cdots & c_{0,0} \end{pmatrix},$$

$$H := J_m \ HBez(a, b) \ = \begin{pmatrix} h_{m-1,0} & \cdots & h_{m-1,m-1} \\ \vdots & \ddots & \vdots \\ h_{0,0} & \cdots & h_{0,m-1} \end{pmatrix}.$$

Now, we can state how to compute the subresultants from Bézout matrices as follows.

Theorem 2. *For polynomials* $a = \sum_{i=0}^{m} a_i y^i$ *and* $b = \sum_{i=0}^{n} b_i y^i$ *in* $\mathbb{B}[y]$, *the* k-*th subresultant of* a, b, *i.e.,* $S_k(a, b)$, *can be obtained from:*

$$(-1)^{(m-1)(m-k-1)/2} a_m^{m-n} S_k(a, b) = \sum_{i=0}^{k} B_{m-k,k-i} \ y^i,$$

where $B_{m-k,i}$ *for* $0 \leq i \leq k$ *denotes the* $(m-k) \times (m-k)$ *minor extracted from the first* $m - k$ *rows, the first* $m - k - 1$ *columns and the* $(m - k + i)$-*th column of* B.

PROOF. [2, Theorem 2.3]

Theorem 3. *For those polynomials* $a, b \in \mathbb{B}[y]$, *the* k-*th subresultant of* a, b, *i.e.,* $S_k(a, b)$, *can be obtained from:*

$$(-1)^{(m-1)(m-k-1)/2} S_k(a, b) = \sum_{i=0}^{k} H_{m-k,k-i} \ y^i,$$

where $H_{m-k,i}$ *for* $0 \leq i \leq k$ *denotes the* $(m-k) \times (m-k)$ *minor extracted from the first* $m - k$ *rows, the first* $m - k - 1$ *columns and the* $(m - k + i)$-*th column of* H.

PROOF. [2, Theorem 2.3]

Abdeljaoued et al. in [2] study further this relation between subresultants and Bézout matrices. Theorem 4 is the main result of this paper.

Theorem 4. *For those polynomials $a, b \in \mathbb{B}[y]$, the k-th subresultant of a, b can be obtained from the following $m \times m$ matrices, where $\tau = (m-1)(m-k-1)/2$:*

$$(-1)^{\tau} a_m^{m-n} S_k(a,b) = (-1)^k \begin{vmatrix} c_{m-1,m-1} & c_{m-1,m-2} & \cdots & \cdots & \cdots & c_{m-1,0} \\ \vdots & \vdots & \cdots\cdots & \vdots \\ c_{k,m-1} & c_{k,m-2} & \cdots\cdots & c_{k,0} \\ & & 1 & -y \\ & & & 1 & -y \\ & & & & \ddots & \ddots \\ & & & & & 1 & -y \end{vmatrix},$$

$$(-1)^{\tau} S_k(a,b) = (-1)^k \begin{vmatrix} h_{m-1,0} & h_{m-1,1} & \cdots\cdots & h_{m-1,m-1} \\ \vdots & \vdots & \cdots\cdots & \vdots \\ h_{k,0} & h_{k,1} & \cdots\cdots & h_{k,m-1} \\ & & 1 & -y \\ & & & 1 & -y \\ & & & & \ddots & \ddots \\ & & & & & 1 & -y \end{vmatrix}.$$

PROOF. [2, Theorem 2.4]

The advantage of this aforementioned method is that one can compute the entire subresultant chain in a bottom-up fashion. This process starts from computing the determinant of matrix H (or B) in Definition 1 to calculate $S_0(a,b)$, the resultant of a, b, and update the last k rows of H (or B) to calculate $S_k(a,b)$ for $1 \le k \le n$.

Example 3. *Consider polynomials $a = -5y^4 x + 3yx - y - 3x + 3$ and $b = -2y^3 x + 3y^3 - x$ in $\mathbb{Z}[x, y]$ where $x < y$. From Definition 2, the Hybrid Bézout matrix of a, b is the matrix A from Example 1 on page 34. Recall from Example 1 that the determinant of this matrix can be calculated using the fraction-free LU decomposition schemes. Theorem 4, for $k = 0$, yields that,*

$$S_0(a,b) = -1763x^7 + 7881x^6 - 19986x^5 + 35045x^4 - 41157x^3$$
$$+ 30186x^2 - 12420x + 2187.$$

For $k = 1$, one can calculate $S_1(a,b)$ from the determinant of:

$$H^{(1)} = \begin{pmatrix} -2x+3 & 0 & 0 & -x \\ 0 & 0 & 11x^2 - 11x + 3 & -3(x-1)(2x-3) \\ 0 & 11x^2 - 11x + 3 & -3(x-1)(2x-3) & 0 \\ 0 & 0 & 1 & -y \end{pmatrix},$$

that is,

$$S_1(a,b) = -242x^5 y + 132x^5 + 847x^4 y - 660x^4 - 1100x^3 y + 1257x^3$$
$$+ 693x^2 y - 1134x^2 - 216xy + 486x + 27y - 81.$$

We can continue calculating subresultants of higher indices with updating matrix
$H^{(1)}$. *For instance, the 2nd and 3rd subresultants are, respectively, from the*
determinant of:

$$H^{(2)} = \begin{pmatrix} -2x+3 & 0 & 0 & -x \\ 0 & 0 & 11x^2 - 11x + 3 & -3(x-1)(2x-3) \\ 0 & 1 & -y & 0 \\ 0 & 0 & 1 & -y \end{pmatrix},$$

and,

$$H^{(3)} = \begin{pmatrix} -2x+3 & 0 & 0 & -x \\ 1 & -y & 0 & 0 \\ 0 & 1 & -y & 0 \\ 0 & 0 & 1 & -y \end{pmatrix},$$

which are,

$$S_2(a, b) = 22yx^3 - 12x^3 - 55yx^2 + 48x^2 + 39yx - 63x - 9y + 27,$$
$$S_3(a, b) = -2y^3x + 3y^3 - x.$$

We further studied the performance of computing subresultants from Theorem 4 in comparison to the Hybrid Bézout matrix in Definition 2 for multivariate polynomials with integer coefficients. In our implementation, we took advantage of the *FFLU* schemes reviewed in Sect. 2 to compute the determinant of these matrices using *smart-pivoting* technique in *parallel*; see Sect. 3.3 for implementation details and results.

3.2   Speculative Bézout Subresultant Algorithms

In Example 3, the Hybrid Bézout matrix was used to compute subresultants of two polynomials in $\mathbb{Z}[x, y]$. We constructed the square matrix H from Definition 1 and updated the last $k \geq 0$ rows following Theorem 4. Thus, the kth subresultant could be directly computed from the determinant of this matrix.

Consider solving systems of polynomial equations by triangular decomposition, and particularly, regular chains. This method uses a *Regular GCD* subroutine (see [8]) which requires the computation of subresultants in a bottom-up fashion: for multivariate polynomials a, b (viewed as univariate in their main variable) compute $S_0(a, b)$, then possibly $S_1(a, b)$, then possibly $S_2(a, b)$, etc., to try and find a regular GCD. This bottom-up approach for computing subresultant chains is discussed in [5].

In the approach explained in the previous section, we would call the determinant algorithm twice for $H^{(0)} := H$ and $H^{(1)}$ to compute S_0, S_1 respectively. Here, we study a speculative approach to compute both S_0 and S_1 at the cost of computing only one of them. This approach can also be extended to compute any two successive subresultants S_k, S_{k+1} for $2 \leq k < \deg(b, x_n)$.

To compute S_0, S_1 of polynomials $a = -5y^4x + 3yx - y - 3x + 3$ and $b = -2y^3x + 3y^3 - x$ in $\mathbb{Z}[x, y]$ from Example 3, consider the $(m + 1) \times m$ matrix, with $m = 4$, derived from the Hybrid Bézout matrix of a, b, $H^{(0,1)} =$

$$\begin{pmatrix} -2x + 3 & 0 & 0 & -x \\ 0 & 0 & 11x^2 - 11x + 3 & -3(x - 1)(2x - 3) \\ 0 & 11x^2 - 11x + 3 & -3(x - 1)(2x - 3) & 0 \\ \mathbf{11x^2 - 11x + 3} & \mathbf{-3(x - 1)(2x - 3)} & \mathbf{0} & \mathbf{0} \\ \mathit{0} & \mathit{0} & \mathit{1} & \mathit{-y} \end{pmatrix}.$$

In this matrix, the first three rows are identical to the first three rows of $H^{(0)}$ and $H^{(1)}$, while the 4th (**bold**) row is the 4th row of $H^{(0)}$ and the 5th (*italicized*) row is the 4th row of $H^{(1)}$. A deeper look into the determinant algorithm reveals that the *Gaussian (row) elimination* for the first three rows in each iteration of the fraction-free LU decomposition is similar in both $H^{(0)}$ and $H^{(1)}$ and the only difference is within the 4th row.

Hence, managing these row eliminations in the fraction-free LU decomposition, we can compute determinants of $H^{(0)}$ and $H^{(1)}$ by using $H^{(0,1)}$ only calling the FFLU algorithm once. Indeed, when this algorithm tries to eliminate the last rows of $H^{(0)}$ and $H^{(1)}$, we should use the last two rows of $H^{(0,1)}$ separately and return two denominators corresponding to S_0, S_1.

We can further extend this speculative approach to compute S_2 and S_3 by updating the matrix $H^{(0,1)}$ to get the $(m + 3) \times m$ matrix $H^{(2,3)} =$

$$\begin{pmatrix} -2x + 3 & 0 & 0 & -x \\ \mathbf{0} & \mathbf{0} & \mathbf{11x^2 - 11x + 3} & \mathbf{-3(x - 1)(2x - 3)} \\ \cancel{0} & \cancel{11x^2 - 11x + 3} & \cancel{-3(x - 1)(2x - 3)} & \cancel{0} \\ \cancel{11x^2 - 11x + 3} & \cancel{-3(x - 1)(2x - 3)} & \cancel{0} & \cancel{0} \\ \mathit{1} & \mathit{-y} & \mathit{0} & \mathit{0} \\ 0 & 1 & -y & 0 \\ 0 & 0 & 1 & -y \end{pmatrix}.$$

Therefore, to calculate subresultants of index 2 and 3, we should respectively consider the 2nd (**bold**) and 5th (*italicized*) rows of $H^{(2,3)}$ in the fraction-free LU decomposition while ignoring the 3rd and 4th (strikethrough) rows. An adaptation of the FFLU algorithm can then modify $H^{(2,3)}$ as follows to return $d_{(2)}$, ignoring the 5th and strikethrough rows.

$$\begin{pmatrix} -2x + 3 & 0 & 0 & -x \\ 0 & -2x + 3 & -y(-2x + 3) & 0 \\ \cancel{0} & \cancel{11x^2 - 11x + 3} & \cancel{-3(x - 1)(2x - 3)} & \cancel{0} \\ \cancel{11x^2 - 11x + 3} & \cancel{-3(x - 1)(2x - 3)} & \cancel{0} & \cancel{0} \\ 1 & -y & 0 & 0 \\ 0 & 0 & -22x^3 + 55x^2 - 39x + 9 & 3(x - 1)(2x - 3)^2 \\ 0 & 0 & -2x + 3 & d_{(2)} \end{pmatrix},$$

where $d_{(2)} = -22x^3y + 12x^3 + 55x^2y - 48x^2 - 39xy + 63x + 9y - 27$ and $S_2 = -d_{(2)}$. Note that the 2nd and 6th rows are swapped to find a proper pivot.

The adapted FFLU algorithm can also modify $H^{(2,3)}$ to rather return $d_{(3)}$, ignoring the 2nd (**bold**) and strikethrough rows,

$$
\begin{pmatrix}
-2x+3 & 0 & 0 & -x \\
\mathbf{0} & \mathbf{0} & \mathbf{11x^2-11x+3} & \mathbf{-3(x-1)(2x-3)} \\
\cancel{0} & \cancel{11x^2-11x+3} & \cancel{-3(x-1)(2x-3)} & \cancel{0} \\
\cancel{11x^2-11x+3} & \cancel{-3(x-1)(2x-3)} & \cancel{0} & \cancel{0} \\
1 & -y(-2x+3) & 0 & x \\
0 & -2x+3 & -y^2(-2x+3) & x \\
0 & 0 & y(-2x+3) & d_{(3)}
\end{pmatrix},
$$

where $d_{(3)} = -2xy^3 + 3y^3 - x$ and $S_3 = d_{(3)}$.

Generally, to compute subresultants of index k and $k+1$, one can construct the matrix $H^{(k,k+1)}$ from the previously constructed $H^{(k-2,k-1)}$ for $k > 1$. This *recycling* of previous information makes computing the next subresultants of index k and $k+1$ much more efficient, and is discussed below. We proceed with an adapted FFLU algorithm over:

- the first $m - k - 1$ rows,
- the **bold** row for computing S_k, or the *italicized* row for computing S_{k+1}, and
- the last k rows

of matrix $H^{(k,k+1)} \in \mathbb{B}^{(m+k)\times k}$ with $\mathbb{B} = \mathbb{Z}[x_1, \ldots, x_v]$,

$$
H^{(k,k+1)} =
\begin{pmatrix}
h_{m-1,0} & h_{m-1,1} & \cdots\cdots\cdots & h_{m-1,m-1} \\
\vdots & \vdots & \cdots\cdots\cdots & \vdots \\
\mathbf{h_{k,0}} & \mathbf{h_{k,1}} & \cdots\cdots\cdots & \mathbf{h_{k,m-1}} \\
h_{k-1,0} & h_{k-1,1} & \cdots\cdots\cdots & h_{k-1,m-1} \\
\vdots & \vdots & \cdots\cdots\cdots & \vdots \\
h_{0,0} & h_{0,1} & \cdots\cdots\cdots & h_{0,m-1} \\
& 1 & -y & \\
& & \ddots & \ddots \\
& & 1 & -y
\end{pmatrix}.
$$

As seen in the last example, the FFLU algorithm, depending on the input polynomials, may create two completely different submatrices to calculate $d_{(2)}$ and $d_{(3)}$. Thus, the cost of computing S_k, S_{k+1} from $H^{(k,k+1)}$ speculatively may not necessarily be less than computing them successively from $H^{(k)}, H^{(k+1)}$ for some $k > 1$.

We improve the performance of computing S_k, S_{k+1} speculatively via *caching*, and then reusing, intermediate data calculated to compute S_{k-2}, S_{k-1} from $H^{(k-2,k-1)}$. In this approach, the adapted FFLU algorithm returns $d_{(k-2)}, d_{(k-1)}$ along with $H^{(k-2,k-1)}$, the reduced matrix $H^{(k-2,k-1)}$ to compute $d_{(k-1)}$, the list of permutation patterns and pivots.

Therefore, we can utilize $H^{(k-2,k-1)}$ to construct $H^{(k,k+1)}$. In addition, if the first $\delta := m - k - 1$ pivots are picked from the first δ rows of $H^{(k-2,k-1)}$, then one can use the first δ rows of the reduced matrix $H^{(k-2,k-1)}$ along with the list of permutation patterns and pivots to perform the first δ row eliminations of $H^{(k,k+1)}$ via recycling the first δ rows of the reduced matrix cached *a priori*.

3.3   Experimentation

In this section, we compare the subresultant algorithms based on (Hybrid) Bézout matrix against the Ducos' subresultant chain algorithm in BPAS and MAPLE 2020. In BPAS, our optimized Ducos' algorithm (denoted OptDucos), is detailed in [5].

Table 4 and Table 5 show the running time of plain and speculative algorithms for randomly generated, non-zero, sparse polynomials $a, b \in \mathbb{Z}[x_1, x_2, \ldots, x_6]$ with $x_6 < \cdots < x_2 < x_1$, $\deg(a, x_1) = \deg(b, x_1) + 1 = d$, and $\deg(a, x_i) = \deg(b, x_i) = 1$ for $2 \le i \le 6$. Table 6 and Table 7 show the running time of plain, speculative and caching subresultant schemes for randomly generated, non-zero, and sparse polynomials $a, b \in \mathbb{Z}[x_1, x_2, \ldots, x_7]$ with $x_7 < \cdots < x_2 < x_1$, $\deg(a, x_1) = \deg(b, x_1) + 1 = d$, and $\deg(a, x_i) = \deg(b, x_i) = 1$ for $2 \le i \le 7$.

Note that the Bézout algorithm in MAPLE computes the resultant of a, b ($S_0(a, b)$) meanwhile both MAPLE's and BPAS's Ducos' algorithm computes the entire subresultant chain. In BPAS, we have the following:

1. Bézout ($\rho = 0$) calculates the resultant ($S_0(a, b)$) via the determinant of Hybrid Bézout matrix of a, b;
2. Bézout ($\rho = 1$) calculates $S_1(a, b)$ following Theorem 4 from the Hybrid Bézout matrix of a, b;
3. SpecBézout ($\rho = 0$) calculates $S_0(a, b), S_1(a, b)$ speculatively using $H^{(0,1)}$;
4. SpecBézout ($\rho = 2$) calculates $S_2(a, b), S_3(a, b)$ speculatively using $H^{(2,3)}$;
5. SpecBézout$_{\text{cached}}$ ($\rho = 2$) calculates $S_2(a, b), S_3(a, b)$ speculatively via $H^{(2,3)}$ and the *cached* information calculated in *SpecBézout* ($\rho = 0$)
6. SpecBézout$_{\text{cached}}$ ($\rho = \text{all}$) calculates the entire subresultant chain using the speculative algorithm and caching.

To compute subresultants from Bézout matrices in MAPLE, we use the command `SubresultantChain( ... , 'representation'='BezoutMatrix')` from the `RegularChains` library. Our Bézout algorithm is up to 3× faster than the MAPLE implementation to calculate only S_0. Moreover, our results show that Bézout algorithms outperform the Ducos' algorithm in both BPAS and MAPLE for sparse polynomials with many variables.

Tables 4 and 6 show that the cost of computing subresultants S_0, S_1 speculatively is comparable to the running time of computing only one of them. Tables 5 and 7 indicate the importance of recycling cached data to compute higher subresultants speculatively. Our Bézout algorithms can calculate all subresultants speculatively in a comparable running time to the Ducos' algorithm.

Table 4. Comparing the execution time (in seconds) of subresultant algorithms based on Bézout matrix for randomly generated, non-zero, sparse polynomials $a, b \in \mathbb{Z}[x_6 < \ldots < x_1]$, $\deg(a, x_1) = \deg(b, x_1) + 1 = d$, and $\deg(a, x_i) = \deg(b, x_i) = 1$ for $2 \le i \le 6$

	MAPLE		BPAS			
d	Bézout $(\rho = 0)$	Ducos	Bézout $(\rho = 0)$	Bézout $(\rho = 1)$	SpecBézout $(\rho = 0)$	OptDucos
10	0.05128	0.03000	0.024299	0.026762	0.032166	0.045270
11	0.06001	0.04574	0.057312	0.068722	0.058843	0.049532
12	0.02515	0.05100	0.007223	0.019530	0.012792	0.061419
13	0.81209	16.81200	0.421278	0.739842	0.594225	9.527660
14	3.14360	112.280	2.414530	3.829530	3.250710	69.957100
15	518.380	7163.30	151.656	779.9240	512.260	3655.820

Table 5. Comparing the execution time (in seconds) of speculative subresultant algorithms for polynomials in Table 4

	BPAS			
d	SpecBézout$(\rho = 0)$	SpecBézout$(\rho = 2)$	SpecBézout$_{cached}(\rho = 2)$	SpecBézout$_{cached}(\rho = all)$
10	0.032166	0.022125	0.016432	0.076283
11	0.058843	0.079425	0.043512	0.193512
12	0.012792	0.010566	0.004148	0.071435
13	0.594225	2.106280	1.535510	7.891180
14	3.250710	8.735510	4.133760	73.59940
15	512.260	953.1170	579.8580	4877.130

Table 6. Comparing the execution time (in seconds) of subresultant algorithms based on Bézout matrix for randomly generated, non-zero, sparse polynomials $a, b \in \mathbb{Z}[x_7 < \ldots < x_1]$, $\deg(a, x_1) = \deg(b, x_1) + 1 = d$, and $\deg(a, x_i) = \deg(b, x_i) = 1$ for $2 \le i \le 7$

	MAPLE		BPAS			
d	Bézout$(\rho = 0)$	Ducos	Bézout$(\rho = 0)$	Bézout$(\rho = 1)$	SpecBézout$(\rho = 0)$	OptDucos
6	0.00098	0.00372	0.001303	0.001427	0.001553	0.002444
7	0.01148	0.43145	0.080210	0.174460	0.095569	0.279023
8	15.1850	34.8540	7.057270	10.834100	8.380050	22.440500
9	74.1390	327.570	36.8450	66.8430	44.7160	194.4860
10	9941.20	inf	4130.980	6278.240	5686.060	14145.30

Table 7. Comparing the execution time (in seconds) of speculative and caching subresultant algorithms for polynomials in Table 6

			BPAS	
d	SpecBézout$_{(\rho\,=\,0)}$	SpecBézout$_{(\rho\,=\,2)}$	SpecBézout$_{\text{cached}(\rho\,=\,2)}$	SpecBézout$_{\text{cached}(\rho\,=\,\text{all})}$
6	0.001553	0.001812	0.001350	0.003519
7	0.095569	0.103801	0.053730	0.213630
8	8.380050	13.10210	5.7240	25.83050
9	44.7160	67.86560	31.12090	136.8930
10	5686.060	8853.10	3856.550	17569.20

As described in Sect. 3.1, polynomial system solving benefits from computing regular GCDs in a bottom-up approach. From a test suite of over 3000 polynomial systems, coming from the literature and collected from MAPLE user-data (see [4, Section 6]) we compare the benefits of (Speculative) Bézout methods for computing subresultants vs BPAS's optimized Ducos algorithm. Table 8 shows this data for some systems of the test suite with at least 5 variables. Table 9 shows systems which are very challenging to solve, requiring at least 50 s. For these hard systems, speculative methods achived a speed-up of up to 1.6× compared to Ducos' method. Note that, in some cases, the regular GCD has high degree and is thus equal to a subresultant of high index. Thus, Ducos' method to compute the entire subresultant chain may be more efficient than repeated calls to the speculative method.

Table 8. Comparing time (in seconds) to solve polynomial systems with **nvar** $\geq$ 5; system names come from a test suite detailed [4]

SysName	OptDucos	Bézout	SpecBézout	SpecBézout$_{\text{cached}}$	OptDucos/ SpecBézout	Bézout/ SpecBézout
Sys2922	7.91041	7.93589	7.95695	7.95698	0.994151	0.997353
Sys2880	5.55801	5.70138	5.46921	5.41538	1.016236	1.042450
Sys2433	8.75830	8.77473	8.75625	8.76812	1.000234	1.002110
Sys2161	1.08153	0.89279	0.56666	0.63128	1.908605	1.575530
Sys2642	8.06066	6.97177	4.89233	3.21021	1.647612	1.425041
Sys2695	3.18267	3.05706	2.98045	2.12872	1.067849	1.025704
Sys2238	8.75708	8.75923	8.75251	8.75813	1.000522	1.000768
Sys2943	6.70348	6.12246	4.54511	4.69512	1.474877	1.347043
Sys1935	4.14390	5.01831	3.01449	1.98466	1.374660	1.664729
Sys2882	2.42182	2.37203	2.38065	2.35716	1.017294	0.996379
Sys2588	4.49268	4.51135	4.49201	4.49792	1.000149	1.004305
Sys2449	1.23251	1.28321	1.24507	1.26588	0.989912	1.030633
Sys2874	6.99887	7.22326	6.99438	7.11027	1.000642	1.032723
Sys2932	6.27556	6.25798	6.31953	6.29113	0.993042	0.990260
Sys2269	1.03128	1.03253	1.03961	1.04012	0.991987	0.993190

Table 9. Comparing time (in seconds) to solve "hard" polynomial systems with nvar ≥ 5; system names come from a test suite detailed [4]

SysName	OptDucos	Bézout	SpecBézout	SpecBézout$_{cached}$	OptDucos/ SpecBézout	Bézout/ SpecBézout
Sys2797	466.4250	425.8670	386.3810	325.1170	1.207163	1.102194
Sys2539	55.8694	55.8531	55.5113	55.4933	1.006451	1.006157
Sys2681	458.6800	458.5810	458.5360	458.5780	1.000314	1.000098
Sys2745	599.8020	599.3290	599.0610	599.2150	1.001237	1.000447
Sys3335	6406.7400	5843.7300	4799.9700	4801.1200	1.334746	1.217451
Sys2703	322.2940	487.0120	485.8170	491.1520	0.663406	1.002460
Sys2000	55.7026	56.1724	56.3106	57.0079	0.989203	0.997546
Sys2877	2127.4900	1914.5200	1253.8200	1247.4400	1.696807	1.526950

References

1. Abdeljaoued, J., Diaz-Toca, G.M., Gonzalez-Vega, L.: Minors of Bézout matrices, subresultants and the parameterization of the degree of the polynomial greatest common divisor. Int. J. Comput. Math. **81**(10), 1223–1238 (2004)
2. Abdeljaoued, J., Diaz-Toca, G.M., González-Vega, L.: Bézout matrices, subresultant polynomials and parameters. Appl. Math. Comput. **214**(2), 588–594 (2009)
3. Asadi, M., et al.: Basic Polynomial Algebra Subprograms (BPAS) (2021). http://www.bpaslib.org
4. Asadi, M., Brandt, A., Moir, R.H.C., Moreno Maza, M., Xie, Y.: Parallelization of triangular decompositions: techniques and implementation. J. Symb. Comput. (2021, to appear)
5. Asadi, M., Brandt, A., Moreno Maza, M.: Computational schemes for subresultant chains. In: Boulier, F., England, M., Sadykov, T.M., Vorozhtsov, E.V. (eds.) CASC 2021. LNCS, vol. 12865, pp. 21–41. Springer, Cham (2021). https://doi.org/10.1007/978-3-030-85165-1_3
6. Bareiss, E.H.: Sylvester's identity and multistep integer-preserving Gaussian elimination. Math. Comput. **22**(103), 565–578 (1968)
7. Bini, D., Pan, V.Y.: Polynomial and Matrix Computations: Fundamental Algorithms. Springer, New York (2012). https://doi.org/10.1007/978-1-4612-0265-3
8. Chen, C., Moreno Maza, M.: Algorithms for computing triangular decomposition of polynomial systems. J. Symb. Comput. **47**(6), 610–642 (2012)
9. Diaz-Toca, G.M., Gonzalez-Vega, L.: Various new expressions for subresultants and their applications. Appl. Algebra Eng. Commun. Comput. **15**(3–4), 233–266 (2004). https://doi.org/10.1007/s00200-004-0158-4
10. Ducos, L.: Optimizations of the subresultant algorithm. J. Pure Appl. Algebra **145**(2), 149–163 (2000)
11. von zur Gathen, J., Gerhard, J.: Modern Computer Algebra, 3rd edn. Cambridge University Press, Cambridge (2013)
12. Geddes, K.O., Czapor, S.R., Labahn, G.: Algorithms for Computer Algebra. Springer, New York (1992). https://doi.org/10.1007/b102438
13. Hou, X., Wang, D.: Subresultants with the Bézout matrix. In: Computer Mathematics, pp. 19–28. World Scientific (2000)

14. Jeffrey, D.J.: LU factoring of non-invertible matrices. ACM Commun. Comput. Algebra **44**(1/2), 1–8 (2010)
15. Kerber, M.: Division-free computation of subresultants using Bézout matrices. Int. J. Comput. Math. **86**(12), 2186–2200 (2009)
16. Lemaire, F., Moreno Maza, M., Xie, Y.: The RegularChains library in MAPLE. ACM SIGSAM Bull. **39**(3), 96–97 (2005)
17. Li, Y.B.: A new approach for constructing subresultants. Appl. Math. Comput. **183**(1), 471–476 (2006)
18. Olver, P.J., Shakiban, C.: Applied Linear Algebra. Prentice Hall, Upper Saddle River (2006)

Application of Symbolic-Numerical Modeling Tools for Analysis of Gyroscopic Stabilization of Gyrostat Equilibria

Andrei V. Banshchikov[✉]

Matrosov Institute for System Dynamics and Control Theory of Siberian Branch
of Russian Academy of Sciences, PO Box 292,
134, Lermontov str., Irkutsk 664033, Russia
bav@icc.ru

Abstract. Using the applied software developed on the basis of the computer algebra system *"Mathematica"* and its functions of symbolic-numerical modeling, the dynamics of the rotational motion along the circular orbit of a satellite-gyrostat in a Newtonian central field of forces is investigated. In accordance with the problem of Lyapunov's stability from the equations of perturbed motion in the first approximation, the regions with an even degree of instability by Poincaré are found in the space of introduced parameters. The paper considers the question of the possibility of gyroscopic stabilization of unstable relative equilibrium positions of the gyrostat, when the vector of the gyrostatic moment of the system is located in one of the planes formed by the principal central axes of inertia. The research results were obtained in a symbolic (analytic) form on a computer and by means of a numerical experiment with the graphic interpretation.

Keywords: Orbital gyrostat · Degree of instability · Gyroscopic stabilization · Parametric analysis · System of inequalities

1 Introduction

The problems of reliability and accuracy of computation, as well as the question of speeding-up of the investigation process can be partially solved if a computer algebra system (CAS) is chosen as a software tool. As can be seen from the publications, more often, there is an approach to using the CAS as a calculator for solving a particular problem. There is another approach when, based on the internal programming language of CAS, an applied software is developed to solve a specific class of problems.

The use of CAS in the problems of celestial mechanics has its own prehistory and is very important for specialists (see, for example, [10,12]). The classical problem of the influence of the structure of forces on the stability of the equilibria of mechanical systems [9] began to develop in the 19th century – the effect of

F. Boulier et al. (Eds.): CASC 2022, LNCS 13366, pp. 51–61, 2022.
https://doi.org/10.1007/978-3-031-14788-3_4

gyroscopic stabilization was discovered. Nevertheless, the problem remains relevant: see, for example, the review in [1] among the large number of publications on this topic. With the help of symbolic computations, the author in the present paper investigates the dynamics of an orbital gyrostat and the question of the possibility of gyroscopic stabilization of its unstable equilibrium positions.

The rigid body with the fixed axis of a statically and dynamically balanced flywheel rotating about that axis with a constant relative angular velocity is a gyrostat. The system moves along the circular Keplerian orbit in a central Newtonian field of forces around the gravitational center. It is accepted that the mutual influence of the motion of the gyrostat about its mass center and the displacement of the latter at a constant angular velocity along the above-mentioned trajectory are neglected. This is a so-called limited formulation of the problem of orbital motion [8].

The stability of the relative equilibrium positions of a satellite-gyrostat for various variants of the positioning of the flywheel rotation axis in its body has been considered by many authors. For example, in [14], the regions are found in the parameter space where there are different numbers of equilibrium orientations of the system, and sufficient conditions for their Lyapunov's stability are obtained from the analysis of the sign-definiteness of the generalized energy integral. The stabilization of the equilibria of an orbital gyrostat (with an arbitrary inertia ellipsoid) using the equations of the first approximation, that is presented here, continues and supplements the studies performed earlier for oblate [2] and prolate [5] axisymmetric gyrostats.

2   Construction of a Symbolic Model and Stability Conditions

For the description of a motion of the system, two rectangular coordinate systems with the poles in the system's mass center O are introduced: $OZ_1Z_2Z_3$ is the orbital coordinate system (OCS), and the coordinate system $Oz_1z_2z_3$ rigidly connected to a body has the axes directed along the principal central axes of inertia of the gyrostat. A, B, and C are the moments of inertia of the system relative to the axes Oz_1, Oz_2, Oz_3, and h_j are the projections (onto the corresponding axis) of a vector of gyrostatic moment of system divided by ω (the module of orbital angular velocity). For the definition of a relative positioning of the OZ_k and Oz_j axes, the directional cosines defined by the aircraft angles α, β, γ are used (see, for example, [14]).

By the construction of a symbolical model, one implies the obtaining of non-linear and linearized differential equations of motion in analytic form in computer memory. The software package (SP) [6] used for this paper is designed for modeling and qualitative analysis in symbolic form of dynamic systems (in particular, the systems of interconnected absolutely rigid bodies). This applied software, the functional description and application technology of which is given in [4, 7], is a set of interactive programs executed in the interpretation mode in the CAS "*Mathematica*" [15] environment.

Consider the position of equilibrium ($\dot\alpha = 0$, $\dot\beta = 0$, $\dot\gamma = 0$) in regard to OCS in general form:

$$\alpha = \alpha_0 = \text{const}, \quad \beta = \beta_0 = \text{const}, \quad \gamma = \gamma_0 = \text{const}. \tag{1}$$

The equations of motion of the satellite-gyrostat with respect to its center of mass in Euler form are widely known (see review [13]). With the help of SP [6], the following results in a symbolic form in computer memory are obtained:

(a) kinetic energy and force function of the approximate Newtonian field of gravitation (as given in [13]);
(b) nonlinear differential equations in the Lagrange form of the second kind describing the motion of an orbital gyrostat;
(c) existence conditions of equilibrium (1).

For example, let us write down the equations determining the relative equilibria of the gyrostat (i.e., conditions (c)):

$$
\begin{cases}
\sin 2\alpha_0 \left(2 \left(2A - B - C\right) \cos^2\beta_0 - (B - C)\left(\cos 2\beta_0 - 3\right) \cos 2\gamma_0 \right) \\
\quad + 4 \left(B - C\right) \cos 2\alpha_0 \sin\beta_0 \sin 2\gamma_0 \ = \ 0, \\[6pt]
\left(B + C - 2\,A + (B - C) \cos 2\gamma_0\right) \sin 2\beta_0 \left(5 - 3 \cos 2\alpha_0\right) \\
\quad + 6 \left(B - C\right) \cos\beta_0 \sin 2\alpha_0 \sin 2\gamma_0 \\
\quad + 8 \left(\sin\beta_0 \left(h_2 \cos\gamma_0 - h_3 \sin\gamma_0\right) - h_1 \cos\beta_0 \right) \ = \ 0, \\[6pt]
\left(B - C\right) \left(\sin 2\gamma_0 \left(3 \cos^2\alpha_0 \cos^2\beta_0 - 3 \sin^2\alpha_0 \sin^2\beta_0\right) \right. \\
\left. + 3 \sin 2\alpha_0 \sin\beta_0 \cos 2\gamma_0\right) + 2 \cos\beta_0 \left(h_2 \sin\gamma_0 + h_3 \cos\gamma_0\right) \ = \ 0.
\end{cases}
\tag{2}
$$

The necessary conditions of stability for the equilibrium can be obtained from the equations of perturbed motion in the first approximation. The linearized equations of perturbed motion in vicinity of (1) look as follows:

$$M\,\ddot q + G\,\dot q + K\,q = 0, \tag{3}$$

where $q = \left(\overline\alpha, \overline\beta, \overline\gamma\right)^T$ is the column vector of deviations of generalized coordinates from the unperturbed motion (1);

$$M = \begin{pmatrix} M_{11} & M_{12} & M_{13} \\ M_{12} & M_{22} & 0 \\ M_{13} & 0 & M_{33} \end{pmatrix}$$ is a positive definite symmetric matrix of kinetic energy;

$$G = \begin{pmatrix} 0 & G_{12} & G_{13} \\ -G_{12} & 0 & G_{23} \\ -G_{13} & -G_{23} & 0 \end{pmatrix}$$ is a skew-symmetric matrix of gyroscopic forces;

$$K = \begin{pmatrix} K_{11} & K_{12} & K_{13} \\ K_{12} & K_{22} & K_{23} \\ K_{13} & K_{23} & K_{33} \end{pmatrix}$$ is a symmetric matrix of potential forces.

Here

$$M_{11} = A\sin^2\beta_0 + (B\cos^2\gamma_0 + C\sin^2\gamma_0)\cos^2\beta_0\,, \quad M_{22} = B\sin^2\gamma_0 + C\cos^2\gamma_0\,,$$
$$M_{12} = (B-C)\cos\beta_0\sin\gamma_0\cos\gamma_0\,, \quad M_{13} = A\sin\beta_0\,, \quad M_{33} = A\,;$$

$$G_{12} = \sin 2\beta_0(A - B\cos^2\gamma_0 - C\sin^2\gamma_0) + h_1\cos\beta_0$$
$$+ \sin\beta_0(h_3\sin\gamma_0 - h_2\cos\gamma_0)\,,$$
$$G_{13} = -\cos\beta_0\,(\cos\beta_0(B-C)\sin 2\gamma_0 + h_2\sin\gamma_0 + h_3\cos\gamma_0)\,,$$
$$G_{23} = -\cos\beta_0\,(A + (C-B)\cos 2\gamma_0) - h_3\sin\gamma_0 + h_2\cos\gamma_0\,;$$

$$K_{11} = \frac{3}{4}\left(\cos 2\alpha_0\big((4A - 2(B+C))\cos^2\beta_0 - (B-C)(\cos 2\beta_0 - 3)\cos 2\gamma_0\big)\right)$$
$$+ 3\,(C-B)\sin 2\alpha_0\sin\beta_0\sin 2\gamma_0\,,$$

$$K_{12} = 3\cos\beta_0\left((B\cos^2\gamma_0 + C\sin^2\gamma_0 - A)\sin 2\alpha_0\sin\beta_0\right.$$
$$\left. + \frac{1}{2}(B-C)\cos 2\alpha_0\sin 2\gamma_0\right),$$

$$K_{13} = \frac{3}{4}(B-C)(4\cos 2\alpha_0\sin\beta_0\cos 2\gamma_0 + \sin 2\alpha_0(\cos 2\beta_0 - 3)\sin 2\gamma_0)\,,$$

$$K_{22} = \frac{1}{4}\big((2A + (C-B)\cos 2\gamma_0 - B - C)(3\cos 2\alpha_0 - 5)\cos 2\beta_0$$
$$+ 3\,(C-B)\sin 2\alpha_0\sin\beta_0\sin 2\gamma_0) + h_1\sin\beta_0 + (h_2\cos\gamma_0 - h_3\sin\gamma_0)\cos\beta_0\,,$$

$$K_{23} = \frac{1}{4}(B-C)\,(6\sin 2\alpha_0\cos\beta_0\cos 2\gamma_0\,(3\cos 2\alpha_0 - 5)\sin 2\beta_0\sin 2\gamma_0)$$
$$- (h_2\sin\gamma_0 + h_3\cos\gamma_0)\sin\beta_0\,,$$

$$K_{33} = \frac{1}{4}(B-C)\,(\cos 2\gamma_0\,(10\cos^2\beta_0 - 3\cos 2\alpha_0\,(\cos 2\beta_0 - 3))$$
$$- 12\sin 2\alpha_0\sin\beta_0\sin 2\gamma_0) + (h_2\cos\gamma_0 - h_3\sin\gamma_0)\cos\beta_0\,.$$

All derivatives in (3) are calculated by the dimensionless time $\tau = \omega t$.

The characteristic equation: $\det(M\lambda^2 + G\lambda + K) = v_3\lambda^6 + v_2\lambda^4 + v_1\lambda^2 + v_0 = 0$ of system (3) contains λ only in even degrees. The stability of equilibrium (1) takes place when all roots with respect to λ^2, being simple, will be real negative numbers. The algebraic conditions providing specified properties of the roots (*necessary conditions of stability*) represent the system of inequalities [11]:

$$\begin{cases} v_3 \equiv \det M > 0\,, \quad v_2 > 0\,, \quad v_1 > 0\,, \quad v_0 \equiv \det K > 0\,, \\ Dis \equiv v_2^2 v_1^2 - 4v_1^3 v_3 - 4v_2^3 v_0 + 18v_3 v_2 v_1 v_0 - 27v_0^2 v_3^2 > 0\,. \end{cases} \tag{4}$$

The first condition in (4) is always satisfied by virtue of the positive definiteness of the kinetic energy matrix. We should note that, if at least one of the conditions in (4) is replaced by a strict contrary inequality, system (3) will be unstable, according to the Lyapunov theorem on instability in the first approximation [9].

Note also that the construction of the symbolic linearized model (3) (i.e., obtaining in the analytical form the elements of the matrices M, G, K), the calculation of the coefficients v_i ($i = \overline{0,3}$) and the discriminant Dis from (4) was also performed using the SP [6].

3   Parametric Analysis

3.1   Investigated Relative Equilibrium Positions

In [10], using the algorithms of constructing Gröbner bases, all the equilibrium positions in regard to the OCS of a satellite-gyrostat are determined analytically or numerically for three special cases. For these cases, the gyrostatic moment vector is in one of the planes formed by the satellite's principal central axes of inertia. For example, in case $h_3 = 0$, $h_1 \neq 0$, and $h_2 \neq 0$, there are the equilibrium orientations (the solutions of system of equations (2)):

$$\begin{cases} \alpha = \alpha_0 = \pi/2, \quad \gamma = \gamma_0 = 0, \\ \beta = \beta_0 = \text{const}: \quad h_2 \sin\beta_0 - \cos\beta_0 \, (h_1 + 4 \, (A - B) \sin\beta_0) = 0; \end{cases} \quad (5)$$

$$\begin{cases} \alpha = \alpha_0 = 0, \quad \gamma = \gamma_0 = 0, \\ \beta = \beta_0 = \text{const}: \quad h_2 \sin\beta_0 - \cos\beta_0 \, (h_1 + (A - B) \sin\beta_0) = 0. \end{cases} \quad (6)$$

Let us parametrize the problem. Without loss of generality, let $h_i > 0$, $(i = 1, 2)$, and $B > A > C$ for definiteness.

Let us introduce dimensionless parameters:

$$H_1 \equiv \frac{h_1}{B}; \quad H_2 \equiv \frac{h_2}{B}; \quad J_A \equiv \frac{A}{B}; \quad J_C \equiv \frac{C}{B}; \quad p_c \equiv \cos\beta_0; \quad p_s \equiv -\sin\beta_0. \quad (7)$$

The values of the parameters belong to the intervals:

$$H_i > 0, \, (i = 1, 2); \quad 1/2 < J_A < 1, \; 1 - J_A < J_C < J_A; \quad (8)$$
$$-1 < p_c < 1, \; \left(p_c \neq 0, \; p_s = \pm\sqrt{1 - p_c^2} \right).$$

3.2   The Gyroscopic Stabilization of Equilibrium (5)

Using (7), let us resolve the equation from (5) with respect to the parameter H_1:

$$H_1 = p_s \left(4 \, (J_A - 1) - \frac{H_2}{p_c} \right). \quad (9)$$

Taking into account notations (7) and expression (9), the equations of motion (3) have the matrices:

$$M = \begin{pmatrix} p_c^2 + J_A p_s^2 & 0 & -J_A p_s \\ 0 & J_C & 0 \\ -J_A p_s & 0 & J_A \end{pmatrix};$$

$$G = \begin{pmatrix} 0 & -2\,(J_A - 1)\,p_c p_s & 0 \\ 2\,(J_A - 1)\,p_c p_s & 0 & H_2 - (J_C + J_A - 1)\,p_c \\ 0 & (J_C + J_A - 1)\,p_c - H_2 & 0 \end{pmatrix};$$

$$K = \begin{pmatrix} 3\,(J_C - p_s^2 - J_A p_c^2) & 0 & -3\,(J_C - 1)\,p_s \\ 0 & K_{22} & 0 \\ -3\,(J_C - 1)\,p_s & 0 & K_{33} \end{pmatrix}, \tag{10}$$

where $K_{22} = H_2/p_c + 4\,(1 - J_A)\,p_c^2$; $K_{33} = H_2 p_c + (1 - J_C)\big(2\,(p_c^2 - p_s^2) - 1\big)$.

The parameter p_s enters the coefficients of the system's characteristic equation only in even degrees. Let us eliminate it, considering $p_c^2 + p_s^2 = 1$. Let us write down these coefficients depending on four parameters J_A, J_C, p_c, H_2 in an explicit form:

$$\begin{aligned}
v_3 &\equiv \det M = J_A J_C p_c^2; \qquad v_2 = H_2^2\,\big(J_A - (J_A - 1)\,p_c^2\big) \\
&\quad + H_2 p_c\,\big((J_A - 1)\,p_c^2\,(6J_A + J_C - 2) - J_A\,(6J_A + J_C - 7)\big) \\
&\quad + p_c^2\,\Big(-(J_A - 1)\,p_c^2\,\big(2\,(3J_A + 1)\,J_C + (1 - 3J_A)^2 - 3J_C^2\big) \\
&\qquad + (J_A\,(3J_A - 2) - 3)\,J_C + 9J_A\,(J_A - 1)^2 + 3J_C^2\,\Big);
\end{aligned}$$

$$\begin{aligned}
v_1 &= H_2^2\,\big(J_A + 3J_C - 3 - 4\,(J_A - 1)\,p_c^2\big) + H_2 p_c\,\big(J_A\,(22 - 19J_C) \\
&\quad - 3J_A^2 + (J_A - 1)\,p_c^2\,(6J_A + 19J_C - 26) - 3\,(J_C - 7)\,(J_C - 1)\big) \\
&\quad + p_c^2\,\big((J_A - 1)\,p_c^2\,(6J_A\,(5 - 7J_C) + 9J_A^2 - 3J_C^2 + 34J_C - 31) \\
&\qquad + 3\,(J_C - 1)\,\big((3J_A - 2)\,J_C + 9\,(J_A - 1)^2\big)\big);
\end{aligned} \tag{11}$$

$$\begin{aligned}
v_0 &\equiv \det K = 3\,\big(H_2 - 4\,(J_A - 1)\,p_c^3\big)\,\big(H_2\,(J_C - 1 - (J_A - 1)\,p_c^2) \\
&\quad + p_c\,(J_C - 1)\,(4\,(J_A - 1)\,p_c^2 - 3J_A - J_C + 4)\big).
\end{aligned}$$

According to the Kelvin–Chetaev's theorems [9], studying the questions on stability of equilibria begins with an analysis of the matrix of potential forces. For applied problems of spacecraft dynamics, one usually sets the distribution of masses in the system, under which the initial matrix of potential forces will be positive definite. Further, due to the influence of dissipative forces, the asymptotic stability of motion is ensured by the Lyapunov theorem. However, potentially unstable systems may also be of interest, for example, because of the possibility of nonstandard situations in orbit.

It is not difficult to show that the principal diagonal first-order minor of the matrix K from (10) on the intervals (8) is negative. Hence, the matrix of potential forces is not positive definite and equilibrium (5) will be unstable. It is known that if the equilibrium position is unstable at potential forces, Kelvin–Chetaev's theorem [9] of the influence of gyroscopic forces tells us that gyroscopic stabilization is possible only for systems with an even degree of instability. The evenness (or oddness) of the degree of instability according to Poincaré is determined by the positivity (or negativity) of the determinant of the matrix of potential forces.

Let us pose the question of the possibility of the gyroscopic stabilization of an unstable equilibrium (5) under a condition $\det K > 0$. With the help of the "*Mathematica*" function:

$$Reduce[\{\,1/2 < J_A < 1,\ 1 - J_A < J_C < J_A,\ -1 < p_c < 1,\ p_c \neq 0,\ H_2 > 0,$$
$$\det K > 0\,\},\{\,J_A,\,J_C,\,p_c,\,H_2\,\},\mathrm{Reals}\,]$$

designed to find the symbolic (analytical) solution of the inequalities systems, the region with an even degree of instability is obtained. Due to the solution bulkiness, its presentation is omitted here. An analysis of the solution obtained allows us to formulate the following conclusion.

Proposition 1. *The region with an even degree of instability for equilibrium (5) with the values of parameters J_A, J_C from (8) lies in the plane $-1 < p_c < 0 \,\wedge\, 0 < H_2 < 2$.*

For the detection of a property of gyroscopic stabilization, it is necessary to find in which part of region with an even degree of instability the remaining inequalities from (4) are fulfilled (except for $v_3 \equiv \det M > 0$, $v_0 \equiv \det K > 0$).

It is not possible to obtain an analytical solution for the entire system of inequalities (4) (with the coefficients v_i ($i = \overline{0,3}$) from (11)) because of the large number of parameters and the complexity of the expressions being analyzed. Therefore, to simplify the analysis, let us move on to symbolic-numerical analysis for fixed values of one or two parameters.

Let two parameters have the following values: $J_A = 51/100$, $J_C = 1/2$. Let us construct the regions with an even degree of instability and of gyroscopic stabilization in the parameter plane p_c, H_2 using the "*Mathematica*" function

$$RegionPlot[\,-1 < p_c < 0 \wedge 0 < H_2 < 2 \wedge v_0 > 0 \wedge v_1 > 0 \wedge v_2 > 0 \wedge Dis > 0,$$
$$\{\,p_c,\,-1,\,0\,\},\{\,H_2,\,0,\,2\,\}]$$

designed for a graphical representation of the solution of the system of inequalities. The result obtained is shown with regions in Fig. 1. The light part of the shaded area in the figure is the region with an even degree of instability. Its darker part determines the parameter values at which gyroscopic stabilization is possible. Outside the selected regions, the system has an odd degree of instability (i.e., $\det K < 0$) and the equilibrium (5) is unstable.

It is noted that as the value of the parameter J_A increases in the interval from (8), the gyroscopic stabilization region narrows and ceases to exist, starting from the value $J_A = 4/5$.

Drawing a conclusion from the symbolic-numerical modeling, we can formulate the following proposition.

Proposition 2. *The unstable equilibrium (5) can be stabilized by gyroscopic forces. Stabilization is possible only for the values of parameters (7) from the intervals:*

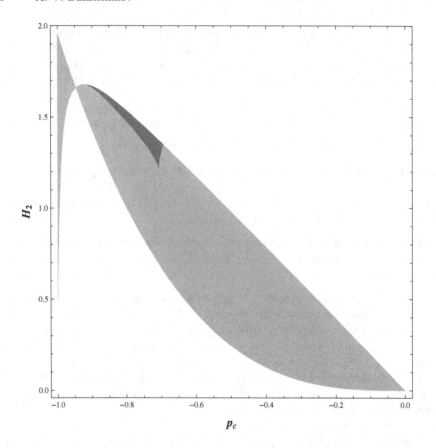

Fig. 1. The region with an even degree of instability and the region of gyroscopic stabilization for $J_A = 51/100$, $J_C = 1/2$

$$\frac{1}{2} < J_A < \frac{4}{5} \wedge 1 - J_A < J_C < J_A \wedge -1 < p_c < 0 \wedge 0 < H_2 < 2.$$

3.3   The Gyroscopic Stabilization of Equilibrium (6)

The equation from (6) in notations (7) is resolved with respect to the parameter H_1 as follows: $H_1 = p_s (J_A - 1 - H_2/p_c)$. Considering the last relation, the matrix of potential forces for the equilibrium (6) takes the form:

$$K = \begin{pmatrix} 3\left(p_s^2 + J_A p_c^2 - J_C\right) & 0 & 3\left(J_C - 1\right) p_s \\ 0 & K_{22} & 0 \\ 3\left(J_C - 1\right) p_s & 0 & K_{33} \end{pmatrix}, \qquad (12)$$

where $K_{22} = H_2/p_c - (J_A - 1)\, p_c^2;$ $K_{33} = H_2 p_c - (J_C - 1)\left(3 + p_c^2\right).$

The principal diagonal first-order minor of the matrix K from (12) on intervals (8) is positive. Based on the structure of the matrix K, a region with an even degree of instability exists only if the conditions $K_{22} < 0$ and $\det K > 0$ are simultaneously satisfied. In this case, the equilibrium (6) will be unstable. The set in the parameter space that satisfies the last inequalities is found using the above-mentioned *Reduce* function and has the following solution:

$$\frac{1}{2} < J_A < 1 \wedge 1 - J_A < J_C < J_A \wedge -1 < p_c < 0$$

$$\wedge \; H_2 > \frac{p_c \left(J_C - 1 \right) \left(4 J_C - J_A \left(p_c^2 + 3 \right) + p_c^2 - 1 \right)}{J_C - 1 - \left(J_A - 1 \right) p_c^2}. \tag{13}$$

The coefficients of the characteristic equation (after the elimination of parameter p_s) become:

$$v_3 \equiv \det M = J_A J_C p_c^2 ; \qquad v_2 = H_2^2 \left(J_A - \left(J_A - 1 \right) p_c^2 \right)$$
$$+ H_2 p_c \left(\left(J_A - 1 \right) p_c^2 \left(J_C - 2 \right) - J_A \left(J_C - 1 \right) \right)$$
$$+ p_c^2 \left(\left(J_A - 1 \right) p_c^2 \left(J_C - 1 \right) + J_C \left(3 + J_A + 3 J_A^2 - 3 \left(J_A + 1 \right) J_C \right) \right);$$

$$v_1 = H_2^2 \left(3 + J_A - 3 J_C + 2 \left(J_A - 1 \right) p_c^2 \right) + H_2 p_c \left(J_A \left(1 - 4 J_C \right) \right.$$
$$\left. + 3 J_A^2 + 3 \left(J_C - 1 \right)^2 - 2 \left(J_A - 1 \right) p_c^2 \left(3 J_A - 2 J_C + 1 \right) \right) \tag{14}$$
$$+ p_c^2 \left(\left(J_A - 1 \right) p_c^2 \left(2 \left(3 J_A + 5 \right) J_C - 3 J_A - 6 J_C^2 - 4 \right) \right.$$
$$\left. - 3 \left(J_C - 1 \right) J_C \left(3 J_A - 3 J_C + 1 \right) \right);$$

$$v_0 \equiv \det K = 3 \left(H_2 - \left(J_A - 1 \right) p_c^3 \right) \left(H_2 \left(\left(J_A - 1 \right) p_c^2 - J_C + 1 \right) \right.$$
$$\left. - p_c \left(J_C - 1 \right) \left(\left(J_A - 1 \right) p_c^2 + 3 J_A - 4 J_C + 1 \right) \right).$$

For the reasons given in Sect. 3.2, let $J_A = 9/10$, $J_C = 4/5$. Let us substitute these values into the coefficients (14) and the discriminant of the characteristic equation. The region of gyroscopic stabilization for equilibrium (6) is found using the function

$$RegionPlot[\; -1 < p_c < 1 \wedge p_c \neq 0 \wedge H_2 > 0 \wedge K_{22} < 0 \wedge \det K > 0$$

$$\wedge \, v_1 > 0 \wedge v_2 > 0 \wedge Dis > 0, \{ p_c, -1, 0 \}, \{ H_2, 0, 1.8 \}]$$

and is shown in Fig. 2. As a result, we can formulate the following

Proposition 3. *The unstable equilibrium (6), whose parameter values belong to intervals (13), can be stabilized by gyroscopic forces.*

It is important to note that the question of the possibility of gyroscopic stabilization is not always resolved positively. In [3] for the case $h_2 = 0$, $h_1 \neq 0$, and $h_3 \neq 0$ the stability of the equilibrium position

$$\begin{cases} \alpha = \alpha_0 = 0, \quad \gamma = \gamma_0 = \pi/2, \\ \beta = \beta_0 = const : \quad h_1 \cos \beta_0 + \sin \beta_0 \left(h_3 + \left(A - C \right) \cos \beta_0 \right) = 0 \end{cases} \tag{15}$$

was considered. In this article, the following proposition has been formulated and proved.

Proposition 4. *The unstable equilibrium (15) for parameters from intervals (8) cannot be stabilized by gyroscopic forces.*

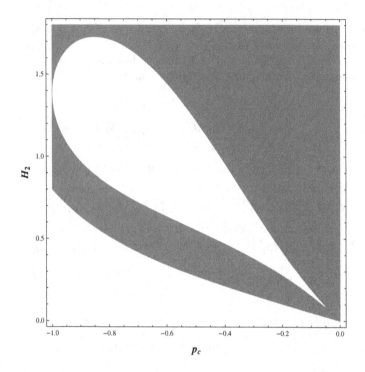

Fig. 2. The region of gyroscopic stabilization when $J_A = 9/10$, $J_C = 4/5$

4   Conclusion

The presented results of modeling and qualitative analysis of differential equations of motion (DEM) of a gyrostat by means of computer algebra indicate that the proposed approach expands our capabilities in the study of multi-parameter problems. In addition, the automatic mode at the stage of obtaining of DEM in symbolic form spares routine computations for specialists and significantly reduces the research time. It is planned to continue the analysis of gyroscopic stabilization of other relative equilibria in order to obtain a more comprehensive qualitative picture of research and compare it with the results obtained earlier [2,5] for an axisymmetric gyrostat.

References

1. Agafonov, S.A.: The stability and stabilization of the motion of non-conservative mechanical systems. J. Appl. Math. Mech. **74**(4), 401–405 (2010)
2. Banshchikov, A.V.: Research on the stability of relative equilibria of oblate axisymmetric gyrostat by means of symbolic-numerical modelling. In: Gerdt, V.P., Koepf, W., Seiler, W.M., Vorozhtsov, E.V. (eds.) CASC 2015. LNCS, vol. 9301, pp. 61–71. Springer, Cham (2015). https://doi.org/10.1007/978-3-319-24021-3_5

3. Banshchikov, A.V.: Symbolic-numerical analysis of the necessary stability conditions for the relative equilibria of an orbital gyrostat. J. Appl. Indust. Math. **14**(2), 213–221 (2020)
4. Banshchikov, A.V., Burlakova, L.A., Irtegov, V.D., Titorenko, T.N.: Symbolic computation in modelling and qualitative analysis of dynamic systems. Comput. Technol. **19**(6), 3–18 (2014). (in Russian)
5. Banshchikov, A.V., Chaikin, S.V.: Analysis of the stability of relative equilibria of a prolate axisymmetric gyrostat by symbolic-numerical modeling. Cosm. Res. **53**(5), 378–384 (2015)
6. Banshchikov, A.V., Irtegov, V.D., Titorenko, T.N.: Software package for modeling in symbolic form of mechanical systems and electrical circuits. Certificate of State Registration of Computer Software. Federal service for intellectual property. No. 2016618253 (2016) (in Russian)
7. Banshchikov, A.V., Vetrov, A.A.: Application of software tools for symbolic description and modeling of mechanical systems. In: Bychkov, I.V. et al. (eds.) CEUR Workshop Proceedings of the 2nd International Workshop on Information, Computation, and Control Systems for Distributed Environments, pp. 33–42 (2020). http://ceur-ws.org/Vol-2638/paper3.pdf
8. Beletskii, V.V.: Motion of an Artificial Satellite Relative to the Center of Mass. Nauka, Moscow (1965).(in Russian)
9. Chetaev, N.G.: The Stability of Motion. Pergamon Press, New York (1961)
10. Gutnik, S.A., Sarychev, V.A.: Application of computer algebra methods for investigation of stationary motions of a gyrostat satellite. Program. Comput. Softw. **43**(2), 90–97 (2017). https://doi.org/10.1134/S0361768817020050
11. Kozlov, V.V.: Stabilization of the unstable equilibria of charges by intense magnetic fields. J. Appl. Math. Mech. **61**(3), 377–384 (1997)
12. Prokopenya, A.N., Minglibayev, M.Z., Mayemerova, G.M.: Symbolic calculations in studying the problem of three bodies with variable masses. Program. Comput. Softw. **40**(2), 79–85 (2014). https://doi.org/10.1134/S036176881402008X
13. Sarychev, V.A.: Problems of orientation of satellites. In: Itogi Nauki i Tekhniki. Series "Space Research", vol. 11, pp. 5–224. VINITI Publication, Moscow (1978). (in Russian)
14. Sarychev, V.A., Mirer, S.A., Degtyarev, A.A.: Dynamics of a gyrostat satellite with the vector of gyrostatic moment in the principal plane of inertia. Cosm. Res. **46**(1), 60–73 (2008)
15. Wolfram, S.: The Mathematica Book, 5th edn. Wolfram Media, Inc., Somerville (2003)

Computer Science for Continuous Data

Survey, Vision, Theory, and Practice of a Computer ~~Algebra~~ *Analysis* System

Franz Brauße[1], Pieter Collins[2], and Martin Ziegler[3(✉)]

[1] University of Manchester, Manchester, UK
[2] Maastricht University, Maastricht, The Netherlands
[3] KAIST, Daejeon, Republic of Korea
`ziegler@kaist.ac.kr`

Abstract. Building on George Boole's work, Logic provides a rigorous foundation for the powerful tools in Computer Science that underlie nowadays ubiquitous processing of discrete data, such as strings or graphs. Concerning continuous data, already Alan Turing had applied "his" machines to formalize and study the processing of real numbers: an aspect of his oeuvre that we transform from theory to practice.

The present essay surveys the state of the art and envisions the future of Computer Science for continuous data: natively, beyond brute-force discretization, based on and guided by and extending classical discrete Computer Science, as bridge between Pure and Applied Mathematics.

1   Introduction and Motivation

Since its early days, Computer Science has enjoyed the support and guidance of Logic, from Theory via Engineering to Practice: recall Alan Turing's 1936 publication preceding nowadays ubiquitous digital computers, or Alonzo Church's Lambda Calculus having led to functional programming languages, or axiomatic structures in Model Theory corresponding to specification of Abstract Data Types, or Hoare Logic for formal program verification—concerning the processing of discrete data, such as graphs or integers or strings.

Continuous data on the other hand commonly arises in Engineering and Science (*natura non facit saltus*) in the form of temperatures and fields; it mathematically includes real numbers, smooth functions, bounded operators, or compact subsets of an abstract metric space. Processing such continuous data has arguably been lacking the foundation and support from Logic in Computer Science that the discrete case is enjoying [11]:

35 years after introduction and hardware standardization of IEEE 754 floating point numbers, mainstream numerics is still governed by this forcible discretization

This work was supported by the National Research Foundation of Korea (grant 2017R1E1A1A03071032) and by the International Research & Development Program of the Korean Ministry of Science and ICT (grant 2016K1A3A7A03950702) and by the ▪ European Union's Horizon 2020 MSCA IRSES project #731143.

F. Boulier et al. (Eds.): CASC 2022, LNCS 13366, pp. 62–82, 2022.
https://doi.org/10.1007/978-3-031-14788-3_5

of continuous data—in spite of violating associative and distributive laws, breaking symmetries, introducing and propagating rounding errors in addition to an involved (and incomplete) axiomatization including NaNs and denormalized numbers.

Deviations between mathematical structures and their hardware counterparts are common also in the discrete realm, such as the wraparound $255 + 1 = 0$ occurring in bytes that led to the "Nuclear Gandhi" programming bug. Therefore nowadays high-level programming languages (like `Java` or `Python`) provide user data types (like `BigInt`) that fully agree with mathematical integers, simulated in software using a variable number of hardware bytes; and advanced discrete data types (such as weighted or labelled graphs) can and do build on that, reliably and efficiently.

The present essay expands on a similar perspective for continuous data types: including real numbers, converging sequences, smooth/integrable functions, bounded operators, compact subsets etc.—exactly, that is, devoid of rounding errors, see Sect. 2. Section 3 discusses imperative programming over such data: with computable semantics including limits, that is, beyond the algebraic realm. Encoding such data over sequences of bits is described in Sect. 4. And Sect. 5 connects discrete complexity theory, with famous classes like P/NP/#P/PSPACE, to operations on continuous data such as integration. Final steps for putting this theory into practice and its applications are collected in Sect. 6.

2   Computable Continuous Data Types

Data types are at the core of Object-Oriented Programming, see Subsect. 2.1. They constitute Computer Science's counterpart to *structures* in Model Theory. This section explains how, unlike in the discrete case, for continuous data types already their specification often poses a challenge—and may require Kleene Logic (Subsect. 2.2), *enrichment* (Subsect. 2.3) and/or *multivaluedness* (Subsect. 2.4) in order to assert mere computability—before proceeding to complexity questions (Sect. 5). Subsects. 2.5 and 2.6 illustrate these with examples.

2.1   Formal Numerical Software Engineering

Formal software engineering of a data type/object proceeds from (i) problem specification via (ii) algorithm design and (iii) analysis to (iv) proof of optimality and finally (v) implementation and (vi) verification/testing in some high-level object-oriented programming language. Depending on the particular endeavour, some of these stages can of course be kept informal of skipped entirely. Item (iv) here refers to Computational Complexity Theory [49], and implies a (meta) "loop": If the algorithm designed (ii) and analyzed (iii) is not optimal (iv), then start over designing a more efficient one (ii).

Note how (ii) implicitly supposes that the problem specified in (i) actually does admit an algorithmic solution—which in the discrete realm is usually the case. In the real setting, however, any computable function must necessarily be

     i) problem specification
     ii) algorithm design and
     iii) algorithm analysis
     iv) proof of optimality — or repeat from (ii)
     v) implementation and
     vi) verification/testing.

Fig. 1. Six stages of full-fledged Formal Software Engineering. In practice some may be omitted depending on the particular endeavour under consideration.

continuous[1] [63, §2.2+§3.2+§4.3]; hence a naïve problem specification (such as of finding the kernel of a given real matrix) easily results in algorithmic unsolvability. Thus the need arises for another (meta) "loop" in Numerical Software Engineering: if (ii) fails, start over from (i).

2.2 Kleene Logic Data Type, Generalized Sierpiński Topology

The sign function is discontinuous and thus uncomputable. More precisely, a real test like "$x > 0$?" may take more runtime when x is close to zero—and in case $x = 0$ fail to terminate at all:

Fact 1. *Real inequality is "complete" for the Halting Problem $H \subseteq \mathbb{N}$ in the following sense [63, Exercise 4.2.9]:*

a) *For every computable real sequence $\bar{x} = (x_j)$, the set $\{j \in \mathbb{N} : x_j > 0\}$ is computably reducible to H.*
b) *There exists a computable sequence $\bar{x}$ of non-negative real numbers such that said set coincides with H.*

Note that fixing a computable real sequence makes the problem independent of encoding issues, as the input consists only of an integer index. The thus non-uniform Item a) follows from the following uniform claim with respect to any of the many equivalent ways of encoding real numbers [63, §4.1]:

c) *The partial sign function* sign $: \mathbb{R} \setminus \{0\} \to \{-1, +1\}$ *is computable; same for comparison* $< : \mathbb{R}^2 \setminus \{(x, x) : x \in \mathbb{N}\} \to \{ff, tt\}$, *where ff and tt denote computational counterparts to Booleans TRUE and FALSE.*

A mathematically undefined expression (like $1/0$) is sometimes denoted to "have" value $\bot$; comparison "$x > 0$" on the other hand is defined mathematically also in case $x = 0$, but not computationally so. The latter is captured by Kleene Logic $\mathbb{K}$ including, in addition to classical Booleans tt and ff, as third value uk for (mathematically defined but computationally) "unknown". Equip $\mathbb{K}$ with the generalized Sierpiński's topology

$$\{\emptyset, \{tt\}, \{ff\}, \{ff, tt\}, \{uk, tt, ff\}\}$$

---

[1] Arguably this also applies to the discrete case, where every function is trivially continuous.

and note that this non-Hausdorff topology fails to separate uk from the other elements. Accordingly, a logical expression with mathematical value uk fails to evaluate computationally.

$\mathbb{K}$ thus serves as "lazy" data type that can store unevaluated, computationally partial predicates: such as "$x > 0$" for every $x \in \mathbb{R}$, as well as any other promise problem. Recall [19] that a (discrete) promise problem P is a disjoint pair $P^+, P^- \subseteq \mathbb{N}$, such that a query "$m \in P?$" answers tt in case $m \in P^+$ and answers $f\!\!f$ in case $m \in P^-$ and gives no answer uk in case $m \notin P^+ \cup P^-$.

2.3   Enrichment/Promises

Promises generalize from decision to function problems, motivated as follows: Topology requires that any non-constant function $f : X \to \mathbb{Z}$ from a connected domain X to the discrete set of integers must be discontinuous. This easily prevents naïve problems from being computable, such as the matrix rank function, or the multiplicities of degenerate eigenvalues. On the other hand providing—in addition to the original continuous data—a suitable integer as input often does render such a problem computable [71]. For example, by Fact 1c), the real sign function is computable on $X_0 := \mathbb{R} \setminus \{0\}$, and on $X_1 := \{0\}$ trivially so. See Subsect. 2.5 below for more examples.

In Constructive Mathematics such an effect is well-known as *enrichment* [36, p. 238/239]; elsewhere also as *advice* [1,8]. It amounts to proceeding from total but discontinuous $f : X \to Y$ to a partial function

$$\tilde{f} : \subseteq X \times \mathbb{Z} \ni (x, k) \mapsto f(x) \in Y \tag{1}$$

for some suitable—and now non-connected—domain $\tilde{X} := \mathrm{dom}(\tilde{f}) \subseteq X \times \mathbb{Z}$ whose projection $X_k := \{x : \exists k : (x, k) \in \tilde{X}\}$ covers X. Put differently, the accompanying argument k entails the *promise* that the "main" input x belongs to the subset $\tilde{X}_k := \{x : (x, k) \in \tilde{X}\}$.

2.4   Multivaluedness/Non-extensionality

Although any computable function must necessarily be continuous, this constraint can be avoided by considering relations, that is, by dropping extensionality. Relations mathematically capture search problems, where a query $x \in X$ has not necessarily one unique answer $y = f(x)$, but a range of possible answers $y \in F(x) \subseteq Y$.

In case the domain X is discrete/countable, namely when arguments $x \in X$ are finitely encoded and read in finite time, then any (deterministic) computation of such a relation F actually computes a *selection*, that is, a function $f \subseteq F$. However in the continuous setting, multivaluedness is well-known unavoidable [42].

Mathematically one may identify the relation F with the single-valued total function $F : X \ni x \mapsto \{y \in Y \mid (x, y) \in F\}$ from X to the powerset 2^Y; but the preferable notation of a *multifunction* $f :\subseteq X \rightrightarrows Y$ emphasizes that not every $y \in F(x)$ needs to occur as output. Another important reason to consider multifunctions $f : X \rightrightarrows Y$ distinct from relations $f \subseteq X \times Y$ is related to compactness: Generalizing continuity for single-valued functions, call such a multifunction f *compact-valued* if $f[Z] \subseteq Y$ is compact for every compact $Z \subseteq X$.

A function problem $f : X \rightarrow Y$ becomes "easier" when *restricting* arguments to $x \in X'$ for some $X' \subset X$, that is, when proceeding to $f' = f|_{X'}$. A search problem $F : X \rightrightarrows Y$ additionally becomes "easier" when increasing the range of possible answers, that is, when proceeding to some $F' \subseteq X \rightrightarrows Y$ satisfying $F'(x) \supseteq F(x)$ for every $x \in \text{dom}(F')$. Such F' is also called a *restriction* of F. Note that, unlike in the single-valued case, F' need not be a subset of F when considered as graphs.

2.5   Examples

Example 2. The Archimedian Property of real numbers states that, to every $x \in \mathbb{R}$, there exists some $k \in \mathbb{Z}$ with $k \geq x$.

Skolemization yields a function $k : \mathbb{R} \rightarrow \mathbb{Z}$ with $\forall x : k(x) \geq x$. The least such function is known as *rounding up* $x \mapsto \lceil x \rceil$ and discontinuous. In fact any such function must be discontinuous and hence uncomputable.

On the other hand the original property formulation suggests formalization as a search (rather than function) problem Arch : $\mathbb{R} \rightrightarrows \mathbb{Z}$. And indeed this relaxation becomes computable as follows:

Obtain some rational input approximation to the argument $x \in \mathbb{R}$ up to error 2^{-0} and round it up, exploiting that integer fractions can be operated on exactly. Note that a different rational approximation to the same argument x up to error 2^{-0} can yield a different output, i.e., violate extensionality.

Similarly, no integer rounding *function* is computable; whereas the following *multi*function is:

$$\text{Round} : \mathbb{R} \ni x \mapsto \{k \in \mathbb{Z} \mid x - 1 < k < x + 1\} \subseteq \mathbb{Z} \tag{2}$$

Example 3. The Fundamental Theorem of Algebra states that, to every monic univariate complex degree-d polynomial $c_0 + c_1 \cdot Z + \cdots + c_{d-1} \cdot Z^{d-1} + Z^d$ decomposes into linear factors $(Z - z_1) \cdots (Z - z_d)$.

This suggests formalization as a mapping

$$F : \mathbb{C}^d \ni (c_0, \ldots, c_{d-1}) \mapsto (z_1, \ldots, z_d) \in \mathbb{C}^d.$$

However note that no order on the roots $z_1, \ldots, z_d$ can be imposed mathematically; hence F should naturally be considered as multifunction.

Moreover it turns out that, similarly to Example 2, multivalued F is computable while no single-valued selection of F is [59].

Example 4. Consider the problem of computing a basis of the kernel of a real matrix A given by its entries. Note that this problem is already multivalued problem to begin with, since such a basis is usually far from unique.

Gaussian elimination involves pivot search and thus tests for real inequality—which are uncomputable: recall Fact 1. Indeed already the (unique) cardinality of a (non-unique) basis depends discontinuously on the matrix entries.

However enriching input A with said cardinality $= \text{rank}(A) \in \mathbb{N}$ does render such a basis computable [71].

Example 5. Consider the problem of computing the spectral decomposition, formalized as computing an (!) eigenvector basis, to a given symmetric real matrix $A \in \mathbb{R}^{d \times d}$.

According to Example 3, a tuple of eigenvalues, repeated according to their multiplicities, can be computed via the characteristic polynomial. However the integer-valued multiplicities themselves are discontinuous and uncomputable in their dependence on the matrix entries. Moreover computing an eigenvector basis, although non-unique/multivalued, is impossible; whereas enriching input A with the number $k \in \{1, \ldots, d\}$ of distinct eigenvalues renders the problem computable [71, Theorem 11]. Moreover such d-fold advice turns out as optimal [70, Theorem 46].

We remark that computing eigen*spaces* is possible, when equipping the latter with the right topology.

Example 6. Alternative to the partial sign function from Fact 1, the following total but multivalued so-called *soft* test [67, §6] is computable as well [7, p. 491]:

$$
\begin{aligned}
\text{``}x <_n 0\text{''} \quad = \quad & tt \quad \text{in case } x < -2^{-n}, \\
& f\!f \quad \text{in case } x > 2^{-n}, \text{ and} \quad (3) \\
& \text{either } tt \text{ or } f\!f \quad \text{in case } -2^{-n} \leq x \leq 2^{-n}.
\end{aligned}
$$

Example 7. Fix promise problems $P_0, P_1, \ldots, P_{d-1} \subseteq \mathbb{N}$ with the aforementioned computational semantics "$(m \in P_j) \in \mathbb{K}$". Consider the partial multi-valued mapping choose $: \mathbb{K}^d \rightrightarrows \{0, 1, \ldots, d-1\}$ assigning to $((m \in P_0), \ldots, (m \in P_{d-1}))$ to some j such that $(m \in P_j) = tt$. This is computable!

2.6   More Continuous Data Types

Real numbers are arguably the most basic continuous structure in Calculus; vector, sequence, and function spaces for instance build on top of them. Similarly, having turned real numbers into a computable data type ("level 0") enables now turning more advanced spaces from Mathematics into computable ones—using the aforementioned techniques to deal with discontinuities: enrichment (Subsect. 2.3) and multivaluedness (Subsect. 2.4). Specifically polynomials and matrix operations have been discussed in Subsect. 2.5 above. Note that each such object can be described with finitely many real numbers: "level 1".

2) Sequence spaces ℓ^p are the next level, each element consisting of countably infinitely many real numbers. A plethora of investigations [9, 34, 43, 45] provide guidance on suitable enrichment (such as integer bounds on the norm) to turn them into computable data types.

3) Power series can be identified with their germs/coefficient sequences in appropriately enriched sequence spaces [28, §3.1]; and analytic functions are local power series—of which finitely many suffice to "cover" any fixed compact subset of their domain [28, §3.2].

4) The hyperspace of non-empty compact subsets of Euclidean space is computably closed under union and under image of continuous functions [63, §6.2], but not under (even promised non-empty) intersection [63, Exercise 5.1.15]. The hyperspace of *convex* compact subsets does satisfy this, and additional, computational closure properties [39].

5) Space of probability measures [24, 44, 57] and subspace of Haar measures on compact groups [51].

6) Spaces of continuous [63, §6.1], of smooth [28], and of integrable functions [29, 60]; equipped with operations like (anti or weak) derivative, or trace.

7) Differential geometry, that is, the hyperspace of closed (smooth) manifolds equipped with (smooth) tensor fields on them.

3   New Numerical Programming

Since the early days of automated digital processing in assembly code, programming has made tremendous progress. Nowadays high-level languages provide both convenience/intuition and soundness/reliability—regarding discrete data, such as integers or strings.

Numerical programming differs from this classical realm in that the underlying data type intrinsically incurs errors, namely from rounding. Tracing and bounding the propagation of such deviations is up to the user programmer, and the involved IEEE 754 standard makes reliable coding inconvenient. Practitioners therefore often imagine operating on real (instead of floating point) numbers. This implicit approach thus trades convenience for reliability.

Algebraic numbers can be processed exactly, and the formal verification of algebraic programs [48] may build on Tarski's decidability of the First-Order Theory of this algebraically closed field (although the latter excludes the exponential as well as many other important analytic functions in Science and Engineering and Calculus [6]).

Computable Analysis [63] on the other hand does provide a realistic characterization of un/computable real functions beyond the algebraic realm. It is however based on the (type-2) Turing machine model: theoretically important but practically inconvenient for programming, not to mention formal verification.

Subsection 3.1 recalls an equivalent but convenient imperative model of computation called ERC that comes as close to, and thus provides a sound formalization of, common implicit conceptions underlying numerical programming. Previous and future ways for implementing it on actual digital computers are discussed in Subsect. 3.2. ERC modifies the semantics of real comparison, and Subsection 3.3 illustrates its use with some basic example algorithms. Subsection 3.4 provides a road map of continuous data types to next implement in ERC.

3.1   Analytic Programming

The preprint [10] formalizes a (proof-of-concept) imperative programming language called ERC supporting a data type REAL that agrees with the mathematical structure $\mathbb{R}$, exactly. Its semantics is carefully designed to capture common conceptions (sometimes implicitly) underlying numerical coding, while achieving "Turing-completeness" over the reals: Any function *realizable* in ERC is computable in the sense of Computable Analysis—and vice versa. ERC thus combines the structural benefits of Computable Analysis (such as closure under composition and including transcendental functions) with the intuitive convenience and practical pervasion of object-oriented imperative programming to replace the hassles of Turing machines.

Paradigm 8. *A mathematical partial function $f :\subseteq \mathbb{R} \to \mathbb{R}$ is realized in ERC as a (multi-)function of type $\mathbb{Z} \times \mathbb{R} \rightrightarrows \mathbb{R}$: It receives its real argument x exactly, as well as a separate integer[2] parameter $p \to -\infty$, and must eventually return some approximation to $y = f(x)$ up to absolute error $2^p \to 0$. To this end during intermediate calculations, it may use arithmetic operations free of rounding errors. The "result" of a possibly partial comparison "$x > y$" according to Fact 1(c) can be stored in a logic variable of type KLEENEAN (Subsect. 2.2), and can be evaluated safely using the multivalued operation from Example 7 to yield a total program.*

The discrepancy between exact argument and approximate return value might suspect to void closure under composition [68, p. 325]; however in combination with the modified partial semantics of comparison (Fact 1c), Computable Analysis does assert closure under composition [64]. In particular an ERC program expressing real function f as above may, in addition to using arithmetic operations, call another ERC representing some other real function g with exact argument z to receive and continue processing real return value $w = g(z)$ exactly, i.e., without having to worry about error propagation.

Remark 9. The error bound 2^p ($\mathbb{Z} \ni p \to -\infty$) is preferable over, say, $1/|p|$:

1. It reflects that π has been approximated up to $2^{-\text{billions}}$ [26].
2. It renders the underlying logic decidable; see Subsect. 3.5.
3. It yields numerical characterizations of popular discrete complexity classes; see Subsect. 5.1.

A similar but more practically fleshed-out programming language for continuous data has been devised in [3].

3.2   Implementations

Superficially, operating on real (including transcendental) numbers exactly as postulated by ERC might seem technically infeasible: within finite time, only

---

[2] One could replace it with some real error bound $\varepsilon > 0$.

finite information can be processed. This reproach is valid in operational semantics, where (the order of) user commands correspond to finite blocks of machine instructions executed in the same order. However functional programming regularly removes this implicit condition, and adapting that relaxation to the imperative setting does enable the above semantics in user space [38,46,69].

Technically speaking, operations can be realized "exactly" by actually processing approximations of variable but finite precision; precision chosen automatically, for any given n, such that (a) the output accuracy attained after initial and propagated errors is $\leq 1/2^n$, and (b) the program flow remains indistinguishable from hypothetical exact calculations.

Both conditions can be implemented by object-oriented overloading the operations involving continuous data. One approach executes the user program symbolically, recording expressions of all variables' contents in dependence on the initial arguments—expressions which thus can be evaluated in any finite initial precision and with error propagation whenever needed for (a) output or (b) to decide/branch on a comparison between two distinct reals. Another approach trades runtime for memory namely, instead of recording and silently re-evaluating symbolic expressions, silently re-executes the user program repeatedly in increasing but finite initial precision until sufficient to (b) decide all comparisons/branches and (a) attain the desired output accuracy. Note that either approach requires the user program to be devoid of side effects, as common in functional programming and "desirable" (but now mandatory) in the imperative setting.

3.3 Example ERC Programs

Paradigm 8 was designed to yield a rigorous and real Turing-complete semantics of analytic programming closest to numerical intuition. Thus it requires only little adaptation to replace classical but uncomputable tests with their replacements, as exemplified in this subsection:

Example 10. The Soft Test $x <_n 0$ from Example 6 can be expressed in ERC:

$$\mathsf{choose}\left(x > 2^{-n+1} , \ x < 2^{-n} \right) \ ieq \ 1$$

Example 11. The multivalued rounding "function" from Eq. 2 after Example 2 can be represented in ERC as follows:

```
INTEGER Round(x:REAL);
LET k:INTEGER=0;
WHILE choose( x<1 , x>1/2 ) = 1
   DO  k:=k+1;   x:=x-1;   ENDWHILE;
WHILE choose( x>-1 , x<-1/2 ) = 1
   DO  k:=k-1;   x:=x+1;   ENDWHILE;
RETURN k;
```

Note that the loop bodies are executed roughly $|x|$ times, that is, exponential in the binary length of the output. See [10, §2.5] for a more efficient version sufficing with a linear number of loop iterations.

Classically, bisection is employed for finding the root of a function $f : [a; b] \to \mathbb{R}$. Rigorously speaking however, this may fail in case the chosen mid-point $c = (a + b)/2$ already happens to coincide with the root of the function: In this case $f(c) = 0$, the test of sign $f(c)$ to decide which of the two sub-intervals to proceed with, fails to terminate; recall Fact 1. Instead, *trisection* with overlapping sign conditions has been suggested [20, p. 336] and can be represented in ERC as follows:

Example 12. **Program** Trisection$(p :$ INTEGER, $f :$ REAL $\to$ REAL)
let $a :$ REAL $= 0$; **let** $b :$ REAL $= 1$;
while choose$\big(\imath(p) \gg b - a, \, b - a \gg \imath(p - 1)\big) = 1$
    let $a' :$ REAL $= b/3 + 2 \times a/3$;   **let** $b' :$ REAL $= 2 \times b/3 + a/3$;
    if choose$\big(0 \gg f(a') \times f(b), \, 0 \gg f(a) \times f(b')\big) = 1$
        then $b := b'$ **else** $a := a'$ **end if**;
end while;   **return** a

Gaussian Elimination, as reported in Example 4, requires and suffices with receiving the rank of the given matrix as additional input in order to become computable. A rigorous implementation using full pivoting can be found at http://github.com/realcomputation/iRRAMx.

Example 13. Common implementations drawing the Mandelbrot Set lack reliability for two reasons: First the defining iteration is conducted in floating-point arithmetic with rounding and truncation errors that propagate and make the computed sequence differ from the mathematical one. Secondly, said sequence is calculated for a fixed number N of iterations after which it is considered to not diverge—without actual mathematical justification.

The first rigorous algorithm for computing the Mandelbrot Set (subject to the Hyperbolicity Conjecture) is due to Peter Hertling [21] with first implementation due to Jihoon Hyun:
http://github.com/realcomputation/MANDELBROT.

3.4   Advanced and Upcoming ERC Programs

Some of the computable continuous data types from Subsects. 2.5 and 2.6 already have been implemented in ERC: such as polynomial root-finding (in the sense of Example 3) with QR-algorithm and Wilkinson Shift, or matrix Gaussian Elimination (in the sense of Example 4) with full pivoting [50], or analytic functions [58] or compact metric groups [51] or solution operators to selected PDEs [35]. Future efforts are directed towards similarly implementing:

- the hyperspace of compact subsets of Euclidean space [25];
- spaces of integrable functions defined on a compact Euclidean domain;
- closed manifolds from differential geometry;
- Random sampling of continuous objects (Subsect. 5.2).

See also Subsect. 6.1 below.

3.5 Verification/Testing

According to Murphy's Law of Computing, "Every non-trivial program has at least one bug". Verification and Testing are two major approaches to prevent or at least reduce the number of errors. Beyond heuristical success stories [55], both methods arguably lack logical justification: A program may well work correctly throughout years and on billions of practical test instances yet still contain fundamental flaws, such as *zero-day exploits*. And Gödel Incompleteness/Hilbert's Tenth Problem translates to arithmetic programs which are correct, but whose correctness provably cannot be proven [13].

Relative to such constraints on integer processing, formally verifying the correctness of floating point calculations is possible [4]—but involved and messy: Designed by the Institute of Electrical and Electronics Engineers, the IEEE 754 standard revolves around 1980ies hardware capabilities to support for example NaNs and de/normalized numbers while violating mathematical associate and distributive laws as well as logical completeness.

Real closed fields of characteristic zero on the other hand do admit an intuitive and elegant axiomatization[3], which is furthermore logically complete according to Tarski. More generally, consider the many-sorted structure involving real numbers and Presburger integers connected via the "precision" embedding $\imath : \mathbb{Z} \ni z \mapsto 2^z \in \mathbb{R}$: It can express many (local) properties of ERC programs and, building on work by van den Dries [2,14], is decidable: justifying automatic formal verification. (Whereas other embeddings, such as $\tilde{\imath} : \mathbb{N} \setminus \{0\} \ni n \mapsto 1/n \in \mathbb{R}$, lead to Gödel undecidability.) Recall Remark 9 and see §4.1 in arXiv:1608.05787 for details.

4 Coding Theory

Digital computers and Turing machines naturally operate on sequences of bits; processing any other data, such as integers or graphs, needs first fixing encodings for input and output. In the discrete case, this is usually straightforward and/or complexity-theoretically inessential (up to polynomial time, say). However concerning continuous data, already real numbers suggest various encodings with surprisingly different algorithmic properties: ranging from the computably "unreasonable" binary expansion via qualitatively to polynomially and even linearly complexity-theoretically "reasonable" signed-digit expansion. But how to distinguish between un/suitable encodings of other spaces common in Calculus and Numerics, such as Sobolev?

This (meta) question has long been answered regarding qualitative computability: *admissibility* [37,56] is a crucial condition for an encoding of a space X to be "reasonable". Following this conception, encodings are partial surjective mappings (historically called *representations*) from Cantor space onto X; and said mapping is required to be (a) sequentially continuous and (b) maximal with respect to sequentially continuous reduction [63, §3.2]. Admissible encodings are

---

[3] Without the second-order property of being topologically complete.

guaranteed to exist for a large class of topological spaces, and to be Cartesian closed. And for (precisely) these does the sometimes so-called *Main Theorem* hold: which characterizes continuity of functions by the continuity of mappings translating codes, so-called *realizers*.

Subsection 4.1 summarizes the preprint [41] on quantitatively/complexity-theoretically refining said qualitative/computable admissibility from topological to metric spaces. Further tailoring such efficient encoding to spaces with additional structure is discussed in Subsect. 4.2.

4.1 Quantitative Coding Theory of Compact Metric Spaces

[41] develops a generic approach to refine qualitative computability over topological spaces to quantitative complexity over metric spaces. It strengthens the notion of unqualified admissibility to *polynomial* and to *linear admissibility*. Informally speaking, the latter two require a representation to be (a) almost "optimally" continuous (namely linearly/polynomially *relative* to the space's entropy) and (b) maximal with respect to *relatively* linearly/polynomially continuous reductions.

A large class of spaces is shown to admit a quantitatively admissible representation, including a generalization of the signed-digit encoding; and quantitatively admissible representations exhibit a quantitative strengthening of the qualitative *Main Theorem*, namely now characterizing quantitative continuity of functions by quantitative continuity of realizers. Quantitative admissibility thus provides the desired criterion for complexity-theoretically "reasonable" encodings.

The contribution then rephrases quantitative admissibility as quantitative continuity of both the representation and of its set-valued inverse. For the latter purpose, it adapts from [52] a new notion of sequential continuity for multifunctions. By establishing a quantitative continuous selection theorem for multifunctions between compact ultrametric spaces, it extends the above quantitative *Main Theorem* from functions to multifunctions aka search problems. Higher-type complexity is captured by generalizing Cantor's (and Baire's) ground space for encodings to other (compact) *ultra*metric spaces.

4.2 Encoding Advanced Spaces in Analysis

Structures expanding on compact metric spaces support operations beyond the metric. Making also these computable (subject to admissible encodings) is discussed in Subsect. 2.6 above. Minimizing their computational cost in turn relies on further refining the encodings from Subsect. 4.1: Work in progress develops and compares tailored representations for spaces of (say, square) integrable functions, and for Sobolev spaces of weakly differentiable functions. Such spaces underlie the mathematical theory of partial differential equations [62], and are thus required for the following complexity considerations:

5   Complexity Theory of Continuous Data

Over the past decades, Numerics has devised a myriad of methods: for efficiently computing algebraic and transcendental constants and functions, for solving ordinary and differential and partial differential equations, for optimization under constraints etc.

The efficiency of such a method can often be shown optimal by comparison to the quantitative stability of the problem it solves. When small perturbations of the input lead to large changes in output, algorithms must necessarily process and operate on high-precision data, incurring a large number of bit manipulations.

Function maximization and Riemann integration are stable; yet information-theoretically their approximation up to guaranteed absolute error $1/2^n$ depends on exponentially many sample points already in the smooth case; recall Remark 9. This demonstrates information theory as another method for rigorous lower complexity bounds in Numerics.

But what if the function is fixed, so that only the precision parameter $n \in \mathbb{N}$ remains as input? Subsect. 5.1 reports on surprising connections of numerical problems in this setting to *unary* classical (i.e., discrete) complexity classes. Subsection 5.2 addresses the question of adapting randomization from the discrete to the continuous setting.

5.1   Computational Complexity of Continuous Data

For any fixed polynomial-time computable real function $f : [0; 1] \to \mathbb{R}$, Harvey Friedman and Ker-I Ko had observed that its maximum $\max(f)$ can be computed relative to an NP_1 oracle and its definite integral $\int f$ relative to a $\#\mathsf{P}_1$ oracle [17,31,33]. Recall that the latter denote restrictions of the famous complexity classes NP and $\#\mathsf{P}$ to inputs $n \in \mathbb{N}$ encoded in unary; and maximizing and integrating/counting a *given, discrete* (e.g., Boolean) function are well-known complete for these respective classes.

Parametric maximization is the problem of maximizing $f : [0; 1] \to \mathbb{R}$ not on the entire interval, but on the subinterval $[0; x]$ for a given real number $x \leq 1$. Similarly, computing the indefinite integral $\int_0^x f(y)\, dy$ involves two arguments: real x and integer precision parameter n. These turn out to be computable in polynomial time relative to NP oracles and $\#\mathsf{P}$ oracles, respectively. And, perhaps surprisingly, this is optimal: there exist (even smooth) polynomial-time computable real functions such that polynomial-time algorithms for parametric maximization and indefinite integration yields polynomial-time solutions to NP and $\#\mathsf{P}$, respectively [32].

Thus, perhaps contrary to intuition, proceeding from discrete to smooth instances does not help (enough) in the rigorous sense of computational complexity theory to proceed from $\mathsf{NP}/\#\mathsf{P}$ to polynomial time.

$\#\mathsf{P}_1$-"completeness" of definite integration generalizes from the real unit interval with respect to the Lebesgue measure to a large class of compact metric groups with respect the Haar measure [51]. Akitoshi Kawamura's 2010 breakthrough result [27] similarly characterizes PSPACE via solving 1D smooth ordinary differential equations. And recent contributions relate #P to solving two linear prototype PDEs, namely (elliptic) Poisson [30] and (parabolic) Heat Equation [35].

Next up on the to-do list is a complexity-theoretic classification of the (hyperbolic) linear Wave Equation, and of the non-linear Navier-Stokes Equation. Subject the Millennium Prize Problem, Navier-Stokes maintains regularity and its solutions remain in classical spaces of continuously differentiable functions with their established coding and computability and complexity theory [61]. But regarding the Wave Equation, its regularity theory is well-established to require Sobolev spaces for computability investigations [53,66]; and Subsect. 4.2 develops the quantitative coding theory necessary for complexity considerations.

5.2 Algorithmic Random Sampling of Continuous Data

Monte Carlo algorithms date back to the Manhattan Project, and randomization has since evolved into an important technique in Computer Science: building up from random bits to random integers, random real numbers etc. Like every subset of natural numbers giving rise to a decision problem, every probability measure gives rise to three conceptually distinct computational problems: (a) evaluating, (b) integrating, and (c) random sampling.

Under mild assumptions, evaluation (a) and integration (b) are known computably equivalent [65]; see also the many works of Hoyrup. The general relation of (a) and (b) to (c) random sampling however seems open so far regarding computability. This includes generalizing the real case with Lebesgue measure to other Haar measures on compact groups, cmp. Subsect. 2.6. More generally consider the problem of computably sampling elements from a separable but not necessarily (sigma-)compact space, such as the Wiener space [40].

Following computability, the natural next question is concerned with computational complexity. Recall Subsect. 5.1 that, over the reals, (iii) sampling takes polynomial time while (ii) integrating characterizes #P. Both generalize from the real unit interval to convex bodies in Euclidean space [12,15]. Beyond the continuous Lebesgue measure, integration remains #P-hard for singular measures [16] but becomes algorithmically easy for discrete (e.g., Dirac) measures.

Question 14. Is there a probability measure space where sampling is significantly harder than integration?

6 From Theory to Applications via Practice

Sections 2 to 5 have expanded on four central concepts from classical computer science and how to extend them to the continuous setting. Key examples illustrate how this has been achieved or is currently in progress or what to approach next: to provide proofs-of-concept and opportunities to gather experience and guidance, to pave the path. The present section explores and details ways to finally flesh out between and beyond said case studies, to turn the theory into practice, and to pursue applications.

The present section explores and details ways for finally fleshing out between and beyond said case studies, to turn the theory into practice, and to pursue applications: Subsect. 6.1 envisions thus growing a rich software library of continuous data types. The question of inputting and outputting smooth vector fields is addressed in Subsect. 6.3. Combining such human-computer interface with the software library for *analytical* computing complements common Computer *Algebra* Systems, see Subsect. 6.4. Subsection 6.5 promotes its benefits to Experimental Mathematics.

6.1   Software Library

The rigorous paradigm of Analytic Programming (Sect. 3) finally allows to extend the six stages (i)–(vi) of Formal Software Engineering (Subsect. 2.1) from discrete to continuous problems. It thus enables creating a collection of abstract data types that build up from basic real numbers to the structures of Advanced Calculus, reliably. Above we have illustrated selected stages of this process (Fig. 1) with independent examples, such as: specification (Subsect. 2.6), efficient coding (Sect. 4), complexity (Subsect. 5.1), imperative implementation (Subsect. 3.4), and verification/testing (Subsecti. 3.5).

After completing these proofs-of-concepts comes extending, for each of the above case-study data type levels (0) and (1) from Subsect. 2.6, the example stage from the demonstration to range full-stack from (i) to (vi); and then similarly applying formal Numerical Software Engineering from specification (i) to verification/testing (vi) for the advanced structures in Calculus on levels (2) to (7). This yields a gradually growing collection of reliable data types with algorithmically optimal methods in agreement with constructive proofs. Specifically the following four examples, formulated abstractly and generically in Mathematics, translate equally universally to algorithms using overloading:

Example 15. Equations are often solved by means of iterations:

a) The multiplicative inverse $y = x^{-1}$ can be computed as solution to $1/y - x = 0$ by Newton's method $y_{n+1} = y_n \cdot (2 - x \cdot y_n)$: generically in many rings, such as for example of matrices or operators.
b) Similarly, the square root $y = \sqrt{x}$ as solution of the equation $x^2 = y$ is also often computed by means of Newton iterations aka Babylonian method $y_{n+1} = (y_n + x \cdot y_n^{-1})/2$: again generically in many rings.
c) Picard's method for solving ODEs amounts to iterations according to Banach's Fixedpoint Theorem in a suitable space of smooth functions.
d) Solutions to Navier-Stokes' nonlinear PDE are also mathematically shown to exist [18, §2] and being computable [61] by means of iterations in some space of integrable functions.

6.2   Hardware Acceleration

Being Turing-complete over the reals, Analytic Programming (Sect. 3) hides but must and can build on processing ordinary variable precision approximations. IEEE 754 floating point numbers have fixed precision, but enjoy a constant-factor acceleration from hardware support—compared to software solutions. Similarly accelerating ERC will thus combine the best of both worlds: reliability and efficiency.

Previous work has already managed to beat the highly optimized software library MPFRin quadruple precision by instead operating on pairs of hardware doubles [23, 25]. Alternative approaches may explore SIMD parallel processing of many single precision floating point numbers on a GPU; or may develop dedicated FPGAs/ASICs for multiprecision processing. In both cases, the communication bottleneck to the main CPU/memory requires outsourcing complex operations and sequences on once transferred data.

This endeavour naturally proceeds with the support of and collaboration with Electrical Engineering.

6.3   User Interface

Processing is the middle part of the IPO model, whose extension from discrete to continuous data have been discussed above. The first and last part of IPO refer to input and output. Historical human-computer-interfaces like keyboard/printer can input/output symbolic data, and are thus suitable for Computer Algebra Systems manipulating expressions: one way of representing functions, but lacking intuition. Intuitively and interactively "grabbing" and "pulling" is supported by common graphical user interfaces, based on mouse devices for input and monitors for output—but these are limited to 2D.

VR glasses can visualize 3D, but doing so for opaque non-scalar fields is challenging to put it mildly; and motion sensing game controllers (like Kinect or Nintendo Switch Pro) allow for "grabbing" and "pulling" in 3D, but they do not support "twisting", i.e., they cannot edit vortices.

Thus arises the need to develop a user interface for input and output of real functions "living" in higher dimensions, such as scalar (e.g., temperature) fields in 2D and 3D, or vector (e.g., force) fields. Its core challenge is for a haptic data glove that, conversely to detecting user motions in space (as mentioned already supported by existing models), can also *exercise* free forces, i.e., to pull/drag the user's hand in any direction and magnitude: allowing to "feel" (as opposed/complement to "view" in VR) vector fields. Moreover, in order to both feel and modify vortices of vector fields, the glove will be able to both sense and exercise twisting motions according to any rotation vector.

This endeavour naturally proceeds with the support of and collaboration with Mechanical Engineering.

6.4   Computer *Analysis* System

Combining the software library from Subsect. 6.1 with an interactive user interface (cmp. Subsect. 6.3) yields a Computer *Analysis* System: complementing contemporary Computer *Algebra* Systems, either standalone or—preferably—as seamless extension to a suitable open system like OSCAR. Here each abstract data type naturally turns into a *package* (interface). The plan is for further integration with some theorem proof assistant, such as Coq/HOL.

6.5   Experimental Transcendental Mathematics

The rise of Computer *Algebra* Systems has truly boosted experimental approaches to discrete branches of Mathematics; see for instance the works of Shalosh B. Ekhad. Computer-assisted proofs of statements in continuous Mathematics are a rising field, with breakthroughs concerning for example the Kepler Conjecture or Smale's 14th Problem. But these contributions remain isolated, with each approach computationally tailored (e.g., whether using hardware floats, or MPFR, and at which precision) to the particular problem: challenging for good reasons [5,47], and far from the convenience and turnkey approaches available in the discrete realm.

The software library from Subsect. 6.1 will remedy this deficiency, supporting reliable off-the-shelf computations for example in transcendental number theory: by putting a variety of theoretical algorithms into practice [22,54] and by spurring the development of new ones.

References

1. Ambos-Spies, K., Brandt, U., Ziegler, M.: Real benefit of promises and advice. In: Bonizzoni, P., Brattka, V., Löwe, B. (eds.) CiE 2013. LNCS, vol. 7921, pp. 1–11. Springer, Heidelberg (2013). https://doi.org/10.1007/978-3-642-39053-1_1
2. Avigad, J., Yin, Y.: Quantifier elimination for the reals with a predicate for the powers of two. Theor. Comput. Sci. **370**(1–3), 48–59 (2007). https://doi.org/10.1016/j.tcs.2006.10.005
3. Bauer, A.: Clerical. https://github.com/andrejbauer/clerical (2017)
4. Boldo, S., Jourdan, J.H., Leroy, X., Melquiond, G.: Verified compilation of floating-point computations. J. Autom. Reason. **54**(2), 135–163 (2015)
5. Bornemann, F., Laurie, D., Wagon, S., Waldvogel, J.: The SIAM 100-Digit Challenge. SIAM (2004). http://www.siam.org/books/100digitchallenge/
6. Brattka, V.: The emperor's new recursiveness: the epigraph of the exponential function in two models of computability. In: Ito, M., Imaoka, T. (eds.) Words, Languages & Combinatorics III, pp. 63–72. World Scientific Publishing, Singapore (2003), iCWLC 2000, Kyoto, Japan, March 14–18 (2000)
7. Brattka, V., Hertling, P.: Feasible real random access machines. J. Complex. **14**(4), 490–526 (1998)
8. Brattka, V., Pauly, A.: Computation with advice. In: Zheng, X., Zhong, N. (eds.) CCA 2010, Proceedings of the Seventh International Conference on Computability and Complexity in Analysis. Electronic Proceedings in Theoretical Computer Science, vol. 24, pp. 41–55 (2010)

9. Brattka, V., Schröder, M.: Computing with sequences, weak topologies and the axiom of choice. In: Ong, L. (ed.) CSL 2005. LNCS, vol. 3634, pp. 462–476. Springer, Heidelberg (2005). https://doi.org/10.1007/11538363_32

10. Brauße, F., et al.: Semantics, logic, and verification of "exact real computation". Tech. rep., arXiv (2021)

11. Braverman, M., Cook, S.A.: Computing over the reals: foundations for scientific computing. Notice AMS **53**(3), 318–329 (2006)

12. Cho, J., Park, S., Ziegler, M.: Computing periods .... In: Proceedings of the WAL-COM: Algorithms and Computation - 12th International Conference, WALCOM 2018, Dhaka, Bangladesh, 3–5 March 2018, pp. 132–143 (2018). https://doi.org/10.1007/978-3-319-75172-6_12

13. Cook, S.A.: Soundness and completeness of an axiom system for program verification. SIAM J. Comput. **7**(1), 70–90 (1978). https://doi.org/10.1137/0207005

14. Dries, L.v.d.: The field of reals with a predicate for the powers of two. Manus. Math.**54**, 187–196 (1986), http://eudml.org/doc/155108

15. Dyer, M.E., Frieze, A.M., Kannan, R.: A random polynomial time algorithm for approximating the volume of convex bodies. J. ACM **38**(1), 1–17 (1991). https://doi.org/10.1145/102782.102783

16. Férée, H., Ziegler, M.: On the computational complexity of positive linear functionals on $\mathcal{C}[0; 1]$. In: Kotsireas, I.S., Rump, S.M., Yap, C.K. (eds.) MACIS 2015. LNCS, vol. 9582, pp. 489–504. Springer, Cham (2016). https://doi.org/10.1007/978-3-319-32859-1_42

17. Friedman, H.: The computational complexity of maximization and integration. Adv. Math. **53**, 80–98 (1984). https://doi.org/10.1016/0001-8708(84)90019-7

18. Giga, Y., Miyakawa, T.: Solutions in l^r of the Navier-stokes initial value problem. Arch. Ration. Mech. Anal. **89**(3), 267–281 (1985)

19. Goldreich, O.: On promise problems: a survey. In: Goldreich, O., Rosenberg, A.L., Selman, A.L. (eds.) Theoretical Computer Science. LNCS, vol. 3895, pp. 254–290. Springer, Heidelberg (2006). https://doi.org/10.1007/11685654_12

20. Hertling, P.: Topological complexity with continuous operations. J. Complex. **12**, 315–338 (1996). https://doi.org/10.1006/jcom.1996.0021

21. Hertling, P.: Is the Mandelbrot set computable? Math. Log. Q. **51**(1), 5–18 (2005)

22. Hertling, P., Spandl, C.: Computing a solution of Feigenbaum's functional equation in polynomial time. Log. Methods Comput. Sci. **10**(4), 4:7, 9 (2014). https://doi.org/10.2168/LMCS-10(4:7)2014

23. Hida, Y., Li, X.S., Bailey, D.H.: Library for double-double and quad-double arithmetic. Tech. rep, Lawrence Berkeley National Laboratory (2007)

24. Hoyrup, M., Rute, J.: Computable measure theory and algorithmic randomness. In: Handbook of Computability and Complexity in Analysis. TAC, pp. 227–270. Springer, Cham (2021). https://doi.org/10.1007/978-3-030-59234-9_7

25. Jiman, H.: Real computation: from computability via efficiency to practice. M.sc. thesis, School of Computing (2021)

26. Kanada, Y.J.: 計算機による円周率計算 (特集 円周率 π). Math. Cult. **1**(1), 72–83 (2003). Calculation of circumferential ratio by computer

27. Kawamura, A.: Lipschitz continuous ordinary differential equations are polynomial-space complete. Comput. Complex. **19**(2), 305–332 (2010). https://doi.org/10.1007/s00037-010-0286-0

28. Kawamura, A., Müller, N., Rösnick, C., Ziegler, M.: Computational benefit of smoothness: parameterized bit-complexity of numerical operators on analytic functions and Gevrey's hierarchy. J. Complex. **31**(5), 689–714 (2015). https://doi.org/10.1016/j.jco.2015.05.001

29. Kawamura, A., Steinberg, F., Ziegler, M.: Towards computational complexity theory on advanced function spaces in analysis. In: Beckmann, A., Bienvenu, L., Jonoska, N. (eds.) CiE 2016. LNCS, vol. 9709, pp. 142–152. Springer, Cham (2016). https://doi.org/10.1007/978-3-319-40189-8_15

30. Kawamura, A., Steinberg, F., Ziegler, M.: On the computational complexity of the Dirichlet problem for Poisson's equation. Math. Struct. Comput. Sci. **27**(8), 1437–1465 (2017). https://doi.org/10.1017/S096012951600013X

31. Ko, K.I.: The maximum value problem and NP real numbers. J. Comput. Syst. Sci. **24**, 15–35 (1982)

32. Ko, K.I.: Complex. Theory Real Funct. Progress in Theoretical Computer Science, Birkhäuser, Boston (1991)

33. Ko, K.I., Friedman, H.: Computational complexity of real functions. Theoret. Comput. Sci. **20**, 323–352 (1982)

34. Køber, P.K.: Uniform domain representations of ℓ_p-spaces. Math. Log. Q. **180**(2), 180–205 (2007)

35. Koswara, I., Pogudin, G., Selivanova, S., Ziegler, M.: Bit-complexity of solving systems of linear evolutionary partial differential equations. In: Santhanam, R., Musatov, D. (eds.) CSR 2021. LNCS, vol. 12730, pp. 223–241. Springer, Cham (2021). https://doi.org/10.1007/978-3-030-79416-3_13

36. Kreisel, G., Macintyre, A.: Constructive logic versus algebraization, I. In: Troelstra, A., van Dalen, D. (eds.) The L. E. J. Brouwer Centenary Sympos. Studies in Logic and the Foundations of Mathematics, vol. 110, pp. 217–260. North-Holland, Amsterdam (1982), (Noordwijkerhout, June 8–13 1981)

37. Kreitz, C., Weihrauch, K.: Theory of representations. Theoret. Comput. Sci. **38**, 35–53 (1985)

38. Lambov, B.: RealLib: an efficient implementation of exact real arithmetic. Math. Struct. Comput. Sci. **17**, 81–98 (2007)

39. Le Roux, S., Ziegler, M.: Singular coverings and non-uniform notions of closed set computability. In: Dillhage, R., Grubba, T., Sorbi, A., Weihrauch, K., Zhong, N. (eds.) Proceedings of the Fourth International Conference on Computability and Complexity in Analysis (CCA 2007). Electronic Notes in Theoretical Computer Science, vol. 202, pp. 73–88. Elsevier (2008), CCA 2007, Siena, Italy, 6–18 June 2007

40. Lee, H.: Random sampling of continuous objects. Ph.D. thesis, School of Computing (2020)

41. Lim, D., Ziegler, M.: Quantitative coding and complexity theory of continuous data. Tech. rep., arXiv (2021)

42. Luckhardt, H.: A fundamental effect in computations on real numbers. Theoret. Comput. Sci. **5**(3), 321–324 (1977)

43. McNicholl, T.H.: A note on the computable categoricity of ℓ^p spaces. In: Beckmann, A., Mitrana, V., Soskova, M. (eds.) CiE 2015. LNCS, vol. 9136, pp. 268–275. Springer, Cham (2015). https://doi.org/10.1007/978-3-319-20028-6_27

44. Mori, T., Tsujii, Y., Yasugi, M.: Computability of probability distributions and characteristic functions. Log. Methods Comput. Sci. **9**, 3:9, 11 (2013). https://doi.org/10.2168/LMCS-9(3:9)2013

45. Mostowski, A.: On computable sequences. Fundam. Math. **44**, 37–51 (1957)

46. Müller, N.T.: The iRRAM: exact arithmetic in C++. In: Blanck, J., Brattka, V., Hertling, P. (eds.) CCA 2000. LNCS, vol. 2064, pp. 222–252. Springer, Heidelberg (2001). https://doi.org/10.1007/3-540-45335-0_14

47. Nakao, M.T., Plum, M., Watanabe, Y.: Numerical Verification Methods and Computer-Assisted Proofs for Partial Differential Equations. Springer Series in Computational Mathematics, Springer (2019). https://doi.org/10.1007/978-981-13-7669-6

48. Neumann, E., Ouaknine, J., Worrell, J.: On ranking function synthesis and termination for polynomial programs. In: Konnov, I., Kovács, L. (eds.) 31st International Conference on Concurrency Theory, CONCUR 2020, September 1–4, 2020, Vienna, Austria (Virtual Conference). LIPIcs, vol. 171, pp. 15:1–15:15. Schloss Dagstuhl - Leibniz-Zentrum für Informatik (2020). https://doi.org/10.4230/LIPIcs.CONCUR.2020.15

49. Papadimitriou, C.H.: Computational Complexity. Addison-Wesley (1994)

50. Park, S., Ziegler, M.: Reliable degenerate matrix diagonalization. Tech. Rep. CS-TR-2018-415, KAIST (2018)

51. Pauly, A., Seon, D., Ziegler, M.: Computing Haar measures. In: Fernández, M., Muscholl, A. (eds.) 28th EACSL Annual Conference on Computer Science Logic, CSL 2020, January 13–16, 2020, Barcelona, Spain. LIPIcs, vol. 152, pp. 34:1–34:17. Schloss Dagstuhl - Leibniz-Zentrum für Informatik (2020). https://doi.org/10.4230/LIPIcs.CSL.2020.34

52. Pauly, A., Ziegler, M.: Relative computability and uniform continuity of relations. J. Log. Anal. **5**(7), 1–39 (2013)

53. Pour-El, M.B., Richards, J.I.: The wave equation with computable initial data such that its unique solution is not computable. Advances in Math. **39**, 215–239 (1981)

54. Rettinger, R.: Bloch's constant is computable. J. Univ. Comput. Sci. **14**(6), 896–907 (2008)

55. Ryu, S., Park, J., Park, J.: Toward analysis and bug finding in javascript web applications in the wild. IEEE Softw. **36**(3), 74–82 (2019). https://doi.org/10.1109/MS.2018.110113408

56. Schröder, M.: Admissible representations in computable analysis. In: Beckmann, A., Berger, U., Löwe, B., Tucker, J.V. (eds.) CiE 2006. LNCS, vol. 3988, pp. 471–480. Springer, Heidelberg (2006). https://doi.org/10.1007/11780342_48

57. Schröder, M.: Admissibly Represented Spaces and Qcb-Spaces. In: Handbook of Computability and Complexity in Analysis. TAC, pp. 305–346. Springer, Cham (2021). https://doi.org/10.1007/978-3-030-59234-9_9

58. Selivanova, S., Steinberg, F., Thies, H., Ziegler, M.: Exact real computation of solution operators for linear analytic systems of partial differential equations. In: Boulier, F., England, M., Sadykov, T.M., Vorozhtsov, E.V. (eds.) CASC 2021. LNCS, vol. 12865, pp. 370–390. Springer, Cham (2021). https://doi.org/10.1007/978-3-030-85165-1_21

59. Specker, E.: The fundamental theorem of algebra in recursive analysis. In: Dejon, B., Henrici, P. (eds.) Constructive Aspects of the Fundamental Theorem of Algebra, pp. 321–329. Wiley-Interscience, London (1969)

60. Steinberg, F.: Complexity theory for spaces of integrable functions. Logical Methods in Computer Science 13(3), Paper No. 21, 39 (2017). https://doi.org/10.23638/LMCS-13(3:21)2017

61. Sun, S.-M., Zhong, N., Ziegler, M.: Computability of the solutions to Navier-Stokes equations via effective approximation. In: Du, D.-Z., Wang, J. (eds.) Complexity and Approximation. LNCS, vol. 12000, pp. 80–112. Springer, Cham (2020). https://doi.org/10.1007/978-3-030-41672-0_7

62. Triebel, H.: Theory of Function Spaces I, II, III. Birkhäuser (1983, 1992, 2006). https://doi.org/10.1007/978-3-0346-0416-1

63. Weihrauch, K.: Computable Analysis. Springer, Berlin (2000). https://doi.org/10.1007/978-3-642-56999-9
64. Weihrauch, K.: The computable multi-functions on multi-represented sets are closed under programming. J. Univ. Comput. Sci. **14**(6), 801–844 (2008)
65. Weihrauch, K., Tavana-Roshandel, N.: Representations of measurable sets in computable measure theory. Logical Methods Comput. Sci. **10**, 3:7,21 (2014). https://doi.org/10.2168/LMCS-10(3:7)2014
66. Weihrauch, K., Zhong, N.: Is wave propagation computable or can wave computers beat the Turing machine? Proc. Lond. Math. Soc. **85**(2), 312–332 (2002)
67. Yap, C., Sagraloff, M., Sharma, V.: Analytic root clustering: a complete algorithm using soft zero tests. In: Bonizzoni, P., Brattka, V., Löwe, B. (eds.) CiE 2013. LNCS, vol. 7921, pp. 434–444. Springer, Heidelberg (2013). https://doi.org/10.1007/978-3-642-39053-1_51
68. Yap, C.K.: On guaranteed accuracy computation. In: Geometric Computation, pp. 322–373. World Scientific Publishing, Singapore (2004)
69. Yu, J., Yap, C., Du, Z., Pion, S., Brönnimann, H.: The design of Core 2: a library for exact numeric computation in geometry and algebra. In: Fukuda, K., Hoeven, J., Joswig, M., Takayama, N. (eds.) ICMS 2010. LNCS, vol. 6327, pp. 121–141. Springer, Heidelberg (2010). https://doi.org/10.1007/978-3-642-15582-6_24
70. Ziegler, M.: Real computation with least discrete advice: a complexity theory of nonuniform computability with applications to effective linear algebra. Ann. Pure Appl. Logic **163**(8), 1108–1139 (2012). https://doi.org/10.1016/j.apal.2011.12.030
71. Ziegler, M., Brattka, V.: Computability in linear algebra. Theoret. Comput. Sci. **326**(1–3), 187–211 (2004)

Computational Aspects of Equivariant Hilbert Series of Canonical Rings for Algebraic Curves

Hara Charalambous[1] (ID), Kostas Karagiannis[2,3](✉) (ID), Sotiris Karanikolopoulos[2], and Aristides Kontogeorgis[2] (ID)

[1] Department of Mathematics, School of Sciences,
Aristotle University of Thessaloniki, 54124 Thessaloniki, Greece
hara@math.auth.gr

[2] Department of Mathematics, National and Kapodistrian University of Athens,
Panepistimioupolis, 15784 Athens, Greece
kontogar@math.uoa.gr

[3] Department of Mathematics, University of Manchester, Manchester M13 9PL, UK
konstantinos.karagiannis@manchester.ac.uk

Abstract. We study computational aspects of the problem of decomposing finite group actions on graded modules arising in arithmetic geometry, in the context of ordinary representation theory. We provide an algorithm to compute the equivariant Hilbert series of automorphisms acting on canonical rings of projective curves, using the formulas of Chevalley and Weil. Further, we apply our results on Fermat curves, determine explicitly the respective equivariant Hilbert series and extend the computation to the short exact sequence that arises from Petri's Theorem. Finally, we implement the above computations in Sage.

Keywords: Hilbert series · Group actions · Holomorphic differentials · Fermat curves

1 Introduction

1.1 Equivariant Hilbert Series

One of the most fundamental problems in representation theory of finite groups is that of decomposing representations into direct sums of indecomposables. Namely, given a finite group G acting on a vector space V over an arbitrary field k, the problem amounts to determining, for each indecomposable representation $W \in \mathrm{Ind}(G)$ over k, a natural number $n_{W,V}$ such that $V = \bigoplus_{W \in \mathrm{Ind}(G)} n_{W,V} W$. In the context of modular representation theory, that is, if

This research is co-financed by Greece and the European Union (European Social Fund- ESF) through the Operational Programme ≪Human Resources Development, Education and Lifelong Learning 2014-2020≫ in the context of the project "On the canonical ideal of algebraic curves" (MIS 5047968).

F. Boulier et al. (Eds.): CASC 2022, LNCS 13366, pp. 83–102, 2022.
https://doi.org/10.1007/978-3-031-14788-3_6

the characteristic of the ground field is positive and divides the order of G, there are several complications that make the general case of this problem practically impossible; however, if $\mathrm{char}(k) = 0$ or $\mathrm{char}(k) = p \nmid |G|$, every indecomposable representation is irreducible, and there is a direct approach using character theory

$$n_{W,V} = \langle \chi_V, \chi_W \rangle := \frac{1}{|G|} \sum_{g \in G} \chi_V(g)\overline{\chi_W(g)},$$

where χ_V denotes the character of the representation $\rho : G \to \mathrm{GL}(V)$.

The above technique can be also used when one generalizes the objects acted on, from vector spaces V over k to modules M over some k-algebra R. Historically, a case of particular interest is that of finite groups acting as automorphisms on polynomial rings: the study of their G-structure is essentially the main motivation behind the development of invariant theory, a subject whose origins date back to Hilbert's fourteenth problem. The next level of abstraction dictates to consider, instead of a polynomial ring, an arbitrary graded, Noetherian k-algebra $R = \bigoplus_{d=0}^{\infty} R_d$ acted upon by a finite group G. Since each graded piece R_d is a vector space over k, one can apply the techniques of the first paragraph to obtain for each $d \in \mathbb{N}$ and each $W \in \mathrm{Irr}(G)$, natural numbers $n_{W,d}$ such that $R_d = \bigoplus_{W \in \mathrm{Irr}(G)} n_{W,d} W$. The decomposition of R is then given by

$$R = \bigoplus_{d=0}^{\infty} R_d = \bigoplus_{d=0}^{\infty} \bigoplus_{W \in \mathrm{Irr}(G)} n_{W,d} W = \bigoplus_{W \in \mathrm{Irr}(G)} \sum_{d=0}^{\infty} n_{W,d} W.$$

One obtains for each $W \in \mathrm{Irr}(G)$ a generating function for the sequence $\{n_{W,d}\}_{d=0}^{\infty}$

$$H_{R,W}(T) = \sum_{d=0}^{\infty} n_{W,d} T^d.$$

By studying the convergence of $H_{R,W}(T)$, the infinite information of the action of G on R, which is infinite dimensional over k, can be packaged in a finite sequence $\{H_{R,W}(T) \mid W \in \mathrm{Irr}(G)\}$ which is called *the equivariant Hilbert series* of the pair (R, G). The best understood case is, again, that of polynomial rings: if $R = \mathrm{Sym}(V)$ is the symmetric algebra of a finite dimensional k-vector space V, Molien's theorem [15, Theorem 2.1] says that

$$H_{R,W}(T) = \frac{\dim W}{|G|} \sum_{g \in G} \frac{\overline{\chi_W(g)}}{\det(\mathrm{Id}_V - gT)}. \tag{1}$$

Of course, hoping to obtain an analogous formula for arbitrary graded, Noetherian k-algebras R, is unrealistic, unless one has some concrete information on the action of G on R. Since graded, Noetherian k-algebras arise as homogeneous coordinate rings of projective varieties, this can be achieved by switching the viewpoint towards algebraic geometry.

1.2   Petri's Theorem

From now on we assume that k is algebraically closed. Let X be a smooth, projective curve of genus g over k. Recall that X does not come a priori with a fixed embedding into projective space; however, it is well known that explicit projective embeddings can be constructed using (very ample) line bundles on X. Of all possible projective embeddings of X, there is one that stands out as *canonical*: that determined by the cotangent bundle Ω_X, referred to also as the sheaf of holomorphic differentials on X. It is given by

$$X \to \mathbb{P}\left(H^0(X, \Omega_X)\right) \cong \mathbb{P}_k^{g-1}, \quad P \mapsto [\omega_1(P) : \cdots : \omega_g(P)],$$

where $\{\omega_1, \ldots, \omega_g\}$ denotes a k-basis for the global sections $H^0(X, \Omega_X)$.

To see that this construction gives an embedding, we rephrase the above in the algebraic language. Recall that the homogeneous coordinate ring of the projectivization $\mathbb{P}\left(H^0(X, \Omega_X)\right)$ is the symmetric algebra $\mathrm{Sym}\left(H^0(X, \Omega_X)\right)$, which may be identified with a polynomial ring in g variables. The canonical embedding is then determined by the so-called canonical map, as ensured by the following classic theorem [14] due to Max Noether, Federigo Enriques and Karl Petri.

Theorem 1. *If X is not hyperelliptic and has genus $g \geq 4$, the canonical map*

$$\phi : S := \mathrm{Sym}\left(H^0(X, \Omega_X)\right) \to S_X := \bigoplus_{m=0}^{\infty} H^0(X, \Omega_X^{\otimes m}),$$

is surjective. Its kernel I_X, the canonical ideal, is generated in degrees 2 and 3.

Quoting from [3, Section 2, §3], the *canonical ring* S_X "is the homogeneous coordinate ring of the canonically embedded curve X". Any action of a finite group G on X induces an action on S_X, and thus, we may seek a formula for its equivariant Hilbert series. Assuming that $\mathrm{char}(k) = p \nmid |G|$, we may use Molien's formula to compute the respective series for S and thus obtain the equivariant Hilbert series for the canonical ideal I_X. It is worth noting that these calculations are the starting point in computing the action of G on the minimal graded resolution of S_X as an S-module. The latter is well-studied in the non-equivariant case mainly due its connection to Green's syzygy conjecture [5]; we hope that this work will shed some light to possible generalizations in the equivariant case.

The main results of this paper are:

1. General formulas (Theorem 3) and an algorithm (Algorithm 2) that gives the equivariant Hilbert series of S_X for arbitrary curves X.
2. Explicit formulas (Theorem 4) for the equivariant Hilbert series of S_X when X is a Fermat curve.
3. A Sage [16] program[1,2] that computes, when X is a Fermat curve:
   (a) $\{H_{S_X, V}(T) : V \in \mathrm{Irr}(G)\}$, by implementing the formulas of Theorem 4.
   (b) $\{H_{S, V}(T) : V \in \mathrm{Irr}(G)\}$, by implementing Molien's formula.

---

[1] http://users.uoa.gr/ kontogar/Code/EquivariantSage.ipynb.
[2] http://users.uoa.gr/ kontogar/Code/EquivariantSage.pdf.

(c) $\{H_{I_X,V}(T) : V \in \mathrm{Irr}(G)\}$, subtracting the two above results.

We remark that similar results were obtained in our preprint [1] using different techniques. We have verified computationally that the two approaches lead
to the same answers; a concrete theoretical proof involves complicated calculations, however we can indicatively provide the reader with one, i.e., for one of
the irreducible representations, upon request.

2   Equivariant Hilbert Series of Canonical Rings

Let X be a smooth, projective curve of genus g over an algebraically closed field k
of arbitrary characteristic $p \geq 0$. Let G be a finite subgroup of its automorphism
group $\mathrm{Aut}_k(X)$ of order $|G|$ not divisible by p. For $m \geq 1$, we denote by $\Omega_X^{\otimes m}$
the sheaf of holomorphic m-differentials on X and by W_m the k-vector space
$H^0(X, \Omega_X^{\otimes m})$ of its global sections. By the Riemann-Roch Theorem [6, IV.1.3],

$$\dim_k W_m = \begin{cases} 1 & \text{, if } m = 0 \\ g & \text{, if } m = 1 \\ (2m-1)(g-1) & \text{, if } m \geq 2 \end{cases}$$

it is further well known that the action of G on X induces an action on W_m for all
$m \geq 1$. Let $\mathrm{Irr}(G)$ denote the group of irreducible representations of G over k; the
isomorphism class of each W_m, viewed as a kG-module, is uniquely determined
by a collection of integers $\{N_{V,m}\}_{V \in \mathrm{Irr}(G)}$ such that $W_m = \bigoplus_{V \in \mathrm{Irr}(G)} N_{V,m}V$.
The classic approach to computing the integers $N_{V,m}$ goes as follows.

Algorithm 1: Computing the multiplicities $N_{V,m}$.

Inputs:

   1. The character table $[\chi_V(g)]_{\substack{V \in \mathrm{Irr}(G) \\ g \in G}}$ of G over k.

   2. The action of G on the closed points of X.

   3. A k-basis $\mathbf{b}_m = \{f(x,y)dx^{\otimes m}\}$ for W_m.

Output: A list of integers $\{N_{V,m}\}_{V \in \mathrm{Irr}(G)}$ such that $W_m = \bigoplus_{V \in \mathrm{Irr}(G)} N_{V,m}V$.
Method:

   1. For each $g \in G$
      (a) Compute the matrix $\rho(g)$, given by the action of g on the basis $\mathbf{b}_m$.
      (b) Produce a list $\{\chi_{W_m}(g) : g \in G\}$ where $\chi_{W_m}(g) = \mathrm{Trace}\,(\rho(g))$.
   2. For each $V \in \mathrm{Irr}(G)$, compute

$$N_{V,m} = \frac{1}{|G|} \sum_{g \in G} \chi_{W_m}(g)\overline{\chi_V(g)}.$$

The downside of the above algorithm comes from input (3), in that there does
not exist a general method to compute explicit bases for the k-vector spaces

$W_m = H^0(X, \Omega_X^{\otimes m})$. Even in the few cases in which bases are known, one of which is that of Fermat curves that we will study in Sect. 3, the sums in step (2) can in practice become rather difficult to compute, see for example our proof of [1, Theorem 20]. An alternative approach, exploited with great success by many authors, see for example [2,4] and [8], is to express the multiplicities $N_{V,m}$ in terms of the ramification data of the action of G on X. The resulting formulas are much easier to use, both in terms of the input required and in terms of computational complexity; however, as is usually the case in such situations, they require some familiarity with technical aspects of arithmetic geometry, which we briefly recall here. For more details the reader may refer to [6, Chapter IV], [12, Chapters 4 & 10], or [13].

From now on, we assume that the characteristic of k is either 0 or does not divide $|G|$. Let $Y = X/G$ be the quotient of X by the action of G. The quotient map $\pi : X \to Y$ is a non-constant, regular morphism of curves of degree $|G|$, so that the number of points in a generic fiber $\pi^{-1}(Q)$, $Q \in Y$ is equal to $|G|$. There exists a finite set of points $Q \in Y$ for which the fiber $\pi^{-1}(Q)$ has cardinality strictly less than $|G|$, called the *branch locus* of π and denoted by $\mathscr{B}$. The ramification locus of π is $\mathscr{R} = \pi^{-1}(\mathscr{B}) \subseteq X$. By [7, Theorem 11.49], the decomposition group of a point $P \in X$ is the cyclic group $G_P = \{\sigma \in G : \sigma(P) = P\}$ and its order is called the *ramification index* of $P \in X$. Since the ramification index is the same for all points in the orbit of $P \in X$, we denote it by e_Q, where $Q = \pi(P) \in Y$. The cyclic group G_P has e_Q-many, distinct, one-dimensional irreducible representations, determined by their characters. Fix ζ_{e_Q} to be a primitive e_Q-th root of unity; the irreducible characters of G_P are all of the of the form χ_P^d, $1 \le d \le e_Q$, where χ_P, is the fundamental character at the point P, that is, the character obtained by letting G_P act on a local uniformizer u_P at P considered modulo u_P^2. The monodromy element σ_P is a generator of G_P such that $\sigma_P(u_P) = \zeta_{e_Q} u_P$. For each irreducible representation V of G, we denote by $n_{d,Q,V}$ the multiplicity of the irreducible character χ_P^d in the decomposition of the restricted representation $\mathrm{Res}_{G_P}^G(V)$, i.e., $n_{d,Q,V} = \langle \chi_P^d, \mathrm{Res}_{G_P}^G(\chi_V) \rangle$. We summarize the above in the table below.

Table 1. Notation for the ramification data of the action of G on X

$Y = X/G$	Quotient of X by the action of G
$\mathscr{R}$	Ramification locus of $\pi : X \to Y$
$\mathscr{B}$	Branch locus $\pi : X \to Y$
e_Q	Ramification index at $Q \in \mathscr{B}$
$G_P = \{\sigma \in G : \sigma(P) = P\}$	decomposition group at $P \in \mathscr{R}$
σ_P	monodromy generator of G_P, $P \in \mathscr{R}$
$\{\chi_P^d : 0 \le d \le e_Q - 1\}$	irreducible characters of $G_P \cong \mathbb{Z}/e_Q\mathbb{Z}$
$\{n_{d,Q,V} : Q \in \mathscr{B}, 0 \le d \le e_Q - 1, V \in \mathrm{Irr}(G)\}$	multiplicities of χ_P^d in $\mathrm{Res}_{G_P}^G(V)$

The following result gives an explicit formula for the multiplicities $N_{V,m}$.

Theorem 2 (Chevalley-Weil [2]). *For each* $V \in \mathrm{Irr}(G)$, *we have that*

$$N_{V,m} = E_{V,m} + (2m-1)(g_Y - 1) \dim V$$

$$+ \sum_{Q \in \mathscr{B}} \sum_{d=0}^{e_Q - 1} \left((m-1)\left(1 - \frac{1}{e_Q}\right) + \left\langle \frac{m-1-d}{e_Q} \right\rangle \right) n_{d,Q,V},$$

where $\mathscr{B}, e_Q$ *and* $n_{d,Q,V}$ *are given in Table 1,* g_Y *is the genus of* Y,

$$E_{V,m} = \begin{cases} N_{V^*,1} & , \text{ if } m = 0 \ (V^* \text{ denotes the dual of } V) \\ 1 & , \text{ if } m = 1 \text{ and } V \text{ is the trivial representation} \\ 0 & , \text{ otherwise.} \end{cases}$$

and $\langle x \rangle = x - \lfloor x \rfloor$ *denotes the fractional part of* x.

Remark 1. For a proof of the above, see [4, Th. 3.8 & Rem. 3.9] The authors compute the multiplicity of V in the equivariant Euler characteristic $[H^0(X, \Omega_X^{\otimes m})] - [H^1(X, \Omega_X^{\otimes m})]$. The formula for $E_{V,m}$, which is the multiplicity in $H^1(X, \Omega_X^{\otimes m})$, follows from the Riemann-Roch theorem combined with Serre's duality. It is worth mentioning that the above result was generalized in [8] to the weakly ramified case.

Theorem 3. *The equivariant Hilbert series of* $S_X = \bigoplus_m H^0(X, \Omega_X^{\otimes m})$

$$H_{S_X,V}(T) = \sum_{m=0}^{\infty} N_{V,m} T^m$$

of an irreducible representation V *of* G *is given by the rational function*

$$H_{S_X,V}(T) = N_{V^*,1} + \delta_V T + \frac{3T-1}{(1-T)^2}(g_Y - 1)\dim V + \frac{T}{(1-T)^2} \dim V |\mathscr{B}|$$

$$- \frac{1}{1-T} \sum_Q \frac{f'_{Q,V}(1)}{e_Q} - \frac{T}{1-T} \sum_Q \frac{f_{Q,V}(T)}{1-T^{e_Q}},$$

where $\delta_V = 1$ *for* $V = V_{\mathrm{triv}}$ *and 0 otherwise,*

$$f_{Q,V}(T) = \sum_{d=0}^{e_Q - 1} n_{d,Q,V} T^d$$

and $|\mathscr{B}|$ *denotes the cardinality of the branch locus of the cover* $X \to X/G$.

Our computations will be in two steps. Write

$$H_{S_X,V}(T) = N_{V^*,1} + \delta_V T + F_V(T) + G_V(T),$$

where

$$F_V(T) = \sum_{m=0}^{\infty} \left((2m-1)(g_Y - 1)\dim V + \sum_Q \sum_{d=0}^{e_Q-1} (m-1)\left(1 - \frac{1}{e_Q}\right) n_{d,Q,V} \right) T^m$$

$$G_V(T) = \sum_{m=0}^{\infty} \sum_Q \sum_{d=0}^{e_Q-1} n_{d,Q,V} \left\langle \frac{m-d-1}{e_Q} \right\rangle T^m. \tag{2}$$

Lemma 1.

$$F_V(T) = \frac{3T-1}{(1-T)^2}(g_Y - 1)\dim V + \frac{2T-1}{(1-T)^2}\dim V \sum_Q \left(1 - \frac{1}{e_Q}\right).$$

Proof. The result follows from the well-known formulas

$$\sum_{m=0}^{\infty}(2m-1)T^m = \frac{3T-1}{(1-T)^2}, \quad \text{and} \quad \sum_{m=0}^{\infty}(m-1)T^m = \frac{2T-1}{(1-T)^2},$$

as well as the fact that $\sum_{d=0}^{e_Q-1} n_{d,Q,V} = \dim V$.

To compute $G_V(T)$, we first prove the following auxiliary lemma.

Lemma 2. *For $A \in \mathbb{Z}$ and $1 < e \in \mathbb{N}$, we have that*

$$\sum_{m=0}^{\infty} \left\langle \frac{m+A}{e} \right\rangle T^m = \frac{T}{e(1-T)^2} + \frac{v_A}{e(1-T)} - \frac{T^{e-v_A}}{(1-T^e)(1-T)},$$

where v_A is the remainder of the division of A by e.

Proof. Recall that $\langle x \rangle = x - \lfloor x \rfloor$. Write $m = \pi e + v$ and $A = \pi_A e + v_A$, for $0 \le v, v_A < e$ and $\pi, \pi_A \in \mathbb{Z}$. Then

$$\left\langle \frac{m+A}{e} \right\rangle = \frac{v + v_A}{e} - \left\lfloor \frac{v + v_A}{e} \right\rfloor,$$

and thus

$$\sum_{m=0}^{\infty} \left\langle \frac{m+A}{e} \right\rangle T^m = \sum_{v=0}^{e-1} \sum_{\pi=0}^{\infty} \left(\frac{v + v_A}{e} - \left\lfloor \frac{v + v_A}{e} \right\rfloor \right) (T^e)^\pi T^v$$

$$= \frac{1}{1-T^e} \sum_{v=0}^{e-1} \left(\frac{v + v_A}{e} - \left\lfloor \frac{v + v_A}{e} \right\rfloor \right) T^v.$$

Next, we remark that since

$$\left\lfloor \frac{v + v_A}{e} \right\rfloor = \begin{cases} 0 & \text{if } 0 \le v + v_A \le e-1 \\ 1 & \text{if } e \le v + v_A < 2e \end{cases},$$

we have that

$$\sum_{v=0}^{e-1}\left(\frac{v+v_A}{e}-\left\lfloor\frac{v+v_A}{e}\right\rfloor\right)T^v = \sum_{v=0}^{e-1}\frac{v+v_A}{e}T^v - \sum_{v=e-v_A}^{e-1}T^v$$

$$= \frac{1}{e}\sum_{v=0}^{e-1}vT^v + \frac{v_A}{e}\sum_{v=0}^{e-1}T^v - T^{e-v_A}\sum_{v=0}^{v_A-1}T^v.$$

Each of the three sums is given by

$$\frac{1}{e}\sum_{v=0}^{e-1}vT^v = \frac{eT^{e+1}-T^{e+1}-eT^e+T}{e(1-T)^2} = -\frac{T^e}{(1-T)}+\frac{T(1-T^e)}{e(1-T)^2},$$

$$\frac{v_A}{e}\sum_{v=0}^{e-1}T^v = \frac{v_A(1-T^e)}{e(1-T)},$$

$$T^{e-v_A}\sum_{v=0}^{v_A-1}T^v = T^{e-v_A}\frac{1-T^{v_A}}{1-T} = \frac{T^{e-v_A}}{1-T}-\frac{T^e}{1-T}.$$

Observe that the first term of the first sum cancels out with the second term of the third sum, and thus

$$\sum_{m=0}^{\infty}\left\langle\frac{m+A}{e}\right\rangle T^m = \frac{1}{1-T^e}\left(\frac{T(1-T^e)}{e(1-T)^2}+\frac{v_A(1-T^e)}{e(1-T)}-\frac{T^{e-v_A}}{1-T}\right)$$

$$= \frac{T}{e(1-T)^2}+\frac{v_A}{e(1-T)}-\frac{T^{e-v_A}}{(1-T^e)(1-T)}.$$

Corollary 1. Let $G_V(T)$ be as in Eq. (2) and $f_{Q,V}(T) = \sum_{d=0}^{e_Q-1}n_{d,Q,V}T^d$. Then

$$G_V(T) = \frac{\dim V \cdot T}{(1-T)^2}\sum_Q\frac{1}{e_Q}+\frac{\dim V}{1-T}\sum_Q\left(1-\frac{1}{e_Q}\right)$$

$$-\frac{1}{1-T}\sum_Q\frac{f'_{Q,V}(1)}{e_Q}-\frac{T}{1-T}\sum_Q\left(\frac{f_{Q,V}(T)}{1-T^{e_Q}}\right).$$

Proof. Observe that if $0 \le d \le e_Q - 1$, the remainder of the division of $A = -d-1$ by e_Q is $v_A = e_Q - d - 1$. Thus, applying Lemma 2 for $A = -d - 1$ we obtain

$$\sum_{m=0}^{\infty} \sum_{Q} \sum_{d=0}^{e_Q-1} n_{d,Q,V} \left\langle \frac{m-d-1}{e_Q} \right\rangle T^m = \sum_{Q} \sum_{d=0}^{e_Q-1} n_{d,Q,V} \sum_{m=0}^{\infty} \left\langle \frac{m-d-1}{e_Q} \right\rangle T^m$$

$$= \sum_{Q} \sum_{d=0}^{e_Q-1} n_{d,Q,V} \left(\frac{T}{e_Q(1-T)^2} + \frac{e_Q - d - 1}{e_Q(1-T)} - \frac{T^{d+1}}{(1-T^{e_Q})(1-T)} \right)$$

$$= \sum_{Q} \sum_{d=0}^{e_Q-1} \frac{n_{d,Q,V} T}{e_Q(1-T)^2} + \left(1 - \frac{1}{e_Q} \right) \frac{n_{d,Q,V}}{1-T} - \frac{n_{d,Q,V} d}{e_Q(1-T)} - \frac{n_{d,Q,V} T^{d+1}}{(1-T^{e_Q})(1-T)}$$

$$= \sum_{Q} \frac{f_{Q,V}(1) T}{e_Q(1-T)^2} + \left(1 - \frac{1}{e_Q} \right) \frac{f_{Q,V}(1)}{1-T} - \frac{f'_{Q,V}(1)}{e_Q(1-T)} - \frac{f_{Q,V}(T) T}{(1-T^{e_Q})(1-T)}.$$

Using again the fact that $f_{Q,V}(1) = \sum_{d=0}^{e_Q-1} n_{d,Q,V} = \dim V$ gives the desired result. $\square$

Proof (of Theorem 3). Let $F_V(T)$ and $G_V(T)$ be as in Eq. (2). By Lemma 1 and Corollary 1 we have that

$$F_V(T) = \frac{3T-1}{(1-T)^2} (g_Y - 1) \dim V + \frac{2T-1}{(1-T)^2} \dim V \sum_{Q} \left(1 - \frac{1}{e_Q} \right),$$

$$G_V(T) = \frac{\dim V \cdot T}{(1-T)^2} \sum_{Q} \frac{1}{e_Q} + \frac{\dim V}{1-T} \sum_{Q} \left(1 - \frac{1}{e_Q} \right)$$

$$- \frac{1}{1-T} \sum_{Q} \frac{f'_{Q,V}(1)}{e_Q} - \frac{T}{1-T} \sum_{Q} \left(\frac{f_{Q,V}(T)}{1 - T^{e_Q}} \right).$$

Adding the second term of $F_V(T)$ to the second term of $G_V(T)$ gives

$$\frac{(2T-1)\dim V}{(1-T)^2} \sum_{Q} \left(1 - \frac{1}{e_Q} \right) + \frac{\dim V}{1-T} \sum_{Q} \left(1 - \frac{1}{e_Q} \right)$$

$$= \frac{\dim V \cdot T}{(1-T)^2} \sum_{Q} \left(1 - \frac{1}{e_Q} \right) = \frac{\dim V \cdot T}{(1-T)^2} \# \mathscr{B} - \frac{\dim V \cdot T}{(1-T)^2} \sum_{Q} \frac{1}{e_Q},$$

and the last term above cancels out with the first term of $G_V(T)$. Thus

$$F_V(T) + G_V(T) = \frac{3T-1}{(1-T)^2} (g_Y - 1) \dim V + \frac{T}{(1-T)^2} \dim V |\mathscr{B}|$$

$$- \frac{1}{1-T} \sum_{Q} \frac{f'_{Q,V}(1)}{e_Q} - \frac{T}{1-T} \sum_{Q} \frac{f_{Q,V}(T)}{1 - T^{e_Q}}.$$

As a corollary we obtain the below algorithm.

Algorithm 2: Computing the equivariant Hilbert series $\{H_{S_X,V}(T) : V \in \mathrm{Irr}(G)\}$.

Inputs:
1. The character table $[\chi_V(g)]_{\substack{V \in \mathrm{Irr}(G) \\ g \in G}}$ of G over k.
2. The action of G on the closed points of X.

Output: A list of rational functions $\{H_V(t) : V \in \mathrm{Irr}(G)\}$
Method:

1. Compute the ramification locus $\mathcal{R}$ and the branch locus $\mathcal{B}$ of $\pi : X \to Y$.
2. Compute g_Y using the Riemann-Hurwitz formula [17, Theorem 3.4.13].
3. For each $Q \in \mathcal{B}$
   (a) Compute the ramification index e_Q.
   (b) For each $V \in \mathrm{Irr}(G)$ and each $0 \leq d \leq e_Q - 1$ compute

$$n_{d,Q,V} = \langle \chi_P^d, \mathrm{Res}_{G_P}^G(\chi_V) \rangle \text{ and } f_{Q,V}(T) = \sum_d n_{d,Q,V} T^d.$$

4. For each $V \in \mathrm{Irr}(G)$, compute $H_{S_X,V}(T)$ using Theorem 3.

There are two advantages of Algorithm 2 over Algorithm 1. Firstly it can be used in the cases in which explicit k-bases for polydifferentials are not known; secondly the inner products of step 3(b) are taken over the decomposition groups G_P which are strictly smaller than the full automorphism group G. On the other hand, its disadvantages are that one needs to compute the ramification data of the cover $\pi : X \to Y$, a problem which is wide open in its full generality, and that computing the multiplicities $n_{d,Q,V}$ is not always a straightforward task. We shall demonstrate how this is done in the next section by applying our results to Fermat curves.

3   The Case of Fermat Curves

Let F_n be a Fermat curve with affine model $x^n + y^n + 1 = 0$, defined over an algebraically closed field k of characteristic $p \geq 0$. We assume that $n \geq 4$, $p > 3$ and $n-1$ is not a power of p. To describe the automorphism group $G = \mathrm{Aut}_k(X)$, we write

$$A := \mathbb{Z}/n\mathbb{Z} \times \mathbb{Z}/n\mathbb{Z} = \{\sigma_{\alpha,\beta} : 0 \leq \alpha, \beta \leq n - 1\}$$
$$S_3 = \langle s, t : s^3 = t^2 = 1, \ tst = s^{-1} \rangle = \{1, s, s^2, t, st, ts\}.$$

and note that S_3 acts on A by conjugation as:

$h \in S_3$	s	s^2	t	ts	st
$h^{-1}\sigma_{\alpha,\beta}h$	$\sigma_{\beta-\alpha,-\alpha}$	$\sigma_{-\beta,\alpha-\beta}$	$\sigma_{-\alpha,\beta-\alpha}$	$\sigma_{\beta,\alpha}$	$\sigma_{\alpha-\beta,-\beta}$

Remark 2. An automorphism $\sigma : F_n \to F_n$ acts on functions $f \in k(F_n)$ by $\sigma(f) = f \circ \sigma^{-1}$. The group acts on the left on points, so $(\sigma_1 \sigma_2)P = \sigma_1(\sigma_2 P)$, and the action on functions satisfies $(\sigma_1 \sigma_2 f) = f \circ (\sigma_1 \sigma_2)^{-1} = f \circ \sigma_2^{-1} \circ \sigma_1^{-1} = \sigma_1(\sigma_2 f)$.

In [11] and [18] the authors prove that F_n has genus $g = \frac{(n-1)(n-2)}{2}$, automorphism group $G = A \rtimes S_3$ and that the action of G on the function field $k(F_n)$, i.e., the field $k(x, y)$ subject to the equation $x^n + y^n + 1 = 0$, is given by

$g \in G$	$\sigma_{\alpha,\beta}$	s	s^2	t	ts	st
$g(x,y)$	$(\zeta_n^\alpha x, \zeta_n^\beta y)$	$\left(\frac{y}{x}, \frac{1}{x}\right)$	$\left(\frac{1}{y}, \frac{x}{y}\right)$	$\left(\frac{1}{x}, \frac{y}{x}\right)$	(y, x)	$\left(\frac{x}{y}, \frac{1}{y}\right)$

where ζ_n is a fixed primitive n-th root of unity. The above gives us the second required input item for Algorithm 2. Regarding the first, we use the character table of G that was computed in [1, Proposition 3]. Recall that S_3 has three irreducible representations: the trivial representation, the sign representation and the standard representation, denoted by $\rho_{\text{triv}}, \rho_{\text{sgn}}$ and ρ_{stan} respectively.

Proposition 1. *The irreducible representations of G are given in the table below,*

Rep.	Degree	Character $\chi(\sigma_{\alpha,\beta}x)$, where $x \in S_3$
$\theta_{\frac{\nu n}{3}, \frac{\nu n}{3}, \rho}$	1	$\zeta^{\frac{\nu n}{3}(\alpha+\beta)} \chi_\rho(x)$
$\theta_{\frac{\nu n}{3}, \frac{\nu n}{3}, \rho_{\text{stan}}}$	2	$\zeta^{\frac{\nu n}{3}(\alpha+\beta)} \chi_{\text{stan}}(x)$
$\theta_{\kappa,\kappa,\rho}$	3	$\begin{cases} \zeta^{\kappa(\alpha+\beta)} + \zeta^{\kappa(\alpha-2\beta)} + \zeta^{\kappa(\beta-2\alpha)} & , \text{ if } x = 1 \\ \zeta^{\kappa(\alpha+\beta)} \chi_\rho(x) & , \text{ if } x = ts \\ \zeta^{\kappa(\alpha-2\beta)} \chi_\rho(x) & , \text{ if } x = t \\ \zeta^{\kappa(\beta-2\alpha)} \chi_\rho(x) & , \text{ if } x = st \\ 0 & , \text{ if } x = s, s^2 \end{cases}$
$\theta_{\kappa,\lambda,\rho_{\text{triv}}}$	6	$\begin{cases} \begin{pmatrix} \zeta^{\kappa\alpha+\lambda\beta} + \zeta^{-(\kappa+\lambda)\alpha+\kappa\beta} + \zeta^{\lambda\alpha-(\kappa+\lambda)\beta} + \\ \zeta^{\lambda\alpha+\kappa\beta} + \zeta^{-(\kappa+\lambda)\alpha+\lambda\beta} + \zeta^{\kappa\alpha-(\kappa+\lambda)\beta} \end{pmatrix} & , \text{ if } x = 1 \\ 0 & , \text{ if } x \neq 1 \end{cases}$

where $\nu \in \{0, 1, 2\}$, $\rho \in \{\rho_{\text{triv}}, \rho_{\text{sgn}}\}$, $\kappa, \lambda \in \mathbb{Z}/n\mathbb{Z}$, $\kappa, \lambda \neq \frac{\nu n}{3}, \kappa \neq \lambda, \kappa \neq -2\lambda, \lambda \neq -2\kappa$ and the representations corresponding to $\kappa, \lambda \in \{\frac{n}{3}, \frac{2n}{3}\}$ appear only when $3 \mid n$.

In what follows, we fix all primitive roots of unity to be compatible with the chosen ζ_n, in the sense that if $n \mid i$, then ζ_i must satisfy $\zeta_i^{i/n} = \zeta_n$, whereas if $i \mid n$ then $\zeta_i = \zeta_n^{n/i}$.

Proposition 2. *The quotient F_n/G is isomorphic to $\mathbb{P}_k^1$, the branch locus of $F_n \to \mathbb{P}_k^1$ consists of three points P_∞, P_1, P_0. The points $Q_\infty = (\zeta_{2n}, 0), Q_1 =$*

$(1, \sqrt[n]{-2}), Q_0 = (\zeta_{6n}^4, \zeta_{6n}^2)$ of F_n lie above each of the three mentioned points, and their isotropy groups and monodromy generators are given in the following table.

point	group	monodromy
$Q_\infty = (\zeta_{2n}, 0)$	$\mathbb{Z}/2n\mathbb{Z}$	$\sigma_{1,1}t$
$Q_1 = (1, \sqrt[n]{-2})$	$\mathbb{Z}/2\mathbb{Z}$	t
$Q_0 = (\zeta_{6n}^4, \zeta_{6n}^2)$	$\mathbb{Z}/3\mathbb{Z}$	$\sigma_{-1,-1}s^2$

Proof. The proof can be found at the appendix.

The above implies that, in the notation of Theorem 3, we have $g_Y = 0$ and $|\mathscr{B}| = 3$. Thus the third and fourth term of $H_{S_X,V}$ simplify as follows

$$\frac{3T-1}{(1-T)^2}(g_Y - 1)\dim V + \frac{T}{(1-T)^2}\dim V|\mathscr{B}| = \frac{\dim V}{(1-T)^2}.$$

Theorem 4. *With the above notation, we have that*

$$H_{S_X,V}(T) = N_{V^*,1} + \delta_V T + \frac{\dim V}{(1-T)^2} - \frac{1}{1-T}\sum_Q \frac{f'_{Q,V}(1)}{e_Q} - \frac{T}{1-T}\sum_Q \frac{f_{Q,V}(T)}{1-T^{e_Q}},$$

where $\delta_V = 1$ for $V = V_{\mathrm{triv}}$ and 0 otherwise, the polynomials $f_{Q,V}(T)$ are given in the table below

$V \in \mathrm{Irr}(G)$	$f_{Q_\infty,V}(T)$	$f_{Q_0,V}(T)$	$f_{Q_1,V}(T)$
$\theta_{0,0,\rho_{\mathrm{triv}}}$	1	1	1
$\theta_{0,0,\rho_{\mathrm{sgn}}}$	T^n	1	T
$\theta_{\frac{n}{3},\frac{n}{3},\rho_{\mathrm{triv}}}$	$T^{\frac{4n}{3}}$	T	1
$\theta_{\frac{n}{3},\frac{n}{3},\rho_{\mathrm{sgn}}}$	$T^{\frac{n}{3}}$	T	T
$\theta_{\frac{2n}{3},\frac{2n}{3},\rho_{\mathrm{triv}}}$	$T^{\frac{2n}{3}}$	T^2	1
$\theta_{\frac{2\nu n}{3},\frac{2\nu n}{3},\rho_{\mathrm{sgn}}}$	$T^{\frac{5n}{3}}$	T^2	T
$\theta_{\frac{\nu n}{3},\frac{\nu n}{3},\rho_{\mathrm{stan}}}$	$T^{\frac{\nu n}{3}} + T^{n+\frac{\nu n}{3}}$	$1 + T + T^2 - T^\nu$	$1+T$
$\theta_{\kappa,\kappa,\rho_{\mathrm{triv}}}$	$T^\kappa + T^{n+\kappa} + T^{[-2\kappa]_n}$	$1 + T + T^2$	$2+T$
$\theta_{\kappa,\kappa,\rho_{\mathrm{sgn}}}$	$T^\kappa + T^{n+\kappa} + T^{[-2\kappa]_n+n}$	$1 + T + T^2$	$1+2T$
$\theta_{\kappa,\lambda,\rho_{\mathrm{triv}}}$	$T^\kappa + T^\lambda + T^{n+\kappa} + T^{n+\lambda}$ $+ T^{[-(\kappa+\lambda)]_n} + T^{[-(\kappa+\lambda)]_n+n}$	$2(1 + T + T^2)$	$3+3T$

and $[x]_n$ denotes the smallest non-negative remainder of the division of x by n.

The proof of the above will be given separately for each of $f_{Q_\infty,V}$, $f_{Q_0,V}$, $f_{Q_1,V}$, by considering all irreducible representations of Proposition 1. To do so, one needs to calculate first the multiplicities $n_{d,Q,V}$, $Q \in \{Q_\infty, Q_0, Q_1\}$ as follows:

1. For each Q, write $G_Q = \{\sigma_Q^i : 0 \le i \le e_Q\}$ where σ_Q is the local monodromy and e_Q is the ramification index, both taken from Proposition 2.
2. For each $0 \le i \le e_Q$, find $\sigma_{\alpha_i,\beta_i} \in A$ and $x_i \in S_3$ such that $\sigma_Q^i = \sigma_{\alpha_i,\beta_i} x_i$.
3. Fix a primitive root of unity ζ_{e_Q} compatible with ζ_n as discussed above.
4. Compute $n_{d,Q,V} = \langle \mathrm{Res}_{G_Q}^G (\chi_V), \chi_Q^d \rangle = \sum_{i=0}^{e_Q-1} \chi_V (\sigma_P^i) \zeta_{e_Q}^{-id}$.

3.1 The Polynomials $f_{Q_\infty,V}(T)$

By Proposition 2, $Q_\infty = (\zeta_{2n}, 0)$ and G_{Q_∞} is generated by the monodromy element $\sigma_{1,1}t$. Since $(\sigma_{1,1}t)^2 = \sigma_{0,1}$, we have that

$$(\sigma_{1,1}t)^{2k} = \sigma_{0,k} \text{ and } (\sigma_{1,1}t)^{2k+1} = \sigma_{1,k+1}t, \quad \text{for } 0 \le k \le n-1,$$

and thus, for $0 \le d \le 2n-1$, we have that

$$n_{d,Q_\infty,V} = \langle \mathrm{Res}_{G_{Q_\infty}}^G (\chi_V), \chi_{Q_\infty}^d \rangle$$
$$= \frac{1}{2n} \sum_{\ell=0}^{n-1} \zeta_{2n}^{2\ell(-d)} \chi_V(\sigma_{0,\ell}) + \frac{1}{2n} \sum_{\ell=0}^{n-1} \zeta_{2n}^{(2\ell+1)(-d)} \chi_V(\sigma_{1,\ell+1}t).$$

- When $V = \theta_{\frac{\nu n}{3},\frac{\nu n}{3},\rho}$, $\rho \in \{\rho_{\mathrm{triv}}, \rho_{\mathrm{sgn}}\}$, Proposition 1 gives

$$\chi_V(\sigma_{0,\ell}) = \zeta_n^{\frac{\nu n\ell}{3}} = \zeta_{2n}^{\frac{2\nu n\ell}{3}} \text{ and } \chi_V(\sigma_{1,\ell+1}t) = \zeta_n^{\frac{\nu n(\ell+2)}{3}} \chi_\rho(t) = \zeta_{2n}^{\frac{2\nu n(\ell+2)}{3}} \chi_\rho(t).$$

Thus, we compute

$$n_{d,Q_\infty,V} = \frac{1}{2n} \sum_{\ell=0}^{n-1} \zeta_{2n}^{2\ell(-d)} \zeta_{2n}^{\frac{2\nu n\ell}{3}} + \frac{1}{2n} \sum_{\ell=0}^{n-1} \zeta_{2n}^{(2\ell+1)(-d)} \zeta_{2n}^{\frac{2\nu n(\ell+2)}{3}} \chi_\rho(t)$$
$$= \frac{1}{2n} \sum_{\ell=0}^{n-1} \zeta_{2n}^{2\ell(\frac{\nu n}{3}-d)} + \frac{1}{2n} \sum_{\ell=0}^{n-1} \zeta_{2n}^{2\ell(\frac{\nu n}{3}-d)+\left(\frac{4\nu n}{3}-d\right)} \chi_\rho(t)$$
$$= \begin{cases} \frac{1}{2} + \frac{1}{2}\zeta_{2n}^{\frac{4\nu n}{3}-d} \chi_\rho(t) & , \text{ if } n \mid \frac{\nu n}{3} - d \\ 0 & , \text{ otherwise.} \end{cases}$$

The only two values of d such that $0 \le d \le 2n-1$ and $n \mid \frac{\nu n}{3} - d$ are $d = \frac{\nu n}{3}$ and $d = n + \frac{\nu n}{3}$. So

$$f_{Q_\infty,V}(T) = \left(\frac{1}{2} + \frac{1}{2}\zeta_{2n}^{\nu n} \chi_\rho(t)\right) T^{\frac{\nu n}{3}} + \left(\frac{1}{2} + \frac{1}{2}\zeta_{2n}^{(\nu-1)n} \chi_\rho(t)\right) T^{n+\frac{\nu n}{3}}.$$

Notice that for $\nu \in \{0,1,2\}$ we always have that $\{\zeta_{2n}^{\nu n}, \zeta_{2n}^{(\nu-1)n}\} = \{-1,1\}$.

- When $V = \theta_{\frac{\nu n}{3}, \frac{\nu n}{3}, \rho_{\text{stan}}}$, Proposition 1 gives

$$\chi_V(\sigma_{0,\ell}) = 2\zeta_{2n}^{\frac{2\nu n\ell}{3}} \quad \text{and} \quad \chi_V(\sigma_{1,\ell+1}t) = 0.$$

We compute as above

$$n_{d,Q_\infty,V} = \frac{1}{n}\sum_{\ell=0}^{n-1} \zeta_{2n}^{2\ell\left(\frac{\nu n}{3}-d\right)} + 0 = \begin{cases} 1 & , \text{if } n \mid \frac{\nu n}{3} - d \\ 0 & , \text{otherwise} \end{cases},$$

and thus

$$f_{Q_\infty,V}(T) = T^{\frac{\nu n}{3}} + T^{n+\frac{\nu n}{3}}.$$

- When $V = \theta_{\kappa,\kappa,\rho}$, $\rho \in \{\rho_{\text{triv}}, \rho_{\text{sgn}}\}$, Proposition 1 gives

$$\chi_V(\sigma_{0,\ell}) = 2\zeta_n^{\kappa\ell} + \zeta_n^{-2\kappa\ell} = 2\zeta_{2n}^{2\kappa\ell} + \zeta_{2n}^{-4\kappa\ell} \quad \text{and}$$
$$\chi_V(\sigma_{1,\ell+1}t) = \zeta_n^{-\kappa(2\ell+1)}\chi_\rho(t) = \zeta_{2n}^{-2\kappa(2\ell+1)}\chi_\rho(t).$$

We then have

$$n_{d,Q_\infty,V} = \frac{1}{n}\sum_{\ell=0}^{n-1}\zeta_{2n}^{2\ell(\kappa-d)} + \frac{1}{2n}\sum_{\ell=0}^{n-1}\zeta_{2n}^{-2\ell(2\kappa+d)} + \frac{1}{2n}\sum_{\ell=0}^{n-1}\zeta_{2n}^{-2\ell(2\kappa+d)}\zeta_{2n}^{-(2\kappa+d)}\chi_\rho(t)$$

$$= \begin{cases} 1 & , \text{if } n \mid \kappa - d \\ \frac{1}{2} + \frac{1}{2}\zeta_{2n}^{-(2\kappa+d)}\chi_\rho(t) & , \text{if } n \mid 2\kappa + d \\ 0 & \text{otherwise.} \end{cases}$$

The first case gives $d = \kappa$ or $d = n + \kappa$, while the second gives $d = [-2k]_n$ or $d = n + [-2k]_n$ so

$$f_{Q_\infty,V}(T) = T^\kappa + T^{n+\kappa} + \left(\frac{1}{2} + \frac{1}{2}\chi_\rho(t)\right)T^{[-2\kappa]_n} + \left(\frac{1}{2} - \frac{1}{2}\chi_\rho(t)\right)T^{[-2\kappa]_n+n}.$$

- Finally, for $V = \theta_{\kappa,\lambda,\rho_{\text{triv}}}$ we have that

$$\chi_V(\sigma_{0,\ell}) = 2\zeta_{2n}^{2\lambda\ell} + 2\zeta_{2n}^{2\kappa\ell} + 2\zeta_{2n}^{-2(\kappa+\lambda)\ell} \quad \text{and} \quad \chi_V(\sigma_{1,\ell+1}t) = 0.$$

Thus

$$n_{d,Q_\infty,V} = \frac{1}{n}\sum_{\ell=0}^{n-1}\left(\zeta_{2n}^{2\ell(\lambda-d)} + \zeta_{2n}^{2\ell(\kappa-d)} + \zeta_{2n}^{-2\ell(\kappa+\lambda+d)}\right)$$

$$= \begin{cases} 1 & , \text{if } n \mid \lambda - d \text{ or } n \mid \kappa - d \text{ or } n \mid \kappa + \lambda + d \\ 0 & , \text{otherwise} \end{cases}$$

and so

$$f_{Q_\infty,V}(T) = T^\kappa + T^\lambda + T^{n+\kappa} + T^{n+\lambda} + T^{[-(\kappa+\lambda)]_n} + T^{[-(\kappa+\lambda)]_n+n}.$$

3.2 The Polynomials $f_{Q_0,V}(T)$

By Proposition 2, $Q_0 = (\zeta_{6n}^4, \zeta_{6n}^2)$ and G_{Q_0} is generated by the monodromy element $\sigma_{-1,-1}s^2$. For $d = 0, 1, 2$ we have

$$n_{d,Q_0,V} = \langle \mathrm{Res}_{G_{Q_0}}^{G}(\chi_V), \chi_{Q_0}^d \rangle$$

$$= \frac{1}{3}\left(\chi_V(1) + \zeta_3^{-d}\chi_V(\sigma_{-1,-1}s) + \zeta_3^{2(-d)}\chi_V\left((\sigma_{-1,-1}s^2)^2\right)\right).$$

- For both $V = \theta_{\frac{\nu n}{3}, \frac{\nu n}{3}, \rho}$, $\rho \in \{\rho_{\mathrm{triv}}, \rho_{\mathrm{sgn}}\}$ we have

$$n_{d,Q_0,V} = \frac{1}{3}\left(1 + \zeta_3^{-d}\zeta_n^{-\frac{2\nu n}{3}} + \zeta_3^{2(-d)}\zeta_n^{-\frac{4\nu n}{3}}\right) = \frac{1}{3}\left(1 + \zeta_3^{-d-2\nu} + \zeta_3^{-2d-4\nu}\right)$$

$$= \begin{cases} 1 & \text{, if } d \equiv -2\nu \bmod 3 \equiv \nu \bmod 3 \\ 0 & \text{, otherwise.} \end{cases}$$

and so $f_{Q_0,V}(T) = T^{[\nu]_3} = T^\nu$.

- When $V = \theta_{\frac{\nu n}{3}, \frac{\nu n}{3}, \rho_{\mathrm{stan}}}$ we have

$$n_{d,Q_0,V} = \frac{1}{3}\left(2 - \zeta_3^{-d-2\nu} - \zeta_3^{-2d-4\nu}\right) = \frac{1}{3}\left(3 - 1 - \zeta_3^{-d-2\nu} - \zeta_3^{-2d-4\nu}\right)$$

$$= \begin{cases} 1 & \text{, if } d \not\equiv -2\nu \bmod 3 \\ 0 & \text{, otherwise} \end{cases}$$

and so $f_{Q_0,V}(T) = 1 + T + T^2 - T^{[-2\nu]_3} = 1 + T + T^2 - T^\nu$.

- When $V = \theta_{\kappa,\kappa,\rho}$, $\rho \in \{\rho_{\mathrm{triv}}, \rho_{\mathrm{sgn}}\}$, we have $n_{d,Q_0,V} = 1$ for $d \in \{0,1,2\}$ and so $f_{Q_0,V} = 1 + T + T^2$.
- When $V = \theta_{\kappa,\lambda,\rho_{\mathrm{triv}}}$, $n_{d,Q,V} = 2$ for $d \in \{0,1,2\}$ and so $f_{Q_0,V} = 2(1+T+T^2)$.

3.3 The Polynomials $f_{Q_1,V}(T)$

By Proposition 2, $Q_1 = (1, \sqrt[n]{-2})$ and G_{Q_1} is generated by the monodromy element t. For $d \in \{0,1\}$ we have

$$n_{d,Q_1,V} = \langle \mathrm{Res}_{G_{Q_1}}^{G}(\chi_V), \chi_{Q_1}^d \rangle$$

$$= \frac{1}{2}\left(\chi_V(1) + (-1)^{-d}\chi_V(t)\right) = \frac{1}{2}\left(\dim V + (-1)^d\chi_V(t)\right).$$

- When $V = \theta_{\frac{\nu n}{3}, \frac{\nu n}{3}, \rho}$, $\rho \in \{\rho_{\mathrm{triv}}, \rho_{\mathrm{sgn}}, \rho_{\mathrm{stan}}\}$, we get

$$f_{Q_1,V}(T) = \begin{cases} 1 & \text{, if } \rho = \rho_{\mathrm{triv}} \\ T & \text{, if } \rho = \rho_{\mathrm{sgn}} \\ 1 + T & \text{, if } \rho = \rho_{\mathrm{stan}}. \end{cases}$$

- When $V = \theta_{\kappa,\kappa,\rho}$, $\rho \in \{\rho_{\mathrm{triv}}, \rho_{\mathrm{sgn}}\}$, we get

$$f_{Q_1,V}(T) = \begin{cases} 2 + T & \text{, if } \rho = \rho_{\mathrm{triv}} \\ 1 + 2T & \text{, if } \rho = \rho_{\mathrm{sgn}}. \end{cases}$$

- Finally, when $V = \theta_{\kappa,\lambda,\rho_{\mathrm{triv}}}$, $f_{Q_1,V}(T) = 3 + 3T$.

4 Implementation and Examples

Let F_n be a Fermat curve over k, with the assumptions on n and k as in the previous section. By Petri's Theorem 1, there exists a short exact sequence

$$0 \to I_X := \ker \phi \hookrightarrow S := \mathrm{Sym}\left(H^0(X, \Omega_X)\right) \overset{\phi}{\twoheadrightarrow} S_X := \bigoplus_{m=0}^{\infty} H^0(X, \Omega_X^{\otimes m}) \to 0,$$

which is split over kG, since $\mathrm{char}(k) \nmid |G|$.

In this section, we present our Sage [16] program[3,4], which, as mentioned in the introduction, computes for each $V \in \mathrm{Irr}(G)$:

1. $H_{S_X,V}(T)$, by implementing the formulas of Theorem 4.
2. $H_{S,V}(T)$, by implementing Molien's formula (see Eq. 1).
3. $H_{I_X,V}(T) = H_{S,V}(T) - H_{S_X,V}(T)$, by subtracting the two above results.

The computation of $H_{S_X,V}(T)$ follows by Theorem 4. First we implement the difference $H_{S_X,V}(T) - N_{V^*,1} - \delta_V T$ for each $V \in \mathrm{Irr}(G)$. Then we read each N_{V^*1} from the implementation of $H_{S_X,V^*}(T) - N_{V,1} - \delta_{V^*} T$ and add T if V is trivial. We remark that implementing the algorithm for $n = 6$ we retrieve same results as in [1, Table 1, pg. 18], where we computed the kG-structure of $H^0(X, \Omega_X^{\otimes m})$ using an alternative approach. For example, the series for $V = \theta_{0,1,\mathrm{triv}}$ is

$$H_{S_X,V}(T) = \frac{T^3}{T^5 - 2T^4 + T^3 + T^2 - 2T + 1}.$$

To implement Molien's formula, it is required to input the character table of G and the representation $G \to \mathrm{GL}\left(H^0(X, \Omega_X)\right)$. The former is taken directly from Proposition 1, while the latter was implemented using the action of G on a basis $\{\omega_{i,j}\}$ of $H^0(X, \Omega_X)$ which we computed in [1, Prop. 6]:

$\sigma_{\alpha,\beta}(\omega_{i,j})$	$s(\omega_{i,j})$	$t(\omega_{i,j})$	$ts(\omega_{i,j})$	$st(\omega_{i,j})$	$s^2(\omega_{i,j})$
$\zeta^{\alpha(i+1)+\beta(j+1)}\omega_{i,j}$	$\omega_{n-3-(i+j),i}$	$-\omega_{n-3-(i+j),j}$	$-\omega_{j,i}$	$-\omega_{i,n-3-(i+j)}$	$\omega_{j,n-3-(i+j)}$

The output is much more complicated than $H_{S_X,V}(T)$: for instance when $n = 6$ and $V = \theta_{0,1,\mathrm{triv}}$ we obtain a rational function with numerator of degree 18 and denominator of degree 30.

The final step is to compute the equivariant Hilbert series of I_X using Petri's Theorem 1. For $n = 6$ and $V = \theta_{0,1,\mathrm{triv}}$, $H_{I_X,V}(T) = H_{S,V}(T) - H_{S_X,V}(T)$ has power series expansion

$$8T^3 + 20T^4 + 49T^5 + 130T^6 + 319T^7 + 667T^8 + 1363T^9 + 2557T^{10} + \text{higher order terms.}$$

The interpretation is that the representation $\theta_{0,1,\mathrm{triv}}$ appears, for example, 2557 times in the decomposition of the degree 10 graded piece of I_X into irreducible summands.

---

[3] http://users.uoa.gr/ kontogar/Code/EquivariantSage.ipynb.
[4] http://users.uoa.gr/ kontogar/Code/EquivariantSage.pdf.

5   Appendix - The Ramification Data of Fermat Curves

In this section we give the details for the proof of Proposition 2. We shall work over $k = \mathbb{C}$ for simplicity, even though the arguments are valid over any algebraically closed field of characteristic prime to the order of G. Recall that all roots of unity are fixed, as per the discussion preceding Proposition 2.

The Fermat curve can be seen as double Kummer cover of the projective line $\mathbb{P}^1$. We will work with Galois extensions of the corresponding function fields and in this way we have the Kummer extension of function fields $\mathbb{C}(F_n)/\mathbb{C}(x)$, where $\mathbb{C}(F_n)$ is the extension obtained by the rational function field $\mathbb{C}(x)$ by adjoining the quantity $y = (-1-x^n)^{\frac{1}{n}}$. Then we can consider the cyclic extension of function fields $\mathbb{C}(x)/\mathbb{C}(x^n)$. The ramification in such extensions is well known, see for example [9, 10], namely there is ramification in the cover $\mathbb{C}(F_n)/\mathbb{C}(F_n)^A$ over the points $x^n = -1, x^n = 0, x^n = \infty$, where $A = \mathbb{Z}/n\mathbb{Z} \times \mathbb{Z}/n\mathbb{Z}$.

Since $G = \text{Aut}(F_n) = A \rtimes S_3$, the Galois extension $\mathbb{C}(F_n)/\mathbb{C}(F_n)^G$ corresponding to the cover $F_n \to F_n/G$ has the intermediate subfield $\mathbb{C}(F_n)^A = \mathbb{C}(x^n)$, and $\mathbb{C}(F_n)^A/\mathbb{C}(F_n)^G$ is Galois with Galois group the symmetric group S_3. Moreover, the extension $\mathbb{C}(F_n)^A/\mathbb{C}(F_n)^G$ corresponds to a ramified cover $\mathbb{P}^1 \to \mathbb{P}^1$ ramified over three points. Such covers can be explained in terms of the j invariant, see [19]. Indeed, if we set $X = -x^n$ then the group S_3 can be realized by the six Möbius automorphisms:

$$X \mapsto \left\{ X, \frac{1}{X}, 1 - X, \frac{1}{1-X}, \frac{X}{1-X}, \frac{X-1}{X} \right\}.$$

The fixed points of these maps are given in the following table:

transform	order	equation	fixed points
$\frac{1}{X}$	2	$X^2 - 1 = 0$	$1, -1$
$1 - X$	2	$2X - 1$	$\frac{1}{2}, \infty$
$\frac{X}{X-1}$	2	$X^2 - 2X = 0$	$0, 2$
$\frac{1}{1-X}, \frac{X-1}{X}$	3	$X^2 - X + 1 = 0$	$\zeta_6, \frac{1}{\zeta_6}$

and the function

$$j(X) = \frac{4}{27} \frac{(X^2 - X + 1)^3}{X^2(X-1)^2}$$

is a generator of the fixed field $\mathbb{C}(X)^{S_3} = \mathbb{C}(F_n)^G = \mathbb{C}(j)$. The fixed points of the S_3-cover $\mathbb{P}^1 \to \mathbb{P}^1$ are $P_{(j=0)}, P_{(j=1)}, P_{(j=\infty)}$. The map j maps the fixed points as follows:

X		$j(X)$
$0, 1, \infty$	$\longmapsto$	∞
$-1, 2, \frac{1}{2}$	$\longmapsto$	1
$\zeta_6, \frac{1}{\zeta_6}$	$\longmapsto$	0

In Fig. 1 we display the ramification diagram above the point $P_{(j=\infty)}$ and in Fig. 2 the respective diagram above the points $P_{(j=1)}$ and $P_{(j=0)}$. Note that in the first row we denote by $P_{i,i'}$ the i-th ramification point above $P_{(X=i')}$, for $i' \in \{0, 1, \infty\}$, the labels in the vertical lines of the first column indicate the Galois groups, whereas in all other columns they indicate ramification indices.

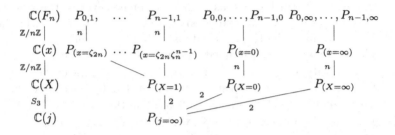

Fig. 1. Ramification diagram for $P_{(j=\infty)}$

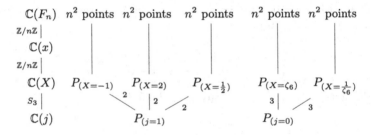

Fig. 2. Ramification diagram for $P_{(j=1)}$ and $P_{(j=\infty)}$

Each of the points $P_{(X=-1)}, P_{(X=2)}, P_{(X=\frac{1}{2})}, P_{(X=\zeta_6)}, P_{(X=\frac{1}{\zeta_6})}$ has n^2 points in the Fermat curve. For instance the point $X = -x^n = \zeta_6$ is lifted to the points (x, y) where $x = (-\zeta_6)^{1/n} = \zeta_n^\ell \zeta_{2n} \zeta_{6n} = \zeta_{6n}^{6\ell+4}$, for $0 \le \ell \le n-1$, and similarly, $y^n = -1 - x^n = -1 + \zeta_6 = \zeta_6^2$, since $\zeta_6^2 - \zeta_6 + 1 = 0$. Therefore, for $0 \le k \le n-1$, $y = \zeta_n^k \zeta_{6n}^2 = \zeta_{6n}^{6k+2}$. This means that the set of points $\{(\zeta_{6n}^{6\ell+4}, \zeta_{6n}^{6k+2}) : 0 \le k, \ell < n\}$ are the n^2 points above the point $P_{(X=\zeta_6)}$.

We will now select an arbitrary point above each $P_{(j=\infty)}, P_{(j=1)}, P_{(j=0)}$ and for each such point we will find the cyclic subgroup and the monodromy element. Recall that by Remark 2, automorphisms $\sigma \in G$ act on functions $f \in \mathbb{C}(F_n)$ by $\sigma(f) = f \circ \sigma^{-1}$.

• Consider the point $Q_\infty = (\zeta_{2n}, 0)$ above $P_{(j=\infty)}$. The isotropy subgroup is a cyclic group of order $2n$. For example we can verify that it is fixed by the element $\sigma_Q = \sigma_{1,1} t$. Further, since $(\sigma_{1,1} t)^2 = \sigma_{0,1}$, we have that

$$(\sigma_{1,1} t)^{2k} = \sigma_{0,k} \quad \text{and} \quad (\sigma_{1,1} t)^{2k+1} = \sigma_{1,k+1} t, \quad \text{for } 0 \le k \le n-1$$

A local uniformizer at Q_∞ is y, which is acted on by $(\sigma_{1,1}t)^2 = \sigma_{0,1}$ by $y \mapsto \zeta_n y$. Hence, the monodromy element at Q_∞ is $\sigma_{1,1}t$.

• Consider the point Q_1 above $P_{(j=1)}$ given by affine coordinates $(1, \sqrt[n]{-2})$, which is fixed by the automorphism t acting on functions as $t(x) = 1/x, t(y) = y/x$. Since the decomposition group at Q_1 is a cyclic group of order 2 the monodromy at Q_1 is the element t.

• A point $Q_0 = (x_0, y_0)$ above $P_{(j=0)}$ is given by $X = \zeta_6$, that is, $x_0^n = -\zeta_6$, therefore $x_0 = (-\zeta_6)^{1/n} = \zeta_n^\ell \zeta_{2n} \zeta_{6n} = \zeta_{6n}^{6\ell+4}$, for $0 \le \ell \le n-1$. Similarly $y_0^n = -1-x^n = -1+\zeta_6 = \zeta_6^2$, since $\zeta_6^2 - \zeta_6 + 1 = 0$. Therefore, $y_0 = \zeta_n^k \zeta_{6n}^2 = \zeta_{6n}^{6k+2}$, for $0 \le k \le n-1$.

Let s be the automorphism acting on functions by $s(x) = y/x$, $s(y) = 1/x$, so that $s^2(x) = 1/y, s^2(y) = x/y$. Observe that the point with coordinates $(x_0, y_0) = (\zeta_{6n}^4, \zeta_{6n}^2)$ is sent by $\sigma_{1,1}s$ to (x_0, y_0). Indeed,

$$(\zeta_{6n}^4, \zeta_{6n}^2) \overset{s}{\longmapsto} (\zeta_{6n}^{2-4}, \zeta_{6n}^{-4}) \overset{\sigma_{1,1}}{\longmapsto} (\zeta_{6n}^4, \zeta_{6n}^2).$$

The function $x - x_0 = x - \zeta_{6n}^4$ is a local uniformizer at (x_0, y_0). By Remark 2, $\sigma_{1,1}s$ acts on functions as $\sigma_{-1,-1}s^2$ and thus

$$\sigma_{-1,-1}s^2(x - \zeta_{6n}^4) = \sigma_{-1,-1}\left(\frac{1}{y} - \zeta_{6n}^4\right) = \frac{\zeta_{6n}^6}{y} - \zeta_{6n}^4 = -\zeta_{6n}^4 \frac{y - \zeta_{6n}^2}{y}$$

$$= -\zeta_{6n}^4 \frac{y - \zeta_{6n}^2}{\zeta_{6n}^2 + (y - \zeta_{6n}^2)}$$

$$= -\zeta_{6n}^2 (y - \zeta_{6n}^2)\left(1 + \sum_{\nu=1}^\infty \frac{-1}{\zeta_{6n}^2}(y - \zeta_{6n}^2)^\nu\right). \tag{3}$$

On the other hand Taylor expansion of the Fermat equation at (x_0, y_0) gives

$$0 = x^n + y^n + 1 = x_0^n + y_0^n + 1 + nx_0^{n-1}(x-x_0) + ny_0^{n-1}(y-y_0)) + \text{higher order terms},$$

that is

$$\zeta_{6n}^{4(n-1)}(x - \zeta_{6n}^4) + \zeta_{6n}^{2(n-1)}(y - \zeta_{6n}^2) \mod \mathfrak{m}_{(x_0,y_0)}^2$$

and this combined with Eq. (3) gives

$$\sigma_{-1,-1}s^2(x - \zeta_{6n}^4) = \zeta_{6n}^2 \zeta_{6n}^{2(n-1)}(x - \zeta_{6n}^4) = \zeta_3(x - \zeta_{6n}^4),$$

i.e., $\sigma_{-1,-1}s^2$ is indeed the monodromy at the point $Q_0 = (x_0, y_0)$.

References

1. Charalambous, H., Karagiannis, K., Karanikolopoulos, S., Kontogeorgis, A.: The equivariant Hilbert series of the canonical ring of Fermat curves. Indagationes Math. (2022). https://doi.org/10.1016/j.indag.2022.06.001
2. Chevalley, C., Weil, A., Hecke, E.: Über das verhalten der integrale 1. gattung bei automorphismen des funktionenkörpers. Abh. Math. Sem. Univ. Hamburg **10**(1), 358–361 (1934). https://doi.org/10.1007/BF02940687

3. Eisenbud, D.: Green's conjecture: an orientation for algebraists. In: Free Resolutions in Commutative Algebra and Algebraic Geometry (Sundance, UT, 1990), Res. Notes Math., vol. 2, pp. 51–78. Jones and Bartlett, Boston (1992)
4. Ellingsrud, G., Lønsted, K.: An equivariant Lefschetz formula for finite reductive groups. Math. Ann. **251**(3), 253–261 (1980). https://doi.org/10.1007/BF01428945
5. Green, M.L.: Koszul cohomology and the geometry of projective varieties. J. Differ. Geom. **19**(1), 125–171 (1984). http://projecteuclid.org/euclid.jdg/1214438426
6. Hartshorne, R.: Algebraic Geometry. Graduate Texts in Mathematics, vol. 52. Springer, New York (1977). https://doi.org/10.1007/978-1-4757-3849-0
7. Hirschfeld, J.W.P., Korchmáros, G., Torres, F.: Algebraic Curves Over a Finite Field. Princeton Series in Applied Mathematics, Princeton University Press, Princeton (2008)
8. Köck, B.: Galois structure of Zariski cohomology for weakly ramified covers of curves. Am. J. Math. **126**(5), 1085–1107 (2004)
9. Kontogeorgis, A.: The group of automorphisms of cyclic extensions of rational function fields. J. Algebra **216**(2), 665–706 (1999)
10. Kontogeorgis, A.I.: The group of automorphisms of the function fields of the curve $x^n + y^m + 1 = 0$. J. Number Theory **72**(1), 110–136 (1998)
11. Leopoldt, H.W.: Über die Automorphismengruppe des Fermatkörpers. J. Number Theory **56**(2), 256–282 (1996)
12. Liu, Q.: Algebraic Geometry and Arithmetic Curves. Oxford Graduate Texts in Mathematics, vol. 6. Oxford University Press, Oxford (2002).Translated from the French by Reinie Erné, Oxford Science Publications
13. Pries, R., Stevenson, K.: A survey of Galois theory of curves in characteristic p. In: WIN–Women in Numbers, Fields Institute Communications, vol. 60, pp. 169–191. American Mathematical Society, Providence (2011). https://doi.org/10.1090/bull/1594
14. Saint-Donat, B.: On Petri's analysis of the linear system of quadrics through a canonical curve. Math. Ann. **206**, 157–175 (1973). https://doi.org/10.1007/BF01430982
15. Stanley, R.P.: Invariants of finite groups and their applications to combinatorics. Bull. Am. Math. Soc. (N.S.) **1**(3), 475–511 (1979). https://doi.org/10.1090/S0273-0979-1979-14597-X
16. Stein, W., et al.: Sage Mathematics Software (Version 8.9). The Sage Development Team (2019). http://www.sagemath.org
17. Stichtenoth, H.: Algebraic Function Fields and Codes. Springer, Berlin (1993)
18. Tzermias, P.: The group of automorphisms of the Fermat curve. J. Number Theory **53**(1), 173–178 (1995)
19. Yoshida, M.: Hypergeometric Functions, My Love. Aspects of Mathematics, E32, Friedr. Vieweg & Sohn, Braunschweig (1997). https://doi.org/10.1007/978-3-322-90166-8

Symbolic-Numeric Algorithm for Calculations in Geometric Collective Model of Atomic Nuclei

Algirdas Deveikis[1], Alexander A. Gusev[2(✉)], Sergue I. Vinitsky[2,3],
Yuri A. Blinkov[3], Andrzej Góźdź[4], Aleksandra Pędrak[5], and Peter O. Hess[6,7]

[1] Vytautas Magnus University, Kaunas, Lithuania
[2] Joint Institute for Nuclear Research, Dubna, Russia
gooseff@jinr.ru
[3] RUDN University, 6 Miklukho-Maklaya, 117198 Moscow, Russia
[4] Institute of Physics, Maria Curie-Skłodowska University, Lublin, Poland
[5] National Centre for Nuclear Research, Warsaw, Poland
[6] Instituto de Ciencias Nucleares, UNAM, Circuito Exterior, C.U., A.P. 70-543,
04510 Mexico D.F., Mexico
[7] Frankfurt Institute for Advanced Studies, 60438 Frankfurt am Main, Germany

Abstract. We developed a symbolic–numeric algorithm involving a set of effective symbolic and numerical procedures for calculations of low lying energy spectra and eigenfunctions of atomic nuclei. The eigenfunctions are expanded over the orthonormal noncanonical $U(5) \supset O(5) \supset O(3)$ basis in Geometric Collective Model. We give implementation of the algorithm and procedures in Wolfram Mathematica. We present benchmark calculations of energy spectrum, quadrupole moment and the reduced upwards transition probability $B(E2)$ for the nucleus ^{186}Os.

Keywords: Orthonormal non-canonical basis · Groups
$U(5) \supset SO(5) \supset SO(3)$ · Irreducible representations · Gram-Schmidt orthonormalization · Geometric Collective Model · Spectral characteristic · Atomic nuclei

1 Introduction

The Bohr–Mottelson (B-M) collective model [2,3] has gained widespread acceptance in calculations of vibrational-rotational quadrupole spectra and electromagnetic transitions in atomic nuclei [4,8,17]. Among others it was applied for such nuclei as: uranium [14], Pt, Os and W isotopes [15]. Some results were also obtained for the super-heavy deformed nuclei [12] where a fit of microscopically derived potential energy surfaces proposed in [9,21–23] has been performed with the help of numerical (FORTRAN) application of the geometric collective model (GCM) [13,20].

Key problems in such numerical large-scale calculations of spectral characteristics of the GCM with the octahedral O_h point symmetry as well as the general Bohr Hamiltonian [18,19] are round-off errors appearing in calculation of high-power polynomials. These polynomials with alternating coefficients and strong numerical cancellations are observed in the Gram–Schmidt orthonormalization of the nonorthogonal set of basis eigenfunctions that we investigated in [7] using both integer and floating point arithmetics implemented in Wolfram Mathematica [25].

In the present paper, we propose some development of effective symbolic procedures for calculations of the spectral characteristic of atomic nuclei in GCM. We give the implementation of the developed procedures in Wolfram Mathematica and performance of benchmark calculations. We analyze round-off errors in calculation of high-power polynomials with alternating coefficients. We show that strong cancellation in Gram–Schmidt orthonormalization usually pose serious problems in numerical calculations [7,14,15,20,26,27].

The structure of the paper is following. In Sect. 2, we describe the statement of the problem separated into subsections corresponding to procedures (subroutines) involving the GCM code. We give the benchmark examples of their execution summing up them in the Tables that show computer memory and execution time with respect to ranges of the quantum numbers involved in the runs: construction of GCM Hamiltonian, construction of orthonormal $U(5) \supset O(5) \supset O(3)$ basis, calculation of β- and γ-dependent matrix elements, and composition of Hamiltonian matrices of algebraic eigenvalue problem. In Sect. 3, benchmark calculations of energy spectrum, quadrupole moment and the reduced upwards transition probability $B(E2)$ for ^{186}Os are presented. Finally, in Sect. 4, the summary of main results and conclusions are given. In Appendices A and B, the sets of input parameters for atomic nuclei and boundary value problem for GCM model are presented.

The CPU times of the benchmark calculations give required estimates for choosing appropriate versions of the presented symbolic-numeric algorithms and programs. The computations were performed with Wolfram Mathematica 10.1 on PC Intel i7-36030QM, CPU 2.40 GHz, RAM 8 GB, 64-bit Windows 8.

2   The Statement of the Problem and Subroutines

Hamiltonian. The classical nuclear collective Hamiltonian constructed in the so called laboratory frame has the general form [20]

$$\hat{H} = \hat{T}(\pi, \alpha) + \hat{V}(\alpha). \tag{1}$$

Quantum description of the collective motions in GCM is performed by using the quadrupole deformation coordinates, $\hat{\alpha}^{[2]} = \alpha_{2m}, m = -2, -1, 0, 1, 2$, and the corresponding conjugate momenta, $\hat{\pi}^{[2]} = \pi_{2m}, m = -2, -1, 0, 1, 2$, subjected to commutation relations $[\hat{\pi}_m^{[2]}, \hat{\alpha}_{m'}^{[2]}] = -\imath\hbar\delta_{mm'}$. The kinetic energy is constructed to contain the two lowest-order terms proportional to the square of the momenta determined in a nonstandard form accepted in [20]:

$$\hat{T} = \frac{1}{B_2}[\hat{\pi} \times \hat{\pi}]^{[0]} + \frac{P_3}{3}\left\{\left[[\hat{\pi} \times \hat{\alpha}]^{[2]} \times \hat{\pi}\right]^{[0]}\right\}, \tag{2}$$

where $\{\ldots\}$ means the sum over all permutations, and B_2 and P_3 are kinetic-energy parameters. For such nonstandard definition of the parameter B_2 with respect to standard one (see Eq. (2)), it will be multiplied by factor $2/\sqrt{5}$. So, in the practice of GCM calculations, the rescaled parameter $\bar{B}_2 = 2B_2/\sqrt{5}$ is really used. The tensor product of spherical tensors $A^{[l_1]}$ and $B^{[l_2]}$ is defined as

$$[A^{[l_1]} \otimes B^{[l_2]}]^{[l]} = \sum_{m_1,m_2} (l_1 m_1 l_2 m_2 | l m) A_{m_1}^{[l_1]} B_{m_2}^{[l_2]},$$

where $(l_1 m_1 l_2 m_2 | l m)$ are SO(3) Clebsch–Gordan coefficients [24]. All terms in the Hamiltonian are coupled to angular momentum 0, i.e., to rotational scalars.

Potential Energy. For the potential energy we use a polynomial expansion up to the sixth order in the deformation variables β and γ specified by the intrinsic deformation coordinates $\hat{a}^{[2]} = a_{2m'}$. The intrinsic frame is defined as coinciding to principal axes of the nucleus. It is determined by a set of three Euler angles $\Omega \in S^3(\Omega)$ and new deformation variable $\alpha_{2m} = \sum_{m'} D_{mm'}^{2*}(\Omega) a_{2m'}$, where $D_{mm'}^{2*}(\Omega)$ denotes the Wigner functions of irreducible representations of SO(3) group [24] (marker * denotes the complex conjugate operation). The choice of principal axes requires the following constraints: $a_{2-2} = a_{22}, a_{2-1} = a_{21} = 0$. The β and γ variables are defined as: $a_{20} = \beta \cos \gamma, a_{22} = (1/\sqrt{2})\beta \sin \gamma$. The potential energy is assumed in the following form:

$$\hat{V}(\beta,\gamma) = \sum_{\rho=2}^{6} \sum_{m=0}^{2} \beta^\rho \cos^m(3\gamma) \hat{V}_{\rho,m}, \tag{3}$$

where potential parameters $\hat{V}_{\rho,m}$ read as:

$$\hat{V}_{2,0} = C_2 \tfrac{1}{\sqrt{5}}; \quad \hat{V}_{3,1} = -C_3 \sqrt{\tfrac{2}{35}}; \quad \hat{V}_{4,0} = C_4 \tfrac{1}{5};$$
$$\hat{V}_{5,1} = -C_5 \sqrt{\tfrac{2}{175}}; \quad \hat{V}_{6,2} = C_6 \tfrac{2}{35}; \quad \hat{V}_{6,0} = D_6 \tfrac{1}{5\sqrt{5}}. \tag{4}$$

Introducing these parameters the potential $\hat{V}(\beta,\gamma)$ takes the form

$$\hat{V}(\beta,\gamma) = C_2 \tfrac{1}{\sqrt{5}}\beta^2 - C_3 \sqrt{\tfrac{2}{35}}\beta^3 \cos(3\gamma) + C_4 \tfrac{1}{5}\beta^4$$
$$-C_5 \sqrt{\tfrac{2}{175}}\beta^5 \cos(3\gamma) + C_6 \tfrac{2}{35}\beta^6 \cos^2(3\gamma) + D_6 \tfrac{1}{5\sqrt{5}}\beta^6. \tag{5}$$

For practical reason, we rescale $\hat{V}_{\rho,m}$ to $V_{\rho,m}$ in oscillator units of length with respect to the β variable using basis parameters of mass B_2' and stiffness C_2':

$$V(\beta,\gamma) = \sum_{\rho=2}^{6} \sum_{m=0}^{2} \beta^\rho \cos^m(3\gamma) V_{\rho,m}, \quad V_{\rho,m} = \hat{V}_{\rho,m} \times \left(\frac{\hbar}{\sqrt{B_2' C_2'}} \right)^{\rho/2}. \tag{6}$$

Basis States and a Range of the Set of Quantum Numbers. We choose as basic functions the eigenfunctions of the five-dimensional harmonic oscillator

$$\hat{H}_5 = \frac{\sqrt{5}}{2B_2'}[\hat{\pi} \times \hat{\pi}]^{[0]} + \frac{\sqrt{5}C_2'}{2}[\hat{\alpha} \times \hat{\alpha}]^{[0]}. \tag{7}$$

Table 1. The degeneracy $d_{\lambda L} = \mu_{\max} - \mu_{\min} + 1$ for a number of L and λ. The first row of the table is formed by the values of λ and the first column of the table is formed by the values of the angular momentum L. The next columns in non empty square contains the degeneracy $d_{\lambda L}$ depending on accessible values of momentum L and seniority λ.

L, λ	5	10	15	20	25	30	35	40	45	50
0		1				1			1	
2	1	1		1	1		1	1		1
5	1	1		1	1		1	1		1
10	1	2	2	2	2	2	2	2	2	2
15		1	3	2	2	3	2	2	3	2
20		1	2	4	4	3	4	4	3	4
25			1	3	4	4	4	4	4	4
30			1	2	4	6	5	5	6	5
35				1	3	4	6	6	5	6

Table 2. The example of calculations of the total number of states defined by quantum numbers $\nu\lambda$ for a number of L up to the specified value of the $\nu_{\max}$. The first row of the table is formed by the values of $\nu_{\max}$ and the first column of the table is formed by the value of the angular momentum L. The next columns contains the total number of states for corresponding values of L and $\nu_{\max}$.

$L, \nu_{\max}$	5	10	15	20	25	30	35	40	45	50
0	5	14	27	44	65	91	120	154	192	234
2	7	22	45	77	117	165	222	287	360	442
5	2	12	30	57	92	135	187	247	315	392
10	1	12	36	72	121	182	256	342	441	552
15	0	2	16	42	81	132	196	272	361	462
20	0	1	12	36	72	121	182	256	342	441
25	0	0	2	16	42	81	132	196	272	361
30	0	0	1	12	36	72	121	182	256	342
35	0	0	0	2	16	42	81	132	196	272

The basis states can be characterized by irreducible representations of the $U(5) \supset O(5) \supset O(3) \supset O(2)$ chain of groups [7]:

- ν is the number of phonons,
- λ is the number of phonons that are not coupled pairwise to zero (seniority),
- L and M are the numbers of the angular momentum and its projection,
- μ is the additional quantum number, denoting the maximal number of phonon triplets coupled to the angular momentum $L = 0$ and counting degenerated states for $L \geq 6$:

$$\nu = 0, 1, 2, \ldots, \nu_{\max}, \quad \lambda = \nu, \nu-2, \ldots, 1 \text{ or } 0, \quad \mu = \mu_{\min}, \mu_{\min}+1, \ldots, \mu_{\max}.$$

$$(8)$$

Here ν_{max} is some chosen as the maximum number of phonons. The range of μ (i.e., μ_{min} and μ_{max}) for given λ and L is determined by inequalities:

$$L/2 \leq \lambda - 3\mu \leq L, \; L = \text{even}, \quad (L+3)/2 \leq \lambda - 3\mu \leq L, \; L = \text{odd}. \quad (9)$$

The solution of inequalities Eqs. (9) gives a range of accessible values of μ at given accessible λ and L:

$$\mu_{min} = \max\left(0, \texttt{Ceiling}\left(\tfrac{\lambda-L}{3}\right)\right), \; \mu_{max} = \texttt{Floor}\left(\frac{\lambda-(L+3(L \bmod 2))/2}{3}\right), \quad (10)$$

where $\texttt{Ceiling}(\mu)$ is the lowest integer but not lower than μ and $\texttt{Floor}(\mu)$ is the largest integer not greater than μ.

2.1   The Representation of the Wave Functions in Coordinate Space

The five-dimensional equation of the B-M collective model (7) in the intrinsic frame $\beta \in R_+^1$ and $\gamma, \Omega \in S^4$ with respect to $\Psi^{int}_{\nu\lambda\mu LM} \in L_2(R_+^1 \otimes S^4)$ with the measure $d\tau = \beta^4 \sin(3\gamma) d\beta d\gamma d\Omega$ reads as

$$\{H^{(BM)} - E^{BM}_\nu\}\Psi^{int}_{\nu\lambda\mu LM} = 0, \; H^{(BM)} = \frac{\hbar^2}{2B_2'}\left(-\frac{1}{\beta^4}\frac{\partial}{\partial\beta}\beta^4\frac{\partial}{\partial\beta} + \frac{\hat{\Lambda}^2}{\beta^2}\right) + \frac{C_2'}{2}\beta^2. \quad (11)$$

Here $E^{BM}_\nu \equiv E^L_\nu = \hbar\omega_2'(\nu + \tfrac{5}{2})$ are the eigenvalues of the five-dimensional harmonic oscillator, $\omega_2' = \sqrt{C_2'/B_2'}$ is the oscillation frequency, $\hbar$ is Planck constant, $\hat{\Lambda}^2$ is the quadratic Casimir operator of O(5) in $L_2(S^4(\gamma, \Omega))$ at nonnegative integers $\nu = 2n_\beta + \lambda$, i.e., at even and nonnegative integers $\nu - \lambda$ determined as

$$(\hat{\Lambda}^2 - \lambda(\lambda+3))\Psi^{int}_{\nu\lambda\mu LM} = 0, \; \hat{\Lambda}^2 = -\frac{1}{\sin(3\gamma)}\frac{\partial}{\partial\gamma}\sin(3\gamma)\frac{\partial}{\partial\gamma} + \sum_{k=1}^{3}\frac{(\hat{\bar{L}}_k)^2}{4\sin^2(\gamma - \tfrac{2}{3}k\pi)}, \quad (12)$$

where the nonnegative integer λ is the seniority (8) and $(\hat{\bar{L}}_k)^2$ are the angular momentum operators of O(3) along the principal axes in intrinsic frame, i.e., with commutator $[\hat{\bar{L}}_i, \hat{\bar{L}}_j] = -\imath\varepsilon_{ijk}\hat{\bar{L}}_k$ [7].

Eigenfunctions $|\nu\lambda\mu LM\rangle$ of the five-dimensional oscillator (7) in the intrinsic frame (11) have the form

$$\Psi^{int}_{\nu\lambda\mu LM}(\beta, \gamma, \Omega) = \langle\beta\gamma\Omega|\nu\lambda\mu LM\rangle = \sum_{K(even)} \Phi^{int}_{\nu\lambda\mu LK}(\beta, \gamma)\mathcal{D}^{(L)*}_{MK}(\Omega), \quad (13)$$

where $\mathcal{D}^{(L)*}_{MK}(\Omega)$ are the orthonormal Wigner functions with measure $d\Omega$,

$$\mathcal{D}^{(L)*}_{MK}(\Omega) = \sqrt{\frac{2L+1}{8\pi^2}}\frac{D^{(L)*}_{MK}(\Omega) + (-1)^L D^{(L)*}_{M,-K}(\Omega)}{1+\delta_{K0}}; \quad (14)$$

summation over K runs even values K in range:

$$K = 0, 2, \ldots, L \qquad \text{for even integer } L : 0 \le L \le L_{\max}, \qquad (15)$$
$$K = 2, \ldots, L - 1 \qquad \text{for odd integer } L : 3 \le L \le L_{\max}.$$

$\Phi^{int}_{\nu\lambda\mu LK}(\beta, \gamma)$ are the nonorthogonal components with overlap $\langle \hat{\phi}^{\lambda\mu' L}(\gamma) | \hat{\phi}^{\lambda\mu L}(\gamma) \rangle$

$$\Phi^{int}_{\nu\lambda\mu LK}(\beta, \gamma) = C_L^{\lambda\mu} F_{\nu\lambda}(\beta) \hat{\phi}_K^{\lambda\mu L}(\gamma), \qquad (16)$$

determined by (17), (18) and normalization factor $C_L^{\lambda\mu} = (\langle \hat{\phi}^{\lambda\mu L}(\gamma) | \hat{\phi}^{\lambda\mu L}(\gamma) \rangle)^{-1/2}$.

2.2 γ-Dependent Part of the Basis States

The components $\hat{\phi}_K^{\lambda\mu L}(\gamma) = (-1)^L \hat{\phi}_{-K}^{\lambda\mu L}(\gamma)$ for even K and $\hat{\phi}_K^{\lambda\mu L}(\gamma) = 0$ for odd L and $K = 0$ as well as for odd K are determined below according to papers [5,6,17,26]. It should be noted that for these components, $L \ne 1$, $|K| \le L$ for $L = $ even and $|K| \le L - 1$ for $L = $ odd:

$$\hat{\phi}_K^{\lambda\mu L}(\gamma) = \sum_{n=0}^{n_{\max}} F_{n\lambda L}^{\sigma\tau\mu}(\gamma) \left[G_{|K|}^{nL}(\gamma) \delta_{L,\text{even}} + \bar{G}_{|K|}^{nL}(\gamma) \delta_{L,\text{odd}} \right]; \qquad (17)$$

$$K = K_{\min}, K_{\min}+2, \ldots, K_{\max}; \qquad K_{\min} = \begin{cases} 0, & L = \text{even}, \\ 2, & L = \text{odd}; \end{cases} \qquad K_{\max} = \begin{cases} L, & L = \text{even}, \\ L-1, & L = \text{odd}; \end{cases}$$

$$n_{\max} = \begin{cases} L/2, & L = \text{even}, \\ (L-3)/2, & L = \text{odd}; \end{cases} \qquad \delta_{L,\text{even}} = \begin{cases} 1, & L = \text{even}, \\ 0, & L = \text{odd}; \end{cases} \qquad \delta_{L,\text{odd}} = \begin{cases} 0, & L = \text{even}, \\ 1, & L = \text{odd}; \end{cases}$$

where $L/2 \le \lambda - 3\mu \le L$ for $L = $ even, and $(L+3)/2 \le \lambda - 3\mu \le L$ for $L = $ odd;

2.3   Wave Function for γ Degree of Freedom $\hat{\phi}_K^{\lambda\mu L}(\gamma)$

Components $\bar{G}_K^{nL}(\gamma)$, $G_K^{nL}(\gamma)$ and $F_{n\lambda L}^{\sigma\tau\mu}(\gamma)$ in Eq. (17) are calculated by

$$\bar{G}_K^{nL}(\gamma) = \sum_{k=3-L,2}^{L-3} \langle L-3, 3, k, K-k | L K \rangle G_{|k|}^{nL-3}(\gamma) \sin 3\gamma (\delta_{K-k,2} - \delta_{K-k,-2});$$

$$G_K^{nL}(\gamma) = (-\sqrt{2})^n \sum_{k=2n-L,2}^{L-2n} \langle L-2n, 2n, k, K-k | L K \rangle S_{|k|}^{(L-2n)/2}(\gamma) S_{|K-k|}^n(-2\gamma);$$

$$S_K^r(\gamma) = \left[\frac{(2r+K)!(2r-K)!}{(4r)!} \right]^{1/2} (\sqrt{6})^r r! \sum_{q=K/2}^{[r/2+K/4]} \left(\frac{1}{2\sqrt{3}} \right)^{2q-K/2}$$
$$\times \frac{1}{(r-2q+K/2)!(q-K/2)!q!} (\cos\gamma)^{r+K/2-2q} (\sin\gamma)^{2q-K/2};$$

$$F_{n\lambda L}^{\sigma\tau\mu}(\gamma) = (-1)^{\mu+\tau-n} 2^{-n/2} \sum_{r=0}^{[(\mu+\tau-n)/2]} C_{rn\lambda L}^{\sigma\tau\mu} 2^{-r} (\cos 3\gamma)^{\mu+\tau-n-2r};$$

$$C_{rn\lambda L}^{\sigma\tau\mu} = \frac{3^n \sigma! \lambda! (-1)^r 2^r (2\mu + 2\tau - 2r + \delta_{L,\text{odd}})!(3r)!}{2^{\mu+n} n! (2\lambda+1)! r! (\mu+\tau-r)!(\mu+\tau-n-2r)!}$$
$$\times \sum_{s=\max(n-\tau,0)}^{\min(\sigma,\lambda,3r-\tau+n)} \frac{(-1)^s 4^s (\tau+s)!(2\lambda+1-2s)!}{s!(\sigma-s)!(\tau-n+s)!(3r-\tau+n-s)!(\lambda-s)!},$$

where $S_K^r(\gamma)$ is taken to be equal 0, if $\sin\gamma = 0$ or $\cos\gamma = 0$, $F_{n\lambda L}^{\sigma\tau\mu}(\gamma)$ is taken to be equal 0, if $\cos 3\gamma = 0$, $C_{rn\lambda L}^{\sigma\tau\mu}$ is taken to be equal 0, if $\mu + \tau - n - 2r < 0$. It has been implemented in Ref. [7].

2.4   Gram–Schmidt Orthogonalization of the Functions $\hat{\phi}_K^{\lambda\mu L}(\gamma)$

Using implementation [7] of orthogonalization of the functions $\hat{\phi}_K^{\lambda\mu L}(\gamma)$ with the Gram–Schmidt method the reduced overlap (a scalar product with integration over γ) is required

$$\langle \hat{\phi}^{\lambda\mu' L}(\gamma)|\hat{\phi}^{\lambda\mu L}(\gamma)\rangle = \int_0^\pi d\gamma \sin(3\gamma) \sum_{K=K_{\min},2}^{K_{\max}} \frac{2\hat{\phi}_K^{\lambda\mu' L}(\gamma)\hat{\phi}_K^{\lambda\mu L}(\gamma)}{1+\delta_{K,0}}. \tag{18}$$

It should be noted that the definition of the reduced overlap integral (18) will be the same for original $\hat{\phi}^{\lambda\mu L}(\gamma)$ as well as for orthogonalized functions $\phi_K^{\lambda\mu L}(\gamma)$.

The degeneracy labelled by μ for the nuclear calculations is small in relevant cases as presented in Table 1, therefore, the original Gram–Schmidt method may be adopted to orthogonalize the functions $\hat{\phi}_K^{\lambda\mu L}(\gamma)$. For large μ, the modified Gram–Schmidt methods will be applied [7].

Application of the Gram–Schmidt method gives the orthogonalized functions

$$\phi_K^{\lambda\mu L}(\gamma) = \hat{\phi}_K^{\lambda\mu L}(\gamma) - \sum_{\mu'=\mu_{\min}}^{\mu-1} \phi_K^{\lambda\mu' L}(\gamma) \frac{\langle \phi^{\lambda\mu' L}(\gamma)|\hat{\phi}^{\lambda\mu L}(\gamma)\rangle}{\langle \phi^{\lambda\mu' L}(\gamma)|\phi^{\lambda\mu' L}(\gamma)\rangle}. \tag{19}$$

This procedure should be applied for all available indexes μ in boundaries given in Eq. (10) and indexes K in boundaries given in Eq. (17).

As produced by the procedure outlined in Eq. (19), the wave functions $\phi_K^{\lambda\mu L}(\gamma)$ are trigonometric polynomials of $\sin(\gamma)$ and $\cos(\gamma)$. For the algebraic integration over the variable γ, it is then sufficient to expand the $\sin(3\gamma)$ and the additional $\cos(3\gamma)$ and to implement the following three definite integrals:

$$\int_0^\pi \sin^{2m}(\gamma)d\gamma = \frac{(2m-1)!!}{2^m m!}\pi, \quad \int_0^\pi \sin^{2m+1}(\gamma)d\gamma = \frac{2^{m+1}m!}{(2m+1)!!}, \quad \int_0^\pi \sin^m(\gamma)\cos(\gamma)d\gamma = 0,$$

for any integer m. For example, the normalization integral for $L = 0$, $\lambda = 27$ and $\mu = 9$ is equal to $\frac{2}{57}$ and shows less than 0.001 sec. computation time on Mathematica. At the same time, direct symbolic integration of this normalization integral takes 436.781 s.

2.5   The Normalized Components $F_{n_\beta}^\lambda(\beta)$

The normalized components $F_{n_\beta}^\lambda(\beta)$ with the number of nodes $n_\beta = (\nu - \lambda)/2$, adapted for calculations of rescaled matrix elements $V(\beta,\gamma)$ from (6), read as

$$F_{n_\beta}^\lambda(\beta) = \sqrt{\frac{2n_\beta!}{\Gamma\left(n_\beta + \lambda + \frac{5}{2}\right)}}\beta^\lambda \exp\left(-\frac{1}{2}\beta^2\right)L_{n_\beta}^{\lambda+\frac{3}{2}}(\beta^2), \tag{20}$$

where $L_{n_\beta}^{\lambda+\frac{3}{2}}(\beta^2)$ is the associated Laguerre polynomial [1].

Table 3. The example of calculations of the matrix elements (27) for a number of L and fixed $\nu_{\max} = 30$. The columns of the table are formed by the value of the angular momentum L, the total number of states $\{\nu\lambda\mu\}$ defined by quantum numbers $\nu\lambda\mu$, the total number of states $\{\lambda\mu\}$ defined by quantum numbers $\lambda\mu$, the total number #MeT of matrix elements (27) in upper triangles of their matrices with $m = 1, 2$, the number #MeN of nonzero matrix elements among #MeT that are given by Eq. (28), the cumulative number #MeZ of angular matrix elements that are calculated equal to 0 among #MeN matrix elements, the maximum memory in MB used to store intermediate data for the current Mathematica session in computation of the overlap integrals, and the CPU time.

L	$\{\nu\lambda\mu\}$	$\{\lambda\mu\}$	#MeT	#MeN	#MeZ	memory	CPU time
0	91	11	132	30	0	0 MB	0.17 s
6	271	37	1406	266	15	3.48 MB	4.09 s
10	326	49	2450	495	80	4.18 MB	28.33 s
15	259	47	2256	534	109	4.45 MB	45.47 s
20	305	62	3906	1010	322	6.26 MB	2.60 min
25	193	50	2550	788	227	6.06 MB	3.08 min
30	174	51	2652	853	138	7.75 MB	5.07 min

In Table 2, we present an example of calculations of the number of functions (20) for a number of L up to the specified value of the $\nu_{\max}$ under condition (9). The presented results show the general tendency: with larger ν, the number of states increases and the calculations involve larger L. If we require larger L the number ν has to be sufficiently large.

2.6   Hamiltonian Matrix Elements and Algebraic Eigenvalue Problem

For the calculation of the matrix elements of the kinetic energy T the gradient formula [11] is applied taking into account the rescaled parameter $\bar{B}_2 = 2B_2/\sqrt{5}$:

$$T^L_{\nu'\lambda'\mu',\nu\lambda\mu} = (-1)^{\frac{|\nu'-\nu|}{2}} \frac{1}{2}\hbar\sqrt{B'_2 C'_2} \frac{1}{\bar{B}_2} \langle\nu'\lambda'|\beta^2|\nu\lambda\rangle\delta_{\lambda',\lambda}\delta_{\mu',\mu} \tag{21}$$

$$-\sqrt{\frac{2}{35}}\hbar^{\frac{3}{2}}(B'_2 C'_2)^{\frac{1}{4}}\frac{P_3}{3}\langle\nu'\lambda'|\beta^3|\nu\lambda\rangle\langle\lambda'\mu'L|\cos(3\gamma)|\lambda\mu L\rangle\left(\delta_{|\nu'-\nu|,1} - 3\delta_{|\nu'-\nu|,3}\right).$$

The potential energy matrix elements V read as

$$V^L_{\nu'\lambda'\mu',\nu\lambda\mu} = \sum_{\rho=2}^{6}\sum_{m=0}^{2} V_{\rho,m}\langle\nu'\lambda'|\beta^\rho|\nu\lambda\rangle\langle\lambda'\mu'L|\cos^m(3\gamma)|\lambda\mu L\rangle. \tag{22}$$

Matrix elements of the quantum Hamiltonian (1) read as

$$H^L_{\nu'\lambda'\mu',\nu\lambda\mu} = T^L_{\nu'\lambda'\mu',\nu\lambda\mu} + V^L_{\nu'\lambda'\mu',\nu\lambda\mu}. \tag{23}$$

The eigenvalues E_n^L and the eigenfunctions Ψ_n^L of the quantum Hamiltonian $H = T + V$ (1) are calculated by solving the Schrödinger equation

$$(H - E_n^L)\Psi_n^L = 0. \tag{24}$$

We seek eigenfunctions Ψ_n^L of Hamiltonian (1) in the form of expansion over the basis functions $\Psi_{\nu\lambda\mu}^{int}(\beta, \gamma, \Omega)$ (13)

$$\Psi_n^L(\beta, \gamma, \Omega) = \sum_{\nu\lambda\mu} \Psi_{\nu\lambda\mu}^{int}(\beta, \gamma, \Omega) D_{\nu\lambda\mu,n}(L). \tag{25}$$

Eigenenergies E_n^L are calculated as an algebraic eigenvalue problem

$$\sum_{\nu\lambda\mu}(H^L_{\nu'\lambda'\mu',\nu\lambda\mu} - \delta_{\nu'\nu}\delta_{\lambda'\lambda}\delta_{\mu'\mu}E_n^L)D_{\nu\lambda\mu,n}(L) = 0. \tag{26}$$

Here $D_{\nu\lambda\mu,n}(L)$ is the eigenvector of Hamiltonian (23) for the n'th state with the angular momentum L. In Eq. (26), indices ν, λ and μ enumerate the total basis. The total number of different collections of (ν, λ and μ) for given L, and up to given $\nu_{\max}$ is the total dimension of the basis. These values are presented in Tables 3, 4, and 7. In Table 7, Dim is this total number of different (ν, λ and μ) for listed L and up to given $\nu_{\max} = 30$, i.e., the dimension of the Hamiltonian matrix.

2.7   Matrix Elements $\langle \lambda'\mu'L| \cos^m(3\gamma)|\lambda\mu L\rangle$

For computation of potential energy matrix elements the matrix elements of powers $m = 0, 1, 2$ of $\cos(3\gamma)$ should be evaluated, that are defined as

$$\langle \lambda'\mu'L| \cos^m(3\gamma)|\lambda\mu L\rangle = \frac{1}{\sqrt{\langle\phi^{\lambda'\mu'L}(\gamma)|\phi^{\lambda'\mu'L}(\gamma)\rangle\langle\phi^{\lambda\mu L}(\gamma)|\phi^{\lambda\mu L}(\gamma)\rangle}}$$
$$\times \int_0^\pi d\gamma \sin(3\gamma) \cos^m(3\gamma) \sum_{K=K_{\min},2}^{K_{\max}} \frac{2\phi_K^{\lambda'\mu'L}(\gamma)\phi_K^{\lambda\mu L}(\gamma)}{1 + \delta_{K,0}}. \tag{27}$$

Here summation boundaries are the same as in Eq. (18). Obviously this integral is equal to $\delta_{\lambda\mu,\lambda'\mu'}$ when $m = 0$. It should be pointed out that only small part of these integrals are not equal to 0. There are useful simple conditions that allow identify the large part of these integrals that are equal to zero. The appropriate selection rules are

$$\lambda + \lambda' + (m \bmod 2) = \text{ odd},$$
$$|\lambda - \lambda'| \le 3n \text{ and } 3n \le \lambda + \lambda', \text{ where } n = m, m - 2, \ldots, 1 \text{ or } 0. \tag{28}$$

Using conditions (28) saves a lot of computation resources and makes it possible to avoid calculation of most of integrals (27) that actually are equal to 0. Nevertheless, these conditions are not precise and some of matrix elements that pass their test may appear to be equal to 0 after their computation. An example

Table 4. The example of calculations of the matrix elements (27) for a number of $\nu_{\max}$ and fixed $L = 18$. The first column of the table is formed by value of the $\nu_{\max}$, other columns are denoted as in Table 3.

$\nu_{\max}$	$\{\nu\lambda\mu\}$	$\{\lambda\mu\}$	#MeT	#MeN	#MeZ	memory	CPU time
10	2	2	6	3	0	0 MB	0.17 s
15	23	12	156	73	2	3.51 MB	5.80 s
20	81	28	812	312	52	3.93 MB	29.14 s
25	181	45	2070	632	170	4.92 MB	1.12 min
30	323	62	3906	953	294	5.97 MB	1.94 min
35	506	78	6162	1253	405	7.08 MB	2.86 min
40	731	95	9120	1577	530	8.50 MB	4.08 min
45	998	112	12656	1898	654	10.02 MB	5.68 min
50	1306	128	16512	2198	765	11.62 MB	7.65 min

of calculations of the matrix elements (27) is presented in Tables 3 and 4. Each evaluation is performed after quitting the Mathematica kernel. $\{\nu\lambda\mu\}$ is the total number of states defined for given $\nu_{\max}$ by quantum numbers in Eq. (8) under conditions in Eqs. (9) and (10); $\{\lambda\mu\}$ is the total number of states defined only by indices λ and μ, and this number is equal to the total number of different pairs of λ and μ among the states $\{\nu\lambda\mu\}$; #MeT – the cumulative number of angular matrix elements in the upper triangles of matrices for $\cos^m(3\gamma)$ with $m = 1, 2$ on states $\{\lambda\mu\}$, here the number of matrix elements with $m = 0$ are not included, since they all are equal to 1 by definition; #MeN is the number of nonzero matrix elements among #MeT that are given by Eq. (28); #MeZ is the cumulative number of angular matrix elements that are evaluated by equal to 0 by direct computation.

2.8   Matrix Elements $\langle \nu'\lambda'|\beta^\rho|\nu\lambda\rangle$

For the first case $|\lambda - \lambda'| \le \rho$, matrix elements $\langle \nu'\lambda'|\beta^\rho|\nu\lambda\rangle$ read as:

$$\int_0^\infty F_{n'_\beta}^{\lambda'}(\beta)\beta^\rho F_{n_\beta}^{\lambda}(\beta)\beta^4 d\beta = \left[\frac{n'_\beta! n_\beta!}{\Gamma(n'_\beta + \lambda' + \frac{5}{2})\Gamma(n_\beta + \lambda + \frac{5}{2})}\right]^{\frac{1}{2}} \tag{29}$$

$$\times (-1)^{n'_\beta + n_\beta} \Gamma\left(\frac{1}{2}(\rho + \lambda' - \lambda + 2)\right)\Gamma\left(\frac{1}{2}(\rho + \lambda - \lambda' + 2)\right)$$

$$\times \sum_\sigma \frac{\Gamma\left(\frac{1}{2}(\rho + \lambda' + \lambda + 5) + \sigma\right)}{\sigma!(n'_\beta - \sigma)!(n_\beta - \sigma)!\Gamma\left(\sigma + \frac{1}{2}(\rho + \lambda' - \lambda) - n_\beta + 1\right)} \frac{1}{\Gamma\left(\sigma + \frac{1}{2}(\rho + \lambda - \lambda') - n'_\beta + 1\right)}$$

the summation bounds for $\rho + \lambda' - \lambda$ even are:

$$\max\left(n'_\beta - (\rho + \lambda - \lambda')/2, n_\beta - (\rho + \lambda' - \lambda)/2, 0\right) \le \sigma \le \min(n'_\beta, n_\beta),$$

the summation bounds for $\rho + \lambda' - \lambda$ odd are:

$$0 \le \sigma \le \min(n'_\beta, n_\beta).$$

Table 5. An example of calculations of the matrix elements over β given by Eqs. (29) and (30) for a number of $\nu_{\max}$ when $\rho = 1, \ldots, 6$. The columns of the table are formed by the value of $\nu_{\max}$, the total number of states $\{\nu\lambda\}$ is defined by the quantum numbers $\nu\lambda$ up to the specified value of $\nu_{\max}$, the total number $\#\mathrm{Me}(\beta)$ of different matrix elements over β, the maximum memory in MB used to store intermediate data for the current Mathematica session in computation of the matrix elements, and the CPU time.

$\nu_{\max}$	$\{\nu\lambda\}$	$\#\mathrm{Me}(\beta)$	memory	CPU time
10	36	495	0 MB	0.30 s
20	121	2690	3.27 MB	1.78 s
40	441	12630	5.70 MB	9.06 s
60	961	29970	5.54 MB	24.26 s
80	1681	54710	5.69 MB	49.47 s
100	2601	86850	5.74 MB	1.44 min

For the second case $|\lambda - \lambda'| > \rho$ and the pair of quantities n'_β, λ' and n_β, λ are interchanged when $\lambda > \lambda'$:

$$\int_0^\infty F_{n'_\beta}^{\lambda'}(\beta)\beta^\rho F_{n_\beta}^{\lambda}(\beta)\beta^4 d\beta = \left[\frac{n'_\beta! n_\beta!}{\Gamma(n'_\beta + \lambda' + \frac{5}{2})\Gamma(n_\beta + \lambda + \frac{5}{2})} \right]^{\frac{1}{2}} \tag{30}$$

$$\times (-1)^{n_\beta} \frac{\Gamma\left(\frac{1}{2}(\rho + \lambda' - \lambda + 2)\right)}{\Gamma\left(\frac{1}{2}(-\rho + \lambda' - \lambda)\right)} \sum_\sigma (-1)^\sigma \frac{\Gamma\left(\frac{1}{2}(\rho + \lambda' + \lambda + 5) + \sigma\right)\Gamma\left(\frac{1}{2}(\lambda' - \lambda - \rho) + n'_\beta - \sigma\right)}{\sigma!(n'_\beta - \sigma)!(n_\beta - \sigma)!\Gamma\left(\sigma + \frac{1}{2}(\rho + \lambda' - \lambda) - n_\beta + 1\right)}$$

the summation bounds for $\rho + \lambda' - \lambda$ even are: $\max\left(n_\beta - (\rho + \lambda' - \lambda)/2, 0\right) \leq \sigma \leq \min(n'_\beta, n_\beta)$, the summation bounds for $\rho + \lambda' - \lambda$ odd are: $0 \leq \sigma \leq \min(n'_\beta, n_\beta)$.

There are selection rules for the matrix elements over the variable β. The matrix elements are equal to zero when

$$|\nu' - \nu| > \rho, \quad \rho \text{ and } |\nu' - \nu| \text{ have unequal parities,}$$
$$|\lambda' - \lambda| > \rho, \quad |\nu' - \nu| \text{ and } |\lambda' - \lambda| \text{ have unequal parities,} \tag{31}$$
$$\rho = 4 \text{ and } |\lambda' - \lambda| \neq 0, \quad \rho = 5 \text{ and } |\lambda' - \lambda| = 5.$$

The formulas of the matrix elements over β Eqs. (29) and (30) are very effective comparing with direct symbolic integration approach. For example, symbolic integration of the matrix element with $n'_\beta = 126, \lambda' = 121, n_\beta = 125, \lambda = 120$, and $\rho = 120$ takes 23.80 s, when Mathematica timing for a computation with Eqs. (29) and (30) returns zero.

In Table 5, we present an example of memory consumption and CPU time of calculations of the matrix elements over β for a number of ν and fixed range of ρ. This interval $\rho = 1, \ldots, 6$ represents all powers of ρ in the expression of the potential energy for the approach adopted in this paper. It should be stressed that the presented procedure is very effective and could be applied for large scale calculations since the quantum numbers managed significantly outperform the ones considered for very large values, e.g., $\lambda \sim 100$ and $\mu \sim 10$.

In Table 6, we present the illustration how the accuracy of calculations depends on the number of significant digits used in computations. The presented results

Table 6. An example of calculations of relative accuracy of the matrix elements over β given by Eq. (30) for a number of ρ when $\nu = 100, \lambda = 70, \nu' = 60, \lambda' = 5$. The first row specifies the number of significant digits used in the corresponding computation. The n.a. indicates that the calculations could not be performed with specified number of significant digits.

ρ, precision	25	26	27	28	32	36	40
1	$2.4 \cdot 10^{-1}$	$1.8 \cdot 10^{-2}$	$3.0 \cdot 10^{-3}$	$3.4 \cdot 10^{-5}$	$3.7 \cdot 10^{-8}$	$1.5 \cdot 10^{-12}$	$2.1 \cdot 10^{-16}$
3	n.a.	$4.8 \cdot 10^{-2}$	$2.4 \cdot 10^{-5}$	$2.4 \cdot 10^{-5}$	$1.3 \cdot 10^{-7}$	$1.9 \cdot 10^{-11}$	$1.7 \cdot 10^{-15}$
5	n.a.	$2.1 \cdot 10^{-3}$	$2.1 \cdot 10^{-3}$	$1.2 \cdot 10^{-3}$	$1.1 \cdot 10^{-7}$	$1.3 \cdot 11^{-10}$	$1.2 \cdot 10^{-16}$
7	n.a.	n.a.	$1.3 \cdot 10^{-1}$	$1.2 \cdot 10^{-3}$	$1.9 \cdot 10^{-7}$	$1.4 \cdot 10^{-10}$	$4.8 \cdot 10^{-15}$
8	n.a.	n.a.	n.a.	$4.5 \cdot 10^{-2}$	$2.4 \cdot 10^{-6}$	$1.4 \cdot 10^{-10}$	$2.2 \cdot 10^{-14}$

gives the background for assertion that large scale calculations of this kind may be performed only symbolically.

3   Benchmark Calculations of GCM for ^{186}Os Nucleus

3.1   The Example of Calculations of Eigenenergies $E_n^{L^\pi}$ (in MeV)

The eigenstates L_n^π are characterized by the angular momentum L, parity $\pi = \pm = (\pm 1)$ [4] and sequence number n for fixed angular momentum starting at the lowest state. The calculated eigenvalues $E_n^{L^\pi}$ of rotational bands of ^{186}Os nucleus are the same as may be produced by the FORTRAN program [20]. In these calculations, the following values of parameters were used: $C_2 = -564.76, C_3 = 733.01, C_4 = 13546., C_5 = -8535.1, C_6 = -41635., D_6 = 0.$, and $C_2' = C2S = 100.$ (in MeV), $B_2 = 112.48$ and $B_2' = B2S = 90.$ (in 10^{-42}MeV s^2), $P_3 = -0.0531$ (in 10^{+42}MeV/s^2), $\hbar = 6.58211828$ (in 10^{-22}MeV s), $\nu_{\max} = NPH = 30$ in expansion of (25). In Table 7, we show a comparison of calculated eigenenergies from algebraic eigenvalue problem (26) and experimental eigenenergies from [15,20]. They are in a good agreement that confirm consistent choice of the parameters of GCM model and our version of the GCM code.

3.2   The Quadrupole Moment Q and Transitions $B(E2)$

The quadrupole operator $Q_m^{(2)}$ is defined as

$$Q_m^{(2)} = \rho_0 R_0^5 \left(\alpha_m^{[2]} - \frac{10}{\sqrt{70\pi}} [\alpha^{[2]} \times \alpha^{[2]}]_m^{[2]} \right), \tag{32}$$

where $\rho_0 = 3Ze/(4\pi R_0^3)$, $R_0 = r_0 A^{1/3}$, $r_0 = 1.1$fm.

The quadrupole moment of nth level with specified L reads as

$$Q_n(L) = \rho_0 R_0^5 \sqrt{\frac{16\pi}{5}} \begin{pmatrix} L & 2 & L \\ -L & 0 & L \end{pmatrix} 10^{-2}$$

$$\times \left(\alpha_{n,n}^{[2]}(L, L) - \frac{10}{\sqrt{70\pi}} [\alpha^{[2]} \times \alpha^{[2]}]_{n,n}^{[2]}(L, L) \right), \tag{33}$$

Table 7. First column shows the labels L_n^π of eigenstates of a given rotational band, where L is the angular momentum, and $\pi = \pm$ is the parity. Dim is a number of components of the eigenvector $D_{\nu\lambda\mu,n}$ in Eq. (26), i.e., Dim is the total number of different (ν, λ and μ) for listed L and up to given $\nu_{\max}$, as well as the dimension of Hamiltonian matrix. Energy calc. are the eigenenergies of algebraic eigenvalue problem, Δ Energy calc.=Energy calc.(L_n^π) - Energy calc.(0_1^+) are the eigenenergies counted of eigenenergy of ground state 0_1^+, Δ Energy exp. are the experimental eigenenergies of rotational bands of ^{186}Os nucleus, all eigenenergies are in MeV.

Level	Code	Dim	Energy calc.	CPU time	Δ Energy calc.	Δ Energy exp.
0_{gs}^+	0_1^+	91	-5.683	5.33 s	0.000	0.000
2_{gs}^+	2_1^+	165	-5.546	23.78 s	0.138	0.137
4_{gs}^+	4_1^+	225	-5.260	58.70 s	0.424	0.433
6_{gs}^+	6_1^+	271	-4.854	1.58 min	0.829	0.867
2_γ^+	2_2^+		-4.937		0.746	0.767
3_γ^+	3_1^+	75	-4.750	4.19 s	0.934	0.910
4_γ^+	4_2^+		-4.596		1.087	1.070
5_γ^+	5_1^+	135	-4.343	17.22 s	1.340	1.275
4_γ^+	4_3^+		-4.174		1.509	1.352
6_γ^+	6_2^+		-4.164		1.520	1.492

where $\begin{pmatrix} L & 2 & L \\ -L & 0 & L \end{pmatrix}$ is 3-j symbol [24]. The reduced upwards transition probability $B(E2)$ is calculated by the expression

$$B_{n_2,n_1}(E2, L_2 \to L_1) = \frac{10^{-4}}{2L_2 + 1}$$

$$\times \left[\rho_0 R_0^5 \left(\alpha_{n_2,n_1}^{[2]}(L_2, L_1) - \frac{10}{\sqrt{70\pi}} [\alpha^{[2]} \times \alpha^{[2]}]_{n_2,n_1}^{[2]}(L_2, L_1) \right) \right]^2. \quad (34)$$

3.3   Matrix Elements $\alpha_{n_2,n_1}^{[2]}(L_2, L_1)$ and $[\alpha^{[2]} \times \alpha^{[2]}]_{n_2,n_1}^{[2]}(L_2, L_1)$

Matrix elements $\alpha_{n_2,n_1}^{[2]}(L_2, L_1)$ and $[\alpha^{[2]} \times \alpha^{[2]}]_{n_2,n_1}^{[2]}(L_2, L_1)$ are given by the following expressions

$$\alpha_{n_2,n_1}^{[2]}(L_2, L_1) = \sqrt{(2L_1 + 1)(2L_2 + 1)} \sqrt{\frac{\hbar}{\sqrt{B_2'C_2'}}}$$

$$\times \sum_{\nu_1\lambda_1\mu_1} \sum_{\nu_2\lambda_2\mu_2} \langle \nu_2\lambda_2 L_2|\beta|\nu_1\lambda_1 L_1\rangle\langle\lambda_2\mu_2 L_2|\alpha^{[2]}|\lambda_1\mu_1 L_1\rangle \quad (35)$$

$$\times D_{\nu_1\lambda_1\mu_1,n_1}(L_1)D_{\nu_2\lambda_2\mu_2,n_2}(L_2),$$

$$[\alpha^{[2]} \times \alpha^{[2]}]_{n_2,n_1}^{[2]}(L_2, L_1) = \sqrt{\frac{1}{7}}\sqrt{(2L_1 + 1)(2L_2 + 1)} \frac{\hbar}{\sqrt{B_2'C_2'}}$$

$$\times \sum_{\nu_1\lambda_1\mu_1} \sum_{\nu_2\lambda_2\mu_2} \langle \nu_2\lambda_2 L_2|\beta^2|\nu_1\lambda_1 L_1\rangle D_{\nu_1\lambda_1\mu_1,n_1}(L_1) \quad (36)$$

$$\times \langle\lambda_2\mu_2 L_2|[\alpha^{[2]} \times \alpha^{[2]}]^{[2]}|\lambda_1\mu_1 L_1\rangle D_{\nu_2\lambda_2\mu_2,n_2}(L_2).$$

Here $D_{\nu_i\lambda_i\mu_i, n_i}(L_i)$ is the eigenvector of the Hamiltonian (23) for the n_ith state with angular momentum L_i from the algebraic eigenvalue problem (26).

3.4   Matrix Elements $\langle\lambda_1\mu_1 L_1|\alpha^{[2]}|\lambda_2\mu_2 L_2\rangle$

Matrix elements of $\alpha^{[2]}$ are calculated by means of the reduced Wigner coefficients in the chain $O(5)\supset O(3)$ [7]

$$\langle\lambda_1\mu_1 L_1|\alpha^{[2]}|\lambda_2\mu_2 L_2\rangle = (-1)^{L_2-L_1}\frac{1}{\sqrt{2L_1+1}}\frac{1}{N}$$

$$\times \sum_{K=-K_s(2)}^{K_s}\sum_{K_1=-K_{1s}(2)}^{K_{1s}}\sum_{K_2=-K_{2s}(2)}^{K_{2s}}\langle 2, K, L_2, K_2|L_1, -K_1\rangle \qquad (37)$$

$$\times \int_0^\pi \phi_{K_1}^{\lambda_1\mu_1 L_1}(\gamma)\phi_K^{\lambda=1\mu=0\,L=2}(\gamma)\phi_{K_2}^{\lambda_2\mu_2 L_2}(\gamma)\sin(3\gamma)d\gamma,$$

where $\langle 2, K, L_2, K_2|L_1, -K_1\rangle$ is Clebsch–Gordan coefficient [24], $\phi_K^{\lambda=1\mu=0L=2}(\gamma)$ are the orthogonalized functions calculated from Eq. (19) at $\lambda = 1, \mu = 0, L = 2$. For all K, the summation bounds and normalization factors N are defined as follows:

$$K_s = \begin{cases} L, & L = \text{even}, \\ L-1, & L = \text{odd}; \end{cases} \quad N = \begin{cases} \langle\lambda_1\mu_1 L_1|\lambda_1\mu_1 L_1\rangle, & (\lambda_1\mu_1 L_1) = (\lambda_2\mu_2 L_2), \\ \sqrt{\langle\lambda_1\mu_1 L_1|\lambda_1\mu_1 L_1\rangle\langle\lambda_2\mu_2 L_2|\lambda_2\mu_2 L_2\rangle}, & \text{otherwise.} \end{cases}$$

The angular brackets $\langle\lambda\mu L|\lambda\mu L\rangle$ here represent the overlap integrals Eq. (18) $\langle\phi^{\lambda\mu L}(\gamma)|\phi^{\lambda\mu L}(\gamma)\rangle$ of the corresponding functions $\phi^{\lambda\mu L}(\gamma)$.

3.5   Matrix Elements $\langle\lambda_1\mu_1 L_1|[\alpha^{[2]}\times\alpha^{[2]}]^{[2]}|\lambda_2\mu_2 L_2\rangle$

Matrix elements of $[\alpha^{[2]}\times\alpha^{[2]}]^{[2]}$ are calculated also by means of the reduced Wigner coefficients

$$\langle\lambda_1\mu_1 L_1|[\alpha^{[2]}\times\alpha^{[2]}]^{[2]}|\lambda_2\mu_2 L_2\rangle = \sqrt{\frac{2}{9(2L_2+1)}}\frac{1}{N}$$

$$\times \sum_{K=-K_s(2)}^{K_s}\sum_{K_1=-K_{1s}(2)}^{K_{1s}}\sum_{K_2=-K_{2s}(2)}^{K_{2s}}\langle L_1, K_1, 2, K|L_2, -K_2\rangle \qquad (38)$$

$$\times \int_0^\pi \phi_{K_1}^{\lambda_1\mu_1 L_1}(\gamma)\phi_K^{\lambda=2\mu=0\,L=2}(\gamma)\phi_{K_2}^{\lambda_2\mu_2 L_2}(\gamma)\sin(3\gamma)d\gamma,$$

where $\phi_K^{\lambda=2\mu=0L=2}(\gamma)$ are the orthogonalized functions calculated from Eq. (19) at $\lambda = 2, \mu = 0, L = 2$.

The selection rules for the matrix elements $\alpha^{[2]}$ and $[\alpha^{[2]}\times\alpha^{[2]}]^{[2]}$ are:

$$\lambda+\lambda_1+\lambda_2 \text{ even}, \ \lambda>|\lambda_1-\lambda_2|, \ \lambda<\lambda_1+\lambda_2, \ L>|L_1-L_2|, \ L<L_1+L_2. \quad (39)$$

The columns of Table 8 are formed by the values of the angular momentum L_2, #MeT is the total number of matrix elements for given L_2 and $L_1 = L_2 - 2, L_2 - 1, L_2$, except for the first row where the $L_1 L_2 = 02, 22, 23$, #MeZ is the number of zero matrix elements that are calculated equal to 0 among the #MeT matrix elements, and the CPU time.

Table 8. The CPU time of calculation of the matrix elements $\langle\lambda_1\mu_1 L_1|\alpha^{[2]}|\lambda_2\mu_2 L_2\rangle$ and $\langle\lambda_1\mu_1 L_1|[\alpha^{[2]}\times\alpha^{[2]}]^{[2]}|\lambda_2\mu_2 L_2\rangle$ for a number of L_2 and fixed $\nu_{\max}=30$.

	$\langle\lambda_1\mu_1 L_1\|\alpha^{[2]}\|\lambda_2\mu_2 L_2\rangle$				$\langle\lambda_1\mu_1 L_1\|[\alpha^{[2]}\times\alpha^{[2]}]^{[2]}\|\lambda_2\mu_2 L_2\rangle$		
L_2	#MeT	#MeZ	CPU time	L_2	#MeT	#MeZ	CPU time
3	59	0	1.80 s	3	77	0	2.56 s
4	113	0	7.86 s	4	167	0	12.06 s
5	72	0	5.80 s	5	105	0	8.16 s
6	213	8	24.42 s	6	317	0	35.66 s
7	152	9	20.33 s	7	225	0	28.64 s
8	333	40	56.45 s	8	490	8	1.34 min
9	253	30	48.70 s	9	377	7	1.14 min
10	466	100	1.91 min	10	690	36	2.74 min

Table 9. Values of the quadrupole moments $Q_n(L)$ (in eb) of ^{186}Os for a number of L and fixed $\nu_{\max}=30$.

n, L	2	4	5	6
1	−1.51	−1.85	0.953	−1.95
2	1.46	−0.517	−0.912	−1.02
3	−0.929	2.13	0.421	0.915

3.6   An Example of Calculations of The $Q_n(L)$ (in eb) of ^{186}Os

The required states are characterized by their angular momentum L and sequence number n for fixed angular momentum starting at the lowest state. The calculated values of the quadrupole moment $Q_n(L)$ (in eb) of ^{186}Os from (33) shown in Table 9 are the same as may be produced by the FORTRAN program [20].

3.7   An Example of Calculations of the $B(E2)$ (in e^2b^2) of ^{186}Os

The states are characterized by their angular momentum L and sequence number n for fixed angular momentum starting at the lowest state. The transitions are indicated as $n_i \to n_j$. The calculated values $B(E2) = B_{n_2,n_1}(E2, L_2 \to L_1)$ (in e^2b^2) of ^{186}Os come from Eq. (34) for a number of $(L_1 L_2)$ transitions and fixed $\nu_{\max}=30$ shown in Table 10 are the same as may be produced by the FORTRAN program [20]. CPU time for calculation of all Q and $B(E2)$ for up to $L = 6$ and with the number of states $n = 3$ is 64 s. (with previously prepared data files for angular matrix elements and eigenvectors of Hamiltonian).

3.8   Finding the Optimal Basis Parameters [20]

As a basis in this code we use the eigenfunctions (13)–(20) of the five-dimensional harmonic oscillator (11), which are respectively parameterized in terms of the

Table 10. Values of the $B(E2) = B_{n_2,n_1}(E2, L_2 \to L_1)$ (in e^2b^2) of ^{186}Os for a number of $(L_1 L_2)$ transitions and fixed $\nu_{\max} = 30$.

transitions	(0,2)	(2,2)	(2,3)
$1 \to 1$	2.99	0.801	0.0207
$2 \to 1$	0.0228	0.0779	1.32
$3 \to 1$	0.00835	0.00182	0.151
$1 \to 2$	0.0389	0.0778	0.00000291
$2 \to 2$	0.249	0.746	0.00131
$3 \to 2$	0.000121	0.0000199	0.0000855
$1 \to 3$	0.00526	0.00182	0.000429
$2 \to 3$	0.248	0.0000199	0.0261
$3 \to 3$	0.0573	0.300	0.00386

Table 11. The values of the phenomenological potential parameters $C_2, C_3, C_4,$ C_5, C_6, D_6, B_2, P_3, Eqs. (4), (5) for $N = 184$ isotones are determined by fitting [9].

	^{298}114	^{300}116	^{302}118	^{304}120	^{306}122	^{308}124
C_2	7579.22	7661.89	7744.29	7826.40	7908.23	7989.76
C_3	$3.25 \cdot 10^{-4}$	$-1.62 \cdot 10^{-3}$	$-1.61 \cdot 10^{-3}$	$-4.83 \cdot 10^{-4}$	$2.20 \cdot 10^{-3}$	$3.51 \cdot 10^{-4}$
C_4	$-2.93 \cdot 10^{-1}$	$1.98 \cdot 10^{-1}$	$1.84 \cdot 10^{-1}$	$-1.16 \cdot 10^{-1}$	$-8.39 \cdot 10^{-1}$	$-3.13 \cdot 10^{-1}$
C_5	$-4.11 \cdot 10^{-3}$	$2.05 \cdot 10^{-2}$	$2.04 \cdot 10^{-2}$	$6.10 \cdot 10^{-3}$	$-2.80 \cdot 10^{-2}$	$-4.44 \cdot 10^{-3}$
C_6	$1.65 \cdot 10^{-4}$	$-7.81 \cdot 10^{-4}$	$-7.72 \cdot 10^{-4}$	$-2.22 \cdot 10^{-4}$	$1.09 \cdot 10^{-3}$	$1.70 \cdot 10^{-4}$
D_6	1.79	-2.08	-1.98	$4.46 \cdot 10^{-1}$	6.15	1.96
B_2	226.573	226.573	226.573	226.573	226.573	226.573
P_3	0	0	0	0	0	0

basis parameters C_2' and B_2'. For a finite set of basic vectors, the parameters have to be chosen to get satisfactory convergence of the calculated energies and B(E2)-values. To find the best set of basis parameters one has to diagonalize a given Hamiltonian (23) and minimize the sum of (lowest) energy eigenvalues E_n^L by varying the basis parameters(see e.g. [16]). Since this procedure is quite time-consuming, we use another scheme that takes much less time and turned out to be also effective: we minimize only the sum of the first NUM diagonal matrix elements of the Hamiltonian for spin $I = 0$ and take B_2' fixed at B_2. The integer variable NUM should be equal to the number of the lowest $L = 0$ basis wave functions which contribute most to the first excited states. (Default: $NUM = 10$). The minimum is found by increasing a do–loop variable S, defined as $S = (C_2'B_2'/\hbar^2)^{1/4}$, successively by 0.5. In the case of failure to find reasonable basis parameters, the program is stopped and should be reruned with changed boundaries for S. In particular, for ^{186}Os: $S = 12.005370$, where $\hbar = 0.6582183$ (in 10^{-22}MeV s) , $B_2' = 90$(in 10^{-42}MeV s^2), $C_2' = 100$ (in MeV).

Table 12. The values of the phenomenological potential parameters, $C_2, C_3, C_4,$ C_5, C_6, D_6, B_2, P_3, Eqs. (4), (5) for ^{152}Sm, ^{154}Sm, ^{186}Os, ^{188}Os, ^{190}Os, ^{194}Pt and ^{196}Pt are determined by fitting [9].

	^{152}Sm	^{154}Sm	^{186}Os	^{188}Os	^{190}Os	^{194}Pt	^{196}Pt
C_2	−422.74	−464.74	−564.76	−398.83	−363.64	−161.58	−169.88
C_3	493.92	311.25	733.01	−380.74	−372.59	368.36	748.09
C_4	7983.60	6454.70	13546.	18295.43	19391.83	2610.08	5704.92
C_5	370.53	88.25	−8535.1	−17660.53	−19246.82	−130535.32	−315802.62
C_6	8279.78	3842.18	−41635.	74725.61	80003.64	583687.93	975896.18
D_6	−28041.74	−17430.86	0	−54507.20	−70794.69	997672.28	1841446.68
B_2	62.714	63.823	112.48	165.514	173.035	203.613	223.380
P_3	0	0	−0.0531	0	0	0	0

Table 13. The values of the phenomenological potential parameters, $C_2, C_3, C_4,$ C_5, C_6, D_6, B_2, P_3, Eqs. (4), (5) for Nobelium isotopes ^{248}No, ^{250}No, ^{252}No, ^{254}No, ^{256}No, and ^{258}No are determined by fitting [9].

	^{248}No	^{250}No	^{252}No	^{254}No	^{256}No	^{258}No
C_2	−742.30	−740.31	−742.30	−820.82	−785.78	−755.83
C_3	308.99	308.17	183.87	307.77	220.57	172.21
C_4	19029.74	18978.78	17371.12	19240.37	18948.17	19152.27
C_5	6261.17	6244.40	8478.34	3008.90	4202.27	3693.95
C_6	6020.57	6004.45	15304.34	7515.07	4820.42	−20225.25
D_6	−39632.38	−39525.07	−11301.10	−58453.68	−42470.76	−20383.43
B_2	226.573	83.289	240.795	226.573	226.573	226.573
P_3	0	0	0	0	0	0

Table 14. The values of the phenomenological potential parameters for Seaborgium isotopes ^{258}Sg, ^{260}Sg, ^{262}Sg, ^{264}Sg, ^{266}Sg, ^{268}Sg, ^{270}Sg, ^{272}Sg are determined by fitting [9].

	^{258}Sg	^{260}Sg	^{262}Sg	^{264}Sg	^{266}Sg	^{268}Sg	^{270}Sg	^{272}Sg
C_2	−889.78	−862.28	−858.57	−707.00	−881.27	−953.97	−948.53	−816.63
C_3	302.98	273.96	156.35	−244.07	−191.92	−135.38	−236.13	−249.84
C_4	21572.37	21479.99	23326.53	15852.44	22306.53	24457.98	27362.39	29458.02
C_5	−181.54	−958.25	603.74	4107.14	1057.65	554.98	−106.83	1252.04
C_6	5756.03	4802.92	−15191.44	5714.98	2767.59	2200.10	7677.81	2858.93
D_6	−91050.30	−85406.47	−92394.99	−42886.06	−87378.21	−101284.80	−124100.10	−136456.34
B_2	226.573	226.573	226.573	226.573	226.573	226.573	226.573	226.573
P_3	0	0	0	0	0	0	0	0

Table 15. The values of the phenomenological potential parameters for Hassium isotopes ^{264}Hs, ^{266}Hs, ^{268}Hs , ^{270}Hs, ^{272}Hs, ^{274}Hs, ^{276}Hs are determined by fitting [9].

	^{264}Hs	^{266}Hs	^{268}Hs	^{270}Hs	^{272}Hs	^{274}Hs	^{276}Hs
C_2	−910.55	−960.16	−957.77	−974.18	−967.13	−892.06	−528.93
C_3	−237.56	−306.53	−305.77	−401.01	−293.31	−366.77	401.95
C_4	23771.37	25603.20	25539.27	25146.02	27982.38	28476.76	31704.06
C_5	870.21	530.39	529.12	465.71	−2668.59	1589.99	26.09
C_6	−8701.36	8602.71	8581.15	20760.67	39469.35	9918.38	27748.59
D_6	−90319.19	−114091.97	−113806.31	−105793.07	−140708.47	−134891.91	−156859.49
B_2	226.573	226.573	226.573	226.573	226.573	226.573	226.573
P_3	0	0	0	0	0	0	0

4   Conclusions

We have developed a symbolic method implemented as a code GCM in the Wolfram Mathematica to compute energy spectrum, quadrupole momentum, and electromagnetic transitions in Geometric Collective Model. The symbolic nature of the developed methods allows one to avoid the numerical round-off errors in the calculation of spectral characteristics (especially close to resonances) of quantum systems under consideration and to study their analytic properties for understanding the dominant symmetries. Efficiency of the elaborated procedures and the code is shown by benchmark calculations of ^{186}Os nucleus and demonstrate quick performance even on a laptop.

The GCM code can be applied to study the properties of super-heavy nuclei using an approach proposed in the papers [9,12]. Sets of the input parameters for some atomic nuclei and super–heavy nuclei are given in Appendix A.

To point out further investigations of the considered GCM model for atomic nuclei in the framework of the Computer Algebra System (CAS) of the boundary value problem (BVP) corresponding to quantum Hamiltonian Eq. (11) is presented in Appendix B. Solution of this problem by the finite element method (FEM) implemented in a suitable CAS code, for example, GCMFEM code [10] gives a possibility to compare GCM results with GCMFEM ones using the alternative FEM reduction of the BVP to algebraic problems and input parameters from Appendix A.

Acknowledgments. The work was partially supported by the RUDN University Strategic Academic Leadership Program, the Bogoliubov–Infeld program, and grant of Plenipotentiary of the Republic of Kazakhstan in JINR. AD is grateful to Prof. A. Góźdź for hospitality during visits in Institute of Physics, Maria Curie-Skłodowska University (UMCS). POH acknowledges financial support from DGAPA-UNAM (IN100421).

A   Appendix. Sets of Input Parameters for Atomic Nuclei

To denote approximately a range of applicability of the GCM code and to make it more friendly for users, we will accompany it by the sets of input files with the values of sets of parameters for atomic nuclei given in the papers [9,12,20].

For example, we present some of them in Tables 11, 12, 13, 14, 15, and 16. In Table 11 the macroscopic potential parameters are given. The value of C_2 is increased as we approach to double closed shell. Even the potential depends more on the quadratic term over β, it is not completely quadratic even if one approaches very close the double closed shell. Because of the great similarity, the authors only depict the PES of the $^{298}114$ and $^{304}120$ in Figs. 29 and 30 in Ref. [9]. The PES is perfectly spherical, thus, the spectrum will be that of a five-dimensional oscillator: The energy scales as $\hbar\sqrt{C_2/B_2}$. The first excited state is a 2^+ state at the energy $\hbar\sqrt{C_2/B_2}$ and at twice this energy, there are three degenerate states with spin and parity 0^+, 2^+ and 4^+. The first $3+$ state is three times the energy of the first 2^+ state. For completeness, in Fig. 31 in Ref. [9], the authors depict the spectrum of the $^{298}114$ nucleus as predicted by the GCM [9].

Table 16. The values of the phenomenological potential parameters for ^{184}W are determined by fitting [12].

C_2	C_3	C_4	C_5	C_6	D_6	B_2	P_3
-521.77	-337.80	14306.01	-502.64	1902.26	-60439.94	112.697	0

The only parameter, which cannot be deduced is the collective mass B_2 of the geometrical model [8]. This parameter has to be adjusted to, e.g., a particular state in the ground state band. Also assuming for neighboring nuclei the same value of B_2 is in general far more accurate than using the Cranking Model. For the case of nuclei in the island of stability, one will use a generic value, i.e., results will scale with B_2 (as it is pointed out in page 128 in Ref. [9]).

B   Appendix. Boundary Value Problem for GCM Model

The equation of geometric collective model (GCM) with respect to components $\Phi_{nK}^L = \Phi_{nK}^L(\beta, \gamma)$ and eigenvalue E_n^L (in MeV), $\bar{B}_2 = 2B_2/\sqrt{5}$ in (10^{-42}MeV s^2) and C_2 in (MeV) are mass and stiffness parameters, variable β in (fm), reads as

$$(T(\beta,\gamma)+T_K^L(\beta,\gamma)+\hat{V}(\beta,\gamma)-E_n^L)\Phi_{nK}^L(\beta,\gamma) = \sum_{K'=K\pm2even} V_{KK'}^L(\beta,\gamma)\Phi_{vK'}^L(\beta,\gamma), \tag{40}$$

$$T(\beta,\gamma) = \frac{\hbar^2}{2\bar{B}_2}\left(-\frac{1}{\beta^4}\frac{\partial}{\partial\beta}\beta^4\frac{\partial}{\partial\beta} - \frac{1}{\beta^2\sin(3\gamma)}\frac{\partial}{\partial\gamma}\sin(3\gamma)\frac{\partial}{\partial\gamma}\right) + \mathcal{K}(\beta,\gamma),$$

$$T_K^L(\beta,\gamma) = +\frac{\hbar^2}{2\bar{B}_2}\left[(L(L+1)-K^2)\left(\frac{2\bar{B}_2}{4J_1}+\frac{2\bar{B}_2}{4J_2}\right) + \frac{K^2 2\bar{B}_2}{2J_3}\right],$$

$$V_{KK'}^L(\bar{\beta},\gamma) = -\frac{\hbar^2}{2\bar{B}_2}\left[\frac{2\bar{B}_2}{8J_1}-\frac{2\bar{B}_2}{8J_2}\right]C_{KK'}^L, \quad C_{KK'}^L = \delta_{K'K-2}C_{KK-2}^L+\delta_{K'K+2}C_{KK+2}^L,$$

$$C_{KK-2}^L = (1+\delta_{K2})^{1/2}[(L+K)(L-K+1)(L+K-1)(L-K+2)]^{1/2},$$

$$C_{KK+2}^L = (1+\delta_{K0})^{1/2}[(L-K)(L+K+1)(L-K-1)(L+K+2)]^{1/2},$$

and the moments of the inertia denoted as $J_k = 4\bar{B}_{(k)}\beta^2\sin^2(\gamma-\frac{2}{3}k\pi)$, where $k = 1, 2, 3$ and $\bar{B}_{(k)} = \bar{B}_2$ is a mass parameter, with potential function $\hat{V}(\beta,\gamma)$ from (3), (4) and (5), and input set of parameters from Tables 11, 12, 13, 14, 15 and 16 in Appendix A, and additional kinetic function $\mathcal{K}(\beta,\gamma)$ determined in [11,14,17,20,23]. The bounded components ϕ_{vK}^L are subjected to homogeneous Neumann or Dirichlet boundary conditions at the boundary points of interval $\gamma = 0$ and $\gamma = \pi/3$ for zero or odd values of L (for details of boundary conditions on interval of the β variable see [18,19,23]), and orthonormalization conditions (see Eq. (15))

$$\int_{\beta=0}^{\beta_{max}}\int_0^{\pi/3}\sum_{K even}\Phi_{n'K}^L(\beta,\gamma)\Phi_{nK}^L(\beta,\gamma)\sin(3\gamma)d\gamma\beta^4 d\beta = \delta_{n'n}. \tag{41}$$

The BVP (40)–(41) will be solved by the FEM implemented in the CAS code.

References

1. Abramowitz, M., Stegun, I.A.: Handbook of Mathematical Functions. Dover, New York (1972). https://dlmf.nist.gov/33.22#vii
2. Bohr, A.: The coupling of nuclear surface oscillations to the motion of individual nucleons. Mat. Fys. Medd. Dan. Vid. Selsk. **26**(14) (1952)
3. Bohr, A., Mottelson, B.: Collective and individual-particle aspects of nuclear structure. Mat. Fys. Medd. Dan. Vid. Selsk. **27**(16) (1953)
4. Bohr, A., Mottelson, B.R.: Nuclear Structure, vol. 2. W A Bejamin Inc., New York; Amsterdam (1970)
5. Chacón, E., Moshinsky, M., Sharp, R.T.: $U(5) \supset O(5) \supset O(3)$ and the exact solution for the problem of quadrupole vibrations of the nucleus. J. Math. Phys. **17**, 668–676 (1976)
6. Chacón, E., Moshinsky, M.: Group theory of the collective model of the nucleus. J. Math. Phys. **18**, 870–880 (1977)
7. Deveikis, A., et al.: Symbolic-numeric algorithm for computing orthonormal basis of $O(5) \times SU(1,1)$ group. In: Boulier, F., England, M., Sadykov, T.M., Vorozhtsov, E.V. (eds.) CASC 2020. LNCS, vol. 12291, pp. 206–227. Springer, Cham (2020). https://doi.org/10.1007/978-3-030-60026-6_12
8. Eisenberg, J.M., Greiner, W.: Nuclear Theory, vol. 1, 3rd edn. North-Holland, Amsterdam (1987)

9. Ermamatov, M.J., Hess, P.O.: Microscopically derived potential energy surfaces from mostly structural considerations. Ann. Phys. **37**, 125–158 (2016)

10. Gusev, A.A., Chuluunbaatar, G., Chuluunbaatar, O., Vinitsky, S.I., Blinkov, Yu.A., Hess, P.O.: Interpolation Hermite Polynomials in Parallelepipeds and FEM Applications Extended Abstract in CASC-2022 and in Mathematics in Computer Science (2022)

11. Hess, P.O.: A gradient formula for the group $U(2l + l)$. J. Phys. G: Nucl. Phys. **4**(3), L59–L63 (1978)

12. Hess, P.O., Ermamatov, M.: In search of a broader microscopic underpinning of the potential energy surface in heavy deformed nuclei. J. Phys. Conf. Ser. **876**, 012012 (2017)

13. Hess, P.O.: The power of the geometrical model of the nucleus. In: Hess, P.O., Stöcker H. (eds.) Walter Greiner Memorial Volume, pp. 183–197. World Scientific, Singapore (2018). https://www.worldscientific.com/worldscibooks/10.1142/10828

14. Hess, P.O., Seiwert, M., Maruhn, J., Greiner, W.: General Collective Model and its Application to $^{238}_{92}$U. Z. Phys. A **296**, 147–163 (1980)

15. Hess, P.O., Maruhn, J., Greiner, W.: The general collective model applied to the chains of Pt, Os and W isotopes. J. Phys. G Nucl. Phys. **7**, 737–769 (1981)

16. Löwdin, P.O.: Studies in perturbation theory. X. Lower bounds to energy eigenvalues in perturbation-theory ground state. Phys. Rev. A **139**, 357–360 (1965)

17. Moshinsky, M., Smirnov, Y.F.: The Harmonic Oscillator in Modern Physics. HAP, Netherlands (1996)

18. Próchniak, L., Zajac, K.K., Pomorski, K., et al.: Collective quadrupole excitations in the $50 < Z$, $N < 82$ nuclei with the general Bohr Hamiltonian. Nucl. Phys. A **648**, 181–202 (1999)

19. Próchniak, L., Rohoziński, S.G.: Quadrupole collective states within the Bohr collective Hamiltonian. J. Phys. G: Nucl. Part. Phys. **36**, 123101 (2009)

20. Troltenier, D., Maruhn, J.A., Hess, P.O.: Numerical application of the geometric collective model. In: Langanke, K., Maruhn, J.A., Konin, S.E. (eds.) Computational Nuclear Physics, vol. 1, pp. 116–139. Springer, Heidelberg (1991). https://doi.org/10.1007/978-3-642-76356-4_6

21. Troltenier, D.: The generalized collective model; Das generalisierte Kollektivmodell. Ph.D. thesis, University of Francfurt, p. 55 (1992)

22. Troltenier, D., Draayer, J.P., Babu, B.R.S., Hamilton, J.H., Ramayya, A.V., Oberacker, V.E.: The ^{108,110,112}Ru isotopes in the generalized collective model. Nucl. Phys. A **601**, 56–68 (1996)

23. Troltenier, D., Maruhn, J.A., Greiner, W., Hess, P.O.: A general numerical solution of collective quadrupole surface motion applied to microscopically calculated potential energy surfaces. Z. Phys. A Hadrons Nuclei **343**, 25–34 (1992)

24. Varshalovitch, D.A., Moskalev, A.N., Hersonsky, V.K.: Quantum Theory of Angular Momentum. Nauka, Leningrad (1975); World Scientific, Singapore (1988)

25. Wolfram Research Inc: Mathematica, Version 13.0.0, Champaign, IL (2022). https://www.wolfram.com/mathematica/

26. Yannouleas, C., Pacheco, J.M.: An algebraic program for the states associated with the $U(5) \supset O(5) \supset O(3)$ chain of groups. Comput. Phys. Commun. **52**, 85–92 (1988)

27. Yannouleas, C., Pacheco, J.M.: Algebraic manipulation of the states associated with the $U(5) \supset O(5) \supset O(3)$ chain of groups: orthonormalization and matrix elements. Comput. Phys. Commun. **54**, 315–328 (1989)

Analyses and Implementations of Chordality-Preserving Top-Down Algorithms for Triangular Decomposition

Mingyu Dong and Chenqi Mou[✉]

LMIB–School of Mathematical Sciences, Beihang University, Beijing 100191, China
{mingyudong,chenqi.mou}@buaa.edu.cn

Abstract. When the input polynomial set has a chordal associated graph, top-down algorithms for triangular decomposition are proved to preserve the chordal structure. Based on these theoretical results, sparse algorithms for triangular decomposition were proposed and demonstrated with experiments to be more efficient in case of sparse polynomial sets. However, existing implementations of top-down triangular decomposition are not guaranteed to be chordality-preserving due to operations which potentially destroy the chordality. In this paper, we first analyze the current implementations of typical top-down algorithms for triangular decomposition in the Epsilon package to identify these chordality-destroying operations. Then modifications are made accordingly to guarantee new implementations of such algorithms are chordality-preserving. In particular, the technique of dynamic checking is introduced to ensure that the modifications also keep the computational efficiency. Experimental results with polynomial sets from biological systems are also reported.

Keywords: Triangular decomposition · Chordal graph · Sparsity · Implementation

1 Introduction

Symbolic computation, also called computer algebra, is an interdisciplinary subject of mathematics and computer science which studies how to solve mathematical problems in terms of symbolic objects by using algorithms and their implementations [11]. As an indispensable method in symbolic computation, triangular decomposition transforms any multivariate polynomial set into finitely many triangular sets or systems which are in the triangular shape with respect to their greatest variables and thus much easier to solve, making operations with polynomial systems like solving them algorithmically feasible [1,16,32,37].

After the introduction of characteristic set, a special kind of triangular set, by Ritt [28,29], solid development on the theories, methods, and algorithms of

This work was partially supported by the National Natural Science Foundation of China (NSFC 11971050) and Beijing Natural Science Foundation (Z180005).

F. Boulier et al. (Eds.): CASC 2022, LNCS 13366, pp. 124–142, 2022.
https://doi.org/10.1007/978-3-031-14788-3_8

triangular decomposition has been witnessed in the last decades [2, 3, 7–9, 12, 13, 17–19, 30, 31, 38], accompanied by many successful applications of triangular decomposition in scientific and engineering areas, e.g., in automated reasoning of geometric theorems [9, 36], stability analysis of biological systems [27, 35], and cryptography [5, 13, 15] etc. Well-known implementations for triangular decomposition include the Epsilon package for top-down triangular decomposition for MAPLE [33], the RegularChains library for regular decomposition in MAPLE [20], the wsolve package for characteristic decomposition [34], and the built-in implementations of triangular decomposition in SINGULAR [14].

This paper focuses on top-down algorithms for triangular decomposition which preserve the chordal structure. The connections between chordal graphs and triangular sets were first established by Cifuentes and Parrilo in their study on the chordal network of polynomial systems by associating a graph to a polynomial set [10]. They also showed that algorithms due to Wang [30, 31] are more efficient when the input polynomial set has a chordal associated graph. Their works inspired Mou and his collaborators to study chordal graphs in top-down algorithms for triangular decomposition: they proved that such algorithms preserve the chordal structure and thus are also sparsity-preserving, explaining the experimental observations by Cifuentes and Parrilo [22]. Then based on these theoretical results top-down algorithms for sparse triangular decomposition were proposed and applied to solve large polynomial systems arising from stability analysis of biological systems [23–25]. Furthermore, algorithms for incremental triangular decomposition and for cylindrical algebraic decomposition were also proved to preserve the chordal structure, leading to more efficient algorithm variants in the sparse case [6, 21].

Though those algorithms for triangular decomposition are proved to be chordality-preserving at the algorithmic level, in real computation with their existing implementations, instances where chordality is destroyed are reported. This means that in the implementations of these top-down algorithms for triangular decomposition, there exist procedures or operations which are against the overall top-down strategy of the algorithm, introducing unwanted relationships between the variables in the polynomial sets. In this paper, we first analyze the current implementations of top-down algorithms for triangular decomposition in the Epsilon package to identify the operations which destroy the chordal structure and then modify them accordingly to have real chordality-preserving implementations for top-down triangular decomposition. Furthermore, we introduce the technique of dynamic checking to make the best use of simplification which speeds up the computation of triangular decomposition considerably while keeping the implementations chordality-preserving. The effectiveness and efficiency of our modified implementations were demonstrated with experiments with benchmark polynomial systems in the ODEbase database for biological models[1].

To our best knowledge, the implementations we present in this paper are the first ones for chordality-preserving top-down triangular decomposition in the community of symbolic computation. It is planned to incorporate these

---

[1] https://odebase.cs.uni-bonn.de/ODEModelApp.

implementations into the next release of the Epsilon package. We believe that the four operations we identify in the original implementations in the Epsilon package to potentially destroy the chordality-preserving property of an implementation of top-down triangular decomposition are also useful as references to those who plan to apply chordal graphs in top-down elimination methods like those for cylindrical algebraic decomposition.

2 Preliminaries

Let $\mathbb{K}$ be a computable field. Denote by $\mathbb{K}[\boldsymbol{x}]$ the polynomial ring in the variables $x_1, \ldots, x_n$ over $\mathbb{K}$. We fix a variable ordering $x_1 < \cdots < x_n$ unless otherwise specified. For a polynomial $F \in \mathbb{K}[\boldsymbol{x}]$, the greatest variable that effectively appears in it is called the *leading variable* of F and denoted by $\mathrm{lv}(F)$. Let $x_k = \mathrm{lv}(F)$. Then F can also be regarded as a univariate polynomial in x_k, with coefficients from $\mathbb{K}[x_1, \ldots, x_{k-1}]$, and accordingly it can be written as $F = \sum_{i=0}^{d_k} C_i x_k^i$, where $C_i \in \mathbb{K}[x_1, \ldots, x_{k-1}]$, $d_k = \deg(F, x_k)$, and $C_{d_k} \neq 0$. The leading coefficient C_{d_k} here is called the *initial* of F, denoted by $\mathrm{ini}(F)$, and plays an important role in the theory of triangular decomposition.

2.1 Triangular Set and Triangular Decomposition

Definition 1. Let $\mathcal{T} = [T_1, \ldots, T_r]$ be an ordered polynomial set in $\mathbb{K}[\boldsymbol{x}]$. If none of $T_1, \ldots, T_r$ is constant and $\mathrm{lv}(T_1) < \cdots < \mathrm{lv}(T_r)$, then $\mathcal{T}$ is called a *triangular set* in $\mathbb{K}[\boldsymbol{x}]$.

Clearly the following polynomial set forms a triangular set in $\mathbb{K}[x_1, \ldots, x_4]$

$$[x_2 + x_1, (x_2^2 - x_1^2 + 2)x_3, (x_3 + x_2)x_4 + x_3 - 1]. \tag{1}$$

One can impose additional conditions on the polynomials and their initials in a triangular set to make it even stronger and have more desirable properties. Commonly used triangular sets include regular sets (or called regular chains) [8, 17, 38], simple sets [3, 26, 31], irreducible sets [32, Sect. 4.1], and normal sets [32, Sect. 5.2], etc.

Let $\mathcal{P}, \mathcal{Q} \subset \mathbb{K}[\boldsymbol{x}]$ be two polynomial sets. We are interested in the zeros defined by $\mathcal{P}$ as equations and $\mathcal{Q}$ as inequations. To be specific, we study the system of equations $P = 0$ and inequations $Q \neq 0$ for all $P \in \mathcal{P}$ and $Q \in \mathcal{Q}$ and denote this system by $\mathcal{P} = 0$ and $\mathcal{Q} \neq 0$ accordingly. Let $\overline{\mathbb{K}}$ be the algebraic closure of $\mathbb{K}$. Then we denote by $\mathsf{Z}(\mathcal{P})$ the common zeros of the polynomials in $\mathcal{P}$ in $\overline{\mathbb{K}}$ and denote $\mathsf{Z}(\mathcal{P}/\mathcal{Q}) := \mathsf{Z}(\mathcal{P}) \setminus \mathsf{Z}(S)$, where $S = \prod_{Q \in \mathcal{Q}} Q$. As one may find, the definition $\mathsf{Z}(\mathcal{P}/\mathcal{Q})$ is indeed the zero set of $\mathcal{P} = 0$ and $\mathcal{Q} \neq 0$.

Definition 2. Let $(\mathcal{T}, \mathcal{U})$ be a pair of polynomial sets in $\mathbb{K}[\boldsymbol{x}]$. Then it is called a *triangular system* if $\mathcal{T}$ is a triangular set, say $\mathcal{T} = [T_1, \ldots, T_r]$, and for each $i = 2, \ldots, r$ and any $\overline{\boldsymbol{x}}_{i-1} \in \mathsf{Z}([T_1, \ldots, T_{i-1}]/\mathcal{U})$, we have $\mathrm{ini}(T_i)(\overline{\boldsymbol{x}}_{i-1}) \neq 0$.

Definition 3. Let $\mathcal{P}$ and $\mathcal{Q}$ be two finite polynomial sets in $\mathbb{K}[\boldsymbol{x}]$. Then the process to compute finite many triangular systems $(\mathcal{T}_1, \mathcal{T}_1), \ldots, (\mathcal{T}_s, \mathcal{U}_s)$ such that $Z(\mathcal{P}/\mathcal{Q}) = \bigcup_{i=1}^{s} Z(\mathcal{T}_i/\mathcal{U}_i)$ is called *triangular decomposition* of $\mathcal{P}$ and $\mathcal{Q}$.

When the triangular set $\mathcal{T}$ in a system $(\mathcal{T}, \mathcal{U})$ is regular, simple, irreducible, or normal, the corresponding triangular system is also called so. Triangular decomposition to different kinds of triangular systems is also named after the resulting triangular systems. For example, in this paper, we are interested in top-down algorithms for regular and simple decomposition which decomposes a polynomial set into regular sets and simple sets, respectively.

Top-down algorithms for triangular decomposition refer to those handle the polynomials in a decreasing order with respect to their leading variables so that when handling the polynomials with a certain leading variable, those with strictly greater leading variables keep the same and newly generated polynomials in the process are only of smaller leading variables. The readers are referred to [23] for a formal definition of top-down triangular decomposition.

2.2 Sparse Triangular Decomposition Based on Chordal Graphs

Consider an undirected graph $G = (V, E)$, where $V := \{x_1, \ldots, x_n\}$ and E are, respectively, the sets of its vertices and edges. An edge in E connecting two vertices x_i and x_j is denoted by (x_i, x_j). Since G is an undirected graph, we have $(x_i, x_j) = (x_j, x_i)$. Let S be a non-empty subset of V. Then the subgraph with its vertices in S and edges consisting of the edges in E whose endpoints belong to S is called the *induced subgraph* of G with respect to S and denoted by $G[S]$. If an induced subgraph $G[S]$ is complete, meaning that all the vertices are connected with edges, then we say that S forms a *clique* in G.

Definition 4. Let $G = (V, E)$ be an undirected graph with $V = \{x_1, \ldots, x_n\}$ and $x_{i_1} < x_{i_2} < \cdots < x_{i_n}$ be an ordering of all the vertices. If for each $j = i_1, i_2, \cdots, i_n$, the set $\{x_j\} \cup \{x_k : x_k < x_j \text{ and } (x_k, x_j) \in E\}$ forms a clique in G, then this vertex ordering is called a *perfect elimination ordering* of G. If G has some perfect elimination ordering, then it is said to be *chordal*.

There exist effective algorithms, e.g., the MCS (maximum cardinality search) algorithm [4], to test whether a given graph is chordal, returning also a perfect elimination ordering in the affirmative case. The specifications of the MCS algorithm are formulated in Algorithm 1 for later references.

Algorithm 1: MCS algorithm $(B, \sigma) := \mathsf{MCS}(G)$

Input: G, a graph

Output: (B, σ): if G is chordal, then $B = \mathsf{True}$ and σ is one perfect elimination ordering of G, otherwise $B = \mathsf{False}$ and $\sigma = \emptyset$

For a non-chordal graph $G = (V, E)$, one can make it chordal by adding a set of new edges E'. The process to find a minimal set E' of edges for G so that $G' := (V, E \cup E')$ is chordal is called *chordal completion* of G. We denote this process by $G' = \mathsf{ChordalComp}(G)$. Here by minimal we mean that any strict subset E'' of E' cannot make $G'' = (V, E \cup E'')$ chordal. The resulting supergraph G' is also called a chordal completion of G if no ambiguity may happen.

In Fig. 1 below, the subgraph (a) is not chordal and (b) is a chordal completion of it with a perfect elimination ordering $x_1 < x_2 < x_3 < x_4 < x_5$. Note that (c) is not a chordal completion of (a) even though it is a chordal graph, for the set of added edges is not minimal.

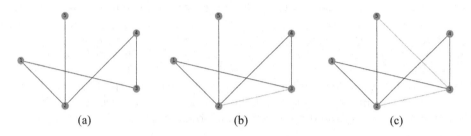

(a)   (b)   (c)

Fig. 1. A chordal graph and its chordal completion (Color figure online)

For a polynomial $F \in \mathbb{K}[\boldsymbol{x}]$, denote by $\mathrm{supp}(F)$ the set of the variables F contains; similarly for a polynomial set $\mathcal{F} \subset \mathbb{K}[\boldsymbol{x}]$ we define $\mathrm{supp}(\mathcal{F}) = \{\mathrm{supp}(F) : F \in \mathcal{F}\}$. In the following way a graph is associated to $\mathcal{F}$.

Definition 5. Let $\mathcal{F} \subset \mathbb{K}[\boldsymbol{x}]$ be a polynomial set. Set $V = \mathrm{supp}(\mathcal{F})$ and $E = \{(x_i, x_j) : \text{there exists } F \in \mathcal{F} \text{ such that } x_i, x_j \in \mathrm{supp}(F)\}$. Then the graph $G = (V, E)$ is called the *associated graph* of $\mathcal{F}$, denoted by $G(\mathcal{F})$.

The associated graph of the polynomial set (1) is illustrated as below in Fig. 2. One can see that the associated graph of a polynomial set $\mathcal{F}$ reflects how the variables in $\mathcal{F}$ are interconnected, and, thus, the definition below for variable sparsity via associated graphs is natural.

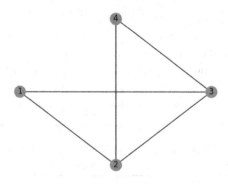

Fig. 2. Associated graph of (1)

Definition 6. Let $\mathcal{F} \subset \mathbb{K}[\boldsymbol{x}]$ be a polynomial set and $G(\mathcal{F}) = (V, E)$ be its associated graph. Then the *variable sparsity* of $\mathcal{F}$ is defined to be $s_v := |E|/\binom{2}{|V|}$.

It is proved that if the input polynomial set has a chordal associated graph and a perfect elimination ordering of this graph is used as the variable ordering, top-down algorithms for triangular decomposition preserve the variable sparsity of the polynomial sets in the process of decomposition and, thus, the following algorithmic framework for sparse triangular decomposition is proposed in [23] and verified to be more efficient for polynomial systems which are sparse with respect to their variables [24]. In Algorithm 2 below, RegDec() represents a top-down algorithm for regular decomposition. It is called an algorithmic framework because by simply replacing RegDec() with any other top-down algorithm for triangular decomposition, say one for simple decomposition, one will have a sparse version of that top-down algorithm.

Algorithm 2: Top-down algorithm for sparse regular decomposition $(\boldsymbol{\sigma}, \psi) = \mathsf{SparseRegDec}(\mathcal{F})$

Input: $\mathcal{F} \in \mathbb{K}[\boldsymbol{x}]$, a sparse polynomial set with respect to the variables
Output: $(\boldsymbol{\sigma}, \Psi)$, a perfect elimination ordering $\boldsymbol{\sigma}$ and triangular
decomposition Ψ of $\mathcal{F}$ with respect to $\boldsymbol{\sigma}$

1 $(B, \boldsymbol{\sigma}) := \mathsf{MCS}(G(\mathcal{F}))$;
2 **if** $B = \mathsf{False}$ **then**
3 $\quad$ $G' := \mathsf{ChordalComp}(G(\mathcal{F}))$;
4 $\quad$ $(B, \boldsymbol{\sigma}) := \mathsf{MCS}(G')$;
5 $\Psi := \mathsf{RegDec}(\mathcal{F}, \boldsymbol{\sigma})$;
6 **return** $(\boldsymbol{\sigma}, \Psi)$;

---

3 Chordality in Top-Down Triangular Decomposition

In this paper, we focus on three typical top-down algorithms for triangular decomposition which have been implemented in the Epsilon package: they are the algorithms for regular decomposition (denoted by RegSer), for simple decomposition (by SimSer), and for triangular decomposition (by TriSer). These algorithms, detailed in [32] and proved to be chordality-preserving [23,25], rely heavily on subresultants for polynomial elimination in the decomposition and, thus, share a similar underlying structure.

We applied sparse algorithms for triangular decomposition in the framework of Algorithm 2 with the implementations in the Epsilon package of these three algorithms to polynomial systems arising from biological models in the database ODEbase in our experiments. For each studied polynomial system $\mathcal{F}$, a finite number of triangular systems $(\mathcal{T}_1, \mathcal{U}_1), \ldots, (\mathcal{T}_s, \mathcal{U}_s)$ will be computed after the triangular decomposition, and we want to check whether all the associated graphs $G(\mathcal{T}_1), \ldots, G(\mathcal{T}_s)$ are indeed subgraphs of the input chordal graph that is either $G(\mathcal{F})$ or its chordal completion. In fact, in our experiments, instances with extra added edges not contained in the input chordal graph were reported.

Example 1. Consider the following polynomial system in Model BM483

$$\{100x_3 - x_1 - 10^{-5}x_1^2x_4 + 10^{-5}x_1x_4 + 0.2x_6, -5 \times 10^{-6}x_2^2x_3 + 5 \times 10^{-6}x_2x_3 + 0.1x_5$$
$$100x_4 - x_2 - 10^{-5}x_2^2x_3 + 10^{-5}x_2x_3 + 0.2x_5, -5 \times 10^{-6}x_1^2x_4 + 5 \times 10^{-6}x_1x_4 + 0.1x_6,$$
$$5 \times 10^{-6}x_2^2x_3 - 5 \times 10^{-6}x_2x_3 - 0.1x_5, 5 \times 10^{-6}x_1^2x_4 - 5 \times 10^{-6}x_1x_4 - 0.1x_6,$$
$$x_1 - 100x_3, x_2 - 100x_4\}.$$

The associated graph of this polynomial set is shown below in Fig. 3 and it is a chordal graph with one perfect elimination ordering $x_1 < x_3 < x_6 < x_4 < x_2 < x_5$.

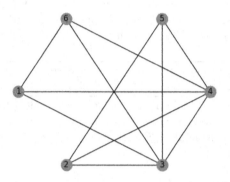

Fig. 3. A chordal associated graph

With this perfect elimination ordering as the variable order, the RegSer function in Epsilon package returns the following two triangular systems

$$([x_1 - 100x_3, x_1^2x_4 - x_1x_4 - 20000x_6, x_2 - 100x_4, 100x_4^2x_1 - x_1x_4 - 20000x_5],$$
$$\{x_1, x_1 - 1\}), ([x_1^2 - x_1, x_1 - 100x_3, x_6, x_2 - 100x_4, 100x_4^2x_1 - x_1x_4 - 20000x_5], \emptyset).$$

Their associated graphs are shown in (a) and (b) of Fig. 4 respectively.

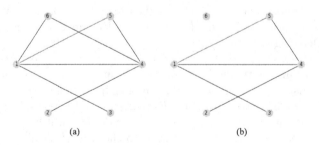

(a)   (b)

Fig. 4. Associated graph (Color figure online)

As one may find, the red edge in these two graphs does not appear in Fig. 3. This means that, even though the top-down algorithm RegSer for regular decomposition is proved theoretically to preserve the chordal structure of the input polynomial set, there may still exist specific operations in the actual implementation of this algorithm which can destroy the chordality-preserving property. In fact, out of 40 polynomial systems we picked from the ODEbase database, our experiments found 7 to produce extra added edges in some associated graphs of the returned triangular systems output by the RegSer function in the Epsilon package.

4   When is the Chordality Destroyed?

In this section, we report our study and analyses on the source codes of the implementations of the functions RegSer, SimSer, and TriSer for top-down triangular decomposition in the Epsilon package. In total, we found the following 4 operations which may destroy the chordality in these implementations.

4.1   Simplifying a Polynomial Set with Its Binomials

Let $\mathcal{P}$ be a polynomial set appearing in the process of triangular decomposition which represents the equations $\mathcal{P} = 0$. In case there exists some simple polynomial in $\mathcal{P}$ like a binomial, we can simplify $\mathcal{P}$ with this simple polynomial. This operation is formulated as Simplify($\mathcal{P}$) as follows.

Simplify($\mathcal{P}$)
1. Consider each polynomial T in the polynomial set $\mathcal{P}$ representing $\mathcal{P} = 0$.
2. If T is a binomial, write $T = c_1 t_1 + c_2 t_2$, with the term t_1 greater than t_2.
3. For any polynomial $P \in \mathcal{P} \setminus \{T\}$, replace every occurrence t_1 in P, if any, with $-\frac{c_2}{c_1} t_2$.

Example 2. Take the polynomial set (1) as input, we have
Input: Simplify($\{x_2 + x_1, (x_2^2 - x_1^2 + 2)x_3, (x_3 + x_2)x_4 + x_3 - 1\}$)
Output: $\{x_2 + x_1, 2x_3, -x_1 x_4 - 1\}$

   Clearly after the operation the output polynomial set is simpler. However, with the associated graph of the output polynomial set shown in Fig. 5, one can find an added edge in red compared with Fig. 2.

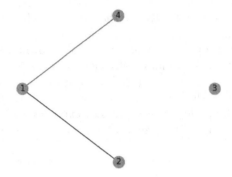

Fig. 5. Associated graph with one added edge after simplification (Color figure online)

Example 3. When computing regular decomposition of the polynomial system in Model BM483 in **ODEbase** database with the function **RegSer** in the Epsilon package with the variable ordering $x_1 < x_3 < x_6 < x_4 < x_2 < x_5$. A triangular set

$$\mathcal{T} = [x_1 - 100x_3, x_2 - 100x_4, 100x_3x_4^2 - x_3x_4 - 200x_5,$$
$$x_1^2x_4 - x_1x_4 + 100000x_1 - 10000000x_3 - 20000x_6]$$

is simplified with **Simplify()**. With the binomial $x_1 - 100x_3$, the substitution $x_3 = x_1/100$ into other polynomials in $\mathcal{T}$ results in a new triangular set

$$\mathcal{T}' = [x_1 - 100x_3, x_2 - 100x_4, 100x_1x_4^2 - x_1x_4 - 20000x_5, x_1^2x_4 - x_1x_4 - 20000x_6].$$

One can find that the vertices x_1 and x_5 appear in $100x_1x_4^2 - x_1x_4 - 20000x_5$ in $\mathcal{T}'$ now, introducing a new edge (x_1, x_5) in $G(\mathcal{T}')$.

4.2   Simplifying a Polynomial System with Binomials

In the process of top-down triangular decomposition, when the handling down to the variable x_{k+1} from x_n has finished, there are many stored polynomial systems $(\mathcal{P}, \mathcal{Q})$ such that for $i = n, \ldots, k+1$, the number of polynomials in $\mathcal{P}$ with leading variable x_i is at most 1. This means that triangular decomposition has been done for $\mathcal{P}$ down to x_{k+1}. At this point, the polynomials in both $\mathcal{P}$ and $\mathcal{Q}$ may be simplified with a simple polynomial $T \in \mathcal{P}$ if $\mathrm{lv}(T) \le x_k$. This operation, formulated as $\mathsf{Filter}((\mathcal{T}, \mathcal{U}), x_k)$ below, is similar to $\mathsf{Simplify}()$ in Sect. 4.1, except that the substitution here is by the other term of the binomial.

$\mathsf{Filter}((\mathcal{P}, \mathcal{Q}), x_k)$
1. Consider each polynomial T in the polynomial set $\mathcal{P}$ representing $\mathcal{P} = 0$.
2. If T is a binomial and $\mathrm{lv}(T) \le x_k$, write $T = c_1t_1 + c_2t_2$, with the term t_1 greater than t_2.
3. For any polynomial P in $\mathcal{P} \cup \mathcal{Q} \setminus \{T\}$, replace every occurrence t_2 in P, if any, with $-\frac{c_1}{c_2}t_1$.

Example 4. The following example shows that this operation can potentially destroy the chordality.

Input: Filter$(({x_4 + x_2, x_3 + x_1 + x_2, x_4^2 - x_2^2 + x_3}, {x_1, x_2, x_4}), x_4)$

Output: $({x_4 + x_2, x_3 + x_1 - x_4, x_3}, {x_1, x_4})$

One can find that both the equation and inequation sets become simpler after this operation. The associated graphs of the input and output polynomial sets are shown in (a) and (b) of Fig. 6 respectively, with the newly added edges colored in red.

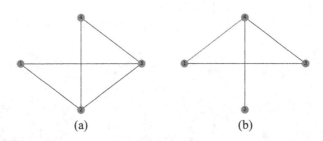

(a)   (b)

Fig. 6. Associated graph with one added edge after simplification (Color figure online)

4.3 Reducing Inequation Polynomials with a Polynomial in the Triangular Set

In top-down triangular decomposition, reduction like the pseudo-division and subresultant (see, e.g., Sects. 1.2–1.3 of [32] for the definitions of these fundamental operations in triangular decomposition) is applied to the polynomials whose leading variables are the one of interest, say x_k. After the reduction, only one polynomial T whose leading variable equals x_k is left, and this polynomial T is an element in the triangular set. Whenever such a polynomial T is found, one can perform reduction on all the current polynomials in Q representing inequations $Q \neq 0$ in the decomposition to simplify Q. This operation is formulated as Reduce(Q, T) below, in which prem(Q, T) computes the pseudo-remainder of Q with respect to T in lv(T).

Reduce(Q, T)
1. For each $Q \in Q$, replace Q with prem(Q, T).

Example 5. We report our experimental results with the function TriSer applied to the polynomial system in Model BM335 in ODEbase database, where the operation Reduce() introduced an extra edge in the process of decomposition.

Figure 7(a) is the associated graph of the input polynomial system in Model BM335 and it is not chordal. Then chordal completion is applied to it with

MCS(), resulting in a chordal graph as (b), with added edges colored in blue, and the following perfect elimination ordering

$$x_2 < x_6 < x_1 < x_5 < x_4 < x_3 < x_7 < x_{23} < x_{12} < x_8 < x_{29} < x_{22} < x_9 < x_{18} < x_{10} < x_{27}$$
$$< x_{11} < x_{25} < x_{13} < x_{26} < x_{28} < x_{14} < x_{21} < x_{24} < x_{17} < x_{16} < x_{20} < x_{19} < x_{15}.$$

Then performing pseudo division of $Q = -943230000000000x_1x_5 - 437100x_8$ in the inequation polynomial set $\mathcal{Q}$ by $T = x_2(101200000000x_1x_5 + 403x_6)$ results in $\mathrm{prem}(Q, T, x_5) = 41317575x_2x_6 - 4808100x_2x_8$. One can find that in the resulting graph shown in (c), an edge in red connecting x_2 and x_8 is added and it is not included in the chordal graph (b).

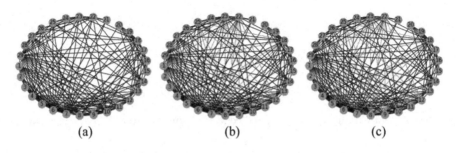

Fig. 7. One added edge with reduction on inequation polynomials (Color figure online)

4.4 Reducing a Triangular System with a Polynomial in the Triangular Set

In top-down triangular decomposition, whenever a triangular system $(\mathcal{T}, \mathcal{U})$ is constructed, one can simplify it by performing pseudo-division on all the polynomials in $\mathcal{T}$ and $\mathcal{U}$ by any polynomial in $\mathcal{T}$. This operation is formulated as ReduceTS$((\mathcal{T}, \mathcal{U}))$ below.

ReduceTS$((\mathcal{T}, \mathcal{U}))$
1. For each polynomial $T \in \mathcal{T}$, replace P with $\mathrm{prem}(P, T)$ for each $P \in \mathcal{T} \cup \mathcal{U} \setminus \{T\}$.

Example 6.
Input: ReduceTS$(([x_2^2 + x_1, x_3^5 - x_1, x_3^{10}x_4 + x_2], \{x_3\}))$
Output: $([x_2^2 + x_1, x_3^5 - x_1, x_1^2x_4 + x_2], \{x_3\})$
    The associated graphs of the input and output polynomial sets are shown in (a) and (b) of Fig. 8 respectively, with one added edge colored in red.

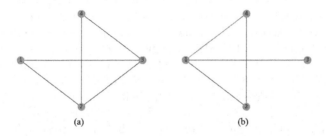

Fig. 8. Associated graph with one added edge after reduction (Color figure online)

4.5   Analysis on the Four Operations

As one can easily find, the occurrences of added edges in the functions in Sects. 4.1–4.2 are all due to simplification via substitution. It is worth mentioning that neither algebraic simplification with binomials in Sects. 4.1–4.2 nor reducing inequation polynomials with a polynomial in the triangular set in Sect. 4.3 appears in the original descriptions of top-down algorithms for triangular decomposition. These operations are found in the implementations of such algorithms only for the efficiency consideration. Similarly, reducing a triangular system is not included in the original descriptions of top-down algorithms for triangular decomposition either. This operation in the implementation is to make a triangular system *perfect*, a stronger notion than our target triangular system. To be short, this operation is for the quality of the output after triangular decomposition. To conclude, the existences of all these four identified "bad" operations in the implementations do not affect the correctness of the implementations.

5   Chordality-Preserving Implementations and Experiments

As analyzed above, all the four identified operations which potentially destroy the chordal structure in triangular decomposition do not affect the correctness of the implementations. Then one straightforward method to construct chordality-preserving implementations of top-down algorithms for triangular decomposition is merely removing the related codes.

5.1   Removing Chordality-Destroying Operations

Take our modifications to the function RegSer for regular decomposition in Epsilon package for example. We simply removed related codes of the four chordality-destroying operations in the implementations, resulting in a new function which we name RegSerC. Then we tested this new function with the benchmark polynomial systems from ODEbase database to see whether it indeed preserves chordality and to compare its efficiency against the original RegSer function.

The experimental results are summarized in Table 1, and all the experiments in this paper were carried out on a Macbook Pro laptop with a 2 GHz quad core i5 CPU and 16 GB 3733 MHz MHz LPDDR4 memory under the operating system MacOS Catalina 10.15.5. In this table, the timings (CPU time in seconds), number of branches in the computed triangular decomposition, and number of branches with added edges compared with the input chordal graph are recorded in the columns "Time", "#Bran", and "#Edge", respectively. In particular, the number "XXX-i" in the column "No." means that the corresponding ID in the ODEbase is BIOMD0000000XXX with the ith perfect elimination ordering (the specific ordering is not provided due to its length) and a dash "—" in the column "Time" means that the corresponding computation does not finish within 2 h.

From this table, we have the following observations: (1) There is no added edge reported with the new function RegSerC for all the finished computation, meaning that (at least) experimentally this function is chordality-preserving; (2) There are considerable efficiency decreases with this new function against the original RegSer one, for all tested systems. Take Model BM332 for example, in total we tested it with 5 perfect elimination orderings: the RegSer function finishes the computation around 312 s on average, while for 4 out of 5 variable orderings, the new RegSerC function cannot finish within 2 h. For the remaining ordering, the computation time with RegSerC is 6.71 times of that with RegSer.

5.2   Further Optimization with Dynamic Checking

Simply removing related codes of chordality-destroying operations can indeed guarantee that the chordality is preserved but unfortunately it also diminishes the efficiency of the implementations. This means that algebraic simplification and reduction are quite effective to improve the computational efficiency. Or in other words, we should keep as much algebraic simplification and reduction as possible in the implementations while still preserving the chordality. Following this strategy we introduce the technique of dynamic checking to test whether some extra edge will be added if some specific algebraic simplification or reduction is performed. This test is in fact quite easy: for example, if algebraic simplification is applied to a polynomial F with a binomial T to have a new polynomial F', then a simple comparison of $G(F')$ to the input chordal graph would tell us whether some extra edge would be added.

As an example, we formulate the technique of dynamic checking with Simplify in Sect. 4.1. Other algebraic simplification and reduction with dynamic checking are the same and we omit their formal descriptions.

Algorithm 3: Algebraic simplification with dynamic checking

Input: A polynomial set $\mathcal{F} \subset \mathbb{K}[\boldsymbol{x}]$, the input chordal graph G
Output: A polynomial set $\mathcal{F}' \subset \mathbb{K}[\boldsymbol{x}]$ after simplification

1 **for** $T \in \mathcal{F}$ **do**
2   **if** T *is a binomial* **then**
3   Write $T = c_1 t_1 + c_2 t_2$, with the term t_1 greater than t_2;
4   **for** $F \in \mathcal{F} \setminus \{T\}$ **do**
5           $F' :=$ polynomial obtained by replacing t_1 in F by $-\frac{c_2}{c_1} t_2$;
6   **if** $G(F') \subseteq G$ **then**
7              $F := F'$;

8 **return** $\mathcal{F}$;

Table 1. Experiments with chordality-preserving top-down implementations for regular decomposition

No.	#Var	RegSer			RegSerCO			RegSerC		
		Time	#Bran	#Edge	Time	#Bran	#Edge	Time	#Bran	#Edge
220-1	58	91.73	1728	288	99.99	1728	0	273.55	1728	0
220-2	58	256.41	5184	1944	307.00	5184	0	450.24	5184	0
332-1	78	328.96	2125	985	376.91	2176	0	—	—	—
332-2	78	248.33	2082	998	272.08	2083	0	—	—	—
332-3	78	244.08	3130	1692	284.24	3160	0	—	—	—
332-4	78	215.94	2719	1030	245.37	2719	0	—	—	—
332-5	78	522.94	3680	1312	642.25	3708	0	2423.99	3732	0
333-1	54	22.82	314	72	25.37	302	0	503.07	319	0
333-2	54	26.52	251	72	28.38	251	0	80.02	299	0
333-3	54	29.85	393	96	35.24	393	0	—	—	
333-4	54	11.93	197	87	14.45	197	0	75.35	260	0
333-5	54	22.97	269	44	26.04	269	0	151.04	366	0
334-1	74	321.06	2223	117	353.48	2223	0	—	—	—
334-2	74	235.91	1640	1081	279.44	1640	0	—	—	—
334-3	74	502.42	3183	1175	544.60	3183	0	—	—	—
334-4	74	257.60	2313	590	274.76	2313	0	1719.56	2552	0
335-1	34	8.22	262	90	8.22	262	0	14.91	205	0
335-2	34	7.03	256	81	7.86	256	0	12.32	196	0
362	34	14.68	500	165	14.80	500	0	30.36	450	0
431-1	27	5.21	66	21	5.29	67	0	18.11	78	0
431-2	27	3.05	42	7	3.45	42	0	10.07	58	0
475-1	23	5.50	40	8	4.43	38	0	8.10	38	0
475-2	23	2.02	42	6	2.20	42	0	4.14	48	0
478-1	33	5.82	67	20	5.88	80	0	9.24	80	0
478-2	33	1.30	34	7	1.44	34	0	1.64	34	0
504-1	75	83.15	142	142	94.04	240	0	2055.42	586	0
504-2	75	112.63	295	284	119.07	319	0	1972.26	2778	0
599-1	30	40.54	46	36	39.75	46	0	—	—	—
599-2	30	—	—	—	—	—	—	—	—	—
599-3	30	30.60	50	30	31.07	50	0	—	—	—

Denote by RegSerCO the new function integrated with algebraic simplification and reduction with dynamic checking. We experimented with RegSerCO with the same polynomial systems, comparing with RegSer and RegSerC, and the experimental results are recorded in Table 1 too. It can be seen that this new function RegSerCO also preserves the chordal structure in the process of triangular decomposition as RegSerC, at the same time, the efficiency loss is under control compared with RegSer: the computation time with RegSerCO is about 1.15 times on average of that with RegSer.

5.3   Chordality-Preserving Implementations for SimSer and TriSer Functions

We did similar modifications to the two functions SimSer and TriSer in the Epsilon package by introducing algebraic simplification and reduction with dynamic

Table 2. Experiments with chordality-preserving top-down implementations for simple decomposition

No.	SimSer			SimSerCO		
	Time	#Bran	#Edge	Time	#Bran	#Edge
220-1	141.06	2016	228	152.78	2016	0
220-2	329.39	6048	2160	395.78	6048	0
332-1	262.13	2082	998	303.69	2083	0
332-2	321.88	3628	2068	405.18	3590	0
333-1	23.53	314	72	24.04	302	0
333-2	27.52	251	83	27.52	251	0
333-3	42.63	450	118	47.25	450	0
333-4	15.98	218	89	18.36	218	0
333-5	31.86	283	54	36.22	283	0
335-1	7.97	262	90	8.85	262	0
335-2	8.03	256	81	8.05	256	0
362-1	26.56	635	213	26.83	636	0
362-2	23.67	874	442	26.64	861	0
362-3	21.99	602	209	22.97	602	0
431-1	7.30	74	23	7.35	72	0
431-2	3.31	45	9	3.26	45	0
431-3	2.76	28	11	2.89	28	0
475-1	5.25	38	4	5.03	36	0
475-2	4.86	42	6	2.86	42	0
475-3	3.94	72	24	4.62	74	0
475-4	2.36	42	18	2.61	42	0
478-1	7.62	97	22	8.00	110	0
478-2	1.64	43	8	1.72	43	0
478-3	1.60	40	0	1.76	40	0

checking to have two new functions SimSerCO and TriSerCO respectively. The results of our experiments with these two new functions are recorded in Tables 2 and 3, respectively.

It can be seen that both these two new functions SimSerCO and TriSerCO preserve the chordality of the input polynomial systems and keep the same level of efficiency compared with the original functions: computation with SimSerCO is 1.17 times on average of that with SimSer (slightly slower), and computation with TriSer is 0.94 times on average of that with TriSer (slightly faster).

Table 3. Experiments with chordality-preserving top-down implementations for triangular decomposition

No.	TriSer			TriSerCO		
	Time	#Bran	#Edge	Time	#Bran	#Edge
220-1	1710.22	1728	288	1566.92	1728	0
220-2	—	—	—	—	—	—
332-1	4193.24	1902	977	3948.40	1883	0
332-2	—	—	—	—	—	—
333-1	48.01	248	72	48.86	236	0
333-2	46.20	241	74	49.54	178	0
333-3	99.65	343	84	105.40	343	0
333-4	19.61	162	85	24.64	162	0
333-5	37.62	196	44	41.06	196	0
335-1	19.57	245	102	20.00	245	0
335-2	17.89	244	81	19.38	244	0
362-1	52.49	393	158	54.59	394	0
362-2	69.67	483	225	71.99	448	0
362-3	49.82	392	163	58.70	389	0
431-1	8.33	64	26	8.78	55	0
431-2	3.47	41	7	3.85	41	0
431-3	2.81	28	12	2.74	26	0
475-1	4.86	40	8	4.54	38	0
475-2	2.24	42	6	2.42	42	0
475-3	3.84	54	26	4.37	54	0
475-4	1.74	42	18	2.09	42	0
478-1	4.86	40	8	4.54	38	0
478-2	2.24	42	6	2.42	42	0
478-3	2.20	38	0	2.13	38	0

6 Concluding Remarks and Future Work

In our experiments with sparse triangular decomposition, instances are found such that existing implementations of top-down triangular decomposition destroy the chordal structure of the input polynomial system, which is inconsistent with the proved theoretical results of the chordality-preserving property of such algorithms. The main contribution of this paper is the real chordality-preserving implementations of top-down triangular decomposition based on the Epsilon package, and they are, to our best knowledge, the first chordality-preserving ones for this kind of triangular decomposition. In order to achieve this, we first analyze the current implementations in the Epsilon package to identify four chordality-destroying operations. Corresponding modifications to these four operations with dynamic checking lead to chordality-preserving implementations. Experimental results with polynomial sets from biological systems show that these implementations are indeed chordality-preserving and their efficiency is comparable to original implementations.

In the future, more implementations for top-down triangular decomposition, like those for irreducible decomposition in which factorization over algebraic field extensions is essential, are planned to be investigated and further transformed into chordality-preserving ones. Furthermore, since the choice of a specific perfect elimination ordering also influences the computational efficiency of sparse triangular decomposition, we also plan to study the underlying reasons for the influence.

Acknowledgments. The authors would like to thank Prof. Dongming Wang for his insightful comments on the implementations in Epsilon package and the referees for their helpful comments resulting in improvements on the previous version of this paper.

References

1. Aubry, P., Lazard, D., Moreno Maza, M.: On the theories of triangular sets. J. Symb. Comput. **28**(1–2), 105–124 (1999)
2. Aubry, P., Moreno Maza, M.: Triangular sets for solving polynomial systems: a comparative implementation of four methods. J. Symb. Comput. **28**(1), 125–154 (1999)
3. Bächler, T., Gerdt, V., Lange-Hegermann, M., Robertz, D.: Algorithmic Thomas decomposition of algebraic and differential systems. J. Symb. Comput. **47**(10), 1233–1266 (2012)
4. Berry, A., Blair, J., Heggernes, P., Peyton, B.: Maximum cardinality search for computing minimal triangulations of graphs. Algorithmica **39**(4), 287–298 (2004)
5. Chai, F., Gao, X.S., Yuan, C.: A characteristic set method for solving Boolean equations and applications in cryptanalysis of stream ciphers. J. Syst. Sci. Complex. **21**(2), 191–208 (2008)
6. Chen, C.: Chordality preserving incremental triangular decomposition and its implementation. In: Bigatti, A.M., Carette, J., Davenport, J.H., Joswig, M., de Wolff, T. (eds.) ICMS 2020. LNCS, vol. 12097, pp. 27–36. Springer, Cham (2020). https://doi.org/10.1007/978-3-030-52200-1_3

7. Chen, C., Golubitsky, O., Lemaire, F., Moreno Maza, M., Pan, W.: Comprehensive triangular decomposition. In: Ganzha, V.G., Mayr, E.W., Vorozhtsov, E.V. (eds.) CASC 2007. LNCS, vol. 4770, pp. 73–101. Springer, Heidelberg (2007). https://doi.org/10.1007/978-3-540-75187-8_7
8. Chen, C., Moreno Maza, M.: Algorithms for computing triangular decompositions of polynomial systems. J. Symb. Comput. **47**(6), 610–642 (2012)
9. Chou, S.-C., Gao, X.-S.: Ritt-Wu's decomposition algorithm and geometry theorem proving. In: Stickel, M.E. (ed.) CADE 1990. LNCS, vol. 449, pp. 207–220. Springer, Heidelberg (1990). https://doi.org/10.1007/3-540-52885-7_89
10. Cifuentes, D., Parrilo, P.: Chordal networks of polynomial ideals. SIAM J. Appl. Algebra Geom. **1**(1), 73–110 (2017)
11. Cox, D., Little, J., O'Shea, D.: Ideals, Varieties, and Algorithms: An Introduction to Computational Algebraic Geometry and Commutative Algebra. Undergraduate Texts in Mathematics, Springer, New York (1997). https://doi.org/10.1007/978-3-319-16721-3
12. Della Dora, J., Dicrescenzo, C., Duval, D.: About a new method for computing in algebraic number fields. In: Caviness, B.F. (ed.) EUROCAL 1985. LNCS, vol. 204, pp. 289–290. Springer, Heidelberg (1985). https://doi.org/10.1007/3-540-15984-3_279
13. Gao, X.S., Huang, Z.: Characteristic set algorithms for equation solving in finite fields. J. Symb. Comput. **47**(6), 655–679 (2012)
14. Greuel, G.M., Pfister, G., Bachmann, O., Lossen, C., Schönemann, H.: A Singular Introduction to Commutative Algebra. Springer, Heidelberg (2002). https://doi.org/10.1007/978-3-662-04963-1
15. Huang, Z., Lin, D.: Attacking bivium and trivium with the characteristic set method. In: Nitaj, A., Pointcheval, D. (eds.) AFRICACRYPT 2011. LNCS, vol. 6737, pp. 77–91. Springer, Heidelberg (2011). https://doi.org/10.1007/978-3-642-21969-6_5
16. Hubert, E.: Notes on triangular sets and triangulation-decomposition algorithms I: polynomial systems. In: Winkler, F., Langer, U. (eds.) SNSC 2001. LNCS, vol. 2630, pp. 1–39. Springer, Heidelberg (2003). https://doi.org/10.1007/3-540-45084-X_1
17. Kalkbrener, M.: A generalized Euclidean algorithm for computing triangular representations of algebraic varieties. J. Symb. Comput. **15**(2), 143–167 (1993)
18. Lazard, D.: A new method for solving algebraic systems of positive dimension. Discret. Appl. Math. **33**(1–3), 147–160 (1991)
19. Lazard, D.: Solving zero-dimensional algebraic systems. J. Symb. Comput. **13**(2), 117–131 (1992)
20. Lemaire, F., Moreno Maza, M., Xie, Y.: The RegularChains library in MAPLE. ACM SIGSAM Bull. **39**(3), 96–97 (2005)
21. Li, H., Xia, B., Zhang, H., Zheng, T.: Choosing the variable ordering for cylindrical algebraic decomposition via exploiting chordal structure. In: Proceedings of ISSAC 2021, pp. 281–288 (2021)
22. Mou, C., Bai, Y.: On the chordality of polynomial sets in triangular decomposition in top-down style. In: Proceedings ISSAC 2018, pp. 287–294 (2018)
23. Mou, C., Bai, Y., Lai, J.: Chordal graphs in triangular decomposition in top-down style. J. Symb. Comput. **102**, 108–131 (2021)
24. Mou, C., Ju, W.: Sparse triangular decomposition for computing equilibria of biological dynamic systems based on chordal graphs. In: IEEE/ACM Transactions Computational Biology and Bioinformatics (2022)

25. Mou, C., Lai, J.: On the chordality of simple decomposition in top-down style. In: Slamanig, D., Tsigaridas, E., Zafeirakopoulos, Z. (eds.) MACIS 2019. LNCS, vol. 11989, pp. 138–152. Springer, Cham (2020). https://doi.org/10.1007/978-3-030-43120-4_12

26. Mou, C., Wang, D., Li, X.: Decomposing polynomial sets into simple sets over finite fields: the positive-dimensional case. Theoret. Comput. Sci. **468**, 102–113 (2013)

27. Niu, W., Wang, D.: Algebraic approaches to stability analysis of biological systems. Math. Comput. Sci. **1**(3), 507–539 (2008)

28. Ritt, J.: Differential Equations from the Algebraic Standpoint. AMS (1932)

29. Ritt, J.: Differential Algebra. AMS (1950)

30. Wang, D.: An elimination method for polynomial systems. J. Symb. Comput. **16**(2), 83–114 (1993)

31. Wang, D.: Decomposing polynomial systems into simple systems. J. Symb. Comput. **25**(3), 295–314 (1998)

32. Wang, D.: Elimination Methods. Texts and Monographs in Symbolic Computation, Springer Science & Business Media, New York (2001). https://doi.org/10.1007/978-3-7091-6202-6

33. Wang, D.: Epsilon: A library of software tools for polynomial elimination. In: Mathematical Software, pp. 379–389. World Scientific (2002)

34. Wang, D.: wsolve: A Maple package for solving system of polynomial equations (2004). http://www.mmrc.iss.ac.cn

35. Wang, D., Xia, B.: Stability analysis of biological systems with real solution classification. In: Proceedings of ISSAC 2005, pp. 354–361 (2005)

36. Wu, W.T.: Basic principles of mechanical theorem proving in elementary geometries. J. Autom. Reason. **2**(3), 221–252 (1986)

37. Wu, W.T.: A zero structure theorem for polynomial-equations-solving and its applications. In: Davenport, J.H. (ed.) EUROCAL 1987. LNCS, vol. 378, pp. 44–44. Springer, Heidelberg (1989). https://doi.org/10.1007/3-540-51517-8_84

38. Yang, L., Zhang, J.: Searching dependency between algebraic equations: an algorithm applied to automated reasoning. In: Artificial Intelligence in Mathematics, pp. 147–156 (1994)

Accelerated Subdivision for Clustering Roots of Polynomials Given by Evaluation Oracles

Rémi Imbach[1] and Victor Y. Pan[2][(✉)]

[1] Université de Lorraine, CNRS, Inria, LORIA, 54000 Nancy, France
remi.imbach@laposte.net
[2] Lehman College and Graduate Center of City University of New York, New York, USA
victor.pan@lehman.cuny.edu

Abstract. In our quest for the design, the analysis and the implementation of a subdivision algorithm for finding the complex roots of univariate polynomials given by oracles for their evaluation, we present sub-algorithms allowing substantial acceleration of subdivision for complex roots clustering for such polynomials. We rely on approximation of the power sums of the roots in a fixed complex disc by Cauchy sums, each computed in a small number of evaluations of an input polynomial and its derivative, that is, in a polylogarithmic number in the degree. We describe root exclusion, root counting, root radius approximation and a procedure for contracting a disc towards the cluster of root it contains, called ε-compression. To demonstrate the efficiency of our algorithms, we combine them in a prototype root clustering algorithm. For computing clusters of roots of polynomials that can be evaluated fast, our implementation competes advantageously with user's choice for root finding, `MPsolve`.

Keywords: Polynomial root finding · Subdivision algorithms · Oracle polynomials

1 Introduction

We consider the

ε-Complex Root Clustering Problem (ε-CRC)
Given: a polynomial $p \in \mathbb{C}[z]$ of degree d, $\varepsilon > 0$
Output: $\ell \leq d$ couples $(\Delta^1, m^1), \ldots, (\Delta^\ell, m^\ell)$ satisfying:
    - the Δ^j's are pairwise disjoint discs of radii $\leq \varepsilon$,
    - for any $1 \leq j \leq \ell$, Δ^j and $3\Delta^j$ contain $m^j > 0$ roots of p,
    - each complex root of p is in a Δ^j for some j.

Victor's research has been supported by NSF Grant CCF 1563942 and PSC CUNY Award 63677 00 51.

F. Boulier et al. (Eds.): CASC 2022, LNCS 13366, pp. 143–164, 2022.
https://doi.org/10.1007/978-3-031-14788-3_9

Here and hereafter *root(s)* stands for *root(s) of* p and are counted with multiplicities, $3\Delta^j$ for the factor 3 concentric dilation of Δ^j, and p is a *Black box polynomial*: its coefficients are not known, but we are given *evaluation oracles*, that is, procedures for the evaluation of p, its derivative p' and hence the ratio p'/p at a point $c \in \mathbb{C}$ with a fixed precision. Such a black box polynomial can come from an experimental process or can be defined by a procedure, for example Mandelbrot's polynomials, defined inductively as

$$\mathtt{Man}_1(z) = z, \quad \mathtt{Man}_k(z) = z\,\mathtt{Man}_{k-1}(z)^2 + 1.$$

$\mathtt{Man}_k(z)$ has degree $d = 2^k - 1$ and d non-zero coefficients but can be evaluated fast, i.e., in $O(k)$ arithmetic operations. Any polynomial given by its coefficients can be handled as a black box polynomial, and the evaluation subroutines for p, p' and p'/p are fast if p is sparse or Mandelbrot-like. One can solve root-finding problems and in particular the ε-CRC problem for black box polynomials by first retrieving the coefficients by means of evaluation-interpolation, e.g., with FFT and inverse FFT, and then by applying the algorithms of [2,4,11,13,19]. Evaluation-interpolation, however, decompresses the representation of a polynomial, which can blow up its input length, in particular, can destroy sparsity. We do not require knowledge of the coefficients of an input polynomial, but instead use evaluation oracles.

Functional root-finding iterations such as Newton's, Weierstrass's (also known as Durand-Kerner's) and Ehrlich's iterations – implemented in MPsolve [4] – can be applied to approximate the roots of black box polynomials. Applying such iterations, however, requires initial points, which the known algorithms and in particular MPsolve obtain by computing root radii, and for that it needs the coefficients of the input polynomial.

Subdivision Algorithms. Let $\mathbf{i}$ stand for $\sqrt{-1}$, $c \in \mathbb{C}$, $c = a + \mathbf{i}b$ and $r, w \in \mathbb{R}$, r and w positive. We call *box* a square complex interval of the form $B(c,w) := [a - \frac{w}{2}, a + \frac{w}{2}] + \mathbf{i}[b - \frac{w}{2}, b + \frac{w}{2}]$ and *disc* $D(c,r)$ the set $\{x \in \mathbb{C} \mid |x - c| \leq r\}$. The *containing disc* $D(B(c,w))$ of a box $B(c,w)$ is $D(c,(3/4)w)$. For a $\delta > 0$ and a box or a disc S, δS denotes factor δ concentric dilation of S.

We consider algorithms based on iterative subdivision of an initial box B_0 (see [2,3,12]) and adopt the framework of [2,3] which relies on two basic subroutines: an *Exclusion Test* (ET) – deciding that a small inflation of a disc contains no root – and a *Root Counter* (RC) – counting the number of roots in a small inflation of a disc. A box B of the subdivision tree is tested for root exclusion or inclusion by applying the ET and RC to $D(B)$, which can fail and return -1 when $D(B)$ has some roots near its boundary circle. In [2], ET and RC are based on the Pellet's theorem, requiring the knowledge of the coefficients of p and shifting the center of considered disc into the origin (*Taylor's shifts*); then Dandelin-Lobachevsky-Gräffe iterations, aka *root-squaring* iterations, enable the following properties for boxes B and discs Δ:

(p1) if $2B$ contains no root, ET applied to $D(B)$ returns 0,

(p2) if Δ and 4Δ contain m roots, RC applied to 2Δ returns m.

(p1) and (p2) bound the depth of the subdivision tree. To achieve quadratic convergence to clusters of roots, [2] uses a complex version of the Quadratic Interval Refinement iterations of J. Abbott [1], aka QIR Abbott iterations, described in details in Algorithm 7 of [3] and, like [12], based on extension of Newton's iterations to multiple roots due to Schröder. [8] presents an implementation of [2] in the C library Ccluster[1], which slightly outperforms MPsolve for initial boxes containing only few roots.

In [6] we applied an ET based on Cauchy sums approximation. It satisfies (p1) and instead of coefficients of p involves $O(\log^2 d)$ evaluations of p'/p with precision $O(d)$ for a disc with radius in $O(1)$; although the output of this ET is only certified if no roots lie on or near the boundary of the input discs, in our extensive experiments it was correct when we dropped this condition.

1.1   Our Contributions

The ultimate goal of our work is to design an algorithm for solving the ε-CRC problem for black box polynomials which would run faster in practice than the known solvers, have low and possibly near optimal Boolean complexity (aka bit complexity). We do not achieve this yet in this paper but rather account for the advances along this path by presenting several sub-routines for root clustering. We implemented and assembled them in an experimental ε-CRC algorithm which outperforms the user's choice software for complex root finding, MPsolve, for input polynomials that can be evaluated fast.

Cauchy ET and RC. We describe and analyze a new RC based on Cauchy sum computations and satisfying property (p2) which only require the knowledge of evaluation oracles. For input disc of radius in $O(1)$, it requires evaluation of p'/p at $O(\log^2 d)$ points with precision $O(d)$ and is based on our ET presented in [6]; the support for its correctness is only heuristic.

Disc Compression. For a set S, let us write $Z(S, p)$ for the set of roots in S and $\#(S, p)$ for the cardinality of $Z(S, p)$; two discs Δ and Δ' are said *equivalent* if $Z(\Delta, p) = Z(\Delta', p)$. We introduce a new sub-problem of ε-CRC:

ε-**Compression into Rigid Disc (ε-CRD)**
Given: a polynomial $p \in \mathbb{C}[z]$ of degree d, $\varepsilon > 0$, $0 < \gamma < 1$,
      a disc Δ s.t. $Z(\Delta, p) \neq \emptyset$ and 4Δ is equivalent to Δ.
Output: a disc $\Delta' \subseteq \Delta$ of radius r' s.t. Δ' is equivalent to Δ and:
     - either $r' \leq \varepsilon$,
     - or $\#(\Delta, p) \geq 2$ and Δ' is at least γ-*rigid*, that is

$$\max_{\alpha, \alpha' \in Z(\Delta', p)} \frac{|\alpha - \alpha'|}{2r'} \geq \gamma.$$

---

[1] https://github.com/rimbach/Ccluster.

The ε-CRD problem can be solved with subdivision and QIR Abbott iteration, but this may require, for an initial disk of radius r, up to $O(\log(r/\max(\varepsilon',\varepsilon)))$ calls to the ET in the subdivision if the radius of convergence of the cluster in Δ for Schröder's iteration is in $O(\varepsilon')$.

Table 1. Runs of `CauchyQIR`, `CauchyComp` and `MPsolve` on Mignotte and Mandelbrot polynomials

		CauchyQIR			CauchyComp			MPsolve
d	$\log_{10}(\varepsilon^{-1})$	t	n	t_N	t	n	t_C	t
Mignotte polynomials, $a = 16$								
1024	5	1.68	30850	0.44	**0.96**	16106	0.27	1.04
1024	10	2.08	30850	0.58	**1.07**	16106	0.37	1.30
1024	50	**2.17**	30850	0.71	2.70	16105	1.96	4.84
2048	5	3.84	62220	0.90	**2.13**	32148	0.51	4.08
2048	10	4.02	62220	1.03	**2.36**	32148	0.70	5.09
2048	50	**4.51**	62220	1.25	5.62	32147	3.78	17.1
Mandelbrot polynomials								
1023	5	10.4	30877	0.86	**6.23**	18701	0.41	27.2
1023	10	10.1	30920	0.91	**6.45**	18750	0.59	30.0
1023	50	10.3	30920	1.06	**8.64**	18713	2.71	45.7
2047	5	24.3	62511	1.95	**15.2**	39296	1.39	229.
2047	10	26.4	62952	2.31	**15.5**	39358	1.71	246.
2047	50	26.1	62952	2.64	**20.4**	39255	6.22	380.

We present and analyze an algorithm solving the ε-CRD problem for $\gamma = 1/8$ based on Cauchy sums approximation and on an algorithm solving the following root radius problem: for a given $c \in \mathbb{C}$, a given non-negative integer $m \le d$ and a $\nu > 1$, find r such that $r_m(c,p) \le r \le \nu r_m(c,p)$ where $r_m(c,p)$ is the smallest radius of a disc centered in c and containing exactly m roots of p. Our compression algorithm requires only $O(\log\log(r/\varepsilon))$ calls to our RC, but a number of evaluations and arithmetic operations increasing linearly with $\log(1/\varepsilon)$.

Experimental Results. We implemented our algorithms[2] within `Ccluster` and assembled them in two algorithms named `CauchyQIR` and `CauchyComp` for solving the ε-CRC problem for black box polynomials. Both implement the subdivision process of [2] with our heuristically correct ET and RC. `CauchyQIR` uses QIR Abbott iterations of [3] (with Pellet's test replaced by our RC), while `CauchyComp` uses our compression algorithm instead of QIR Abbott iterations.

---

[2] they are not publicly realeased yet.

We compare runs of `CauchyQIR` and `CauchyComp` to emphasize the practical improvements allowed by using compression in subdivision algorithms for root finding. We also compare running times of `CauchyComp` and `MPsolve` to demonstrate that subdivision root finding can outperform solvers based on functional iterations for polynomials that can be evaluated fast. `MPsolve` does not cluster roots of a polynomial, but approximate each root up to a given error ε. Below we used the latest version[3] of `MPsolve` and call it with: `mpsolve -as -Ga -j1 -oN` where N stands for $\max(1, \lceil \log_{10}(1/\varepsilon) \rceil)$.

All the timings given below have to be understood as sequential running times on a `Intel(R) Core(TM) i7-8700 CPU @ 3.20GHz` machine with Linux. We highlight with boldface the best running time for each example. We present in Table 1 results obtained for Mandelbrot and Mignotte polynomials of increasing degree d for decreasing error ε. The Mignotte polynomial of degree d and parameter a is defined as

$$\text{Mig}_{d,a}(z) = z^d - 2(2^{\frac{a}{2}-1}z - 1)^2.$$

In Table 1, we account for the running time t for the three above-mentionned solvers. For `CauchyQIR` (resp. `CauchyComp`), we also give the number n of exclusion tests in the subdivision process, and the time t_N (resp. t_C) spent in QIR Abbott iterations (resp. compression). Mignotte polynomials have two roots with mutual distance close to the theoretical separation bound; with the ε used in Table 1, those roots are not separated.

1.2 Related Work

The subdivision root-finders of Weyl 1924, Henrici 1974, Renegar 1987, [3,12], rely on ET, RC and root radii sub-algorithms and heavily use the coefficients of p. Design and analysis of subdivision root-finders for a black box p have been continuing since 2018 in [16] (now over 150 pages), relying on the novel idea and techniques of compression of a disc and on novel ET, RC and root radii sub-algorithms, and partly presented in [5,6,10,14,15], and this paper. A basic tool of Cauchy sum computation was used in [20] for polynomial deflation, but in a large body of our results only Thm. 5 is from [20]; we deduced it in [5,16] from a new more general theorem of independent interest. Alternative derivation and analysis of subdivision in [16] (yielding a little stronger results but presently not included) relies on Schröder's iterations, extended from [12]. The algorithms are analyzed in [10,14–16], under the model for black box polynomial root-finding of [9]. [5,6] complement this study with some estimates for computational precision and Boolean complexity. We plan to complete them using much more space (cf. 46 pages in each of [20] and [3]).[4] Meanwhile we borrowed from [3]

---

[3] `3.2.1` available here: https://numpi.dm.unipi.it/software/mpsolve.

[4] In [20, Sect. 2], called "The result", we read: "The method is involved and many details still need to be worked out. In this report also many proofs will be omitted. A full account of the new results shall be given in a monograph" which has actually never appeared. [3] deduced *a posteriori estimates*, depending on root separation and Mahler's measure, that is, on the roots themselves, not known a priori.

Pellet's RC (involving coefficients), Abbott's QIR and the general subdivision algorithm with connected components of boxes extended from [12,18]. With our novel sub-algorithms, however, we significantly outperform MPsolve for polynomials that can be evaluated fast; all previous subdivision root-finders have never come close to such level. MPsolve relies on Ehrlich's (aka Aberth's) iterations, whose Boolean complexity is proved to be unbounded because iterations diverge for worst case inputs [17], but divergence never occurs in decades of extensive application of these iterations.

1.3   Structure of the Paper

In Sect. 2, we describe power sums and their approximation with Cauchy sums. In Sect. 3, we present and analyze our Cauchy ET and RC. Section 4 is devoted to root radii algorithms and Sect. 5 to the presentation of our algorithm solving the ε-CRD problem. We describe the experimental solvers CauchyQIR and CauchyComp in Sect. 6, numeric results in Sect. 7 and conclude in Sect. 8. We introduce additional definitions and properties in the rest of this section.

1.4   Definitions and Two Evaluations Bounds

Troughout this paper, log is the binary logarithm and for a positive real number a, let $\overline{\log}a = \max(1, \log a)$.

Annuli, Intervals. For $c \in \mathbb{C}$ and positives $r \le r' \in \mathbb{R}$, the annulus $A(c, r, r')$ is the set $\{z \in \mathbb{C} \mid r' \le |z - c| \le r'\}$.

Let $\Box\mathbb{R}$ be the set $\{[a - \frac{w}{2}, a + \frac{w}{2}] \mid a, w \in \mathbb{R}, w \ge 0\}$ of real intervals. For $\Box a = [a - \frac{w}{2}, a + \frac{w}{2}] \in \Box\mathbb{R}$ the center $c(\Box a)$, the width $w(\Box a)$ and the radius $r(\Box a)$ of $\Box a$ are respectively a, w and $w/2$.

Let $\Box\mathbb{C}$ be the set $\{\Box a + i\Box b \mid \Box a, \Box b \in \Box\mathbb{R}\}$ of complex intervals. If $\Box c \in \Box\mathbb{C}$, then $w(\Box c)$ (resp. $r(\Box c)$) is $\max(w(\Box a), w(\Box b))$ (resp $w(\Box c)/2$). The center $c(\Box c)$ of $\Box c$ is $c(\Box a) + ic(\Box b)$.

Isolation and Rigidity of a Disc are defined as follows [12,16].

Definition 1. (Isolation) *Let $\theta > 1$. The disc $\Delta = D(c, r)$ has isolation θ for a polynomial p or equivalently is at least θ-isolated if $Z\left(\frac{1}{\theta}\Delta, p\right) = Z(\theta\Delta, p)$, that is, $Z(A(c, r/\theta, r\theta), p) = \emptyset$.*

Definition 2. (Rigidity) *For a disc $\Delta = D(c, r)$, define*

$$\gamma(\Delta) = \max_{\alpha, \alpha' \in Z(\Delta, p)} \frac{|\alpha - \alpha'|}{2r}$$

and remark that $\gamma(\Delta) \le 1$. We say that Δ has rigidity γ or equivalently is at least γ-rigid if $\gamma(\Delta) \ge \gamma$.

Oracle Numbers and Oracle Polynomials. Our algorithms deal with numbers that can be approximated arbitrarily closely by a Turing machine. We call such approximation automata *oracle numbers* and formalize them through interval arithmetic.

For $a \in \mathbb{C}$ we call *oracle* for a a function $\mathcal{O}_a : \mathbb{N} \to \square\mathbb{C}$ such that $a \in \mathcal{O}_a(L)$ and $r(\mathcal{O}_a(L)) \leq 2^{-L}$ for any $L \in \mathbb{N}$. In particular, one has $|c(\mathcal{O}_a(L)) - a| \leq 2^{-L}$. Let $\mathcal{O}_\mathbb{C}$ be the set of oracle numbers which can be computed with a Turing machine. For a polynomial $p \in \mathbb{C}[z]$, we call *evaluation oracle* for p a function $\mathcal{I}_p : (\mathcal{O}_\mathbb{C}, \mathbb{N}) \to \square\mathbb{C}$, such that if $\mathcal{O}_a$ is an oracle for a and $L \in \mathbb{N}$, then $p(a) \in \mathcal{I}_p(\mathcal{O}_a, L)$ and $r(\mathcal{I}_p(\mathcal{O}_a, L)) \leq 2^{-L}$. In particular, one has $|c(\mathcal{I}_p(\mathcal{O}_a, L)) - p(a)| \leq 2^{-L}$.

Consider evaluation oracles $\mathcal{I}_p$ and $\mathcal{I}_{p'}$ for p and p'. If p is given by $d' \leq d+1$ oracles for its coefficients, one can easily construct $\mathcal{I}_p$ and $\mathcal{I}_{p'}$ by using, for instance, Horner's rule. However for procedural polynomials (*e.g.* Mandelbrot), fast evaluation oracles $\mathcal{I}_p$ and $\mathcal{I}_{p'}$ are built from procedural definitions.

To simplify notations, we let $\mathcal{I}_p(a, L)$ stand for $\mathcal{I}_p(\mathcal{O}_a, L)$. In the rest of the paper, $\mathcal{P}$ (resp. $\mathcal{P}'$) is an evaluation oracle for p (resp. p'); $\mathcal{P}(a, L)$ (resp. $\mathcal{P}'(a, L)$) will stand for $\mathcal{I}_p(\mathcal{O}_a, L)$ (resp. $\mathcal{I}_{p'}(\mathcal{O}_a, L)$).

Two Evaluation Bounds. The lemma below provides estimates for values of $|p|$ and $|p'/p|$ on the boundary of isolated discs. See [7, Appendix A.1] for a proof.

Lemma 3. *Let $D(c, r)$ be at least θ-isolated, $z \in \mathbb{C}$, $|z| = 1$ and g be a positive integer. Let $\mathrm{lcf}(p)$ be the leading coefficient of p. Then*

$$|p(c + rz^g)| \geq |\mathrm{lcf}(p)| \frac{r^d(\theta - 1)^d}{\theta^d} \quad and \quad \left| \frac{p'(c + rz^g)}{p(c + rz^g)} \right| \leq \frac{d\theta}{r(\theta - 1)}.$$

2 Power Sums and Cauchy Sums

Definition 4. (Power sums of the roots in a disc) *The h-th power sum of (the roots of) p in the disc $D(c, r)$ is the complex number*

$$s_h(p, c, r) = \sum_{\alpha \in Z(\Delta, p)} \#(\alpha, p) \alpha^h, \tag{1}$$

where $\#(\alpha, p)$ stands for the multiplicity of α as a root of p.

The power sums $s_h(p, c, r)$ are equal to Cauchy's integrals over the boundary circle $\partial D(c, r)$; by following [20] they can be approximated by Cauchy sums obtained by means of the discretization of the integrals: let $q \geq 1$ be an integer and ζ be a primitive q-th root of unity. When $p(c + r\zeta^g) \neq 0$ for $g = 0, \ldots, q-1$, and in particular when $D(c, r)$ is at least θ-isolated with $\theta > 1$, define the Cauchy sum $\widetilde{s}_h^q(p, c, r)$ as

$$\widetilde{s}_h^q(p, c, r) = \frac{r}{q} \sum_{g=0}^{q-1} \zeta^{g(h+1)} \frac{p'(c + r\zeta^g)}{p(c + r\zeta^g)}. \tag{2}$$

For conciseness of notations, we write s_h for $s_h\,(p, 0, 1)$ and $\tilde{s}_h{}^q$ for $\tilde{s}_h{}^q\,(p, 0, 1)$. The following theorem, proved in [6,20], allows us to approximate power sums by Cauchy sums in $D(0, 1)$.

Theorem 5. *For $\theta > 1$ and integers h, q s.t. $0 \le h < q$ let the unit disc $D(0, 1)$ be at least θ-isolated and contain m roots of p. Then*

$$|\tilde{s}_h{}^q - s_h| \le \frac{m\theta^{-h} + (d - m)\theta^h}{\theta^q - 1}. \tag{3}$$

Fix $e > 0$. If $q \ge \lceil \log_\theta(\frac{d}{e}) \rceil + h + 1$ then $|\tilde{s}_h{}^q - s_h| \le e$. (4)

Remark that $s_0\,(p, c, r)$ is the number of roots of p in $D(c, r)$ and $s_1\,(p, c, r)\,/m$ is their center of gravity when $m = \#\,(D(c, r), p)$.

Next we extend Theorem 5 to the approximation of 0-th and 1-st power sums by Cauchy sums in any disc, and define and analyze our basic algorithm for the computation of these power sums.

2.1   Approximation of the Power Sums

Let $\Delta = D(c, r)$ and define $p_\Delta(z)$ as $p(c + rz)$ so that α is a root of p_Δ in $D(0, 1)$ if and only if $c + r\alpha$ is a root of p in Δ. Following Newton's identities, one has:

$$s_0\,(p, c, r) = s_0\,(p_\Delta, 0, 1), \tag{5}$$
$$s_1\,(p, c, r) = cs_0\,(p_\Delta, 0, 1) + rs_1\,(p_\Delta, 0, 1). \tag{6}$$

Next since $p'_\Delta(z) = rp'(c + rz)$, one has

$$\tilde{s}_h{}^q\,(p, c, r) = \frac{1}{q}\sum_{g=0}^{q-1}\zeta^{g(h+1)}\frac{p'_\Delta(\zeta^g)}{p_\Delta(\zeta^g)} = \tilde{s}_h{}^q\,(p_\Delta, 0, 1)$$

and can easily prove:

Corollary 6. (of theorm 5) *Let $\Delta = D(c, r)$ be at least θ-isolated. Let $q > 1$, $s_0^* = \tilde{s}_0{}^q\,(p, c, r)$ and $s_1^* = \tilde{s}_1{}^q\,(p, c, r)$. Let $e > 0$. One has*

$$|s_0^* - s_0\,(p, c, r)| \le \frac{d}{\theta^q - 1}. \tag{7}$$

If $q \ge \lceil \log_\theta(1 + \frac{d}{e}) \rceil$ then $|s_0^* - s_0\,(p, c, r)| \le e$. (8)

Let Δ contain m roots.

$$|mc + rs_1^* - s_1\,(p, c, r)| \le \frac{rd\theta}{\theta^q - 1}. \tag{9}$$

If $q \ge \lceil \log_\theta(1 + \frac{r\theta d}{e}) \rceil$ then $|mc + rs_1^* - s_1\,(p, c, r)| \le e$. (10)

2.2   Computation of Cauchy Sums

Next we suppose that $D(c,r)$ and q are such that $p(c + r\zeta^g) \neq 0 \; \forall 0 \leq g < q$, so that $\widetilde{s}_h^{\,q}(p,c,r)$ is well defined. We approximate Cauchy sums with evaluation oracles $\mathcal{P}$, $\mathcal{P}'$ by choosing a sufficiently large L and computing the complex interval:

$$\Box\widetilde{s}_h^{\,q}(p,c,r,L) = \frac{r}{q}\sum_{g=0}^{q-1}\mathcal{O}_{\zeta^{g(h+1)}}(L)\frac{\mathcal{P}'(c+r\zeta^g,L)}{\mathcal{P}(c+r\zeta^g,L)}. \tag{11}$$

$\Box\widetilde{s}_h^{\,q}(p,c,r,L)$ is well defined for $L > \max_{0\leq g<q}(-\log_2(p(c+r\zeta^g)))$ and contains $\widetilde{s}_h^{\,q}(p,c,r)$. The following result specifies L for which we obtain that $\mathrm{r}\,(\Box\widetilde{s}_h^{\,q}(p,c,r,L)) \leq e$ for an $e > 0$. See [7, Appendix A.2] for a proof.

Lemma 7. *For strictly positive integer d, reals r and e and $\theta > 1$, let*

$$L(d,r,e,\theta) := \max\left((d+1)\log\frac{\theta}{er(\theta-1)} + \log(26rd),1\right)$$

$$\in O\left(d\left(\log\frac{1}{re} + \log\frac{\theta}{\theta-1}\right)\right).$$

If $L \geq L(d,r,e,\theta)$ then $\mathrm{r}\,(\Box\widetilde{s}_h^{\,q}(p,c,r,L)) \leq e$.

In the sequel let $L(d,r)$ stand for $L(d,r,1/4,2)$.

2.3   Approximating the Power Sums $s_0, s_1, \ldots, s_h$

Our Algorithm 1 computes, for a given integer h, approximations to power sums $s_0, s_1, \ldots, s_h$ (of p_Δ in $D(0,1)$) up to an error e, based on Eqs. (2) and (4).

Algorithm 1 satisfies the following proposition. See [7, Appendix A.3] for a proof.

Proposition 8. *Algorithm 1 terminates for an $L \leq L(d,r,e/4,\theta)$.*
Let $\mathbf{ApproxShs}(\mathcal{P},\mathcal{P}',\Delta,\theta,h,e)$ return $(success,[\Box s_0,\ldots,\Box s_h])$. Let $\Delta = D(c,r)$ and $p_\Delta(z) = p(c+rz)$. If $\theta > 1$, one has:

(a) *If $A(c,r/\theta,r\theta)$ contains no root of p, then $success = true$ and for all $i \in \{0,\ldots,h\}$, $\mathrm{w}\,(\Box s_i) < e$ and $\Box s_i$ contains $s_i\,(p_\Delta,0,1)$.*

(b) *If $e \leq 1$ and $D(c,r\theta)$ contains no root of p then $success = true$ and for all $i \in \{0,\ldots,h\}$, $\Box s_i$ contains the unique integer 0.*

(c) *If $e \leq 1$ and $A(c,r/\theta,r\theta)$ contains no root of p, $\Box s_0$ contains the unique integer $s_0\,(p,c,r) = \#(\Delta,p)$.*

(d) *If $success = false$, then $A(c,r/\theta,r\theta)$ and $D(c,r\theta)$ contain (at least) a root of p.*

(e) *If $success = true$ and $\exists i \in \{0,\ldots,h\}$, s.t. $\Box s_i$ does not contain 0 then $A(c,r/\theta,r\theta)$ and $D(c,r\theta)$ contains (at least) a root of p.*

Algorithm 1. ApproxShs$(\mathcal{P}, \mathcal{P}', \Delta, \theta, h, e)$

Require: $\mathcal{P}, \mathcal{P}'$ evaluation oracles for p and p', s.t. p is monic of degree d. $\Delta = D(c, r)$,
    $\theta \in \mathbb{R}, \theta > 1, h \in \mathbb{N}, h \geq 0, e \in \mathbb{R}, e > 0$.
Ensure: a flag $success \in \{true, false\}$, a vector $[\square s_0, \ldots, \square s_h]$.
1: $e' \leftarrow e/4$, $q \leftarrow \lceil \log_\theta(4d/e) \rceil + h + 1$
2: $\ell \leftarrow \frac{r^d(\theta-1)^d}{\theta^d}$, $\ell' \leftarrow \frac{d\theta}{r(\theta-1)}$
3: $L \leftarrow 1$
4: $[\square s_0, \ldots, \square s_h] \leftarrow [\mathbb{C}, \ldots, \mathbb{C}]$
5: **while** $\exists i \in \{0, \ldots, h\}$ s.t. $\mathrm{w}(\square s_i) \geq e$ **do**
6:     $L \leftarrow 2L$
7:     **for** $g = 0, \ldots, q - 1$ **do**
8:         Compute intervals $\mathcal{P}(c + r\zeta^g, L)$ and $\mathcal{P}'(c + r\zeta^g, L)$
9:         **if** $\exists g \in \{0, \ldots, q-1\}$ s.t. $|\mathcal{P}(c + r\zeta^g, L)| < \ell$ **or** $\left| \frac{\mathcal{P}'(c+r\zeta^g,L)}{\mathcal{P}(c+r\zeta^g,L)} \right| > \ell'$ **then**
10:            **return** $false$, $[\square s_0, \ldots, \square s_h]$
11:        **if** $\exists g \in \{0, \ldots, q-1\}$ s.t. $\frac{\ell}{2} \in |\mathcal{P}(c + r\zeta^g, L)|$ **or** $2\ell' \in \left| \frac{\mathcal{P}'(c+r\zeta^g,L)}{\mathcal{P}(c+r\zeta^g,L)} \right|$ **then**
12:            **continue**
13:        **for** $i = 0, \ldots, h$ **do**
14:            $\square s_i^* \leftarrow \square \widetilde{s_i}^q (p, c, r, L)$ //as in Eq. (11)
15:            $\square s_i \leftarrow \square s_i^* + [-e', e'] + \mathbf{i}[-e', e']$
16: **return** $true$, $[\square s_0, \ldots, \square s_h]$

---

3   Exclusion Test and Root Counters

In this section we define and analyse our base tools for disc exclusion and root counting. We recall in Subsect. 3.1 and Subsect. 3.2 the RC and the ET presented in [6]. In Subsect. 3.3, we propose a heuristic certification of root counting in which the assumed isolation for a disc Δ is heuristically verified by applying sufficiently many ETs on the contour of Δ.

For $d \geq 1, r > 0$ and $\theta > 1$, define

$$C(d, r, e, \theta) := \log(L(d, r, e, \theta)) \log_\theta(d/e) \qquad (12)$$

and $C(d, r) = C(d, r, 1/4, 2)$.

3.1   Root Counting with Known Isolation

For a disc Δ which is at least θ-isolated for $\theta > 1$, Algorithm 2 computes the number m of roots in Δ as the unique integer in the interval of width < 1 obtained by approximating 0-th cauchy sum of p_Δ in the unit disc within error $< 1/2$.

Proposition 9. *Let $\Delta = D(c, r)$. **CauchyRC1**$(\mathcal{P}, \mathcal{P}', \Delta, \theta)$ requires evaluation of $\mathcal{P}$ and $\mathcal{P}'$ at $O(C(d, r, 1, \theta))$ points and $O(C(d, r, 1, \theta))$ arithmetic operations, all with precision less than $L(d, r, 1/4, \theta)$. Let m be the output of the latter call.*

(a) If $A(c, r/\theta, r\theta)$ contains no roots of p then $m = \#(\Delta, p)$.
(b) If $m \neq 0$ then p has a root in the disc $\theta\Delta$.

Algorithm 2. CauchyRC1$(\mathcal{P}, \mathcal{P}', \Delta, \theta)$

Require: $\mathcal{P}, \mathcal{P}'$ evaluation oracles for p and p', s.t. p is monic of degree d. $\Delta = D(c, r)$,
   $\theta \in \mathbb{R}, \theta > 1$.
Ensure: An integer $m \in \{-1, 0, \ldots, d\}$.
 1: $(success, [\square s_0]) \leftarrow$ **ApproxShs**$(\mathcal{P}, \mathcal{P}', \Delta, \theta, 0, 1)$
 2: **if** $success = false$ or $\square s_0$ contains no integer **then**
 3:     **return** -1
 4: **return** the unique integer in $\square s_0$

Proposition 9 is a direct consequence of Proposition 8: in each execution of the while loop in **ApproxShs**$(\mathcal{P}, \mathcal{P}', \Delta, \theta, 0, 1)$, $\mathcal{P}$ and $\mathcal{P}'$ are evaluated at $O(\log_\theta d/e)$ points and the while loop executes an $O(\log(L(d, r, 1, \theta)))$ number of times.

3.2   Cauchy Exclusion Test

We follow [6] and increase the chances for obtaining a correct result for the exclusion of a disc with unknown isolation by approximating the first three power sums of p_Δ in $D(0, 1)$ in Algorithm 3. One has:

Proposition 10. *Let $\Delta = D(c, r)$.* **CauchyET**$(\mathcal{P}, \mathcal{P}', \Delta)$ *requires evaluation of* $\mathcal{P}$ *and* $\mathcal{P}'$ *at* $O(C(d, r))$ *points and* $O(C(d, r))$ *arithmetic operations, all with precision less than* $L(d, r)$. *Let m be the output of the latter call.*

(a) If $D(c, 4r/3)$ contains no roots of p then $m = 0$. Let B be a box so that $2B$ contains no root and suppose $\Delta = D(B)$; then $m = 0$.
(b) If $m \neq 0$ then p has a root in the disc $(4/3)\Delta$.

Algorithm 3. CauchyET$(\mathcal{P}, \mathcal{P}', \Delta)$

Require: $\mathcal{P}, \mathcal{P}'$ evaluation oracles for p and p', s.t. p is monic of degree d. $\Delta = D(c, r)$.
Ensure: An integer $m \in \{-1, 0\}$.
 1: $(success, [\square s_0, \square s_1, \square s_2]) \leftarrow$ **ApproxShs**$(\mathcal{P}, \mathcal{P}', \Delta, 4/3, 2, 1)$
 2: **if** $success = false$ or $0 \notin \square s_0$ or $0 \notin \square s_1$ or $0 \notin \square s_2$ **then**
 3:     **return** -1
 4: **return** 0

3.3   Cauchy Root Counter

We begin with a lemma illustrated in Fig. 1. See [7, Appendix A.4] for a proof.

Lemma 11. *Let $c \in \mathbb{C}$ and $\rho_-, \rho_+ \in \mathbb{R}$. Define $\mu = \frac{\rho_+ + \rho_-}{2}$, $\rho = \frac{\rho_+ - \rho_-}{2}$, $w = \frac{\mu}{\rho}$, $v = \lceil 2\pi w \rceil$ and $c_j = c + \mu e^{j\frac{2\pi i}{v}}$ for $j = 0, \ldots, v - 1$. Then the re-union of the discs $D(c_j, (5/4)\rho)$ covers the annulus $A(c, \rho_-, \rho_+)$.*

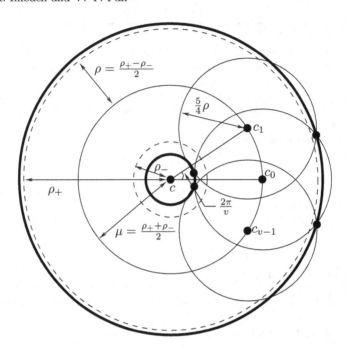

Fig. 1. Illustration for Lemma 11. In bold line, the inner and outer circles of the annulus covered by the v discs $D(c_j, (5/4)\rho)$.

For a disc $D(c, r)$ and a given $a > 1$, we follow Lemma 11 and cover the annulus $A\left(c, r/a, ra\right)$ with v discs of radius $r\frac{5(a-1/a)}{4*2}$ centered at v equally spaced points of the boundary circle of $D(c, r\frac{a+1/a}{2})$. Define

$$f_-(a, \theta) = \frac{1}{2}(a(1 - \frac{5}{4}\theta) + \frac{1}{a}(1 + \frac{5}{4}\theta)) \tag{13}$$

and

$$f_+(a, \theta) = \frac{1}{2}(a(1 + \frac{5}{4}\theta) + \frac{1}{a}(1 - \frac{5}{4}\theta)), \tag{14}$$

then the annulus $A\left(c, rf_-(a, \theta), rf_+(a, \theta)\right)$ covers the θ-inflation of those v discs.

Algorithm 4 counts the number of roots of p in a disc and satisfies:

Proposition 12. *The call* **CauchyRC2**$(\mathcal{P}, \mathcal{P}', \Delta, a)$ *amounts to* $\lceil 2\pi\frac{a^2+1}{a^2-1}\rceil$ *calls to* **CauchyET** *and one call to* **CauchyRC1**.

Let $\Delta = D(c, r)$ *and* A *be the annulus* $A\left(c, rf_-(a, \frac{4}{3}), rf_+(a, \frac{4}{3})\right)$. *Let* m *be the output of the latter call.*

(a) *If* A *contains no root then* $m \geq 0$ *and* Δ *contains* m *roots.*
(b) *If* $m \neq 0$, *then* A *contains a root.*

We state the following corollary.

Algorithm 4. CauchyRC2$(\mathcal{P}, \mathcal{P}', \Delta, a)$

Require: $\mathcal{P}, \mathcal{P}'$ evaluation oracles for p and p', s.t. p is monic of degree d. $\Delta = D(c, r)$.
   $a \in \mathbb{R}, a > 1$.
Ensure: An integer $m \in \{-1, 0, \ldots, d\}$.
   // Verify that Δ is at least a-isolated with **CauchyET**
1: $\rho_- \leftarrow \frac{1}{a}r$, $\rho_+ = ar$.
2: $\rho \leftarrow \frac{\rho_+ - \rho_-}{2}$, $\mu \leftarrow \frac{\rho_+ + \rho_-}{2}$, $w \leftarrow \frac{\mu}{\rho}$, $v \leftarrow \lceil 2\pi w \rceil$, $\zeta \leftarrow \exp(\frac{2\pi i}{v})$
3: **for** $i = 0, \ldots, v - 1$ **do**
4:     $c_i \leftarrow c + \mu\zeta^i$
5:     **if** **CauchyET**$(\mathcal{P}, \mathcal{P}', D(c_i, \frac{5}{4}\rho))$ returns -1 **then**
6:         **return** -1 // $A\left(c, rf_-(a, \frac{4}{3}), rf_+(a, \frac{4}{3})\right)$ contains a root
   // Δ is at least a-isolated according to **CauchyET**
7: **return** **CauchyRC1**$(\mathcal{P}, \mathcal{P}', \Delta, a)$

Corollary 13. (of Proposition *12*) *Let* $\theta = 4/3$ *and* $a = 11/10$. *Remark that*

$$f_-(a, \theta) = \frac{93}{110} > 2^{-1/4} \text{ and } f_+(a, \theta) = \frac{64}{55}.$$

The call **CauchyRC2**$(\mathcal{P}, \mathcal{P}', \Delta, a)$ *amounts to* $\lceil 2\pi\frac{a^2+1}{a^2-1} \rceil = 67$ *calls to* **Cauchy-ET** *for discs of radius* $\frac{21}{176}r \in O(r)$ *and one call to* **CauchyRC1** *for* Δ. *This requires evaluation of* $\mathcal{P}$ *and* $\mathcal{P}'$ *at* $O(C(d, r))$ *points, and* $O(C(d, r))$ *arithmetic operations, all with precision less than* $L(d, r)$.

4   Root Radii Algorithms

4.1   Approximation of the Largest Root Radius

For a monic p of degree d and bit-size $\tau = \log \|p\|_1$, we describe a naive approach to the approximation of the largest modulus r_d of a root of p. Recall Cauchy's bound for such a polynomial: $r_d \leq 1 + 2^\tau$. The procedure below finds an r so that $r_d < r$ and either $r = 1$ or $r/2 < r_d$ when p is given by the evaluation oracles $\mathcal{P}, \mathcal{P}'$.

1: $r \leftarrow 1$, $m \leftarrow -1$
2: **while** $m \leq d$ **do**
3:     $m \leftarrow$ **CauchyRC2**$(\mathcal{P}, \mathcal{P}', D(0, r), 4/3)$
4:     **if** $m < d$ **then**
5:         $r \leftarrow 2r$

As a consequence of Proposition 12 each execution of the while loop terminates and the procedure terminates after no more than $O(\tau)$ execution of the **while** loop. It requires evaluation of $\mathcal{P}$ and $\mathcal{P}'$ at $O(\tau C(d, r))$ points and $O(\tau C(d, r))$ arithmetic operations all with precision less than $L(d, r)$. Its correctness is implied by correctness of the results of **CauchyRC2** which is in turn implied by correctness of the results of **CauchyET**.

4.2   Approximation of the $(d+1-m)$-th Root Radius

For a $c \in \mathbb{C}$ and an integer $m \geq 1$, we call $(d+1-m)$-th root radius from c and write it $r_m(c,p)$ the smallest radius of a disc centered in c and containing exactly m roots of p.

Algorithm 5 approximates $r_m(c,p)$ within the relative error ν. It is based on the RC **CauchyRC2** and reduces the width of an initial interval $[l,u]$ containing $r_m(c,p)$ with a double exponential sieve.

Algorithm 5. RootRadius$(\mathcal{P}, \mathcal{P}', \Delta, m, \nu, \varepsilon)$

Require: $\mathcal{P}$, $\mathcal{P}'$ evaluation oracles for p and p', s.t. p is monic of degree d. A disc $\Delta = D(c,r)$, an integer $m \geq 1$, $\nu \in \mathbb{R}$, $\nu > 1$, and $\varepsilon \in \mathbb{R}$ such that $0 < \varepsilon \leq r/2$
Ensure: $r' > 0$

1: choose a s.t. $\nu^{-\frac{1}{4}} < f_-(a, \frac{4}{3}) < f_+(a, \frac{4}{3}) < 2$ // when $\nu = 2$ take $a = 11/10$
2: $l \leftarrow 0$, $u \leftarrow r$
       // Find a lower bound to $r_{d+1-m}(c,p)$
3: $m' \leftarrow$ **CauchyRC2**$(\mathcal{P}, \mathcal{P}', D(c,\varepsilon), a)$
4: **if** $m' = m$ **then**
5:     **return** ε
6: **else**
7:     $l \leftarrow f_-(a, \frac{4}{3})\varepsilon$
       // Apply double exponential sieve to get $l \leq r_{d+1-m} \leq u \leq \nu l$
8: **while** $l < u/\nu$ **do**
9:     $t \leftarrow (lu)^{\frac{1}{2}}$
10:    $m' \leftarrow$ **CauchyRC2**$(\mathcal{P}, \mathcal{P}', D(c,t), a)$
11:    **if** $m' = m$ **then**
12:       $u \leftarrow t$
13:    **else**
14:       $l \leftarrow f_-(a, \frac{4}{3})t$
15: **return** u

The correctness of Algorithm 5 for given input parameters is implied by correctness of the results of **CauchyRC2** which is in turn implied by correctness of the results of **CauchyET**. Algorithm 5 satisfies the proposition below. See [7, Appendix A.5] for a proof.

Proposition 14. *The call* **RootRadius**$(\mathcal{P}, \mathcal{P}', D(c,r), m, \nu, \varepsilon)$ *terminates after* $O(\log\log(r/\varepsilon))$ *iterations of the while loop. Let* $\Delta = D(c,r)$ *and* r' *be the output of the latter call.*

(a) *If* Δ *contains at least a root of* p *then so does* $D(c, 2r')$.
(b) *If* Δ *contains* m *roots of* p *and* **CauchyRC2** *returns a correct result each time it is called in Algorithm 5, then either* $r' = \varepsilon$ *and* $r_m(c,p) \leq \varepsilon$, *or* $r_m(c,p) \leq r' \leq \nu r_m(c,p)$.

5 A Compression Algorithm

We begin with a geometric lemma illustrated in Fig. 2.

Lemma 15. *Let* $c \in \mathbb{C}$ *and* $r, \varepsilon, \theta \in \mathbb{R}$ *satisfying* $0 < \varepsilon \leq r/2$ *and* $\theta \geq 2$. *Let* $c' \in D(c, \frac{r+\varepsilon}{\theta})$ *and* $u = \max\left(|c - c'| + \frac{r}{\theta}, r\right)$. *Then*

$$D\left(c, \frac{r}{\theta}\right) \subseteq D\left(c', u\right) \subseteq D\left(c, \frac{7}{4}r\right) \subset D(c, r\theta).$$

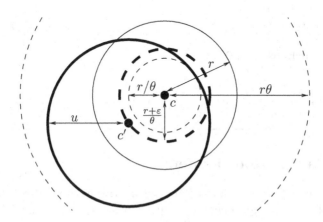

Fig. 2. Illustration for Lemma 15 with $\theta = 2$ and $\varepsilon = r/4$. c' is on the boundary circle of $D(c, (r + \varepsilon)/2)$, and $u := |c - c'| + r/\theta$.

The following lemma is a direct consequence of Lemma 15 because $s_1(p, c, r)/m$ is the center of gravity of the roots of p in $D(c, r)$.

Lemma 16. *Let* $D(c, r)$ *be at least* $\theta \geq 2$-*isolated and contain* m *roots. Let* s_1^* *approximate* $s_1(p, c, r)$ *such that* $|s_1^* - s_1(p, c, r)| \leq \frac{m\varepsilon}{\theta}$ *and* $\varepsilon \leq \frac{r}{2}$. *Then for* $c' = \frac{s_1^*}{m}$ *and* $u = \max\left(|c - c'| + \frac{r}{\theta}, r\right)$, *the disc* $D(c', u)$ *contains the same roots of* p *as* $D(c, r)$.

Algorithm 6 solves the ε-CRD problem for $\gamma = 1/8$. It satisfies the proposition below. See [7, Appendix A.6] for a proof.

Proposition 17. *The call* **Compression**$(\mathcal{P}, \mathcal{P}', \Delta, \varepsilon)$ *where* $\Delta = D(c, r)$ *requires evaluation of* $\mathcal{P}$ *and* $\mathcal{P}'$ *at* $O\left(C(d, \varepsilon) \overline{\log \log} \frac{r}{\varepsilon}\right)$ *points and the same number of arithmetic operations, all with precision less than* $L(d, \varepsilon/4)$. *Let* $m, D(c', r')$ *be the output of the latter call.*

(a) *If* Δ *is at least 2-isolated and* $Z(\Delta, p) \neq \emptyset$, *and if the call to* **RootRadius** *returns a correct result, then* $D(c', r')$ *is equivalent to* Δ, *contains* m *roots of* p *and satisfies: either* $r' \leq \varepsilon$, *or* $D(c', r')$ *is at least 1/8-rigid.*
(b) *If* $m' > 0$ *then* $D(c', 2r')$ *contains at least a root of* p.

Algorithm 6. Compression$(\mathcal{P}, \mathcal{P}', \Delta, \varepsilon)$

Require: $\mathcal{P}$, $\mathcal{P}'$ evaluation oracles for p and p', s.t. p is monic of degree d. A disc $\Delta = D(c, r)$, and a strictly positive $\varepsilon \in \mathbb{R}$.

Ensure: An integer m and a disc $D(c', r')$.

1: $\theta \leftarrow 2$, $\varepsilon' \leftarrow \varepsilon/2\theta$
2: $(success, [\Box s_0, \Box s_1]) \leftarrow \textbf{ApproxShs}(\mathcal{P}, \mathcal{P}', \Delta, \theta, 1, \min(\varepsilon', 1))$
3: **if not** $success$ **or** $\Box s_0$ does not contain an integer > 0 **then**
4:     **return** $-1, \emptyset$
5: $m \leftarrow$ the unique integer in $\Box s_0$
6: **if** $r/2 < \varepsilon$ **then**
7:     **return** $m, D(c, r/2)$
8: $c' \leftarrow c (\Box s_1) /m$ // $|c' - s_1(p, c, r)/m| < \varepsilon/4\theta$
9: **if** $m = 1$ **then**
10:     $m \leftarrow \textbf{CauchyRC1}(\mathcal{P}, \mathcal{P}', D(c', 2\varepsilon'), 2)$
11:     **return** $m, D(c', 2\varepsilon')$
12: $u \leftarrow \max\left(|c - c'| + \frac{r}{\theta}, r\right)$
13: $r' \leftarrow \textbf{RootRadius}(\mathcal{P}, \mathcal{P}', D(c', u), \frac{4}{3}, m, \theta, \varepsilon/2)$
14: **return** $m, D(c', r')$

---

6   Two Cauchy Root Finders

In order to demonstrate the efficiency of the algorithms presented in this paper, we describe here two experimental subdivision algorithms, named CauchyQIR and CauchyComp, solving the ε-CRC problem for oracle polynomials based on our Cauchy ET and RCs. Both algorithms can fail –in the case where **CauchyET** excludes a box of the subdivision tree containing a root – but account for such a failure. Both algorithm adapt the subdivision process described in [2]. CauchyQIR uses QIR Abbott iterations to ensure fast convergence towards clusters of roots. CauchyComp uses ε-compression presented in Sect. 5. In both solvers, the main subdivision loop is followed by a post-processing step to check that the output is a solution of the ε-CRC problem. The main subdivision loop does not involve coefficients of input polynomials but use evaluation oracles instead. However, we use coefficients obtained by evaluation-interpolation in the post-processing step in the case where some output discs contain more than one root. We observe no failure of our algorithms in all our experiments covered in Sect. 7.

6.1   Subdivision Loop

Let B_0 be a box containing all the roots of p. Such a box can be obtained by applying the process described in Subsect. 4.1.

Sub-Boxes, Component and Quadrisection. For a box $B(a + \mathbf{i}b, w)$, let $Children_1(B)$ be the set of the four boxes $\{B((a \pm w/4) + \mathbf{i}(b \pm w/4), w/2)\}$, and

$$Children_n(B) := \bigcup_{B' \in Children_{n-1}(B)} Children_1(B').$$

A box B is a *sub-box* of B_0 if $B = B_0$ or if there exist an $n \geq 1$ s.t. $B \in Children_n(B_0)$. A *component* C is a set of connected sub-boxes of B_0 of equal widths. The *component box* $B(C)$ of a component C is the smallest (square) box subject to $C \subseteq B(C) \subseteq B_0$ minimizing both $\mathcal{R}e(c\,(B(C)))$ and $\mathcal{I}m\,(c\,(B(C)))$. We write $D(C)$ for $D(B(C))$. If S is a set of components (resp. discs) and $\delta > 0$, write δS for the set $\{\delta D(A)\ (\text{resp. } A) \mid A \in S\}$.

Definition 18. *Let Q be a set of components or discs. We say that a component C (resp. a disk Δ) is γ-separated (or γ-sep.) from Q when $\gamma D(C)$ (resp. $\gamma\Delta$) has empty intersection with all elements in Q.*

Remark 19. *Let Q be a set of components and $C \notin Q$ a component. If $Z(\mathbb{C}, p) = Z(\{C\} \cup Q, p)$ and C is 4-separated from Q then $2D(C)$ is at least 2-isolated.*

Subdivision Process. We describe in Algorithm 7 a subdivision algorithm solving the ε-CRC problem. The components in the working queue Q are sorted by decreasing radii of their containing discs. It is parameterized by the flag *compression* indicating whether compression or QIR Abbott iterations have to be used. In QIR Abbott iterations of Algorithm 7 in [3], we replace the Graeffe Pellet test for counting roots in a disc Δ by **CauchyRC2**$(\mathcal{P}, \mathcal{P}', \Delta, 4/3)$. If a QIR Abbott iteration in step 12 fails for input Δ, m, it returns Δ. Steps 20–21 prevent C to artificially inflate when a compression or a QIR Abbott iteration step does not decrease $D(C)$. For a component C, $Quadrisect(C)$ is the set of components obtained by grouping the set of boxes

$$\bigcup_{B \in C} \{B' \in Children_1(B) \mid \mathbf{CauchyET}(\mathcal{P}, \mathcal{P}', D(B')) = -1\}$$

into components.

The **while** loop in steps 4-22 terminates because all our algorithms terminate, and as a consequence of (a) in Proposition 9: any component will eventually be decreased until the radius of its containing disc reaches $\varepsilon/2$.

6.2   Output Verification

After the subdivision process described in steps 1–22 of Algorithm. 7, R is a set of pairs of the form $\{(\Delta^1, m^1), \ldots, (\Delta^\ell, m^\ell)\}$ satisfying, for any $1 \leq j \leq \ell$:

- Δ^j is a disc of radii $\leq \varepsilon$, m^j is an integer ≥ 1,
- Δ^j contains at least a root of p,
- for any $1 \leq j' \leq \ell$ s.t. $j' \neq j$, $3\Delta^j \cap \Delta^{j'} = \emptyset$.

The second property follows from (b) of Proposition 10 and (b) of Proposition 17 when compression is used. Otherwise, remark that a disk Δ in the output of QIR Abbott iteration in step 12 of Algorithm 7 verifies **CauchyRC2**$(\mathcal{P}, \mathcal{P}', \Delta, 4/3) > 0$ and apply (b) of Proposition 12. The third property follows from the **if** statement in step 15 of Algorithm 7. Decompose R as the disjoint union $R_1 \cup R_{>1}$ where R_1 is the subset of pairs (Δ^i, m^i) of R where $m^i = 1$ and $R_{>1}$ is the subset of pairs (Δ^i, m^i) of R where $m^i > 1$, and make the following remark:

Algorithm 7. CauchyRootFinder$(\mathcal{P}, \mathcal{P}', \varepsilon, compression)$

Require: $\mathcal{P}$ and $\mathcal{P}'$ evaluation oracles for p and p', s.t. p is monic of degree d. A
   (strictly) positive $\varepsilon \in \mathbb{R}$, a flag $compression \in \{true, false\}$.
Ensure: A flag $success$ and a list $R = \{(\Delta^1, m^1), \ldots, (\Delta^\ell, m^\ell)\}$
 1: $B_0 \leftarrow$ box s.t. $\#(B, p) = d$ as described in Subsect. 4.1
 2: $Q \leftarrow \{B_0\}$ // Q is a queue of components
 3: $R \leftarrow \{\}$ // R is the empty list of results
 4: **while** Q is not empty **do**
 5:    $C \leftarrow pop(Q)$
 6:    **if** C is 4-separated from Q **then**
 7:       **if** $compression$ **then**
 8:          $m, D(c,r) \leftarrow$ **Compression**$(\mathcal{P}, \mathcal{P}', 2D(C), \varepsilon/2)$
 9:       **else**
10:          $m \leftarrow$ **CauchyRC1**$(\mathcal{P}, \mathcal{P}', 2D(C), 2)$
11:          **if** $m > 0$ **then**
12:             $D(c,r) \leftarrow$ QIR Abbott iteration for $D(C), m$
13:       **if** $m \leq 0$ **then**
14:          **return** $fail, \emptyset$
15:       **if** $r \leq \varepsilon/2$ **and** $D(c, 2r)$ is 3-sep. from $2Q$ **and** is 1-sep. from $6Q$ **then**
16:          $push(R, (D(c, 2r), m))$
17:          **continue**
18:       **else**
19:          $C' \leftarrow$ component containing $D(c,r)$
20:          **if** $C' \subset C$ **then**
21:             $C \leftarrow C'$
22:    $push(Q, Quadrisect(C))$
23: $success \leftarrow$ verify R as described in Subsect. 6.2
24: **return** $success, R$

Remark 20. *If $m^1 + \ldots + m^\ell = d$ and for any $(\Delta^i, m^i) \in R_{>1}$, Δ^i contains exactly m^i roots of p, then R is a correct output for the ε-CRC problem with input p of degree d and ε.*

According to Remark 20, checking that R is a correct output for the ε-CRC problem for fixed input p of degree d and ε amount to check that the m^i's add up to d and that for any $\Delta^i \in R_{>1}$, Δ^i contains exactly m^i roots of p. For this last task, we use evaluation-interpolation to approximate the coefficients of p and then apply the Graeffe-Pellet test of [2].

7 Experiments

We implemented Algorithm 7 in the C library `Ccluster`. Call `CauchyComp` (resp. `CauchyQIR`) the implementation of Algorithm 7 with $compression = true$ (resp. $false$). In the experiments we conducted so far, `CauchyComp` and `CauchyQIR` never failed.

Test Suite. We experimented CauchyComp, CauchyQIR and MPsolve on Mandelbrot and Mignotte polynomials as defined in Sect. 1 as well as Runnel and random sparse polynomials. Let $r = 2$. The Runnel polynomial is defined inductively as

$$\text{Run}_0(z) = 1, \quad \text{Run}_1(z) = z, \quad \text{Run}_{k+1}(z) = \text{Run}_k(z)^r + z\,\text{Run}_{k-1}(z)^{r^2}$$

It has real coefficients, a multiple root (zero), and can be evaluated fast. We generate random sparse polynomials of degree d, bitsize τ and $\ell \geq 2$ nonzero terms as follows, where p_i stands for the coefficient of the monomial of degree i in p: p_0 and p_d are randomly chosen in $[-2^{\tau-1}, 2^{\tau-1}]$, then $\ell - 2$ integers $i_1, \ldots, i_{\ell-1}$ are randomly chosen in $[1, d-1]$ and $p_{i_1}, \ldots, p_{i_{\ell-1}}$ are randomly chosen in $[-2^{\tau-1}, 2^{\tau-1}]$. The other coefficients are set to 0.

Results. We report in Table 1 results of those experiments for Mandelbrot and Mignotte polynomials with increasing degrees and increasing values of $\log_{10}(\varepsilon^{-1})$. We account for the running time t for the three above-mentionned solvers. For CauchyQIR (resp. CauchyComp), we also give the number n of exclusion tests in the subdivision process, and the time t_N (resp. t_C) spent in QIR Abbott iterations (resp. compression).

Our compression algorithm allows smaller running times for low values of $\log_{10}(\varepsilon^{-1})$ because it compresses a component C on the cluster it contains as of $2D(C)$ is 2-isolated, whereas QIR Abbott iterations require the radius Δ to be near the radius of convergence of the cluster for Schröder's iterations.

We report in Table 2 the results of runs of CauchyComp and MPsolve for polynomials of our test suite of increasing degree, for $\log_{10}(\varepsilon^{-1}) = 16$. For random sparse polynomials, we report averages over 10 examples. The column t_V accounts for the time spent in the verification of the output of CauchyComp (see Subsect. 6.2); it is 0 when all the pairs (Δ^j, m^j) in the output verify $m^j = 1$. It is > 0 when there is at least a pair with $m^j > 1$.

The maximum precision L required in all our tests was 106, which makes us believe that our analysis in Proposition 8 is very pessimistic. Our experimental solver CauchyComp is faster than MPsolve for polynomials that can be evaluated fast.

Table 2. Runs of CauchyComp and MPsolve on polynomials of our test suite for $\log_{10}(\varepsilon^{-1}) = 16$

		CauchyComp			MPsolve
d	t	n	t_C	t_V	t
Mandelbrot polynomials					
255	1.31	5007	0.21	0.00	**0.58**
511	**3.25**	10679	0.64	0.00	4.13
1023	**6.47**	18774	0.84	0.00	31.7
2047	**16.2**	39358	2.35	0.00	267.
Runnels polynomials					
341	2.55	4967	0.38	0.00	**0.45**
682	5.66	9392	0.87	0.02	**3.32**
1365	**12.6**	18030	2.00	0.05	26.2
2730	**29.7**	35612	4.26	0.12	236.
Mignotte polynomials, $a = 16$					
256	0.29	4131	0.15	0.00	**0.21**
512	**0.58**	8042	0.27	0.00	0.70
1024	**1.24**	16105	0.55	0.02	2.99
2048	**2.69**	32147	1.05	0.04	11.6
10 randomSparse polynomials with 3 terms and bitsize 256					
767	.902	10791.	.415	0.0	**.602**
1024	**1.35**	15526.	.560	0.0	1.36
1535	**2.04**	21244.	.861	0.0	2.35
2048	**2.98**	30642.	1.16	0.0	4.10
10 randomSparse polynomials with 5 terms and bitsize 256					
2048	4.77	29583.	1.60	0.0	**4.09**
3071	**6.92**	43003.	2.45	0.0	10.0
4096	**9.82**	56659.	3.38	0.0	24.0
6143	**17.7**	86857.	5.40	0.0	44.5
10 randomSparse polynomials with 10 terms and bitsize 256					
3071	11.9	44714.	4.09	0.0	**10.3**
4096	**17.5**	58138.	5.82	0.0	17.6
6143	**29.1**	85451.	8.93	0.0	51.9
8192	**40.6**	116289.	12.4	0.0	66.5

8   Conclusion

We presented, analyzed and verified practical efficiency of two basic subroutines for solving the complex root clustering problem for black box polynomials. One is a root counter, the other one is a compression algorithm. Both algorithms are well-known tools used in subdivision procedures for root finding.

We propose our compression algorithm not as a replacement of QIR Abbott iterations, but rather as a complementary tool: in future work, we plan to use compression to obtain a disc where Schröder's/QIR Abbott iterations would converge fast. The subroutines presented in this paper laid down the path toward a Cauchy Root Finder, that is, an algorithm solving the ε-CRC problem for black box polynomials.

References

1. Abbott, J.: Quadratic interval refinement for real roots. ACM Commun. Comput. Algebra **48**(1/2), 3–12 (2014)
2. Becker, R., Sagraloff, M., Sharma, V., Xu, J., Yap, C.: Complexity analysis of root clustering for a complex polynomial. In: Proceedings of the ACM on International Symposium on Symbolic and Algebraic Computation, pp. 71–78. ACM (2016)
3. Becker, R., Sagraloff, M., Sharma, V., Yap, C.: A near-optimal subdivision algorithm for complex root isolation based on the Pellet test and Newton iteration. J. Symbol. Comput. **86**, 51–96 (2018)
4. Bini, D.A., Robol, L.: Solving secular and polynomial equations: a multiprecision algorithm. J. Comput. Appl. Math. **272**, 276–292 (2014)
5. Imbach, R., Pan, V.Y.: New practical advances in polynomial root clustering. In: Slamanig, D., Tsigaridas, E., Zafeirakopoulos, Z. (eds.) MACIS 2019. LNCS, vol. 11989, pp. 122–137. Springer, Cham (2020). https://doi.org/10.1007/978-3-030-43120-4_11
6. Imbach, R., Pan, V.Y.: New progress in univariate polynomial root finding. In: Proceedings of the 45th International Symposium on Symbolic and Algebraic Computation, pp. 249–256 (2020)
7. Imbach, R., Pan, V.Y.: Accelerated subdivision for clustering roots of polynomials given by evaluation oracles. arXiv preprint 2206.08622 (2022)
8. Imbach, R., Pan, V.Y., Yap, C.: Implementation of a near-optimal complex root clustering algorithm. In: Davenport, J.H., Kauers, M., Labahn, G., Urban, J. (eds.) ICMS 2018. LNCS, vol. 10931, pp. 235–244. Springer, Cham (2018). https://doi.org/10.1007/978-3-319-96418-8_28
9. Louis, A., Vempala, S.S.: Accelerated Newton iteration: roots of black box polynomials and matrix eigenvalues. In: IEEE 57th Annual Symposium on Foundations of Computer Science, pp. 732–740 (2016)
10. Luan, Q., Pan, V.Y., Kim, W., Zaderman, V.: Faster numerical univariate polynomial root-finding by means of subdivision iterations. In: Boulier, F., England, M., Sadykov, T.M., Vorozhtsov, E.V. (eds.) CASC 2020. LNCS, vol. 12291, pp. 431–446. Springer, Cham (2020). https://doi.org/10.1007/978-3-030-60026-6_25
11. Moroz, G.: Fast real and complex root-finding methods for well-conditioned polynomials. arXiv preprint 2102.04180 (2021)
12. Pan, V.Y.: Approximating complex polynomial zeros: modified Weyl's quadtree construction and improved Newton's iteration. J. Complex. **16**, 213–264 (2000)
13. Pan, V.Y.: Univariate polynomials: nearly optimal algorithms for numerical factorization and root-finding. J. Symbol. Comput. **33**, 701–733 (2002)
14. Pan, V.Y.: Old and new nearly optimal polynomial root-finders. In: England, M., Koepf, W., Sadykov, T.M., Seiler, W.M., Vorozhtsov, E.V. (eds.) CASC 2019. LNCS, vol. 11661, pp. 393–411. Springer, Cham (2019). https://doi.org/10.1007/978-3-030-26831-2_26

15. Pan, V.Y.: Acceleration of subdivision root-finding for sparse polynomials. In: Boulier, F., England, M., Sadykov, T.M., Vorozhtsov, E.V. (eds.) CASC 2020. LNCS, vol. 12291, pp. 461–477. Springer, Cham (2020). https://doi.org/10.1007/978-3-030-60026-6_27

16. Pan, V.Y.: New progress in polynomial root-finding. arXiv preprint 1805.12042 (2022)

17. Reinke, B.: Diverging orbits for the Ehrlich-Aberth and the Weierstrass root finders. arXiv preprint 2011.01660 (2020)

18. Renegar, J.: On the worst-case arithmetic complexity of approximating zeros of polynomials. J. Complex. **3**(2), 90–113 (1987)

19. Sagraloff, M., Mehlhorn, K.: Computing real roots of real polynomials. J. Symbol. Comput. **73**, 46–86 (2016)

20. Schönhage, A.: The fundamental theorem of algebra in terms of computational complexity. Manuscript. University of Tübingen, Germany (1982)

On Equilibrium Positions in the Problem of the Motion of a System of Two Bodies in a Uniform Gravity Field

Valentin Irtegov and Tatiana Titorenko[✉]

Institute for System Dynamics and Control Theory SB RAS,
134, Lermontov Street, Irkutsk 664033, Russia
{irteg,titor}@icc.ru

Abstract. In the problem of motion of a system of two rigid bodies connected by a spherical hinge in a uniform gravity field, the conditions for the existence of two- and one-dimensional invariant manifolds are presented, and the manifolds themselves are found with the use of computer algebra tools. From a mechanical point of view, these solutions correspond to equilibrium positions of the system. Their instability in the first approximation is proved.

1 Introduction

This work continues the study [5]. The rotation of the system of two connected rigid bodies S_1 and S_2 (see Fig. 1) in a uniform gravity field is considered. The first body has a fixed point O_1. The bodies are connected by an ideal spherical hinge O_2.

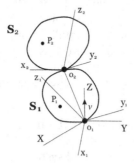

Fig. 1. .

To describe the motion of the mechanical system, the following coordinate systems are introduced: the inertial O_1XYZ (its Z axis with the unit vector $\boldsymbol{\nu}$ is directed vertically upwards), the moving frames $O_1x_1y_1z_1$ and $O_2x_2y_2z_2$ attached rigidly to the bodies S_1 and S_2, respectively. The x_i, y_i, z_i $(i = 1, 2)$ axes are directed along the principal inertia axes of the bodies. The positions of

F. Boulier et al. (Eds.): CASC 2022, LNCS 13366, pp. 165–184, 2022.
https://doi.org/10.1007/978-3-031-14788-3_10

$O_1x_1y_1z_1$ with respect to O_1XYZ and $O_2x_2y_2z_2$ with respect to $O_1x_1y_1z_1$ are defined by Euler's angles $\psi_1, \theta_1, \varphi_1$ and $\psi_2, \theta_2, \varphi_2$.

The mechanical system studied in [5] is characterized as follows: the distribution of mass in the bodies is arbitrary, the connection point O_2 does not lie on the principal axes of inertia of the body S_1, but the centers of masses of the bodies P_1 and P_2 belong to their principal axes of inertia. The equations of motion for the system have been derived with the help of the software package [1] written in the language of computer algebra system (CAS) "Mathematica". First, according to a geometric description of the mechanical system, its characteristic function (the Lagrange function) in symbolic form has been constructed, then, using this function as a starting point, the equations of motion have been obtained. The problem of their qualitative analysis was stated. Within the framework of solving this problem, solutions of the equations corresponding to permanent rotations of the system have been found, and the sufficient conditions of their stability in the sense of Lyapunov have been derived.

In the present work, the above mechanical system is studied in a more general case. We assume that the centers of masses of the bodies do not lie on their principal axes of inertia. The equations of motion of the system are derived analogously to the previous case. The problem of qualitative analysis of the equations is stated. In the present paper, we restrict ourselves by considering equilibrium solutions of the equations. Two techniques are used to find them: from the stationary conditions for the family of the first integrals of the problem, and, directly, from the equations of motion. The stability of the solutions is analyzed on the base of Lyapunov's stability theorems in the first approximation. All computations are performed with the aid of CAS "Mathematica".

As was mentioned in [5], similar problems arise in many applications, e.g., in modelling and the study of dynamical properties of various technical devices and instruments. Such problems are considered, e.g., in [2,3].

The paper is organized as follows. In Sect. 2, the Lagrange function and the equations of motion with their first integrals for the mechanical system in question are given. In Sect. 3, we seek equilibrium positions of the system, using the stationary conditions of the first integrals. In Sect. 3.2, the same problem is solved with the help of the equations of motion. In Sect. 4, the stability of the solutions is analyzed. In Sect. 5, we give a conclusion.

2   The Lagrange Function and the Equations of Motion

The Lagrange function of the mechanical system under consideration derived by the technique [5] has the form: $L = T + U$, where

$$
\begin{aligned}
2T = {} & A_1 p_1^2 + B_1 q_1^2 + C_1 r_1^2 + A_2(b_{11}p_1 + p_2 + b_{12}q_1 + b_{13}r_1)^2 \\
& + B_2(b_{21}p_1 + b_{22}q_1 + b_{23}r_1 + q_2)^2 \\
& + C_2(b_{31}p_1 + b_{32}q_1 + b_{33}r_1 + r_2)^2 + m_2(d_1^2 + d_2^2 + d_3^2) \\
& + 2m_2 \Big[a_2 \left[(b_{31}p_1 + b_{32}q_1 + b_{33}r_1 + r_2)(b_{23}d_1 + b_{22}d_2 + b_{21}d_3) \right.
\end{aligned}
$$

$$-(b_{21}p_1 + b_{22}q_1 + b_{23}r_1 + q_2)(b_{33}d_1 + b_{32}d_2 + b_{31}d_3)]$$
$$-b_2[(b_{31}p_1 + b_{32}q_1 + b_{33}r_1 + r_2)(b_{13}d_1 + b_{12}d_2 + b_{11}d_3)$$
$$-(b_{11}p_1 + b_{12}q_1 + b_{13}r_1 + p_2)(b_{33}d_1 + b_{32}d_2 + b_{31}d_3)]$$
$$+c_2[(b_{21}p_1 + b_{22}q_1 + b_{23}r_1 + q_2)(b_{13}d_1 + b_{12}d_2 + b_{11}d_3)$$
$$-(b_{11}p_1 + b_{12}q_1 + b_{13}r_1 + p_2)(b_{23}d_1 + b_{22}d_2 + b_{21}d_3)]\Big],$$

$$U = -g\Big[m_1(a_1a_{13} + b_1a_{23} + c_1a_{33}) + m_2\Big(a_{13}(a_2b_{11} + b_2b_{21} + c_2b_{31} + s_1)$$
$$+a_{23}(a_2b_{12} + b_2b_{22} + c_2b_{32} + s_2) + a_{33}(a_2b_{13} + b_2b_{23} + c_2b_{33} + s_3)\Big)\Big]$$

are the kinetic energy and the force function of the system, respectively.

Here $d_1 = p_1s_2 - q_1s_1$, $d_2 = r_1s_1 - p_1s_3$, $d_3 = q_1s_3 - r_1s_2$; a_i, b_i, c_i $(i = 1, 2)$ are the coordinates of the centers of masses of the bodies; s_1, s_2, s_3 are the coordinates of the connection point O_2; A_i, B_i, C_i $(i = 1, 2)$ are the principal moments of inertia of the bodies; m_1, m_2 are the masses of the bodies; g is the acceleration due to gravity; $p_i = \dot\psi_i \sin\varphi_i \sin\theta_i + \dot\theta_i \cos\varphi_i$, $q_i = \dot\psi_i \cos\varphi_i \sin\theta_i - \dot\theta_i \sin\varphi_i$, $r_i = \dot\varphi_i + \dot\psi_i \cos\theta_i$ are the projections of the vector of angular velocity of the body S_i onto the axes $O_ix_iy_iz_i$; $\alpha = \|a_{kl}\|$, $\beta = \|b_{kl}\|$ are the cosine matrices 3×3 of angles between the axes O_1XYZ and $O_1x_1y_1z_1$, and the axes $O_1x_1y_1z_1$ and $O_2x_2y_2z_2$, respectively. Their elements are related to Euler's angles $\psi_i, \theta_i, \varphi_i$ as follows:

$$a_{kl} = \zeta_{kl}^{(1)}, \quad b_{kl} = \zeta_{kl}^{(2)} \quad (k, l = 1, 2, 3), \text{ where}$$
$$\zeta_{11}^{(i)} = \cos\varphi_i \cos\psi_i - \cos\theta_i \sin\varphi_i \sin\psi_i,$$
$$\zeta_{12}^{(i)} = \cos\psi_i \cos\theta_i \sin\varphi_i + \cos\varphi_i \sin\psi_i, \quad \zeta_{13}^{(i)} = \sin\varphi_i \sin\theta_i,$$
$$\zeta_{21}^{(i)} = -\cos\psi_i \sin\varphi_i - \cos\varphi_i \cos\theta_i \sin\psi_i, \tag{1}$$
$$\zeta_{22}^{(i)} = \cos\varphi_i \cos\psi_i \cos\theta_i - \sin\varphi_i \sin\psi_i,$$
$$\zeta_{23}^{(i)} = \cos\varphi_i \sin\theta_i, \quad \zeta_{31}^{(i)} = \sin\psi_i \sin\theta_i, \quad \zeta_{32}^{(i)} = -\cos\psi_i \sin\theta_i,$$
$$\zeta_{33}^{(i)} = \cos\theta_i \quad (i = 1, 2).$$

The equations of motion produced by the package according to the formulae

$$\frac{d}{dt}\left(\frac{\partial L}{\partial \omega^{(i)}}\right) = \frac{\partial L}{\partial \omega^{(i)}} \times \omega^{(i)} + \frac{\partial L}{\partial \alpha^{(i)}} \times \alpha^{(i)} + \frac{\partial L}{\partial \beta^{(i)}} \times \beta^{(i)} + \frac{\partial L}{\partial \gamma^{(i)}} \times \gamma^{(i)},$$

$$\dot\alpha^{(i)} = \alpha^{(i)} \times \omega^{(i)}, \quad \dot\beta^{(i)} = \beta^{(i)} \times \omega^{(i)}, \quad \dot\gamma^{(i)} = \gamma^{(i)} \times \omega^{(i)} \quad (i = 1, 2),$$

where $\omega^{(i)} = (p_i, q_i, r_i)$, $\alpha^{T(1)} = (a_{11}, a_{21}, a_{31})$, $\beta^{T(1)} = (a_{12}, a_{22}, a_{32})$, $\gamma^{T(1)} = (a_{13}, a_{23}, a_{33})$, $\alpha^{T(2)} = (b_{11}, b_{21}, b_{31})$, $\beta^{T(2)} = (b_{12}, b_{22}, b_{32})$, $\gamma^{T(2)} = (b_{13}, b_{23}, b_{33})$, are written as:

$$[A_1 + A_2b_{11}^2 + B_2b_{21}^2 + b_{31}(C_2b_{31} + 2(a_2b_{23} - b_2b_{13})m_2s_2)$$
$$+m_2(s_2(2(b_2b_{11} - a_2b_{21})b_{33} + 2c_2(b_{13}b_{21} - b_{11}b_{23}) + s_2)$$
$$+2(b_2(b_{12}b_{31} - b_{11}b_{32}) + a_2(b_{21}b_{32} - b_{22}b_{31}) + c_2(b_{11}b_{22} - b_{12}b_{21}))s_3 + s_3^2)]\dot p_1$$
$$+[A_2b_{11} + m_2((b_2b_{32} - c_2b_{23})s_2 + (c_2b_{22} - b_2b_{32})s_3)]\dot p_2$$

$$+[A_2b_{11}b_{12} + B_2b_{21}b_{22} + C_2b_{31}b_{32} + (b_2b_{13} - a_2b_{23})b_{31}m_2s_1$$
$$+(a_2b_{21}b_{32} - b_2b_{11}b_{32} + c_2(b_{11}b_{23} - b_{13}b_{21}))m_2s_1 - m_2(b_2(b_{13}b_{32} - b_{12}b_{32})$$
$$+a_2(b_{22}b_{32} - b_{23}b_{32}) + c_2(b_{12}b_{23} - b_{13}b_{22}) + s_1)s_2]\,\dot{q}_1$$
$$+[B_2b_{21} + m_2((c_2b_{13} - a_2b_{32})s_2 + (a_2b_{32} - c_2b_{12})s_3)]\,\dot{q}_2$$
$$+[A_2b_{11}b_{13} + B_2b_{21}b_{23} + C_2b_{31}b_{32} + (a_2b_{22}b_{31} + b_2(b_{11}b_{32} - b_{12}b_{31}))\,m_2s_1$$
$$-(a_2b_{21}b_{32} + c_2(b_{11}b_{22} - b_{12}b_{21}))m_2s_1 - m_2(b_2(b_{13}b_{32} - b_{12}b_{32})$$
$$+a_2(b_{22}b_{32} - b_{23}b_{32}) + c_2(b_{12}b_{23} - b_{13}b_{22}) + s_1)s_3]\,\dot{r}_1$$
$$+[C_2b_{31} + m_2((a_2b_{23} - b_2b_{13})s_2 + (b_2b_{12} - a_2b_{22})s_3)]\,\dot{r}_2 + \Phi_1 = 0,$$
$$[A_2b_{11}b_{12} + B_2b_{21}b_{22} + C_2b_{31}b_{32} + m_2s_1(b_2b_{13} - a_2b_{23})b_{31}$$
$$+m_2s_1\,(a_2b_{21}b_{33} - b_2b_{11}b_{33} + c_2(b_{11}b_{23} - b_{13}b_{21}))$$
$$-m_2s_2\,(b_2(b_{13}b_{32} - b_{12}b_{33}) + a_2(b_{22}b_{33} - b_{23}b_{32})$$
$$+c_2(b_{12}b_{23} - b_{13}b_{22}) + s_1)]\,\dot{p}_1$$
$$+[A_2b_{12} + m_2(s_1(c_2b_{23} - b_2b_{33}) + s_3(b_2b_{31} - c_2b_{21}))]\,\dot{p}_2$$
$$+[B_1 + A_2b_{12}^2 + B_2b_{22}^2 + b_{32}(C_2b_{32} + 2m_2s_1\,(b_{13}b_2 - a_2b_{23}))$$
$$+m_2(s_1(2(a_2b_{22}b_{33} - b_2b_{12}b_{33} + c_2(b_{12}b_{23} - b_{13}b_{22})) + s_1)$$
$$+2(b_2(b_{12}b_{31} - b_{11}b_{32}) + a_2(b_{21}b_{32} - b_{22}b_{31}) + c_2(b_{11}b_{22} - b_{12}b_{21}))s_3 + s_3^2)]\,\dot{q}_1$$
$$+[B_2b_{22} + m_2(s_1(a_2b_{33} - c_2b_{13}) + s_3(b_{11}c_2 - a_2b_{31}))]\,\dot{q}_2$$
$$+[A_2b_{12}b_{13} + B_2b_{22}b_{23} + C_2b_{32}b_{33} + m_2s_2(b_2(b_{11}b_{32} - b_{12}b_{31})$$
$$+a_2(b_{22}b_{31} - b_{21}b_{32}) + c_2(b_{12}b_{21} - b_{11}b_{22})) - m_2s_3(b_2(b_{11}b_{33} - b_{13}b_{31})$$
$$+a_2(b_{23}b_{31} - b_{21}b_{33}) + c_2(b_{13}b_{21} - b_{11}b_{23}) + s_2)]\,\dot{r}_1$$
$$+[C_2b_{32} + m_2(s_1(b_2b_{13} - a_2b_{23}) + s_3(a_2b_{21} - b_2b_{11}))]\,\dot{r}_2 + \Phi_2 = 0,$$
$$[A_2b_{11}b_{13} + B_2b_{21}b_{23} + C_2b_{31}b_{33} + m_2s_1\,(a_2b_{22}b_{31} + b_2(b_{11}b_{32} - b_{12}b_{31}))$$
$$-m_2s_1\,(a_2b_{21}b_{32} + c_2(b_{11}b_{22} - b_{12}b_{21})) - m_2s_3\,(b_2(b_{13}b_{32} - b_{12}b_{33})$$
$$+a_2(b_{22}b_{33} - b_{23}b_{32}) + c_2(b_{12}b_{23} - b_{13}b_{22}) + s_1)]\,\dot{p}_1$$
$$+[A_2b_{13} + m_2(s_1(b_2b_{32} - c_2b_{22}) + s_2(c_2b_{21} - b_2b_{31}))]\,\dot{p}_2$$
$$+[A_2b_{12}b_{13} + B_2b_{22}b_{23} + C_2b_{32}b_{33} + m_2s_2\,(b_2(b_{11}b_{32} - b_{12}b_{31})$$
$$+a_2(b_{22}b_{31} - b_{21}b_{32}) + c_2(b_{12}b_{21} - b_{11}b_{22})) - m_2s_3\,(b_2(b_{11}b_{33} - b_{13}b_{31})$$
$$+a_2(b_{23}b_{31} - b_{21}b_{33}) + c_2(b_{13}b_{21} - b_{11}b_{23}) + s_2)]\,\dot{q}_1$$
$$+[B_2b_{23} + m_2(s_1(c_2b_{12} - a_2b_{32}) + s_2(a_2b_{31} - c_2b_{11}))]\,\dot{q}_2$$
$$+[C_1 + A_2b_{13}^2 + B_2b_{23}^2 + C_2b_{33}^2 + m_2(s_1(2(b_2(b_{13}b_{32} - b_{12}b_{33})$$
$$+a_2(b_{22}b_{33} - b_{23}b_{32}) + c_2(b_{12}b_{23} - b_{13}b_{22})) + s_1) + 2(b_2(b_{11}b_{33} - b_{13}b_{31})$$
$$+a_2(b_{23}b_{31} - b_{21}b_{33}) + c_2(b_{13}b_{21} - b_{11}b_{23}))s_2 + s_2^2)]\,\dot{r}_1$$
$$+[C_2b_{33} + m_2(s_1(a_2b_{22} - b_2b_{12}) + s_2(b_2b_{11} - a_2b_{21}))]\,\dot{r}_2 + \Phi_3 = 0,$$
$$[A_2b_{11} + m_2(s_2(b_2b_{33} - c_2b_{23}) + s_3(c_2b_{22} - b_2b_{32}))]\,\dot{p}_1 + A_2\dot{p}_2$$
$$+[A_2b_{12} + m_2(s_1(c_2b_{23} - b_2b_{33}) + s_3(b_2b_{31} - c_2b_{21}))]\,\dot{q}_1$$
$$+[A_2b_{13} + m_2(s_1(b_2b_{32} - c_2b_{22}) + s_2(c_2b_{21} - b_2b_{31}))]\,\dot{r}_1 + \Phi_4 = 0,$$
$$[B_2b_{21} + m_2(s_2(c_2b_{13} - a_2b_{33}) + s_3(a_2b_{32} - c_2b_{12}))]\,\dot{p}_1$$
$$+[B_2b_{22} + m_2(s_1(a_2b_{33} - c_2b_{13}) + s_3(c_2b_{11} - a_2b_{31}))]\,\dot{q}_1 + B_2\dot{q}_2$$
$$+[B_2b_{23} + m_2(s_1(c_2b_{12} - a_2b_{32}) + s_2(a_2b_{31} - c_2b_{11}))]\,\dot{r}_1 + \Phi_5 = 0,$$
$$[C_2b_{31} + m_2(s_2(a_2b_{23} - b_2b_{13}) + s_3(b_2b_{12} - a_2b_{22}))]\,\dot{p}_1$$
$$+[C_2b_{32} + m_2(s_1(b_2b_{13} - a_2b_{23}) + s_3(a_2b_{21} - b_2b_{11}))]\,\dot{q}_1$$
$$+[C_2b_{33} + m_2(s_1(a_2b_{22} - b_2b_{12}) + s_2(b_2b_{11} - a_2b_{21}))]\,\dot{r}_1 + C_2\dot{r}_2 + \Phi_6 = 0;$$

$$(2)$$

$$\dot{a}_{11} = a_{21}r_1 - a_{31}q_1,\quad \dot{a}_{12} = a_{22}r_1 - a_{32}q_1,\quad \dot{a}_{13} = a_{23}r_1 - a_{33}q_1,$$
$$\dot{a}_{21} = a_{31}p_1 - a_{11}r_1,\quad \dot{a}_{22} = a_{32}p_1 - a_{12}r_1,\quad \dot{a}_{23} = a_{33}p_1 - a_{13}r_1, \qquad (3)$$
$$\dot{a}_{31} = a_{11}q_1 - a_{21}p_1,\quad \dot{a}_{32} = a_{12}q_1 - a_{22}p_1,\quad \dot{a}_{33} = a_{13}q_1 - a_{23}p_1,$$

$$\dot{b}_{11} = b_{21}r_1 - b_{31}q_1, \ \dot{b}_{12} = b_{22}r_1 - b_{32}q_1, \ \dot{b}_{13} = b_{23}r_1 - b_{33}q_1,$$
$$\dot{b}_{21} = b_{31}p_1 - b_{11}r_1, \ \dot{b}_{22} = b_{32}p_1 - b_{12}r_1, \ \dot{b}_{23} = b_{33}p_1 - b_{13}r_1, \qquad (4)$$
$$\dot{b}_{31} = b_{11}q_1 - b_{21}p_1, \ \dot{b}_{32} = b_{12}q_1 - b_{22}p_1, \ \dot{b}_{33} = b_{13}q_1 - b_{23}p_1.$$

Here Φ_i $(i = 1, \ldots, 6)$ are the quadratic polynomials of p_j, q_j, r_j $(j = 1, 2)$. These are rather cumbersome and are given in Appendix.

Equations (2)–(4) admit the following first integrals.

- The integrals of energy and kinetic moment:

$$H = T - U = h, \ V = \frac{\partial L}{\partial \omega^{(1)}} \cdot \gamma^{(1)} = c,$$

  where h and c are some constants.
- The geometric integrals:

$$
\begin{aligned}
V_1 &= a_{11}^2 + a_{21}^2 + a_{31}^2 = 1, & V_7 &= b_{11}^2 + b_{21}^2 + b_{31}^2 = 1, \\
V_2 &= a_{12}^2 + a_{22}^2 + a_{32}^2 = 1, & V_8 &= b_{12}^2 + b_{22}^2 + b_{32}^2 = 1, \\
V_3 &= a_{13}^2 + a_{23}^2 + a_{33}^2 = 1, & V_9 &= b_{13}^2 + b_{23}^2 + b_{33}^2 = 1, \\
V_4 &= a_{11}a_{12} + a_{21}a_{22} + a_{31}a_{32} = 0, & V_{10} &= b_{11}b_{12} + b_{21}b_{22} + b_{31}b_{32} = 0, \\
V_5 &= a_{11}a_{13} + a_{21}a_{23} + a_{31}a_{33} = 0, & V_{11} &= b_{11}b_{13} + b_{21}b_{23} + b_{31}b_{33} = 0, \\
V_6 &= a_{12}a_{13} + a_{22}a_{23} + a_{32}a_{33} = 0, & V_{12} &= b_{12}b_{13} + b_{22}b_{23} + b_{32}b_{33} = 0.
\end{aligned}
\qquad (5)
$$

We pose the problem of finding the stationary solutions and invariant manifolds (IMs) [4] for Eqs. (2)–(4) and the investigation of their stability.

3 Finding Stationary Solutions and IMs

It is shown [4] that stationary solutions and IMs can be obtained from both the stationary conditions of the problem's first integrals and, directly, from the equations of motion. Let us use the first technique.

3.1 The Usage of Stationary Conditions

Compose the linear combination of the integrals

$$2\Omega = 2\lambda_0 H - \lambda_1 V_3 - \lambda_2 V_7 - \lambda_3 V_8 - \lambda_4 V_9 - 2(\lambda_5 V_{10} + \lambda_6 V_{11} + \lambda_7 V_{12})$$

and write down the necessary conditions of extremum for the integral Ω with respect to the variables $p_1, p_2, q_1, q_2,\ r_1, r_2, a_{13}, a_{23}, a_{33}, b_{11}, b_{12}, b_{13}, b_{21}, b_{22}, b_{23}, b_{31}, b_{32}, b_{33}$:

$$
\begin{aligned}
&\partial\Omega/\partial p_i = 0, \ \partial\Omega/\partial q_i = 0, \ \partial\Omega/\partial r_i = 0 \ (i = 1, 2), \\
&\partial\Omega/\partial a_{k3} = 0, \ \partial\Omega/\partial b_{kl} = 0 \ (k, l = 1, 2, 3).
\end{aligned}
\qquad (6)
$$

The variables $a_{11}, a_{12}, a_{21}, a_{22}, a_{31}, a_{32}$ are sought with the help of the integrals V_1, V_2, V_j $(j = 4, 5, 6)$ under the corresponding values of a_{13}, a_{23}, a_{33}.

Solutions of Eqs. (6) allow us to determine the stationary solutions and IMs of the differential equations (2)–(4) corresponding to the integral Ω. As was mentioned before, we are interested in equilibrium solutions. In order to find them, we put $p_i = q_i = r_i = 0$ $(i = 1, 2)$ in (6). The equations take the form:

$$
\begin{aligned}
&\lambda_0 g\,(a_1 m_1 + m_2(a_2 b_{11} + b_2 b_{21} + c_2 b_{31} + s_1)) - \lambda_1 a_{13} = 0, \\
&\lambda_0 g\,(b_1 m_1 + m_2(a_2 b_{12} + b_2 b_{22} + c_2 b_{32} + s_2)) - \lambda_1 a_{23} = 0, \\
&\lambda_0 g(c_1 m_1 + m_2(a_2 b_{13} + b_2 b_{23} + c_2 b_{33} + s_3)) - \lambda_1 a_{33} = 0, \\
&\lambda_0 a_2 g\, m_2 a_{13} - \lambda_2 b_{11} - \lambda_5 b_{12} - \lambda_6 b_{13} = 0, \\
&\lambda_0 a_2 g\, m_2 a_{23} - \lambda_3 b_{12} - \lambda_5 b_{11} - \lambda_7 b_{13} = 0, \\
&\lambda_0 a_2 g\, m_2 a_{33} - \lambda_4 b_{13} - \lambda_6 b_{11} - \lambda_7 b_{12} = 0, \\
&\lambda_0 b_2 g\, m_2 a_{13} - \lambda_2 b_{21} - \lambda_5 b_{22} - \lambda_6 b_{23} = 0, \\
&\lambda_0 b_2 g\, m_2 a_{23} - \lambda_3 b_{22} - \lambda_5 b_{21} - \lambda_7 b_{23} = 0, \\
&\lambda_0 b_2 g\, m_2 a_{33} - \lambda_4 b_{23} - \lambda_6 b_{21} - \lambda_7 b_{22} = 0, \\
&\lambda_0 c_2 g\, m_2 a_{13} - \lambda_2 b_{31} - \lambda_5 b_{32} - \lambda_6 b_{33} = 0, \\
&\lambda_0 c_2 g\, m_2 a_{23} - \lambda_3 b_{32} - \lambda_5 b_{31} - \lambda_7 b_{33} = 0, \\
&\lambda_0 c_2 g\, m_2 a_{33} - \lambda_4 b_{33} - \lambda_6 b_{31} - \lambda_7 b_{32} = 0.
\end{aligned}
\tag{7}
$$

The resulting system is multiparametric (20 parameters) that leads to cumbersome expressions in the process of computation.

For the polynomials of system (7), under the following constraints on the parameters of the problem

$$
a_1 = -\frac{m_2 s_1}{m_1}, \; b_1 = -\frac{m_2 s_2}{m_1}, \; c_1 = -\frac{m_2 s_3}{m_1}
$$

and the relations $b_{21} = b_{12}, b_{31} = b_{13}, b_{32} = b_{23}$ (we assume that the cosine matrix β is symmetric), a Gröbner basis was constructed with respect to an eliminating ordering of the variables with the help of the built-in function

$GroebnerBasis[polys_1, \{b_{11}, b_{33}\}, \{\lambda_2, \lambda_3, \lambda_4, \lambda_5, \lambda_6, \lambda_7, a_{13}, a_{23}, a_{33}\},$

CoefficientDomain $\rightarrow$ RationalFunctions,

MonomialOrder $\rightarrow$ EliminationOrder]

Here $polys_1$ is the list of the polynomials of system (7).

After the transformation of the basis into a lexicographical one by means of

$GroebnerBasis[polys_2, \{\lambda_2, \lambda_3, \lambda_4, \lambda_5, \lambda_6, \lambda_7, a_{13}, a_{23}, a_{33}, b_{11}, b_{33}\},$

CoefficientDomain $\rightarrow$ RationalFunctions],

where $polys_2$ is the list of polynomials obtained at the previous step, we have a system dividing into two subsystems. One of them is presented below.

$$
\begin{aligned}
&c_2(a_2 b_{12} + b_2(b_{22} - b_{33})) + c_2^2 b_{23} - b_2(a_2 b_{13} + b_2 b_{23}) = 0, \\
&a_2^2 b_{12} - b_2(b_{12} b_2 + b_{13} c_2) + a_2(b_2(b_{22} - b_{11}) + c_2 b_{23}) = 0, \\
&\lambda_1 b_2 a_{33} - \lambda_0 g m_2 c_2(a_2 b_{12} + b_2 b_{22} + c_2 b_{23}) = 0, \\
&\lambda_0 g m_2\,(a_2 b_{12} + b_2 b_{22} + b_{23} c_2) - a_{23} \lambda_1 = 0, \\
&\lambda_0 g m_2 a_2(a_2 b_{12} + b_2 b_{22} + c_2 b_{23}) - \lambda_1 b_2 a_{13} = 0, \\
&b_2 c_2 g^2 m_2^2 \lambda_0^2 - \lambda_1 \lambda_7 = 0, \; a_2 c_2 g^2 m_2^2 \lambda_0^2 - \lambda_1 \lambda_6 = 0,
\end{aligned}
\tag{8}
$$

$$a_2 b_2 g^2 \lambda_0^2 m_2^2 - \lambda_1 \lambda_5 = 0, \quad c_2^2 g^2 m_2^2 \lambda_0^2 - \lambda_1 \lambda_4 = 0,$$
$$b_2^2 g^2 \lambda_0^2 m_2^2 - \lambda_1 \lambda_3 = 0, \quad a_2^2 g^2 m_2^2 \lambda_0^2 - \lambda_1 \lambda_2 = 0.$$

All computations have been performed on a computer with an Intel Core i7 CPU (3.6 GHz) and 32 GB of RAM. The total computation time is 46 s.

Then, expressions (5) are added to the first five equations of system (8) and a lexicographical basis for the polynomials of the resulting system with respect to $a_{11} > a_{12} > a_{21} > a_{22} > a_{31} > a_{13} > a_{23} > a_{33} > b_{11} > b_{33} > b_{13} > b_{22} > b_{23} > \lambda_1$ is constructed. Again we obtain a system splitting into two subsystems. Below, both subsystems are represented.

$$(a_2^2 + b_2^2 + c_2^2) g^2 m_2^2 \lambda_0^2 - \lambda_1^2 = 0,$$
$$a_2^2 b_{12}^2 + b_2^2 (b_{12}^2 + b_{23}^2) + c_2^2 b_{23}^2 + 2(a_2 c_2 b_{12} b_{23} \mp a_2 b_2 b_{12} \mp b_2 c_2 b_{23}) = 0,$$
$$-a_2 b_{12} - b_2 (b_{22} \mp 1) - c_2 b_{23} = 0,$$
$$a_2^2 (b_2 b_{13} + c_2 b_{12}) + b_2^2 (a_2 b_{23} + c_2 b_{12}) + c_2^2 (a_2 b_{23} + b_2 b_{13}) \mp 2 a_2 b_2 c_2 = 0,$$
$$-a_2^2 (a_2 b_{12} - b_2 (b_{33} \pm 1) + c_2 b_{23}) + b_2^2 (c_2 b_{23} - a_2 b_{12}) + b_2 c_2^2 (b_{33} \mp 1) = 0,$$
$$-a_2^2 b_2 (b_{11} \mp 1) + b_2^2 (c_2 b_{23} - a_2 b_{12}) + c_2^2 (a_2 b_{12} - b_2 (b_{11} \pm 1) + c_2 b_{23}) = 0,$$
$$(a_2^2 + b_2^2 + c_2^2) g m_2 \lambda_0 a_{33} \mp c_2 \lambda_1 = 0,$$
$$\pm b_2 \lambda_1 - (a_2^2 + b_2^2 + c_2^2) g m_2 \lambda_0 a_{23} = 0, \tag{9}$$
$$\pm a_2 \lambda_1 - (a_2^2 + b_2^2 + c_2^2) g m_2 \lambda_0 a_{13} = 0,$$
$$(a_2^2 + b_2^2)(a_{31}^2 + a_{32}^2 - 1) + c_2^2 (a_{31}^2 + a_{32}^2) = 0,$$
$$a_2^2 (a_{22}^2 + a_{32}^2 - 1) + (b_2 a_{22} + c_2 a_{32})^2 = 0,$$
$$(a_2^2 + b_2^2)(a_{21} + a_{32}(a_{22} a_{31} - a_{21} a_{32})) + b_2 c_2 a_{31} + c_2^2 a_{32}(a_{22} a_{31} - a_{21} a_{32}) = 0,$$
$$a_2 a_{12} + b_2 a_{22} + c_2 a_{32} = 0,$$
$$a_2 (a_2^2 a_{11} + a_2 c_2 a_{31} + b_2^2 a_{11})(a_{32}^2 - 1) + [b_2 (a_2^2 + b_2^2) a_{22} a_{31} + c_2^2 (a_2 a_{11} a_{32}$$
$$+ b_2 a_{22} a_{31}) + c_2 (b_2^2 a_{31} + c_2^2 a_{31}) a_{32}] a_{32} = 0.$$

Next, we find $\lambda_1 = \pm \sqrt{a_2^2 + b_2^2 + c_2^2}\, m_2 g \lambda_0$ from the first equation of (9). Under the above values of λ_1, the latter 13 equations of each of the subsystems together with equations $p_i = q_i = r_i = 0$ $(i = 1, 2)$ and the relations $b_{21} = b_{12}$, $b_{31} = b_{13}, b_{32} = b_{23}$ determine the four IMs of codimension 22 of the differential equations (2)–(4). It is verified by direct computation according to the definition of IM. Let us consider one of these IMs, e.g., the one defined by the equations

$$p_1 = 0,\ p_2 = 0,\ q_1 = 0,\ q_2 = 0,\ r_1 = 0,\ r_2 = 0,$$
$$b_{21} - b_{12} = 0,\ b_{31} - b_{13} = 0, b_{32} - b_{23} = 0,$$
$$a_2^2 b_{12}^2 + b_2^2 (b_{12}^2 + b_{23}^2) + c_2^2 b_{23}^2 + 2(a_2 c_2 b_{12} b_{23} + a_2 b_2 b_{12} + b_2 c_2 b_{23}) = 0,$$
$$-a_2 b_{12} - b_2 (b_{22} + 1) - c_2 b_{23} = 0,$$
$$a_2^2 (b_2 b_{13} + c_2 b_{12}) + b_2^2 (a_2 b_{23} + c_2 b_{12}) + c_2^2 (a_2 b_{23} + b_2 b_{13}) + 2 a_2 b_2 c_2 = 0,$$
$$-a_2^2 (a_2 b_{12} - b_2 (b_{33} - 1) + c_2 b_{23}) + b_2^2 (c_2 b_{23} - a_2 b_{12}) + b_2 c_2^2 (b_{33} + 1) = 0, \tag{10}$$
$$-a_2^2 b_2 (b_{11} + 1) + b_2^2 (c_2 b_{23} - a_2 b_{12}) + c_2^2 (a_2 b_{12} - b_2 (b_{11} - 1) + c_2 b_{23}) = 0,$$
$$\sqrt{a_2^2 + b_2^2 + c_2^2}\, a_{33} + c_2 = 0,$$
$$\sqrt{a_2^2 + b_2^2 + c_2^2}\, a_{23} + b_2 = 0,$$
$$\sqrt{a_2^2 + b_2^2 + c_2^2}\, a_{13} + a_2 = 0,$$
$$(a_2^2 + b_2^2)(a_{31}^2 + a_{32}^2 - 1) + c_2^2 (a_{31}^2 + a_{32}^2) = 0,$$

$$a_2^2(a_{22}^2 + a_{32}^2 - 1) + (b_2 a_{22} + c_2 a_{32})^2 = 0,$$
$$(a_2^2 + b_2^2)(a_{21} + a_{32}(a_{22}a_{31} - a_{21}a_{32})) + b_2 c_2 a_{31} + c_2^2 a_{32}(a_{22}a_{31} - a_{21}a_{32}) = 0,$$
$$a_2 a_{12} + b_2 a_{22} + c_2 a_{32} = 0,$$
$$a_2(a_2^2 a_{11} + a_2 c_2 a_{31} + b_2^2 a_{11})(a_{32}^2 - 1) + [b_2(a_2^2 + b_2^2)a_{22}a_{31} + c_2^2(a_2 a_{11}a_{32}$$
$$+ b_2 a_{22}a_{31}) + c_2(b_2^2 a_{31} + c_2^2 a_{31})a_{32}]a_{32} = 0.$$

The differential equations $\dot{a}_{32} = 0$, $\dot{b}_{12} = 0$ on this IM have the following family of solutions:

$$a_{32} = a_{32}^0 = \text{const}, \; b_{12} = b_{12}^0 = \text{const}. \tag{11}$$

From a geometric point of view, Eqs. (10) in the space R^{24} define a two-dimensional surface whose points correspond to the fixed points of the phase space of the system under study.

Next, let us find λ_j $(j = 2, \ldots, 7)$ from the latter six equations of (8) when $\lambda_1 = \sqrt{a_2^2 + b_2^2 + c_2^2}\, m_2 g \lambda_0$ and substitute them into the integral Ω. Having added a combination of the integrals V_1, V_2 to the resulting expression, we have:

$$\Omega_1 = \lambda_0 \Big[H - \frac{1}{2}\sqrt{a_2^2 + b_2^2 + c_2^2}\, g m_2 V_3 - \frac{g m_2 (a_2^2 V_7 + b_2^2 V_8 + c_2^2 V_9)}{2\sqrt{a_2^2 + b_2^2 + c_2^2}}$$
$$- \frac{g m_2 (a_2 b_2 V_{10} + a_2 c_2 V_{11} + b_2 c_2 V_{12})}{\sqrt{a_2^2 + b_2^2 + c_2^2}} \Big] - \lambda_8 V_1^2 - \lambda_9 V_2^2. \tag{12}$$

Using the maps of an atlas on IM (10), e.g.,

$$p_1 = 0, \; p_2 = 0, \; q_1 = 0, \; q_2 = 0, \; r_1 = 0, \; r_2 = 0,$$
$$a_{11} = \frac{a_2 c_2 z_1 \pm b_2 z a_{32}}{(a_2^2 + b_2^2)\, z^{1/2}}, \; a_{12} = -\frac{a_2 c_2 a_{32} \mp b_2 z_1}{a_2^2 + b_2^2}, a_{13} = -a_2 z^{-1/2},$$
$$a_{21} = \frac{b_2 c_2 z_1 \mp a_2 z a_{32}}{(a_2^2 + b_2^2)\, z^{1/2}}, \; a_{22} = -\frac{b_2 c_2 a_{32} \pm a_2 z_1}{a_2^2 + b_2^2}, \; a_{23} = -b_2 z^{-1/2},$$
$$a_{31} = -z_1 z^{-1/2}, \; a_{33} = -c_2 z^{-1/2},$$
$$b_{11} = -\frac{a_2(a_2 + b_2 b_{12}) + c_2 z_2}{a_2^2 + c_2^2}, \; b_{13} = \frac{a_2 z_2 - (a_2 + b_2 b_{12})\, c_2}{a_2^2 + c_2^2},$$
$$b_{21} = b_{12}, \; b_{22} = \frac{c_2 z_2 - b_2(a_2 b_{12} + b_2)}{b_2^2 + c_2^2}, \; b_{23} = -\frac{a_2 c_2 b_{12} + b_2(c_2 + z_2)}{b_2^2 + c_2^2},$$
$$b_{31} = \frac{a_2 z_2 - (a_2 + b_2 b_{12})\, c_2}{a_2^2 + c_2^2}, \; b_{32} = -\frac{a_2 c_2 b_{12} + b_2(c_2 + z_2)}{b_2^2 + c_2^2},$$
$$b_{33} = \frac{1}{(a_2^2 + c_2^2)(b_2^2 + c_2^2)}[a_2^2(b_2(b_2 + a_2 b_{12}) - c_2 z_2)$$
$$+ b_2(a_2(b_2^2 + 2c_2^2)b_{12} + b_2 c_2 z_2) - c_2^4],$$

it is not difficult to show that the integral Ω_1 takes a stationary value on this IM. Here $z_1 = [(1 - a_{32}^2)(a_2^2 + b_2^2) - c_2^2 a_{32}^2]^{1/2}$, $z_2 = [c_2^2 - b_{12}(a_2^2 b_{12} + 2a_2 b_2 + (b_2^2 + c_2^2)b_{12})]^{1/2}$, $z = a_2^2 + b_2^2 + c_2^2$.

Equations (10) together with (11) allow one to obtain up to the eight families of solutions of the equations of motion (2)–(4). We represent one of them, e.g.,

$$p_1 = 0, \; p_2 = 0, \; q_1 = 0, \; q_2 = 0, \; r_1 = 0, \; r_2 = 0,$$

$$a_{11} = \frac{a_2 c_2 D_1 + a_{32}^0 b_2 D}{(a_2^2 + b_2^2) D^{1/2}}, \quad a_{12} = -\frac{a_2 a_{32}^0 c_2 - b_2 D_1}{a_2^2 + b_2^2}, \quad a_{13} = -a_2 D^{-1/2},$$

$$a_{21} = \frac{b_2 c_2 D_1 - a_2 a_{32}^0 D}{(a_2^2 + b_2^2) D^{1/2}}, \quad a_{22} = -\frac{a_{32}^0 b_2 c_2 + a_2 D_1}{a_2^2 + b_2^2}, \quad a_{23} = -b_2 D^{-1/2},$$

$$a_{31} = -D_1 D^{-1/2}, \quad a_{32} = a_{32}^0, \quad a_{33} = -c_2 D^{-1/2},$$

$$b_{11} = -\frac{a_2(a_2 + b_2 b_{12}^0) + c_2 D_2}{a_2^2 + c_2^2}, \quad b_{12} = b_{12}^0, \quad b_{13} = \frac{a_2 D_2 - (a_2 + b_2 b_{12}^0) c_2}{a_2^2 + c_2^2},$$

$$b_{21} = b_{12}^0, \quad b_{22} = \frac{c_2 D_2 - b_2(a_2 b_{12}^0 + b_2)}{b_2^2 + c_2^2}, \quad b_{23} = -\frac{a_2 c_2 b_{12}^0 + b_2(c_2 + D_2)}{b_2^2 + c_2^2},$$

$$b_{31} = \frac{a_2 D_2 - (a_2 + b_{12}^0 b_2) c_2}{a_2^2 + c_2^2}, \quad b_{32} = -\frac{a_2 c_2 b_{12}^0 + b_2(c_2 + D_2)}{b_2^2 + c_2^2},$$

$$b_{33} = \frac{1}{(a_2^2 + c_2^2)(b_2^2 + c_2^2)} [a_2^2(b_2(b_2 + a_2 b_{12}^0) - c_2 D_2)$$
$$+ b_2(a_2(b_2^2 + 2c_2^2) b_{12}^0 + b_2 c_2 D_2) - c_2^4]. \tag{13}$$

The rest of the solutions differs from the above by the signs of the expressions. Here a_{32}^0 and b_{12}^0 are the parameters of the families, $D_1 = [(1 - a_{32}^{0^2})(a_2^2 + b_2^2) - a_{32}^{0^2} c_2^2]^{1/2}$, $D_2 = [c_2^2 - b_{12}^0(a_2^2 b_{12}^0 + 2 a_2 b_2 + b_{12}^0(b_2^2 + c_2^2))]^{1/2}$, $D = a_2^2 + b_2^2 + c_2^2$. Solutions (13) are real, in particular, when the following conditions hold:

$$a_2 \neq 0, \; b_2 \neq 0, \; c_2 \neq 0 \text{ and } -\sigma_1 \leq a_{32}^0 \leq \sigma_1 \text{ and } -\sigma_2 \leq b_{12}^0 \leq \sigma_2,$$

where $\sigma_1 = \sqrt{(a_2^2 + b_2^2) D^{-1}}$, $\sigma_2 = \sqrt{(a_2^2 + c_2^2)(b_2^2 + c_2^2)} D^{-1}$, $\sigma = a_2 b_2 D^{-1}$.

It is easy to verify by direct computation that the integral Ω_1 also takes a stationary value on the elements of the family of solutions (13). From a mechanical point of view, the elements of this family correspond to equilibria of the mechanical system under consideration.

By means of relations (1), the family of solutions (13) can be represented in the initial variables (Euler's angles). Since these solutions are rather cumbersome we give here the expressions obtained under some constraints on the parameters of the problem. For instance, under the following conditions

$$b_{12}^0 = -\frac{2 a_2 b_2}{2 a_2^2 + b_2^2}, \quad c_2 = a_2,$$

the one-parametric families of solutions correspond to the family of solutions (13) in the initial variables:

$$\varphi_1 = \pm \arccos \left(\pm \frac{b_2}{\sqrt{a_2^2 + b_2^2}} \right), \quad \psi_1 = \pm \arccos \left(\pm \frac{a_{32}^0 \sqrt{2 a_2^2 + b_2^2}}{\sqrt{a_2^2 + b_2^2}} \right),$$

$$\theta_1 = \mp \arccos\left(-\frac{a_2}{\sqrt{2a_2^2 + b_2^2}}\right), \quad \varphi_2 = \mp \arccos\left(\pm\frac{2a_2}{\sqrt{4a_2^2 + b_2^2}}\right),$$

$$\psi_2 = \mp \arccos\left(\mp\frac{2a_2}{\sqrt{4a_2^2 + b_2^2}}\right), \quad \theta_2 = \mp \arccos\left(-\frac{2a_2^2}{2a_2^2 + b2^2}\right), \tag{14}$$

where a_{32}^0 is the parameter of the families.

When $b_{12}^0 = \frac{2a_2b_2}{a_2^2 + 2b_2^2}$, $c_2 = b_2$, we have the families of solutions:

$$\varphi_1 = \pm \arccos\left(\pm\frac{b_2}{\sqrt{a_2^2 + b_2^2}}\right), \quad \psi_1 = \pm \arccos\left(\pm\frac{a_{32}^0\sqrt{a_2^2 + 2b_2^2}}{\sqrt{a_2^2 + b_2^2}}\right),$$

$$\theta_1 = \mp \arccos\left(-\frac{a_2}{\sqrt{a_2^2 + 2b_2^2}}\right), \quad \varphi_2 = \mp \arccos\left(\pm\frac{a_2^2 + b_2^2}{\sqrt{(a_2^2 + b_2^2)^2 + a_2^2 b_2^2}}\right),$$

$$\psi_2 = \mp \arccos\left(\mp\frac{a_2^2 + b_2^2}{\sqrt{(a_2^2 + b_2^2)^2 + a_2^2 b_2^2}}\right),$$

$$\theta_2 = \mp \arccos\left(a_2^2\left(\frac{1}{a_2^2 + 2b_2^2} - \frac{2}{a_2^2 + b_2^2}\right)\right). \tag{15}$$

Substituting expressions (14) and (15) into the equations of motion (2)–(4) written in Euler's variables shows that these equations are identically satisfied.

3.2   The Usage of the Equations of Motion

When $p_i = q_i = r_i = 0$ $(i = 1, 2)$, the equations of motion (2) take the form:

$$
\begin{aligned}
&a_{33}(b_1 m_1 + m_2(a_2 b_{12} + b_2 b_{22} + c_2 b_{32} + s_2)) \\
&-a_{23}(c_1 m_1 + m_2(a_2 b_{13} + b_2 b_{23} + c_2 b_{33} + s_3)) = 0, \\
&a_{13}(c_1 m_1 + m_2(a_2 b_{13} + b_2 b_{23} + c_2 b_{33} + s_3)) \\
&-a_{33} m_2(a_2 b_{11} + b_2 b_{21} + c_2 b_{31} + s_1) - m_1 a_1 a_{33} = 0, \\
&a_{23}(a_1 m_1 + a_{23} m_2(a_2 b_{11} + b_2 b_{21} + c_2 b_{31} + s_1)) \\
&-a_{13}(b_1 m_1 + m_2(a_2 b_{12} + b_2 b_{22} + c_2 b_{32} + s_2)) = 0, \\
&b_2(a_{13} b_{31} + a_{23} b_{32} + a_{33} b_{33}) - c_2(a_{13} b_{21} + a_{23} b_{22} + a_{33} b_{23}) = 0, \\
&c_2(a_{13} b_{11} + a_{23} b_{12} + a_{33} b_{13}) - a_2(a_{13} b_{31} + a_{23} b_{32} + a_{33} b_{33}) = 0, \\
&a_2(a_{13} b_{21} + a_{23} b_{22} + a_{33} b_{23}) - b_2(a_{13} b_{11} + a_{23} b_{12} + a_{33} b_{13}) = 0.
\end{aligned} \tag{16}
$$

System (16) is multiparametric (11 parameters).

We assume that the elements of the cosine matrix β are related as follows:

$$
\begin{aligned}
&b_1 m_1 + m_2(a_2 b_{12} + b_2 b_{22} + c_2 b_{32} + s_2) = 0, \\
&a_1 m_1 + m_2(a_2 b_{11} + b_2 b_{21} + c_2 b_{31} + s_1) = 0, \\
&b_2 b_{31} - c_2 b_{21} = 0, \quad c_2 b_{12} - a_2 b_{32} = 0, \quad a_2 b_{22} - b_2 b_{12} = 0.
\end{aligned} \tag{17}
$$

Add relations (17) and (5) to Eqs. (16) and construct a lexicographical basis with respect to $a_{11} > a_{12} > a_{21} > a_{22} > a_{23} > a_{31} > a_{13} > a_{33} > b_{11} > b_{12} >$

$b_{13} > b_{21} > b_{22} > b_{23} > b_{31} > b_{32} > b_{33} > s_1 > s_2 > s_3$ for the polynomials of the resulting system. As a result, we have the system of equations dividing into two subsystems. Both subsystems are given below:

$$
\begin{aligned}
& -c_1 m_1 - m_2 s_3 = 0, \quad -a_1 m_1 - m_2 s_1 = 0, \\
& (b_1 m_1 + m_2 s_2)^2 - (a_2^2 + b_2^2 + c_2^2) m_2^2 = 0, \\
& b_2^2 - (b_2^2 + c_2^2) b_{33}^2 = 0, \\
& -b_1 c_2 m_1 - m_2((a_2^2 + b_2^2 + c_2^2) b_{32} + c_2 s_2) = 0, \\
& a_2^2 c_2^2 - (b_2^2 + c_2^2)(a_2^2 + b_2^2 + c_2^2) b_{31}^2 = 0, \\
& b_2 b_{23} + c_2 b_{33} = 0, \\
& b_1 b_2 m_1 + (a_2^2 + b_2^2 + c_2^2) m_2 b_{22} + b_2 m_2 s_2 = 0, \\
& b_2 b_{31} - c_2 b_{21} = 0, \quad b_{13} = 0, \\
& -a_2^2 b_{12} m_2 - b_{12}(b_2^2 + c_2^2) m_2 - a_2(b_1 m_1 + m_2 s_2) = 0, \\
& -a_2 c_2 b_{11} - (b_2^2 + c_2^2) b_{31} = 0, \\
& a_{33} = 0, \quad a_{13} = 0, \quad 1 - a_{31}^2 - a_{32}^2 = 0, \quad a_{23} \pm 1 = 0, \\
& a_{22} = 0, \quad a_{21} = 0, \quad a_{12}^2 + a_{32}^2 - 1 = 0, \\
& a_{11} + (a_{12} a_{31} - a_{11} a_{32}) a_{32} = 0.
\end{aligned}
\tag{18}
$$

From the first three equations of system (18), we find the constraints on the parameters of the problem

$$
s_1 = -\frac{a_1 m_1}{m_2}, \quad s_2 = -\frac{b_1 m_1 \pm \sqrt{a_2^2 + b_2^2 + c_2^2}\, m_2}{m_2}, \quad s_3 = -\frac{c_1 m_1}{m_2}
$$

under which the latter 17 equations of the system together with the relations $p_i = q_i = r_i = 0$ $(i = 1, 2)$ determine the four one-dimensional IMs of the equations of motion (2)–(4). It is easy to verify by direct computation according to the definition of IM. Below, the equations of one of these IMs are represented.

$$
\begin{aligned}
& p_1 = 0, \; p_2 = 0, \; q_1 = 0, \; q_2 = 0, \; r_1 = 0, \; r_2 = 0, \\
& b_2^2 - (b_2^2 + c_2^2) b_{33}^2 = 0, \\
& -b_1 c_2 m_1 - m_2(a_2^2 + b_2^2 + c_2^2) b_{32} + c_2(b_1 m_1 + \sqrt{a_2^2 + b_2^2 + c_2^2}\, m_2) = 0, \\
& a_2^2 c_2^2 - (b_2^2 + c_2^2)(a_2^2 + b_2^2 + c_2^2) b_{31}^2 = 0, \\
& b_2 b_{23} + c_2 b_{33} = 0, \\
& b_1 b_2 m_1 + (a_2^2 + b_2^2 + c_2^2) m_2 b_{22} - b_2(b_1 m_1 + \sqrt{a_2^2 + b_2^2 + c_2^2}\, m_2) = 0, \\
& b_2 b_{31} - c_2 b_{21} = 0, \quad b_{13} = 0, \\
& (a_2 \sqrt{a_2^2 + b_2^2 + c_2^2} - a_2^2 b_{12} - b_{12}(b_2^2 + c_2^2)) m_2 = 0, \\
& -a_2 c_2 b_{11} - (b_2^2 + c_2^2) b_{31} = 0, \\
& a_{33} = 0, \quad a_{13} = 0, \quad 1 - a_{31}^2 - a_{32}^2 = 0, \quad a_{23} - 1 = 0, \\
& a_{22} = 0, \quad a_{21} = 0, \quad a_{12}^2 + a_{32}^2 - 1 = 0, \\
& a_{11} + (a_{12} a_{31} - a_{11} a_{32}) a_{32} = 0.
\end{aligned}
\tag{19}
$$

The differential equation $\dot{a}_{32} = 0$ on IM (19) has the family of solutions:

$$
a_{32} = a_{32}^0 = \text{const.}
\tag{20}
$$

Thus, from a geometrical point of view, Eqs. (19) in the space R^{24} define a curve whose points correspond to the fixed points of the phase space of the system under study.

Applying the technique [5], we find the combination of the integrals

$$2\Omega_2 = 2\lambda_0 H - \lambda_2 V_1^2 - \lambda_3 V_2^2 - \lambda_4 V_3^2 - \lambda_5 V_7^2$$
$$-gm_2\lambda_0\sqrt{a_2^2 + b_2^2 + c_2^2}\, V_8 - \lambda_7 V_9^2$$

which takes a stationary value on IM (19). It can be verified by direct computation, using the maps of an atlas on this IM, e.g.,

$$p_1 = 0,\ p_2 = 0,\ q_1 = 0,\ q_2 = 0,\ r_1 = 0,\ r_2 = 0,$$
$$a_{11} = \mp a_{32},\ a_{12} = \mp\sqrt{1 - a_{32}^2},\ a_{13} = 0,\ a_{21} = 0,$$
$$a_{22} = 0,\ a_{23} = 1,\ a_{31} = -\sqrt{1 - a_{32}^2},\ a_{33} = 0,$$
$$b_{11} = \bar{D}\bar{D}_1,\ b_{12} = a_2\bar{D},\ b_{13} = 0,$$
$$b_{21} = -a_2b_2\bar{D}\bar{D}_1^{-1},\ b_{22} = b_2\bar{D},\ b_{23} = \mp c_2\bar{D}_1^{-1},$$
$$b_{31} = -a_2c_2\bar{D}\bar{D}_1^{-1},\ b_{32} = c_2\bar{D},\ b_{33} = \pm b_2\bar{D}_1^{-1}.$$

Here $\bar{D} = (a_2^2 + b_2^2 + c_2^2)^{-1/2}$, $\bar{D}_1 = (b_2^2 + c_2^2)^{1/2}$.

Equations (19) together with (20) allow one to obtain up to the eight families of solutions for the equations of motion (2)–(4). One of them is given by:

$$p_1 = 0,\ p_2 = 0,\ q_1 = 0,\ q_2 = 0,\ r_1 = 0,\ r_2 = 0,$$

$$a_{11} = -a_{32}^0,\ a_{12} = -\sqrt{1 - a_{32}^{0\,2}},\ a_{13} = 0,\ a_{21} = 0,\ a_{22} = 0,$$

$$a_{23} = 1,\ a_{31} = -\sqrt{1 - a_{32}^{0\,2}},\ a_{32} = a_{32}^0,\ a_{33} = 0,$$

$$b_{11} = \frac{\sqrt{b_2^2 + c_2^2}}{\sqrt{a_2^2 + b_2^2 + c_2^2}},\ b_{12} = \frac{a_2}{\sqrt{a_2^2 + b_2^2 + c_2^2}},\ b_{13} = 0,$$

$$b_{21} = -\frac{a_2b_2}{\sqrt{(b_2^2 + c_2^2)(a_2^2 + b_2^2 + c_2^2)}},\ b_{22} = \frac{b_2}{\sqrt{a_2^2 + b_2^2 + c_2^2}},$$

$$b_{23} = -\frac{c_2}{\sqrt{b_2^2 + c_2^2}},\ b_{31} = -\frac{a_2c_2}{\sqrt{(b_2^2 + c_2^2)(a_2^2 + b_2^2 + c_2^2)}},$$

$$b_{32} = \frac{c_2}{\sqrt{a_2^2 + b_2^2 + c_2^2}},\ b_{33} = \frac{b_2}{\sqrt{b_2^2 + c_2^2}}. \tag{21}$$

Here a_{32}^0 is the parameter of the family.

The integral Ω_2 takes a stationary value on the elements of the family of solutions (21). From a mechanical point of view, the elements of this family correspond to equilibria of the mechanical system under consideration.

The following families of solutions correspond to the family of solutions (21) in the initial variables:

$$\varphi_1 = 0,\ \psi_1 = -\arccos(-a_{32}^0),\ \theta_1 = \frac{\pi}{2},\ \varphi_2 = 0,$$

$$\psi_2 = \arccos\left(\frac{\sqrt{b_2^2 + c_2^2}}{\sqrt{a_2^2 + b_2^2 + c_2^2}}\right),\ \theta_2 = -\arccos\left(\frac{b_2}{\sqrt{b_2^2 + c_2^2}}\right);$$

$$\varphi_1 = \pm\pi,\ \psi_1 = \arccos(a_{32}^0),\ \theta_1 = -\frac{\pi}{2},\ \varphi_2 = \pm\pi,$$

$$\psi_2 = -\arccos\left(-\frac{\sqrt{b_2^2 + c_2^2}}{\sqrt{a_2^2 + b_2^2 + c_2^2}}\right), \ \theta_2 = \arccos\left(\frac{b_2}{\sqrt{b_2^2 + c_2^2}}\right).$$

4   On the Stability of Solutions

The integrals Ω_1 and Ω_2 taking stationary values both on IMs (10) and (19) and the elements of the families of solutions (13) and (21) can be used to obtain the sufficient conditions of their stability by the Routh–Lyapunov method [6]. In such a way, the sufficient conditions of stability for the permanent rotation of the system of two bodies were derived in [5]. In this work, such approach did not allow us to solve the question of stability for the solutions. The stability analysis of solutions (13) and (21) has been performed by the linear approximation method [7].

First, let us investigate the family of solutions (21). This problem is solved on the IM given by the equations [5]:

$$V_1 - 1 = 0, \ V_2 - 1 = 0, \ V_4 - 1 = 0, \ V_5 - 1 = 0, \ V_6 - 1 = 0,$$
$$b_{11}b_{12} + b_{21}b_{22} + b_{31}b_{32} = 0, \ b_{11}b_{13} + b_{21}b_{23} + b_{31}b_{33} = 0, \tag{22}$$
$$b_{12}b_{13} + b_{22}b_{23} + b_{32}b_{33} = 0, \ b_{13} - b_{21}b_{32} + b_{22}b_{31} = 0.$$

We write the equations of motion (2)–(4) on IM (22). To do this, the variables $a_{11}, a_{12}, a_{21}, a_{22}, a_{31}, b_{11}, b_{13}, b_{23}, b_{33}$ are eliminated from them with the help of (22). The differential equations in the Poisson form take the form:

$$\dot{a}_{32} = \frac{a_{32}a_{33}(a_{23}p_1 - a_{13}q_1) + \sqrt{(a_{13}^2 + a_{23}^2)(1 - a_{32}^2) - a_{32}^2a_{33}^2}\,(a_{13}p_1 + a_{23}q_1)}{a_{13}^2 + a_{23}^2},$$

$$\dot{a}_{13} = a_{23}r_1 - a_{33}q_1, \ \dot{a}_{23} = a_{33}p_1 - a_{13}r_1, \ \dot{a}_{33} = a_{13}q_1 - a_{23}p_1,$$

$$\dot{b}_{21} = b_{31}p_2 + \frac{(b_{21}b_{22} + b_{31}b_{32})\,r_2}{b_{12}}, \ \dot{b}_{31} = -\frac{b_{12}b_{21}p_2 + b_{21}b_{22}q_2 + b_{31}b_{32}q_2}{b_{12}},$$

$$\dot{b}_{12} = b_{22}r_2 - b_{32}q_2, \ \dot{b}_{22} = b_{32}p_2 - b_{12}r_2, \ \dot{b}_{32} = b_{12}q_2 - b_{22}p_2.$$

The differential equations in the Lagrange form are rather cumbersome and are not given here.

The following family of solutions

$$p_1 = 0, \ p_2 = 0, \ q_1 = 0, \ q_2 = 0, \ r_1 = 0, \ r_2 = 0,$$
$$a_{32} = a_{32}^0, \ a_{13} = 0, \ a_{23} = 1, \ a_{33} = 0,$$
$$b_{12} = \frac{a_2}{\sqrt{a_2^2 + b_2^2 + c_2^2}}, \ b_{21} = -\frac{a_2b_2}{\sqrt{(b_2^2 + c_2^2)(a_2^2 + b_2^2 + c_2^2)}},$$
$$b_{22} = \frac{b_2}{\sqrt{a_2^2 + b_2^2 + c_2^2}}, \ b_{31} = -\frac{a_2c_2}{\sqrt{(b_2^2 + c_2^2)(a_2^2 + b_2^2 + c_2^2)}},$$
$$b_{32} = \frac{c_2}{\sqrt{a_2^2 + b_2^2 + c_2^2}} \tag{23}$$

corresponds to solutions (21) on IM (22).

Let us consider the special case $a_2 = b_2 = c_2$. Taking into account the above restrictions, we write the differential equations linearized in the neighbourhood of the elements of the family of solutions (23):

$$\left(A_1 + \frac{1}{6}(4A_2 + B_2 + C_2) + \frac{(b_1^2 + c_1^2)m_1^2}{m_2} - 3c_2^2 m_2\right)\dot{y}_{10} + \sqrt{2}\left(\frac{A_2}{\sqrt{3}} - \sqrt{2}z_1\right)\dot{y}_{11}$$

$$+\left(\frac{2A_2 - B_2 - C_2}{3\sqrt{2}} - \frac{a_1 b_1 m_1^2}{m_2}\right)\dot{y}_{12} + \left(z_1 - \frac{B_2}{\sqrt{6}}\right)\dot{y}_{13} + \left(\frac{B_2 - C_2}{2\sqrt{3}} - \frac{a_1 c_1 m_1^2}{m_2}\right)\dot{y}_{14}$$

$$+\left(z_1 - \frac{C_2}{\sqrt{6}}\right)\dot{y}_{15} + \sqrt{\frac{3}{2}}\, c_2 m_2 g(y_9 - y_7) = 0,$$

$$\left(\frac{2A_2 - B_2 - C_2}{3\sqrt{2}} - \frac{a_1 b_1 m_1^2}{m_2}\right)\dot{y}_{10} + \left(\frac{A_2}{\sqrt{3}} + \sqrt{2}\, a_1 c_2 m_1\right)\dot{y}_{11}$$

$$+\left(\frac{1}{3}(A_2 + 3B_1 + B_2 + C_2) + \frac{(a_1^2 + c_1^2)m_1^2}{m_2}\right)\dot{y}_{12} + \left(\frac{B_2}{\sqrt{3}} - z_2\right)\dot{y}_{13}$$

$$+\left(\frac{C_2 - B_2}{\sqrt{6}} - \frac{b_1 c_1 m_1^2}{m_2}\right)\dot{y}_{14} + \left(\frac{C_2}{\sqrt{3}} + z_3\right)\dot{y}_{15} = 0,$$

$$\left(\frac{B_2 - C_2}{2\sqrt{3}} - \frac{a_1 c_1 m_1^2}{m_2}\right)\dot{y}_{10} + \left(\frac{C_2 - B_2}{\sqrt{6}} - \frac{b_1 c_1 m_1^2}{m_2}\right)\dot{y}_{12} + \frac{\sqrt{6}z_1 - B_2}{\sqrt{2}}\dot{y}_{13}$$

$$+\left(\frac{1}{2}(B_2 + 2C_1 + C_2) + \frac{(a_1^2 + b_1^2)m_1^2}{m_2} - 3c_2^2 m_2\right)\dot{y}_{14} + \frac{C_2 - \sqrt{6}z_1}{\sqrt{2}}\dot{y}_{15}$$

$$+\frac{c_2 m_2 g}{\sqrt{2}}(y_7 + y_9 - 2y_5) = 0,$$

$$\sqrt{2}\left(\frac{A_2}{\sqrt{3}} - \sqrt{2}z_1\right)\dot{y}_{10} + A_2\dot{y}_{11} + \left(\frac{A_2}{\sqrt{3}} + \sqrt{2}\, a_1 c_2 m_1\right)\dot{y}_{12}$$

$$+c_2 m_2 g\left(\sqrt{2}y_4 - y_7 + y_9\right) = 0,$$

$$\left(z_1 - \frac{B_2}{\sqrt{6}}\right)\dot{y}_{10} + \left(\frac{B_2}{\sqrt{3}} - z_2\right)\dot{y}_{12} + B_2\dot{y}_{13} + \frac{\sqrt{6}z_1 - B_2}{\sqrt{2}}\dot{y}_{14}$$

$$+c_2 m_2 g\left(\frac{\sqrt{3}\,y_2 - y_4}{\sqrt{2}} + y_5 - y_9\right) = 0,$$

$$\left(z_1 - \frac{C_2}{\sqrt{6}}\right)\dot{y}_{10} + \left(\frac{C_2}{\sqrt{3}} + z_3\right)\dot{y}_{12} + \frac{C_2 - \sqrt{6}\,z_1}{\sqrt{2}}\dot{y}_{14} + C_2\dot{y}_{15}$$

$$+c_2 m_2 g\left(y_7 - \frac{\sqrt{3}\,y_2 - y_4}{\sqrt{2}} - y_5\right) = 0,$$

$$\dot{y}_1 - \sqrt{1 - a_{32}^{02}}\, y_{12} = 0, \quad \dot{y}_2 - y_{14} = 0, \quad \dot{y}_3 = 0, \quad \dot{y}_4 + y_{10} = 0,$$

$$\dot{y}_6 + \frac{y_{11} + 2y_{15}}{\sqrt{6}} = 0, \quad \dot{y}_8 - \frac{y_{11} + 2y_{13}}{\sqrt{6}} = 0, \quad \dot{y}_5 + \frac{y_{13} - y_{15}}{\sqrt{3}} = 0,$$

$$\dot{y}_7 + \frac{y_{15} - y_{11}}{\sqrt{3}} = 0, \quad \dot{y}_9 + \frac{y_{11} - y_{13}}{\sqrt{3}} = 0. \tag{24}$$

Here y_i $(i = 1, \ldots, 15)$ are the deviations from the unperturbed motion, $z_1 = c_2(b_1 m_1 + \sqrt{3}c_2 m_2)/\sqrt{2}$, $z_2 = c_2 m_1(a_1 + \sqrt{3}c_1)/\sqrt{2}$, $z_3 = c_2 m_1(\sqrt{3}c_1 - a_1)/\sqrt{2}$. The characteristic equation of system (24) is

$$|A\lambda + B| = \lambda^7(f_0\lambda^8 + f_2\lambda^6 + f_4\lambda^4 + f_6\lambda^2 + f_8) = 0,$$

where A, B are the matrices of the 15th order: A is the matrix composed of the coefficients of the derivatives of system (24), B is the matrix of the coefficients of y_i; f_0, f_2, f_4, f_6, f_8 are the expressions of $a_1, b_1, c_1, c_2, g, A_l, B_l, C_l, m_l (l = 1, 2)$. These are rather cumbersome and are not represented here.

Let us find the number of linearly independent eigenvectors corresponding to the multiple root $\lambda = 0$. To do this, we compute the rank of the matrix $A\lambda + B$ when $\lambda = 0$, using the built-in function *MatrixRank*. The rank r is 10 (the "Maple" function *Rank* gives the same result). The number of linearly independent eigenvectors is $15 - r = 5$. The multiple of the root $\lambda = 0$ is 7. Thus, the Jordan form of the matrix of linear system (24) is non-diagonal. The instability of the elements of the family under study in the linear approximation thus follows.

In an analogous way, the instability in the linear approximation for the elements of the family of solutions (13) in the special case $c_2 = b_2 = a_2$, $b_{12}^0 = -2/3$, $s_1 = -s_2$ was proved.

5 Conclusion

In the problem of the rotation of two connected rigid bodies in a uniform gravity field, the Lagrange function and the equations of motion for the mechanical system have been derived in a symbolic form with the help of the software package written in the CAS "Mathematica" language. Using the Gröbner basis method, the stationary solutions and IMs of the equations have been found. From a mechanical point of view, these solutions correspond to equilibria of the mechanical system. Their instability in the linear approximation has been proved. Many questions concerning the stability of the obtained solutions remain yet unresolved. The analysis of the dynamics of the mechanical system in other force fields is also of interest. They will be addressed in our future work.

Appendix

$$\Phi_1 = q_1 r_2[(A_2 - B_2)(b_{12}b_{21} + b_{11}b_{22}) + b_{33}C_2 - m_2((b_2(b_{12}-b_{23}b_{31} + b_{21}b_{33})$$
$$-a_2(b_{22} + b_{13}b_{31} - b_{11}b_{33}))s_1 - (b_2(b_{11} - b_{23}b_{32} + b_{22}b_{33})$$
$$-a_2(b_{21} + b_{13}b_{32} - b_{12}b_{33}))s_2)]$$
$$+p_2 q_1[A_2 b_{13} + (b_{22}b_{31} + b_{21}b_{32})(B_2 - C_2) + m_2((b_2(b_{32} - b_{13}b_{21}$$
$$+b_{11}b_{23}) - c_2(b_{22} + b_{13}b_{31} - b_{11}b_{33}))s_1 - (b_2(b_{31} - b_{13}b_{32} + b_{12}b_{23})$$
$$-c_2(b_{21} + b_{13}b_{32} - b_{12}b_{33}))s_2)]$$
$$-p_1 r_1[A_2 b_{11}b_{12} + B_2 b_{21}b_{22} + C_2 b_{31}b_{32} - m_2((b_2(b_{11}b_{33} - b_{13}b_{31})$$
$$+a_2(b_{23}b_{31} - b_{21}b_{33}) + c_2(b_{13}b_{21} - b_{11}b_{23}))s_1 + (b_2(b_{13}b_{32} - b_{12}b_{33})$$
$$+a_2(b_{22}b_{33} - b_{23}b_{32}) + c_2(b_{12}b_{23} - b_{13}b_{22}) + s_1)s_2)]$$

$$+q_1q_2[B_2b_{23} - (b_{12}b_{31} + b_{11}b_{32})(A_2 - C_2) + m_2((a_2(b_{13}b_{21} - b_{11}b_{23} - b_{32})$$
$$+c_2(b_{12} - b_{23}b_{31} + b_{21}b_{33}))s_1 + (a_2(b_{31} - b_{13}b_{22} + b_{12}b_{23})$$
$$-c_2(b_{11} - b_{23}b_{32} + b_{22}b_{33}))s_2)]$$
$$+p_2q_2[(B_2 - A_2)b_{31} + m_2((b_{13}b_2 + a_2b_{23})s_2 - (b_{12}b_2 + a_2b_{22})s_3)]$$
$$+p_2r_2[(A_2 - C_2)b_{21} + m_2((a_2b_{33} + b_{13}c_2)s_2 - (a_2b_{32} + b_{12}c_2)s_3)]$$
$$+q_2r_2[(C_2 - B_2)b_{11} + m_2((b_2b_{33} + b_{23}c_2)s_2 - (b_2b_{32} + b_{22}c_2)s_3)]$$
$$+p_2r_1[(b_{23}b_{31} + b_{21}b_{33})(B_2 - C_2) - A_2b_{12} + m_2((b_2(b_{12}b_{21} - b_{11}b_{22} + b_{33})$$
$$-c_2(b_{23} - b_{12}b_{31} + b_{11}b_{32}))s_1 + (c_2(b_{21} + b_{13}b_{32} - b_{12}b_{33})$$
$$-b_2(b_{31} - b_{13}b_{22} + b_{12}b_{23}))s_3)]$$
$$-m_2[c_2(p_2^2 + q_2^2)(b_{33}s_2 - b_{32}s_3) + b_2(p_2^2 + r_2^2)(b_{23}s_2 - b_{22}s_3)$$
$$+a_2(q_2^2 + r_2^2)(b_{13}s_2 - b_{12}s_3)]$$
$$-2p_1r_2[-b_{11}(A_2 - B_2)b_{21} + m_2((a_2(b_{13}b_{31} - b_{11}b_{33}) + b_2(b_{23}b_{31} - b_{21}b_{33}))s_2$$
$$+(a_2(b_{11}b_{32} - b_{12}b_{31}) + b_2(b_{21}b_{32} - b_{22}b_{31}))s_3)]$$
$$+r_1r_2[(A_2 - B_2)(b_{13}b_{21} + b_{11}b_{23}) - C_2b_{32} + m_2((a_2(b_{23} - b_{12}b_{31} + b_{11}b_{32})$$
$$-b_2(b_{13} + b_{22}b_{31} - b_{21}b_{32}))s_1 + (a_2(b_{21} + b_{13}b_{32} - b_{12}b_{33}) + b_2(b_{11} - b_{23}b_{32}$$
$$+b_{22}b_{33}))s_3)]$$
$$+2p_1p_2[(B_2 - C_2)b_{21}b_{31} + m_2((b_2(b_{13}b_{21} - b_{11}b_{23}) + (b_{13}b_{31} - b_{11}b_{33})c_2)s_2$$
$$+(b_2(b_{11}b_{22} - b_{12}b_{21}) + c_2(b_{11}b_{32} - b_{12}b_{31}))s_3)]$$
$$-2p_1q_2[A_2b_{11}b_{31} - b_{11}b_{31}C_2 + m_2((a_2(b_{13}b_{21} - b_{11}b_{23}) + c_2(b_{21}b_{33} - b_{23}b_{31}))s_2$$
$$+(a_2(b_{11}b_{22} - b_{12}b_{21}) + c_2(b_{22}b_{31} - b_{21}b_{32}))s_3)]$$
$$-q_2r_1[B_2b_{22} + (b_{13}b_{31} + b_{11}b_{33})(A_2 - C_2) + m_2((a_2(b_{12}b_{21} - b_{11}b_{22} + b_{33})$$
$$-c_2(b_{13} + b_{22}b_{31} - b_{21}b_{32}))s_1 + (a_2(b_{13}b_{22} - b_{12}b_{23} - b_{31}) + c_2(b_{11} - b_{23}b_{32}$$
$$+b_{22}b_{33}))s_3)]$$
$$+p_1q_1[A_2b_{11}b_{13} + B_2b_{21}b_{23} + b_{31}b_{33}C_2 - m_2((b_2(b_{12}b_{31} - b_{11}b_{32})$$
$$+a_2(b_{21}b_{32} - b_{22}b_{31}) + c_2(b_{11}b_{22} - b_{12}b_{21}))s_1 + (b_2(b_{13}b_{32} - b_{12}b_{33})$$
$$+a_2(b_{22}b_{33} - b_{23}b_{32}) + c_2(b_{12}b_{23} - b_{13}b_{22}) + s_1)s_3)]$$
$$+q_1^2[A_2b_{12}b_{13} + B_2b_{22}b_{23} + b_{32}b_{33}C_2 + m_2((b_2(b_{11}b_{32} - b_{12}b_{31})$$
$$+a_2(b_{22}b_{31} - b_{21}b_{32}) + c_2(b_{12}b_{21} - b_{11}b_{22}))s_2 - (b_2(b_{11}b_{33} - b_{13}b_{31})$$
$$+a_2(b_{23}b_{31} - b_{21}b_{33}) + c_2(b_{13}b_{21} - b_{11}b_{23}) + s_2)s_3)]$$
$$-r_1^2[A_2b_{12}b_{13} + B_2b_{22}b_{23} + C_2b_{32}b_{33} - m_2((b_2(b_{12}b_{31} - b_{11}b_{32})$$
$$+a_2(b_{21}b_{32} - b_{22}b_{31}) + c_2(b_{11}b_{22} - b_{12}b_{21}))s_2 + (b_2(b_{11}b_{33} - b_{13}b_{31})$$
$$+a_2(b_{23}b_{31} - b_{21}b_{33}) + c_2(b_{13}b_{21} - b_{11}b_{23}) + s_2)s_3)]$$
$$+q_1r_1[C_1 - B_1 - A_2(b_{12}^2 - b_{13}^2) - B_2(b_{22}^2 - b_{23}^2) - C_2(b_{32}^2 - b_{33}^2)$$
$$+m_2(s_2(2(b_2(b_{11}b_{33} - b_{13}b_{31})$$
$$+a_2(b_{23}b_{31} - b_{21}b_{33}) + (b_{13}b_{21} - b_{11}b_{23})c_2) + s_2) + 2(b_2(b_{11}b_{32} - b_{12}b_{31})$$
$$+a_2(b_{22}b_{31} - b_{21}b_{32}) + c_2(b_{12}b_{21} - b_{11}b_{22}))s_3 - s_3^2)]$$
$$+g[m_1(b1a_{33} - c1a_{23}) + m_2(a_{33}(a_2b_{12} + b_2b_{22} + c_2b_{32} + s_2)$$
$$-a_{23}(a_2b_{13} + b_2b_{23} + c_2b_{33} + s_3))],$$

$$\Phi_2 = p_1r_2[(A_2 - B_2)(b_{12}b_{21} + b_{11}b_{22}) - C_2b_{33} + m_2((a_2(b_{13}b_{31} - b_{11}b_{33} - b_{22})$$
$$+b_2(b_{12} + b_{23}b_{31} - b_{21}b_{33}))s_1 + (a_2(b_{21} - b_{13}b_{32} + b_{12}b_{33}) - b_2(b_{11} + b_{23}b_{32}$$
$$-b_{22}b_{33}))s_2)]$$
$$+p_1p_2[(b_{22}b_{31} + b_{21}b_{32})(B_2 - C_2) - A_2b_{13} + m_2((c_2(b_{22} - b_{13}b_{31} + b_{11}b_{33})$$
$$-b_2(b_{13}b_{21} - b_{11}b_{23} + b_{32}))s_1 + (b_2(b_{13}b_{22} - b_{12}b_{23} + b_{31}) - c_2(b_{21} - b_{13}b_{32}$$
$$+b_{12}b_{33}))s_2)]$$
$$-p_1q_2[B_2b_{23} + (A_2 - C_2)(b_{12}b_{31} + b_{11}b_{32}) - m_2((a_2(b_{13}b_{21} - b_{11}b_{23} + b_{32})$$
$$-c_2(b_{12} + b_{23}b_{31} - b_{21}b_{33}))s_1 - (a_2(b_{13}b_{22} - b_{12}b_{23} + b_{31}) - c_2(b_{11} + b_{23}b_{32}$$
$$-b_{22}b_{33}))s_2)]$$

$$+q_1r_1[A_2b_{11}b_{12} + B_2b_{21}b_{22} + C_2b_{31}b_{32} + m_2((a_2(b_{21}b_{33} - b_{23}b_{31})$$
$$+b_2(b_{13}b_{31} - b_{11}b_{33}) - c_2(b_{13}b_{21} - b_{11}b_{23}))s_1 - (a_2(b_{22}b_{33} - b_{23}b_{32})$$
$$+b_2(b_{13}b_{32} - b_{12}b_{33}) + c_2(b_{12}b_{23} - b_{13}b_{22}) + s_1)s_2)]$$
$$+r_2^2m_2[(a_2b_{13} + b_2b_{23})s_1 - (a_2b_{11} - b_2b_{21})s_3]$$
$$-p_2q_2[(A_2 - B_2)b_{32} + m_2((a_2b_{23} + b_2b_{13})s_1 - (a_2b_{21} + b_2b_{11})s_3)]$$
$$+q_2^2m_2[(a_2b_{13} + c_2b_{33})s_1 - (a_2b_{11} + c_2b_{31})s_3]$$
$$+p_2^2m_2[(b_2b_{23} + c_2b_{33})s_1 - (b_2b_{21} + c_2b_{31})s_3]$$
$$+2q_1r_2[(A_2 - B_2)b_{12}b_{22} - m_2((a_2(b_{12}b_{33} - b_{13}b_{32}) + b_2(b_{22}b_{33} - b_{23}b_{32}))s_1$$
$$+(a_2(b_{11}b_{32} - b_{12}b_{31}) + b_2(b_{21}b_{32} - b_{22}b_{31}))s_3)]$$
$$+r_1r_2((A_2 - B_2)(b_{13}b_{22} + b_{12}b_{23}) + C_2b_{31} + m_2((a_2(b_{23} - b_{12}b_{31} + b_{11}b_{32})$$
$$-b_2(b_{13} + b_{22}b_{31} - b_{21}b_{32}))s_2 - (a_2(b_{22} - b_{13}b_{31} + b_{11}b_{33}) - b_2(b_{12} + b_{23}b_{31}$$
$$-b_{21}b_{33}))s_3))$$
$$+p_2r_2[(A_2 - C_2)b_{22} - m_2((a_2b_{33} + b_{13}c_2)s_1 - (a_2b_{31} + c_2b_{11})s_3)]$$
$$-q_2r_2((B_2 - C_2)b_{12} + m_2((b_2b_{33} + c_2b_{23})s_1 - (b_2b_{31} + c_2b_{21})s_3))$$
$$-2q_1q_2[(A_2 - C_2)b_{12}b_{32} + m_2((a_2(b_{12}b_{23} - b_{13}b_{22}) + c_2(b_{23}b_{32} - b_{22}b_{33}))s_1$$
$$+(a_2(b_{11}b_{22} - b_{12}b_{21}) + c_2(b_{22}b_{31} - b_{21}b_{32}))s_3)]$$
$$+p_2r_1[A_2b_{11} + (B_2 - C_2)(b_{23}b_{32} + b_{22}b_{33}) + m_2((b_2(b_{12}b_{21} - b_{11}b_{22} + b_{33})$$
$$-c_2(b_{23} - b_{12}b_{31} + b_{11}b_{32}))s_2 - (b_2(b_{13}b_{21} - b_{11}b_{23} + b_{32}) + c_2(b_{22} - b_{13}b_{31}$$
$$+b_{11}b_{33}))s_3)]$$
$$+q_2r_1[B_2b_{21} - (b_{13}b_{32} + b_{12}b_{33})(A_2 - C_2) - m_2((a_2(b_{12}b_{21} - b_{11}b_{22} + b_{33})$$
$$-c_2(b_{13} + b_{22}b_{31} - b_{21}b_{32}))s_2 - (a_2(b_{13}b_{21} - b_{11}b_{23} + b_{32}) - c_2(b_{12} + b_{23}b_{31}$$
$$-b_{21}b_{33}))s_3)]$$
$$+2p_2q_1[b_{22}b_{32}(B_2 - C_2) + m_2((b_2(b_{12}b_{23} - b_{13}b_{22}) + c_2(b_{12}b_{33} - b_{13}b_{32}))s_1$$
$$+(b_2(b_{11}b_{22} - b_{12}b_{21}) + c_2(b_{11}b_{32} - b_{12}b_{31}))s_3)]$$
$$+r_1^2[A_2b_{11}b_{13} + B_2b_{21}b_{23} + C_2b_{31}b_{33} + m_2((a_2(b_{22}b_{31} - b_{21}b_{32})$$
$$+b_2(b_{11}b_{32} - b_{12}b_{31}) + c_2(b_{12}b_{21} - b_{11}b_{22}))s_1 - (a_2(b_{22}b_{33} - b_{23}b_{32})$$
$$+b_2(b_{13}b_{32} - b_{12}b_{33}) + c_2(b_{12}b_{23} - b_{13}b_{22}) + s_1)s_3)]$$
$$-p_1^2(A_2b_{11}b_{13} + B_2b_{21}b_{23} + C_2b_{31}b_{33} - m_2((a_2(b_{21}b_{32} - b_{22}b_{31})$$
$$+b_2(b_{12}b_{31} - b_{11}b_{32}) + c_2(b_{11}b_{22} - b_{12}b_{21}))s_1 + (a_2(b_{22}b_{33} - b_{23}b_{32})$$
$$+b_2(b_{13}b_{32} - b_{12}b_{33}) + c_2(b_{12}b_{23} - b_{13}b_{22}) + s_1)s_3))$$
$$+p_1q_1(-A_2b_{12}b_{13} - B_2b_{22}b_{23} - C_2b_{32}b_{33} + m_2((a_2(b_{21}b_{32} - b_{22}b_{31})$$
$$+b_2(b_{12}b_{31} - b_{11}b_{32}) + c_2(b_{11}b_{22} - b_{12}b_{21}))s_2 + (a_2(b_{23}b_{31} - b_{21}b_{33})$$
$$+b_2(b_{11}b_{33} - b_{13}b_{31}) + c_2(b_{13}b_{21} - b_{11}b_{23}) + s_2)s_3))$$
$$+p_1r_1[A_1 - C_1 + A_2(b_{11}^2 - b_{13}^2) + B_2(b_{21}^2 - b_{23}^2) + C_2(b_{31}^2 - b_{33}^2)$$
$$-m_2((2(a_2(b_{22}b_{33} - b_{23}b_{32}) + b_2(b_{13}b_{32} - b_{12}b_{33}) + c_2(b_{12}b_{23} - b_{13}b_{22}))$$
$$+s_1)s_1 - (2(a_2(b_{21}b_{32} - b_{22}b_{31}) + b_2(b_{12}b_{31} - b_{11}b_{32}) + c_2(b_{11}b_{22} - b_{12}b_{21}))$$
$$+s_1)s_1 + s_3)s_3)] - g[m_1(a_1a_{33} - c_1a_{13}) - m_2(a_{13}(a_2b_{13} + b_2b_{23} + c_2b_{33} + s_3)$$
$$-a_{33}(a_2b_{11} + b_2b_{21} + c_2b_{31} + s_1))],$$

$$\Phi_3 = r_2^2\,m_2[(a_2b_{11} + b_2b_{21})s_2 - (a_2b_{12} + b_2b_{22})s_1]$$
$$+q_2^2\,m_2[(a_2b_{11} + c_2b_{31})s_2 - (a_2b_{12} + c_2b_{32})s_1]$$
$$+p_2^2\,m_2[(b_2b_{21} + c_2b_{31})s_2 - (b_2b_{22} + c_2b_{32})s_1]$$
$$+p_2q_2[(B_2 - A_2)b_{33} + m_2((b_2b_{12} + a_2b_{22})s_1 - (b_2b_{11} + a_2b_{21})s_2)]$$
$$+2r_1r_2[b_{13}(A_2 - B_2)b_{23} + m_2((a_2(b_{13}b_{32} - b_{12}b_{33}) + b_2(b_{23}b_{32} - b_{22}b_{33}))s_1$$
$$+(a_2(b_{11}b_{33} - b_{13}b_{31}) + b_2(b_{21}b_{33} - b_{23}b_{31}))s_2)]$$
$$+p_2r_2[b_{23}(A_2 - C_2) + m_2((a_2b_{32} + c_2b_{12})s_1 - (a_2b_{31} + c_2b_{11})s_2)]$$
$$+q_2r_2[(C_2 - B_2)b_{13} + m_2((b_2b_{32} + b_{22}c_2)s_1 - (b_2b_{31} + b_{21}c_2)s_2)]$$
$$-2p_2r_1[(C_2 - B_2)b_{23}b_{33} + m_2((b_2(b_{13}b_{22} - b_{12}b_{23}) + c_2(b_{13}b_{32} - b_{12}b_{33}))s_1$$

$$+(b_2(b_{11}b_{23} - b_{13}b_{21}) + c_2(b_{11}b_{33} - b_{13}b_{31}))s_2)]$$
$$+2q_2r_1\,[(C_2 - A_2)b_{13}b_{33} + m_2((a_2(b_{13}b_{22} - b_{12}b_{23}) + c_2(b_{22}b_{33} - b_{23}b_{32}))s_1$$
$$+(a_2(b_{11}b_{23} - b_{13}b_{21}) + c_2(b_{23}b_{31} - b_{21}b_{33}))s_2)]$$
$$+p_1^2\,[A_2b_{11}b_{12} + B_2b_{21}b_{22} + C_2b_{31}b_{32} + m_2((b_2(b_{13}b_{31} - b_{11}b_{33})$$
$$+a_2(b_{21}b_{33} - b_{23}b_{31}) + c_2(b_{11}b_{23} - b_{13}b_{21}))s_1 - (b_2(b_{13}b_{32} - b_{12}b_{33})$$
$$+a_2(b_{22}b_{33} - b_{23}b_{32}) + c_2(b_{12}b_{23} - b_{13}b_{22}) + s_1)s_2)]$$
$$-q_1^2\,[A_2b_{11}b_{12} + B_2b_{21}b_{22} + C_2b_{31}b_{32} - m_2((b_2(b_{11}b_{33} - b_{13}b_{31})$$
$$+a_2(b_{23}b_{31} - b_{21}b_{33}) + c_2(b_{13}b_{21} - b_{11}b_{23}))s_1 + (b_2(b_{13}b_{32} - b_{12}b_{33})$$
$$+a_2(b_{22}b_{33} - b_{23}b_{32}) + c_2(b_{12}b_{23} - b_{13}b_{22}) + s_1)s_2)]$$
$$+q_1r_2\,[(A_2 - B_2)(b_{13}b_{22} + b_{12}b_{23}) - C_2b_{31} + m_2((b_2(b_{13} - b_{22}b_{31} + b_{21}b_{32})$$
$$-a_2(b_{23} + b_{12}b_{31} - b_{11}b_{32}))s_2 + (a_2(b_{22} + b_{13}b_{31} - b_{11}b_{33})$$
$$-b_2(b_{12} - b_{23}b_{31} + b_{21}b_{33}))s_3)]$$
$$+p_1r_2\,[(A_2 - B_2)(b_{13}b_{21} + b_{11}b_{23}) + C_2b_{32} + m_2((b_2(b_{13} - b_{22}b_{31} + b_{21}b_{32})$$
$$-a_2(b_{23} + b_{12}b_{31} - b_{11}b_{32}))s_1 + (a_2(b_{21}$$
$$-b_{13}b_{32} + b_{12}b_{33}) - b_2(b_{11} + b_{23}b_{32} - b_{22}b_{33}))s_3)]$$
$$+p_2q_1\,[(b_{23}b_{32} + b_{22}b_{33})(B_2 - C_2) - A_2b_{11} + m_2((b_2(b_{12}b_{21} - b_{11}b_{22} - b_{33})$$
$$+c_2(b_{23} + b_{12}b_{31} - b_{11}b_{32}))s_2 + (b_2(b_{32} - b_{13}b_{21} + b_{11}b_{23})$$
$$-c_2(b_{22} + b_{13}b_{31} - b_{11}b_{33}))s_3)]$$
$$+p_1p_2\,[A_2b_{12} + (B_2 - C_2)(b_{23}b_{31} + b_{21}b_{33}) + m_2((c_2(b_{23} + b_{12}b_{31} - b_{11}b_{32})$$
$$-b_2(b_{33} - b_{12}b_{21} + b_{11}b_{22}))s_1 + (b_2(b_{13}b_{22} - b_{12}b_{23} + b_{31})$$
$$-c_2(b_{21} - b_{13}b_{32} + b_{12}b_{33}))s_3)]$$
$$-q_1q_2\,[B_2b_{21} + (A_2 - C_2)(b_{13}b_{32} + b_{12}b_{33}) - m_2((a_2(b_{33} - b_{12}b_{21} + b_{11}b_{22})$$
$$-c_2(b_{13} - b_{22}b_{31} + b_{21}b_{32}))s_2 + (a_2(b_{13}b_{21} - b_{11}b_{23} - b_{32})$$
$$+c_2(b_{12} - b_{23}b_{31} + b_{21}b_{33}))s_3)]$$
$$+p_1q_2(B_2b_{22} - (A_2 - C_2)(b_{13}b_{31} + b_{11}b_{33}) + m_2((a_2(b_{33} - b_{12}b_{21} + b_{11}b_{22})$$
$$-c_2(b_{13} - b_{22}b_{31} + b_{21}b_{32}))s_1 - (a_2(b_{31} + b_{13}b_{22} - b_{12}b_{23})$$
$$-c_2(b_{11} + b_{23}b_{32} - b_{22}b_{33}))s_3))$$
$$-q_1r_1\,[A_2b_{11}b_{13} + B_2b_{21}b_{23} + C_2b_{31}b_{33} - m_2((a_2(b_{21}b_{32} - b_{22}b_{31})$$
$$+b_2(b_{12}b_{31} - b_{11}b_{32}) + c_2(b_{11}b_{22} - b_{12}b_{21}))s_1 + (a_2(b_{22}b_{33} - b_{23}b_{32})$$
$$+b_2(b_{13}b_{32} - b_{12}b_{33}) + c_2(b_{12}b_{23} - b_{13}b_{22}) + s_1)s_3)]$$
$$+p_1r_1\,[A_2b_{12}b_{13} + B_2b_{22}b_{23} + C_2b_{32}b_{33} + m_2((a_2(b_{22}b_{31} - b_{21}b_{32})$$
$$+b_2(b_{11}b_{32} - b_{12}b_{31}) + c_2(b_{12}b_{21} - b_{11}b_{22}))s_2 - (a_2(b_{23}b_{31} - b_{21}b_{33})$$
$$+b_2(b_{11}b_{33} - b_{13}b_{31}) + c_2(b_{13}b_{21} - b_{11}b_{23}) + s_2)s_3)]$$
$$-p_1q_1\,[A_1 - B_1 - A_2(b_{11}^2 - b_{12}^2) + B_2(b_{21}^2 - b_{22}^2) + C_2(b_{31}^2 - b_{32}^2)$$
$$-m_2((2(a_2(b_{22}b_{33} - b_{23}b_{32}) + b_2(b_{13}b_{32} - b_{12}b_{33}) + c_2(b_{12}b_{23} - b_{13}b_{22})) + s_1)s_1$$
$$+(2(b_2(a_2(b_{21}b_{33} - b_{23}b_{31}) + b_{13}b_{31} - b_{11}b_{33}) + c_2(b_{11}b_{23} - b_{13}b_{21})) - s_2)s_2)]$$
$$+g(m_1(a_1a_{23} - b_1a_{13}) + m_2(a_{23}(a_2b_{11} + b_2b_{21} + c_2b_{31} + s_1)$$
$$-a_{13}(a_2b_{12} + b_2b_{22} + c_2b_{32} + s_2))),$$

$$\Phi_4 = (A_2 - B_2 + C_2)(b_{21}p_1 + b_{22}q_1 + b_{23}r_1)\,r_2$$
$$-(A_2 + B_2 - C_2)(b_{31}p_1 + b_{32}q_1 + b_{33}r_1)\,q_2 + (C_2 - B_2)\,q_2r_2$$
$$+r_1^2\,[(C_2 - B_2)b_{23}b_{33} + m_2((b_2(b_{13}b_{22} - b_{12}b_{23}) + c_2(b_{13}b_{32} - b_{12}b_{33}))s_1$$
$$+(b_2(b_{11}b_{23} - b_{13}b_{21}) + c_2(b_{11}b_{33} - b_{13}b_{31}))s_2)]$$
$$-p_1q_1\,[(B_2 - C_2)(b_{22}b_{31} + b_{21}b_{32}) - m_2((b_2(b_{13}b_{21} - b_{11}b_{23})$$
$$+c_2(b_{13}b_{31} - b_{11}b_{33}))s_1 + (b_2(b_{12}b_{23} - b_{13}b_{22}) + c_2(b_{12}b_{33} - b_{13}b_{32}))s_2)]$$
$$+p_1^2\,[(C_2 - B_2)b_{21}b_{31} + m_2((b_2(b_{11}b_{23} - b_{13}b_{21}) + c_2(b_{11}b_{33} - b_{13}b_{31}))s_2$$
$$+(b_2(b_{12}b_{21} - b_{11}b_{22}) + c_2(b_{12}b_{31} - b_{11}b_{32}))s_3)]$$
$$-q_1^2\,[(B_2 - C_2)b_{22}b_{32} - m_2((b_2(b_{13}b_{22} - b_{12}b_{23}) + c_2(b_{13}b_{32} - b_{12}b_{33}))s_1$$

$$+(b_2(b_{12}b_{21} - b_{11}b_{22}) + c_2(b_{12}b_{31} - b_{11}b_{32}))s_3)]$$
$$-q_1 r_1 \left[(B_2 - C_2)(b_{23}b_{32} + b_{22}b_{33}) - m_2((b_2(b_{11}b_{22} - b_{12}b_{21})\right.$$
$$+c_2(b_{11}b_{32} - b_{12}b_{31}))s_2 + (b_2(b_{13}b_{21} - b_{11}b_{23}) + (b_{13}b_{31} - b_{11}b_{33})c_2)s_3)]$$
$$-p_1 r_1 \left[(b_{23}b_{31} + b_{21}b_{33})(B_2 - C_2) - m_2((b_2(b_{11}b_{22} - b_{12}b_{21})\right.$$
$$+c_2(b_{11}b_{32} - b_{12}b_{31}))s_1 + (b_2(b_{12}b_{23} - b_{13}b_{22}) + c_2(b_{12}b_{33} - b_{13}b_{32}))s_3)]$$
$$+gm_2(b_2(a_{13}b_{31} + a_{23}b_{32} + a_{33}b_{33}) - c_2(a_{13}b_{21} + a_{23}b_{22} + a_{33}b_{23})),$$

$$\Phi_5 = (A_2 + B_2 - C_2)(b_{31}p_1 + b_{32}q_1 + b_{33}r_1)\,p_2$$
$$+(A_2 - B_2 - C_2)(b_{11}p_1 + b_{12}q_1 + b_{13}r_1)\,r_2 + (A_2 - C_2)\,p_2 r_2$$
$$+r_1^2 \left[(A_2 - C_2)\,b_{13}b_{33} + m_2((a_2(b_{12}b_{23} - b_{13}b_{22}) + c_2(b_{23}b_{32} - b_{22}b_{33}))s_1\right.$$
$$+(a_2(b_{13}b_{21} - b_{11}b_{23}) + c_2(b_{21}b_{33} - b_{23}b_{31}))s_2)]$$
$$+p_1 q_1 \left[(A_2 - C_2)(b_{12}b_{31} + b_{11}b_{32}) + m_2((a_2(b_{11}b_{23} - b_{13}b_{21})\right.$$
$$+c_2(b_{23}b_{31} - b_{21}b_{33}))s_1 + (a_2(b_{13}b_{22} - b_{12}b_{23}) + c_2(b_{22}b_{33} - b_{23}b_{32}))s_2)]$$
$$+q_1^2 \left[b_{12}b_{32}(A_2 - C_2) + m_2((a_2(b_{12}b_{23} - b_{13}b_{22}) + c_2(b_{23}b_{32} - b_{22}b_{33}))s_1\right.$$
$$+(a_2(b_{11}b_{22} - b_{12}b_{21}) + c_2(b_{22}b_{31} - b_{21}b_{32}))s_3)]$$
$$+p_1^2 \left[(A_2 - C_2)\,b_{11}b_{31} + m_2((a_2(b_{13}b_{21} - b_{11}b_{23}) + c_2(b_{21}b_{33} - b_{23}b_{31}))s_2\right.$$
$$+(a_2(b_{11}b_{22} - b_{12}b_{21}) + c_2(b_{22}b_{31} - b_{21}b_{32}))s_3)]$$
$$+q_1 r_1 \left[(A_2 - C_2)(b_{13}b_{32} + b_{12}b_{33}) + m_2((a_2(b_{12}b_{21} - b_{11}b_{22})\right.$$
$$+c_2(b_{21}b_{32} - b_{22}b_{31}))s_2 + (a_2(b_{11}b_{23} - b_{13}b_{21}) + c_2(b_{23}b_{31} - b_{21}b_{33}))s_3)]$$
$$+p_1 r_1 \left[(A_2 - C_2)(b_{13}b_{31} + b_{11}b_{33}) + m_2((a_2(b_{12}b_{21} - b_{11}b_{22})\right.$$
$$+c_2(b_{21}b_{32} - b_{22}b_{31}))s_1 + (a_2(b_{13}b_{22} - b_{12}b_{23}) + c_2(b_{22}b_{33} - b_{23}b_{32}))s_3)]$$
$$-gm_2 \left[a_2(a_{13}b_{31} + a_{23}b_{32} + a_{33}b_{33}) - c_2(a_{13}b_{11} + a_{23}b_{12} + a_{33}b_{13})\right],$$

$$\Phi_6 = -(A_2 - B_2 - C_2)(b_{11}p_1 + b_{12}q_1 + b_{13}r_1)\,q_2$$
$$-(A_2 - B_2 + C_2)(b_{21}p_1 + b_{22}q_1 + b_{23}r_1)\,p_2 - (A_2 - B_2)\,p_2 q_2$$
$$-r_1^2 \left[(A_2 - B_2)b_{13}b_{23} - m_2((a_2(b_{12}b_{33} - b_{13}b_{32}) + b_2(b_{22}b_{33} - b_{23}b_{32}))s_1\right.$$
$$+(a_2(b_{13}b_{31} - b_{11}b_{33}) + b_2(b_{23}b_{31} - b_{21}b_{33}))s_2)]$$
$$-p_1 q_1 \left[(A_2 - B_2)(b_{12}b_{21} + b_{11}b_{22}) - m_2((a_2(b_{11}b_{33} - b_{13}b_{31})\right.$$
$$+b_2(b_{21}b_{33} - b_{23}b_{31}))s_1 + (a_2(b_{13}b_{32} - b_{12}b_{33}) + b_2(b_{23}b_{32} - b_{22}b_{33}))s_2)]$$
$$-q_1^2 \left[(A_2 - B_2)b_{12}b_{22} - m_2((a_2(b_{12}b_{33} - b_{13}b_{32}) + b_2(b_{22}b_{33} - b_{23}b_{32}))s_1\right.$$
$$+(a_2(b_{11}b_{32} - b_{12}b_{31}) + b_2(b_{21}b_{32} - b_{22}b_{31}))s_3)]$$
$$-p_1^2 \left[(A_2 - B_2)b_{11}b_{21} - m_2((a_2(b_{13}b_{31} - b_{11}b_{33}) + b_2(b_{23}b_{31} - b_{21}b_{33}))s_2\right.$$
$$+(a_2(b_{11}b_{32} - b_{12}b_{31}) + b_2(b_{21}b_{32} - b_{22}b_{31}))s_3)]$$
$$-q_1 r_1 \left[(A_2 - B_2)(b_{13}b_{22} + b_{12}b_{23}) - m_2((a_2(b_{12}b_{31} - b_{11}b_{32})\right.$$
$$+b_2(b_{22}b_{31} - b_{21}b_{32}))s_2 + (a_2(b_{11}b_{33} - b_{13}b_{31}) + b_2(b_{21}b_{33} - b_{23}b_{31}))s_3)]$$
$$-p_1 r_1 \left[(A_2 - B_2)(b_{13}b_{21} + b_{11}b_{23}) - m_2((a_2(b_{12}b_{31} - b_{11}b_{32})\right.$$
$$+b_2(b_{22}b_{31} - b_{21}b_{32}))s_1 + (a_2(b_{13}b_{32} - b_{12}b_{33}) + b_2(b_{23}b_{32} - b_{22}b_{33}))s_3)]$$
$$+gm_2(a_2(a_{13}b_{21} + a_{23}b_{22} + a_{33}b_{23}) - b_2(a_{13}b_{11} + a_{23}b_{12} + a_{33}b_{13})).$$

References

1. Banshchikov, A.V., Burlakova, L.A., Irtegov, V.D., Titorenko, T.N.: Software Package LinModel for the Analysis of the Dynamics of Large Dimensional Mechanical Systems. Certificate of State Registration of Software Programs. FGU-FIPS. 2008610622 (2008)

2. Burov, A.A.: Linear invariant relations in the problem of the motion of a bundle of two bodies. Dokl. Phys. **65**(4), 147–148 (2020). https://doi.org/10.1134/S1028335820040035
3. Gutnik, S.A., Sarychev, V.A.: Application of computer algebra methods to investigation of stationary motions of a system of two connected bodies moving in a circular orbit. Comput. Math. Math. Phys. **60**(1), 74–81 (2020). https://doi.org/10.1134/S0965542520010091
4. Irtegov, V.D., Titorenko, T.N.: About stationary movements of the generalized Kovalevskaya top and their stability. Mech. Solids **54**(1), 81–91 (2019). https://doi.org/10.3103/S0025654419010072
5. Irtegov, V., Titorenko, T.: On the study of the motion of a system of two connected rigid bodies by computer algebra methods. In: Boulier, F., England, M., Sadykov, T.M., Vorozhtsov, E.V. (eds.) CASC 2020. LNCS, vol. 12291, pp. 266–281. Springer, Cham (2020). https://doi.org/10.1007/978-3-030-60026-6_15
6. Lyapunov, A.M.: On permanent helical motions of a rigid body in fluid. Collected Works, USSR Acad. Sci., Moscow-Leningrad **1**, 276–319 (1954). (in Russian)
7. Lyapunov, A.M.: The General Problem of the Stability of Motion. Taylor & Francis, London (1992)

An Interpolation Algorithm for Computing Dixon Resultants

Ayoola Jinadu$^{(\boxtimes)}$ and Michael Monagan

Department of Mathematics, Simon Fraser University,
Burnaby, BC V5A 1S6, Canada
{ajinadu,mmonagan}@sfu.ca

Abstract. Given a system of polynomial equations with parameters, we present a new algorithm for computing its Dixon resultant R. Our algorithm interpolates the monic square-free factors of R one at a time from monic univariate polynomial images of R using sparse rational function interpolation. In this work, we use a modified version of the sparse multivariate rational function interpolation algorithm of Cuyt and Lee.

We have implemented our new Dixon resultant algorithm in Maple with some subroutines coded in C for efficiency. We present timing results comparing our new Dixon resultant algorithm with Zippel's algorithm for interpolating R and a Maple implementation of the Gentleman & Johnson minor expansion algorithm for computing R.

Keywords: Dixon resultant · Parametric polynomial systems · Resultant · Sparse rational function interpolation · Kronecker substitution

1 Introduction

Let $X = \{x_1, x_2, \cdots, x_n\}$ denote the set of variables and let $Y = \{y_1, y_2, \cdots, y_m\}$ be the set of parameters with $n \geq 2$ and $m \geq 0$. Let $\mathcal{F} = \{f_1, f_2, \cdots, f_n\} \subset \mathbb{Q}[X, Y]$ be a parametric polynomial system where f_i is a polynomial in variables X with coefficients in $\mathbb{Q}[Y]$. Let $I = \langle f_1, f_2, \cdots, f_n \rangle$ be the ideal generated by $\mathcal{F}$. The Dixon resultant [5,6] of $\mathcal{F}$ in x_1 is the determinant of the Dixon matrix (see Sect. 2) and it is a polynomial in the elimination ideal $I \cap \mathbb{Q}[Y][x_1]$. It is used to eliminate $n - 1$ variables from a polynomial system in n variables.

Let $R = \sum_{k=0}^{d} r_k(y_1, \cdots, y_m) x_1^k \in \mathbb{Q}[Y][x_1]$ be the Dixon resultant of $\mathcal{F}$ in x_1 where $d = \deg(R, x_1) > 0$. Let $C = \gcd(r_0, \cdots, r_d)$ be the polynomial content of R. In this paper we will compute the monic square-free factors of R. The monic square-free factorization of R is a factorization of the form $\hat{r} \prod_{j=1}^{l} R_j^j$ such that

1. $\hat{r} = C/L$ for some $L \in \mathbb{Q}[Y]$,
2. each R_j is monic and square-free in $\mathbb{Q}(Y)[x_1]$, i.e., $\gcd(R_j, R_j') = 1$, and
3. $\gcd(R_i, R_j) = 1$ for $i \neq j$.

This monic square-free factorization exists and it is unique [8, Section 14.6]. Note, the factors R_j are not necessarily irreducible over $\mathbb{Q}$. The monic square-free part S of R is the product of the monic square-free factors R_j, that is, $S = \prod_{j=1}^{l} R_j$.

© The Author(s), under exclusive license to Springer Nature Switzerland AG 2022
F. Boulier et al. (Eds.): CASC 2022, LNCS 13366, pp. 185–205, 2022.
https://doi.org/10.1007/978-3-031-14788-3_11

In this paper we present a new Dixon resultant algorithm that interpolates the monic square-free factors R_j one at a time and does not interpolate R. We interpolate the R_j's because it is cheap to compute a square-free factorization of a monic image of R and the square-free factorization factors will be consistent from one image to the next with high probability. Interpolating the R_j's instead of R results in a huge gain because all unwanted repeated factors and the polynomial content are removed. The advantage of our algorithm over other known polynomial interpolation algorithms [2, 25] is that the number of polynomial terms in R_j to be interpolated is much less than in R. Furthermore, the number of primes used by our algorithm in the sparse interpolation step when we apply the Chinese remainder theorem is reduced. Thus the number of black box[1] probes required to interpolate the monic square-free factors R_j is much fewer than the number required to interpolate R. We give a real example from [14].

Example 1. *[14, robot arms system, page 17] Let*

$$C = -65536 \left(al^2 + 1\right)^8 l_2^8 \left(al^2 l_2^2 + 2al^2 l_2 l_3 + al^2 l_3^2 + l_2^2 - 2l_2 l_3 + l_3^2\right)^4,$$

polynomial content

$$A_1 = t_1^2 + 1,$$

$$A_2 = (al^2 l_1^2 + 2al^2 l_1 x - al^2 l_2^2 - 2al^2 l_2 l_3 - al^2 l_3^2 + al^2 x^2 + al^2 y^2 + l_1^2 + 2l_1 x - l_2^2$$
$$+ 2l_2 l_3 - l_3^2 + x^2 + y^2)t_1^2 + (-4al^2 l_1 y - 4l_1 y) t_1 + al^2 l_1^2 - 2al^2 l_1 x - al^2 l_2^2$$
$$- 2al^2 l_2 l_3 - al^2 l_3^2 + al^2 x^2 + al^2 y^2 + l_1^2 - 2l_1 x - l_2^2 + 2l_2 l_3 - l_3^2 + x^2 + y^2,$$

$$A_3 = (aa^2 + 2aal_2)t_1^2 + aa^2 - 4aal_1 + 2aal_2 + 4l_1^2 - 4l_1 l_2,$$

$$A_4 = (aa^2 - 2aal_2)t_1^2 + aa^2 - 4aal_1 - 2aal_2 + 4l_1^2 + 4l_1 l_2,$$

where $X = \{t_1, t_2, b_1, b_2\}$ are the variables, t_1 is the main variable and $Y = \{aa, al, l_1, l_2, l_3, x, y\}$ are the parameters. The Dixon resultant R of the robot arms system in t_1 has $6,924,715$ terms in expanded form and it factors as

$$C A_1^{24} A_2^4 A_3^2 A_4^2.$$

Our new Dixon resultant algorithm computes R_1, R_2 and R_3 where $R_1 = A_1$, $R_2 = \mathrm{monic}(A_2, t_1)$ and $R_3 = \mathrm{monic}(A_3 A_4, t_1)$. The largest coefficient of R_1, R_2 and R_3 is the leading coefficient of A_2 which has only 14 terms! Notice that R_1 and R_2 are irreducible over $\mathbb{Q}$ but R_3 is not.

Our motivation to investigate Dixon resultants stems from sets of parametric polynomial systems listed in [11, 12, 14, 15]. Lewis tried to solve these polynomial systems using Gröbner bases and Triangular sets in Maple and Magma, but they often failed badly; they took a very long time to execute and often ran out

---

[1] A black box is a computer program that takes as input a list of integers together with a prime and outputs the evaluation of the represented object modulo the prime. Black box representations are space efficient. The represented object such as a polynomial, a rational function, and a determinant of a matrix of polynomials is assumed to be unknown. A function call to the black box is referred to as a black box probe.

of memory. The failure of these methods is due to the intermediate expression swell caused by the parameters. This led Lewis to develop the Dixon-EDF (Early Detection Factor) algorithm [14] which is a variant of the Gaussian elimination. It is a modified row reduction of the Dixon matrix that factors out the gcd of each pivot row at each step. The Dixon-EDF method is able to detect factors of the Dixon resultant early. One can interrupt it part way to switch to another method. Lewis often switches to the Gentleman & Johnson minor expansion algorithm [7] to finish the computation. The drawback of the Dixon-EDF method is that it is not automatic and expression swell may occur when computing in $\mathbb{Q}[Y, x_1]$.

Our first contribution is a new algorithm that computes the monic square-free factors R_j of R from monic univariate images in x_1 using sparse multivariate rational function interpolation to interpolate the coefficients of R_j in $\mathbb{Q}(Y)$ modulo primes and uses Chinese remaindering and rational number reconstruction [8,18] to recover the rational coefficients of R_j. We have modified the sparse rational function interpolation algorithm of Cuyt and Lee [4] for this purpose. The only interpolation method that has been applied to Dixon resultants that we are aware of was done by Kapur and Saxena in [12]. They used Zippel's sparse interpolation [25] to interpolate R. Zippel's method does $O(m\hat{D}t)$ black box probes for the first image modulo a prime, where m is the number of parameters, $\hat{D} = \deg(R, x_1) + \sum_{i=1}^{m} \deg(R, y_i)$ and $t = \#R$. But one has to recover the integer coefficients of R which may need more primes. Using the support of the result obtained for the first prime, the integer coefficients can be recovered using $O(t)$ probes to the black box for each subsequent prime [25].

Our second contribution is a Maple + C implementation of our algorithm. For our benchmark problem (*Heron5d* system [22]), the Gentleman & Johnson algorithm ran out of space (>64 GB), Zippel's algorithm takes more than 10^5 s and our new algorithm takes 23.12 s on 1 core.

We provide an overview of our Dixon resultant algorithm. Let $\hat{r} \prod_{j=1}^{l} R_j^j$ be the monic square-free factorization of R. For $1 \le j \le l$, our algorithm will compute each R_j in the form

$$R_j = x_1^{d_{T_j}} + \sum_{k=0}^{T_j-1} \frac{f_{jk}(y_1, y_2, \cdots, y_m)}{g_{jk}(y_1, y_2, \cdots, y_m)} x_1^{d_{j_k}} \in \mathbb{Q}(y_1, y_2, \cdots, y_m)[x_1],$$

where $\gcd(f_{jk}, g_{jk}) = 1$, $f_{jk}, g_{jk} \in \mathbb{Q}[y_1, y_2, \cdots, y_m]$ and $d_{T_j} = \deg(R_j, x_1)$.

If more primes are needed to recover the R_j's, one can set up a system of linear equations using the support found with the first prime to solve for the coefficients of f_{jk} and g_{jk} before doing Chinese remaindering. This method costs $O(\sum_{k=1}^{l} \sum_{k=1}^{T_j} d_{j,k}^3)$ arithmetic operations in $\mathbb{Z}_p$, where p is the prime and $d_{j,k} = \#f_{jk} + \#g_{jk}$ is the total number of unknowns in the k-th rational coefficient of R_j. Instead, we reduce this cost to $O(\sum_{j=1}^{l} \sum_{k=1}^{T_j} d_{j,k}^2)$ arithmetic operations in $\mathbb{Z}_p$ as follows. We pick α and β in $\mathbb{Z}_p^m$ at random, a shift $s \in [1, p-2]$ at random and probe the black box to compute

$$G(\alpha^i, x_1, z) := x_1^{d_{T_j}} + \sum_{k=0}^{T_j-1} \frac{f_{jk}(z\beta_1 + \alpha_1^{s+i}, \cdots, z\beta_m + \alpha_m^{s+i})}{g_{jk}(z\beta_1 + \alpha_1^{s+i}, \cdots, z\beta_m + \alpha_m^{s+i})} x_1^{d_{j_k}} \in \mathbb{Z}_p(z)[x_1]$$

for $0 \leq i < N$ and $N = \max_{j=1}^{l} \max_{k=0}^{T_j-1} \{\#f_{jk}, \#g_{jk}\}$. Then for $1 \leq j \leq l$, we collect the $\#f_{jk}$ (or $\#g_{jk}$) rational coefficients modulo p from $G(\alpha^i, x_1, 0)$ and set up a shifted transposed Vandermonde system [9,25] to solve for the coefficients of f_{jk} and g_{jk} for each R_j.

In a preliminary stage of this work, we first designed our algorithm to interpolate the monic square-free part $S = \prod_{j=1}^{l} R_j$ from the monic univariate images of R in x_1. But we discovered that when $l > 1$, interpolating the R_j's instead of S often reduces the number of black box probes required. These savings are realized because there is a further reduction in the number of terms in the largest polynomial coefficient of R_j to be interpolated compared to the monic product S. Also, the same monic univariate images that yield the first monic square-free factor R_1 can re-used for subsequent monic square-free factors in R.

Table 1 contains the number of black box probes required for interpolating S versus interpolating the monic square-free factors R_j one at a time for the *robot arms* problem [14] and it shows a significant reduction in the number of black box probes when the main variable is t_1, t_2 or b_2.

Table 1. Interpolating S versus interpolating the square-free factors R_j

Main variable	t_1	t_2	b_1	b_2
Interpolating square-free part S	$222,301$	$3,137,373$	$116,741$	$5,531,491$
Interpolating square-free-factors R_j one at a time	$19,241$	$1,210,889$	$116,741$	$1,335,853$
Savings in # of probes	$203,060$	$1,926,484$	0	$4,195,638$
# of terms in the largest coefficient of R_j	14	691	85	624
# of terms in the largest coefficient of S	106	$2,200$	85	$2,388$

Notice in column b_1 that both methods used the same number of black box probes. This is because the number of terms in the largest polynomial coefficient of R_j and S are the same. Thus no savings is realized even though the number of the monic square-free factors for this case is more than 1.

Paper Outline

In Sect. 2, we present a Dixon resultant formulation for polynomial systems. In Sect. 3, we give an overview of the rational function interpolation algorithm of Cuyt and Lee [4] and we modify it to use Kronecker substitutions to combat the large prime and unlucky evaluation point problems that occurs when the adopted sparse polynomial algorithm in Cuyt and Lee's method is the Ben-Or/Tiwari sparse polynomial algorithm [2]. Our algorithms without their failure probability bounds are presented in Sect. 4. In Sect. 5, we explain how we evaluate the polynomial entries in the Dixon matrix which is the most expensive part of our algorithm and we compare our new Dixon resultant algorithm with a Maple implementation of the Gentleman & Johnson minor expansion algorithm and a Maple+C implementation of Zippel's algorithm to interpolate R on the parametric polynomial systems from [14,15].

2   Dixon Resultants

Let $\mathbf{x}^{\alpha} = x_1^{\alpha_1} x_2^{\alpha_2} \cdots x_n^{\alpha_n}$ and let $\{\bar{x}_1, \bar{x}_2, \cdots, \bar{x}_n\}$ be a set of new variables. For each $i \in \{0, 1, 2, \cdots, n\}$, we define $\pi_i(\mathbf{x}^{\alpha}) = \bar{x}_1^{\alpha_1} \bar{x}_2^{\alpha_2} \cdots \bar{x}_i^{\alpha_i} x_{i+1}^{\alpha_{i+1}} x_{i+2}^{\alpha_{i+2}} \cdots x_n^{\alpha_n}$ such that $\pi_0(\mathbf{x}^{\alpha}) = \mathbf{x}^{\alpha}$. Extending the map π_i naturally to polynomials, we have

$$\pi_i(f(x_1, x_2, \cdots, x_n)) = f(\bar{x}_1, \bar{x}_2, \cdots, \bar{x}_i, x_{i+1}, x_{i+2} \cdots, x_n).$$

There are three major steps involved in computing the Dixon resultant of a polynomial system. The first step is to construct the cancellation matrix [5,6]. We refer to the determinant of the cancellation matrix as the Dixon polynomial. The Dixon polynomial acts as the link between the cancellation matrix and the Dixon matrix. Although it is important to select the order of the $n-1$ variables to eliminate because the order affects the size and degree of the Dixon polynomial, we do not focus on the optimal order. Further information about the optimal order can be found in [3,17].

Definition 2. *Given a polynomial system $\mathcal{F}$, let $X_e = \{x_2, \cdots, x_n\}$ be the set of variables to be eliminated and let x_1 be the main variable to appear in the Dixon resultant. Let $\overline{X}_e = \{\bar{x}_2, \bar{x}_3 \cdots, \bar{x}_n\}$ be the set of the new variables corresponding to X_e. We define the $n \times n$ cancellation matrix*

$$\mathcal{C} = \begin{pmatrix} \pi_0(f_1(X_e)) & \pi_0(f_2(X_e)) \ldots & \pi_0(f_n(X_e)) \\ \pi_1(f_1(X_e)) & \pi_1(f_2(X_e)) \ldots & \pi_1(f_n(X_e)) \\ \vdots & \vdots & \vdots \\ \pi_{n-1}(f_1(X_e)) & \pi_{n-1}(f_2(X_e)) \ldots & \pi_{n-1}(f_n(X_e)) \end{pmatrix}. \tag{1}$$

Definition 3. *Let $P = \prod_{i=1}^{n-1}(X_{e_i} - \overline{X}_{e_i})$ and let $\Delta_{X_e} = \dfrac{\det(\mathcal{C})}{P}$. We refer to $\Delta_{X_e} \in \mathbb{Q}[Y, x_1][X_e, \overline{X}_e]$ as the Dixon polynomial of $\mathcal{F}$ with respect to X_e.*

The determinant of the cancellation matrix $\det(\mathcal{C})$ is a multiple of the Dixon polynomial Δ_{X_e}. One must not compute Δ_{X_e} by expanding $\det(\mathcal{C})$ then dividing by P because $\det(\mathcal{C})$ which equals $P \times \Delta_{X_e}$, is much bigger than Δ_{X_e}, since there are 2^{n-1} terms in P when P is expanded. Instead, we follow Lewis [16] and create a new cancellation matrix $\hat{\mathcal{C}}$ using the identity

$$\text{Row } (\hat{\mathcal{C}}_1) = \text{Row}(\mathcal{C}_1), \quad \text{Row } (\hat{\mathcal{C}}_{i+1}) = \frac{\text{Row } (\mathcal{C}_{i+1}) - \text{Row } (\mathcal{C}_i)}{X_{e_i} - \overline{X}_{e_i}} \tag{2}$$

for $i = n - 1, n - 2 \cdots 1$ and then compute the determinant of $\hat{\mathcal{C}}$ which produces the Dixon polynomial. The second step in Dixon's method is to construct the Dixon matrix from the Dixon polynomial. To do this, we rewrite the Dixon polynomial as a bilinear form. We give the following definition to formalize this.

Definition 4. *Let $\overline{V}$ be a monomial column vector in variables $\overline{X}_e$ when Δ_{X_e} is viewed as a polynomial in variables $\overline{X}_e$ and let V be a monomial row vector* in X_e when Δ_{X_e} is viewed as a polynomial in variables X_e. A Dixon polynomial $\Delta_{X_e} \in \mathbb{Q}[Y, x_1][X_e, \overline{X}_e]$ can be written in bilinear form as $\Delta_{X_e} = VD\overline{V}$ and matrix D is the Dixon matrix with entries in $\mathbb{Q}[Y, x_1]$. The Dixon resultant $R \in \mathbb{Q}[Y, x_1]$ is the determinant of the Dixon matrix D.

Example 5. Let $\mathcal{F} = \{x_2^2 + x_3^2 - y_3^2, (x_2 - y_1)^2 + x_3^2 - y_2^2, -x_3y_1 + 2x_1\}$ with variables $X = \{x_1, x_2, x_3\}$ and parameters $Y = \{y_1, y_2, y_3\}$. Let $X_e = \{x_2, x_3\}$ be the variables to be eliminated and let $\overline{X}_e = \{\bar{x}_2, \bar{x}_3\}$ be the new variables corresponding to X_e. Using the identity 2, it follows that the cancellation matrix

$$\hat{C} = \begin{bmatrix} x_2^2 + x_3^2 - y_3^2 & (x_2 - y_1)^2 + x_3^2 - y_2^2 & -x_3y_1 + 2x_1 \\ x_2 + \bar{x}_2 & x_2 - 2y_1 + \bar{x}_2 & 0 \\ x_3 + \bar{x}_3 & x_3 + \bar{x}_3 & -y_1 \end{bmatrix}$$

and the Dixon polynomial

$$\Delta_{X_e} = y_1 \left(-2x_2y_1 + y_1^2 - y_2^2 + y_3^2\right) \bar{x}_2 + y_1 \left(x_2y_1^2 - x_2\, y_2^2 + x_2\, y_3^2 - 2y_1\, y_3^2 + 4x_1x_3\right)$$
$$+ y_1 \left(-2x_3y_1 + 4x_1\right) \bar{x}_3.$$

The Dixon polynomial Δ_{X_e} expressed in bilinear form yields

$$VD\overline{V} = \begin{bmatrix} x_2 & x_3 & 1 \end{bmatrix} \begin{bmatrix} -2y_1^2 & 0 & y_1^3 - y_1y_2^2 + y_1y_3^2 \\ 0 & -2y_1^2 & 4x_1y_1 \\ y_1^3 - y_1y_2^2 + y_1y_3^2 & 4x_1y_1 & -2y_1^2y_3^2 \end{bmatrix} \begin{bmatrix} \bar{x}_2 \\ \bar{x}_3 \\ 1 \end{bmatrix}.$$

Finally, the Dixon resultant $R = \det(D)$ is

$$2y_1^4(16x_1^2 + y_1^4 - 2y_1^2y_2^2 - 2y_1^2y_3^2 + y_2^4 - 2y_2^2y_3^2 + y_3^4).$$

An alternative method for constructing the Dixon matrix D can be found in [23,24]. This method also avoids the intermediate expression swell as it constructs D as the product of a transformation matrix F and a Sylvester matrix $\hat{S}$ using an extended recurrence formula.

The last step of the Dixon's method is to compute the determinant of the Dixon matrix D. Unfortunately, the Dixon matrix obtained may be rectangular thus eliminating the possibility of computing its determinant or it may be singular thus providing no information about the solutions of $\mathcal{F}$. Dixon's method was originally designed to compute Dixon resultants of $n + 1$ generic n-degree polynomials in n variables. However, for geometric problems arising in practice, the Dixon resultant is almost always zero because these systems do not have a generic degree shape [11]. These problems were addressed by Kapur, Saxena and Yang in [11]. They proved that the determinant of any maximal minor M of the Dixon matrix D is an element of the elimination ideal $I \cap \mathbb{Q}[Y][x_1]$. Thus, once a Dixon matrix D is constructed, we find any minor of D of maximal rank, and compute its determinant. Hence, the requirement for $\mathcal{F}$ to be generic n-degree in Dixon's method is no longer necessary.

Our idea to select a maximal minor M of a Dixon matrix D proceeds as follows. We pick a 62 bit prime p and choose $\beta \in \mathbb{Z}_p^{m+1}$ at random. Then we compute $B = D(\beta)$ and identify a maximal minor from B in D with high probability. This requires Gaussian elimination over $\mathbb{Z}_p$ only and in contrast to Kapur, Saxena and Yang [11] crucially avoids doing polynomial arithmetic in $\mathbb{Q}[Y, x_1]$.

3 Modified Interpolation Using Kronecker Substitution

Let $f = \sum_{k=1}^{t} a_k M_k(x_1, \cdots, x_n) \in \mathbb{Z}[x_1, \cdots, x_n]$ with $a_k \neq 0$ be a sparse polynomial. The Ben-Or/Tiwari algorithm [2] interpolates f using $2T$ points $\{(2^j, 3^j, \cdots, p_n^j) : 0 \leq j \leq 2T - 1\}$ where p_n is the n-th prime assuming a term bound $T \geq t$ is known. In this work, the Ben-Or/Tiwari algorithm is the preferred polynomial algorithm for the Cuyt and Lee's rational function interpolation algorithm [4] because it requires the fewest number of black box probes.

Let $m_i = M_i(2, 3, \cdots, p_n)$ be the monomial evaluations. The Ben-Or/Tiwari algorithm is done modulo a prime p satisfying $p > \max_{i=1}^{t} m_i \leq p_n^d$ where $d = \deg f$. However, such a prime p may be too large to use machine arithmetic. For example, suppose $n = 8$ and $\deg(f, x_i) = 11$. Then the prime p required by the Ben-Or/Tiwari sparse polynomial algorithm must be larger than $2^{11} 3^{11} \cdots 19^{11} = 7.2 \times 10^{77}$. This is the primary disadvantage of using the Ben-Or/Tiwari algorithm. Also, one has to deal with unlucky evaluation points problem posed by using points $(2^j, 3^j, \cdots, p_n^j)$ in modular GCD algorithms [9].

We avoid these problems in the Cuyt and Lee sparse multivariate rational function interpolation algorithm by using Kronecker substitution to map a multivariate rational function into a univariate rational function and we evaluate at powers of a generator of $\mathbb{Z}_p^*$ instead of powers of prime $(2^j, 3^j, \cdots, p_n^j)$. To invert a Kronecker substitution, we need to know the partial degrees of a multivariate rational function $A = f/g$ for all variables involved.

3.1 Partial Degrees of $A = f/g$ in Each Variable

Let $A = f/g$ be a rational function in variables $y_1, \cdots, y_m$. Let $d_{f_i} \geq \deg(f, y_i)$ and $d_{g_i} \geq \deg(g, y_i)$ be partial degree bounds. Let A be viewed as

$$A = f/g = \frac{\sum_{k=0}^{d_{f_i}} a_k(y_1, \cdots, y_{i-1}, y_{i+1}, y_{i+2}, \cdots, y_m) y_i^k}{\sum_{k=0}^{d_{g_i}} b_k(y_1, \cdots, y_{i-1}, y_{i+1}, y_{i+2}, \cdots, y_m) y_i^k}$$

such that $f, g \in \mathbb{Z}[y_1, y_2, \cdots, y_{i-1}, y_{i+1}, y_{i+2}, \cdots, y_m][y_i]$. Let p be a prime and let z be a new variable. Let $\alpha = (\alpha_1, \cdots, \alpha_{i-1}, \alpha_{i+1}, \cdots, \alpha_m) \in (\mathbb{Z}_p \setminus \{0\})^{m-1}$ be selected at random. To obtain partial degree bounds for each d_{f_i} and d_{g_i}, we use enough distinct points for z selected at random from $\mathbb{Z}_p \setminus \{0\}$ and compute

$$H_i(z) := H_{f_i}/H_{g_i} = A(\alpha_1, \cdots, \alpha_{i-1}, \theta z + \beta, \alpha_{i+1}, \cdots, \alpha_n) \in \mathbb{Z}_p(z)$$

such that $d_{f_i} = \deg(H_{f_i}, z)$ and $d_{g_i} = \deg(H_{g_i}, z)$ where $\beta, \theta \in \mathbb{Z}_p$ are chosen at random. Observe that if $\mathrm{LC}(H_{f_i}, z)(\alpha) = 0$ or $\mathrm{LC}(H_{g_i}, z)(\alpha) = 0$ then the wrong partial degrees would be obtained. For example, let $A = f/g = \frac{(2-y_3)y_1^2 y_2 + y_1}{y_1 + y_2}$ and suppose we want to determine $\deg(A, y_1)$. Let prime $p = 3137$ and let z be a new variable. Let $H_1(z) := H_{f_1}/H_{g_1} = A(\theta z + \beta, \alpha_2, \alpha_3) = \frac{(2-\alpha_3)(\theta z+\beta)^2 \alpha_2 + \theta z + \beta}{\theta z + \beta + \alpha_2}$. Observe that if $\alpha_3 = 2$ then $\mathrm{LC}(H_{f_1}, z)(\alpha_2, 2) = 0$ for any $\beta, \theta, \alpha_2 \in \mathbb{Z}_p$. The wrong partial degree bound of f will be returned in this case since $H_1(z) = A(\theta z + \beta, \alpha_2, 2) = \frac{\theta z + \beta}{\theta z + \alpha_2 + \beta}$. Thus it is important that we pick prime $p \gg \deg f \deg g$ and α randomly.

3.2   Algorithm by Cuyt and Lee

Let $\mathbb{K}$ be a field and let $A = f/g \in \mathbb{K}(y_1, \cdots, y_m)$ be a sparse multivariate rational function with $\gcd(f, g) = 1$. Cuyt and Lee's rational function algorithm [4] reduces interpolation of sparse rational functions to sparse polynomials interpolation. The first step in their algorithm is to introduce a homogenizing variable z to form the auxiliary rational function

$$A(y_1 z, \cdots, y_m z) = \frac{f_0 + f_1(y_1, \cdots, y_m)z + \cdots + f_{\deg f}(y_1, \cdots, y_m)z^{\deg f}}{g_0 + g_1(y_1, \cdots, y_m)z + \cdots + g_{\deg g}(y_1, \cdots, y_m)z^{\deg g}}.$$

In the case when constant terms g_0 and f_0 are both zero, one has to pick $\beta \in (\mathbb{K} \setminus \{0\})^m$ and perform a basis shift to obtain auxiliary rational function $\hat{A}(y_1, \cdots, y_m, z) := A(y_1 z + \beta_1, \cdots, y_m z + \beta_m)$ so that

$$\hat{A}(y_1, \cdots, y_m, z) = \frac{f_0 + f_1(y_1, \cdots, y_m)z + \cdots + f_{\deg f}(y_1, \cdots, y_m)z^{\deg f}}{g_0 + g_1(y_1, \cdots, y_m)z + \cdots + g_{\deg g}(y_1, \cdots, y_m)z^{\deg g}}.$$

The basis shift forces the auxiliary rational function to have non-zero constant terms f_0 and g_0. This is important because their method normalizes on f_0 or g_0. That is they write

$$\hat{A}(y_1, \cdots, y_m, z) = \frac{\frac{f_0}{g_0} + \frac{f_1(y_1, \cdots, y_m)}{g_0}z + \cdots + \frac{f_{\deg f}(y_1, \cdots, y_m)}{g_0}z^{\deg f}}{1 + \frac{g_1(y_1, \cdots, y_m)}{g_0}z + \cdots + \frac{g_{\deg g}(y_1, \cdots, y_m)}{g_0}z^{\deg g}}$$

Thus for a black box rational function $A = f/g$, we interpolate $\hat{A}$ using univariate dense auxiliary rational functions

$$\hat{A}(\alpha^j, z) = \frac{\frac{f_0}{g_0} + \frac{f_1(\alpha^j)}{g_0}z + \cdots + \frac{f_{\deg f}(\alpha)}{g_0}z^{\deg f}}{1 + \frac{g_1(\alpha^j)}{g_0}z + \cdots + \frac{g_{\deg g}(\alpha^j)}{g_0}z^{\deg g}} \in \mathbb{K}(z)$$

for $j = 0, 1, 2, \cdots$. To interpolate $\hat{A}(\alpha^j, z)$ we use $\deg f + \deg g + 2$ black box probes on z. Since the sparsity of $A = f/g$ is destroyed by the basis shift, Cuyt and Lee adjust the coefficients of the lower degree terms in the numerator and denominator of $\hat{A}(\alpha^j, z)$ by the contributions from the higher degree terms before the coefficients interpolation step is performed. We will show how to do this in our Dixon resultant algorithm (See Subroutine Remove-Shift on page 196). Thus using an appropriate sparse polynomial interpolation algorithm such as [2,25], the adjusted coefficients of the auxiliary rational functions produces the desired rational function $A = f/g$ that was represented by a black box.

3.3   Kronecker Substitution

Using a Kronecker substitution in Cuyt and Lee's method, we reduce the problem of interpolating a sparse multivariate rational function into a univariate rational function interpolation.

Definition 6. *Let $\mathbb{K}$ be an integral domain and let $A = f/g \in \mathbb{K}(y_1, \cdots, y_m)$. Let $r = (r_1, r_2, \cdots, r_{m-1}) \in \mathbb{Z}^{m-1}$ with $r_i > 0$. Let $K_r : \mathbb{K}(y_1, \cdots, y_m) \to \mathbb{K}(y)$ be the Kronecker substitution*

$$K_r(A) = \frac{f(y, y^{r_1}, y^{r_1 r_2}, \cdots, y^{r_1 r_2 \cdots r_{m-1}})}{g(y, y^{r_1}, y^{r_1 r_2}, \cdots, y^{r_1 r_2 \cdots r_{m-1}})}.$$

Let $d_i = max\{(\deg f, y_i), \deg(g, y_i)\}$ for $1 \le i \le m$. Provided we choose $r_i > d_i$ for $1 \le i \le m-1$ then K_r is invertible, $g \ne 0$ and $K_r(A) = 0 \iff f = 0$.

Definition 7. *Let $\mathbb{K}$ be a field and let $A = f/g \in \mathbb{K}(y_1, \cdots, y_m)$ such that $\gcd(f, g) = 1$. Let z be the homogenizing variable and let $r = (r_1, \cdots, r_{m-1})$ with $r_i > d_i = max\{(\deg f, y_i), \deg(g, y_i)\}$. Let K_r be the Kronecker substitution and let*

$$F(y, z) = \frac{f(zy, zy^{r_1}, \cdots, zy^{r_1 r_2 \cdots r_{m-1}})}{g(zy, zy^{r_1}, \cdots, zy^{r_1 r_2 \cdots r_{m-1}})} \in \mathbb{K}[y](z).$$

Following the presentation of auxiliary rational functions in [4], we need to guarantee the existence of a constant term in the denominator of $F(y, z)$. Thus we use a basis shift $\beta \in (\mathbb{K} \setminus \{0\})^m$ and instead define an auxiliary rational function

$$F(y, z) := \frac{f(zy + \beta_1, zy^{r_1} + \beta_2, \cdots, zy^{r_1 r_2 \cdots r_{m-1}} + \beta_m)}{g(zy + \beta_1, zy^{r_1} + \beta_2, \cdots, zy^{r_1 r_2 \cdots r_{m-1}} + \beta_m)} \in \mathbb{K}[y](z) \qquad (3)$$

with Kronecker substitution K_r.

Although the degree of the mapped rational function $K_r(A)$ is exponential in y, the degree of the auxiliary functions with Kronecker substitution $F(y, z)$ in z through which $K_r(A)$ is interpolated remains the same. Consequently, the number of terms and the number of probes needed to interpolate $A = f/g$ does not change. To recover the exponents in y we require prime $p > \prod_{i=1}^{m} r_i$.

Example 8. *Let*

$$A = f/g = \frac{y_1 + y_2 + y_3}{y_1 + y_3} \in \mathbb{Z}_{3137}(y_1, y_2, y_3).$$

Observe that $d_i = max\{\deg(f, y_i), \deg(g, y_i)\} = 1$ for $1 \le i \le 2$. Let $r = (2, 2)$ where $r_i > d_i$ and let $\beta = (2, 3, 5)$ be a basis shift. Let $K_r(A) = A(y, y^2, y^4) = \frac{y^4 + y^2 + y}{y^4 + y}$. Then $F(y, z) = A(zy+2, zy^2+3, zy^4+5) = \frac{(y^4 + y^2 + y)z + 10}{(y^4 + y)z + 7} \in \mathbb{Z}_{3137}[y](z)$ is an auxiliary rational function with Kronecker substitution K_r.

3.4   Bad Evaluation Points

Definition 9. *Let p be prime and let $A = f/g \in \mathbb{Z}_p(y_1, \cdots, y_m)$ with $\gcd(f, g) = 1$. Let $\alpha \in \mathbb{Z}_p \setminus \{0\}$ and $\beta \in (\mathbb{Z}_p \setminus \{0\})^m$ with $A(\beta) \in \mathbb{Z}_p$. Let $i \ge 0$ and let*

$$F(y^i, z) := \frac{f_i(y, z)}{g_i(y, z)} = \frac{f(zy^i + \beta_1, zy^{(r_1)i} + \beta_2, \cdots, zy^{(r_1 r_2 \cdots r_{m-1})i} + \beta_m)}{g(zy^i + \beta_1, zy^{(r_1)i} + \beta_2, \cdots, zy^{(r_1 r_2 \cdots r_{m-1})i} + \beta_m)}$$

be the i-th auxiliary rational function with Kronecker substitution K_r. We say that $\alpha \in \mathbb{Z}_p$ is a bad evaluation point if $\deg(f_i(\alpha, z)) < \deg f$ or $\deg(g_i(\alpha, z)) < \deg g$. That is $\mathrm{LC}(f_i, z)(y = \alpha) = 0$ or $\mathrm{LC}(g_i, z)(y = \alpha) = 0$.

Example 10. *Let*

$$A = f/g = \frac{2891y_1 + y_2 + y_3}{y_2^2 + y_1 + y_3} \in \mathbb{Z}_{3137}(y_1, y_2, y_3).$$

Clearly $\gcd(f, g) = 1$. *The rational function* $A = f/g$ *does not have a constant term in the numerator or denominator. Let* $\beta = (5, 2, 3) \in \mathbb{Z}_{3137}$ *serve as the basis shift for* A. *Let* $r = (2, 3)$ *and let* $K_r(A) = A(y, y^2, y^6) = \frac{y^6 + y^2 + 2891y}{y^6 + y^4 + y}$. *Then an auxiliary rational function* $F(y, z)$ *with Kronecker substitution* K_r *is*

$$F(y, z) = \frac{f_1(y, z)}{g_1(y, z)} = \frac{1912 + (y^6 + y^2 + 2891y)z}{12 + y^4 z^2 + (y^6 + 4y^2 + y)z} \in \mathbb{Z}_{3137}[y](z).$$

If $\alpha = 3$ *is randomly picked in* $\mathbb{Z}_{3137}^*$, *then the auxiliary rational function*

$$F(3, z) = \frac{f_1(\alpha, z)}{g_1(\alpha, z)} = \frac{1108}{z^2 + 2217z + 1162} \in \mathbb{Z}_{3137}(z).$$

Thus $\deg(f_1(\alpha, z)) < 1$ *which implies that* $\alpha = 3$ *is a bad evaluation point.*

We avoid bad evaluation points with high probability in our Dixon resultant algorithm by picking any generator $\alpha \in \mathbb{Z}_p^*$ and a random shift $s \in [1, p-2]$ where p is prime and instead compute $F(\alpha^{s+j}, z)$ for $j = 0, 1, 2, \cdots$ [9].

4   The Dixon Resultant Algorithm

For the purpose of description in this paper, we assume that there is one monic square-free factor to be interpolated. That is, our algorithms are presented to interpolate only one square-free factor. The implementation of our algorithm handles more than one monic square-free factor. Let

$$S = x_1^{d_T} + \sum_{k=0}^{T-1} \frac{f_k(y_1, \cdots, y_m)}{g_k(y_1, \cdots, y_m)} x_1^{d_k}. \tag{4}$$

Definition 11. *Let M be a Dixon matrix of polynomials in $\mathbb{Z}[x_1, y_1, \cdots, y_m]$. For our algorithms, a black box $\boldsymbol{BB} : \mathbb{Z}_p^{m+1} \to \mathbb{Z}_p$ is a program that takes a prime p and $\alpha \in \mathbb{Z}_p^{m+1}$ as inputs and outputs $\det(M(\alpha)) \mod p$.*

The implication of the black box representation of $\det(M)$ is that information such as number of terms and variable degrees are unknown. The degree bounds needed are degrees $[d_0, \cdots, d_T]$ as defined in Eq. 4, total degree bounds for the rational function coefficients $\frac{f_k(y_1, \cdots, y_m)}{g_k(y_1, \cdots, y_m)}$ and the maximum partial degrees $\max\left(\max_{k=0}^{T-1}(\deg(f_k, y_i), \deg(g_k, y_i))\right)$ of S with respect to each variable y_i. For lack of space, we will not present the algorithms to compute these degree bounds.

We now present our Dixon resultant algorithm labelled as algorithm Dixon-Res. It calls Algorithms SparseKron, MQRFR and Subroutines PolyInterp, Rat-Fun, Remove-Shift, VanderSolver, BMStep. The MQRFR algorithm is the Maximal Quotient Rational Function Reconstruction algorithm in [13, page 186].

Algorithm 1: DixonRes

Input: A prime p and a black box $\mathbf{BB} : \mathbb{Z}_p^{m+1} \to \mathbb{Z}_p$

Output: A square-free factor $\overline{S} \in \mathbb{Q}[x_1, y_1, \cdots, y_m]$ of R or **FAIL**.

1 Compute T and $d = [d_0, \cdots, d_T]$ as defined in 4 and $\hat{D} = \deg(\det M, x_1)$.

2 Compute $e_k = \deg(f_k) + \deg(g_k) + 2$ for $0 \le k \le T - 1$.

3 Let $e_{\max} = \max_{k=0}^{T-1} \{e_k\}$ and assume that $e_0 \ge e_1, \cdots \ge e_{T-1}$.

4 Compute $D_{y_i} = \max\left(\max_{k=0}^{T-1}(\deg(f_k, y_i), \deg(g_k, y_i))\right)$ for $1 \le i \le m - 1$.

5 Initialize $r_i = D_{y_i} + 1$ for $1 \le i \le m$. // Prime $p > \prod_{j=1}^{m} r_i$.

6 Let $K_r : \mathbb{Z}_p(y_1, \cdots, y_m)[x_1] \to \mathbb{Z}_p(y)[x_1]$ be the Kronecker substitution $K_r(S)$ where S is as defined in 4 and $r = (r_1, r_2, \cdots, r_{m-1})$.

7 Pick a random basis shift $\beta \in (\mathbb{Z}_p \setminus \{0\})^m$ such that $\mathbf{BB}(\beta) \in \mathbb{Z}_p$.

8 Pick a random shift $s \in [1, p-2]$ and any generator α for $\mathbb{Z}_p^*$.

9 Pick $\theta \in \mathbb{Z}_p^{e_{\max}}, \delta \in \mathbb{Z}_p^{\hat{D}+1}$ at random and set $k = 0$.

10 **for** $i = 1, 2, \cdots$ **while** $k \le T - 1$ **do**

11     $\hat{Y}_i \leftarrow (\alpha^{s+i-1}, \alpha^{(s+i-1)r_1}, \cdots \alpha^{(s+i-1)(r_1 r_2 \cdots r_{m-1})})$.

12   Let $Z_i = [\hat{Y}_i \theta_j + \beta \in \mathbb{Z}_p^m : 1 \le j \le e_{\max}]$ be the evaluation points.

13     $H_i \leftarrow \text{PolyInterp}(\mathbf{BB}, Z_i, \delta, d_T, e_{\max})$ // $|H_i| = e_{\max}$

14   **if** $H_i = \mathbf{FAIL}$ return **FAIL** **end**

15   **if** $i \in \{2, 4, 6, 10, 16, 26, \cdots\}$ **then**

16   **for** $j = 1, 2, \ldots, i$ **do**

17           $A_j \leftarrow \text{RatFun}(H_j, \theta, d_k, e_k, p)$ // $A_j = N_j(z)/\hat{N}_j(z) \in \mathbb{Z}_p(z)$.

18   **if** $\deg(N_j, z) \ne \deg(f_k)$ or $\deg(\hat{N}_j, z) \ne \deg(g_k)$ **then**

19   return **FAIL** // α^{s+j-1} is a bad evaluation point

20   **end**

21   **end**

22        $F_k \leftarrow \text{BMStep}([\text{coeff}(N_j, z^{\deg(f_k)}) : 1 \le j \le i], \alpha, s, r)$.

23        $G_k \leftarrow \text{BMStep}([\text{coeff}(\hat{N}_j, z^{\deg(g_k)}) : 1 \le j \le i], \alpha, s, r)$

24   **if** $F_k \ne \mathbf{FAIL}$ and $G_k \ne \mathbf{FAIL}$ **then**

25           $f_k \leftarrow \text{Remove-Shift}(F_k, [\hat{Y}_1, \cdots, \hat{Y}_i], [N_1, \cdots, N_i], \alpha, s, \beta, r)$

26           $g_k \leftarrow \text{Remove-Shift}(G_k, [\hat{Y}_1, \cdots, \hat{Y}_i], [\hat{N}_1, \cdots, \hat{N}_i], \alpha, s, \beta, r)$

27   **if** $f_k \ne \mathbf{FAIL}$ and $g_k \ne \mathbf{FAIL}$ **then**

28              $k \leftarrow k + 1$ // We have interpolated the k-th coefficient of S.

29   **end**

30   **end**

31   **end**

32 **end**

33 $\hat{S} \leftarrow x_1^{d_T} + \sum_{k=0}^{T-1} \frac{f_k(y_1, \cdots, y_m)}{g_k(y_1, \cdots, y_m)} x_1^{d_k}$ // $\hat{S} = S \mod p$ where S is as defined in 4

34 $L \leftarrow \text{LCM}\{g_k \in \mathbb{Z}_p[y_1, y_2, \cdots, y_m] : 0 \le k \le T - 1\}$

35 $\overline{M} \leftarrow \hat{S} \times L \in \mathbb{Z}_p[x_1, y_1, y_2, \cdots, y_m]$. // Clear the denominators

36 Apply rational number reconstruction to the coefficients of $\overline{M} \mod p$ to get $\overline{S}$

37 **if** $\overline{S} \ne \text{FAIL}$ **then**

38   return $\overline{S}$

39 **else**

40     $\overline{S} \leftarrow \text{SparseKron}(\mathbf{BB}, \hat{S}, \overline{M}, \{(\deg f_k, \deg g_k) : 0 \le k \le T - 1\}, e_{\max}, \hat{D}, d_T)$

41   **if** $\overline{S} \ne \text{FAIL}$ **then return** $\overline{S}$ **else return** FAIL **end**

42 **end**

Subroutine 2: Remove-Shift : The effect of the basis shift β is corrected

Input: $\overline{F} \in \mathbb{Z}_p[y_1, \cdots, y_m]$, basis shift $\beta \in (\mathbb{Z}_p \setminus \{0\})^m$, shift $s \in [1, p-2]$ and r which defines the Kronecker substitution K_r.

Input: $[\hat{Y}_j \in \mathbb{Z}_p^m : 1 \le j \le i], [N_j \in \mathbb{Z}_p[z] : 1 \le j \le i]$ and a generator α for $\mathbb{Z}_p^*$

Output: $f_k \in \mathbb{Z}_p[y_1, \cdots, y_m]$ or **FAIL**

1 $(\overline{A}, f_k) \leftarrow (\overline{F}, \overline{F})$
2 Initialize $\Gamma_j = 0$ for $j = 1, 2, \cdots, i$.
3 **for** $d = \deg(\overline{F}) - 1, \deg(\overline{F}) - 2, \cdots, 0$ **do**
4   **if** $\overline{A} \ne 0$ **then**
5   Pick $\theta \in \mathbb{Z}_p^{d+2}$ at random.
6   **for** $j = 1, 2, \cdots, i$ **do**
7   Compute polynomial evaluations :
$$\{Z_{j,t} = \overline{A}(\hat{Y}_{j,1}\theta_t + \beta_1, \cdots, \hat{Y}_{j,m}\theta_t + \beta_m) \mod p : 1 \le t \le d+2\}.$$
8   Interpolate $\overline{W}_j \in \mathbb{Z}_p[z]$ using points $(\theta_t, Z_{j,t} : 1 \le t \le d+2)$.
9          $\Gamma_j \leftarrow \Gamma_j + \overline{W}_j$
10   **end**
11   **end**
12   **if** $d \ne 0$ **then**
13   Compute $P = [\text{coeff}(N_j, z^d) - \text{coeff}(\Gamma_j, z^d) \mod p : 1 \le j \le i]$.
       // The P_j's are adjusted to correct the effect of the basis shift β.//
14   **if** $[P_j = 0 : 1 \le j \le i]$ **then**
15          $\overline{A} = 0$ // There is no monomial of total degree d.
16   **else**
17          $\overline{A} \leftarrow$ BMStep$([P_1, \cdots, P_i], \alpha, s, r)$. // $\overline{A} \in \mathbb{Z}_p[y_1, \cdots, y_m]$.
18   **if** $\overline{A} = $ **FAIL then** return **FAIL end** // More P_j's are needed.
19   **end**
20   **else**
21       $\overline{A} \leftarrow \text{coeff}(N_1, z^0) - \text{coeff}(\Gamma_1, z^0) \mod p$// We get the constant term.
22   **end**
23    $f_k \leftarrow f_k + \overline{A}$.
24 **end**
25 **return** f_k.

Subroutine 3: BMStep

Input: $P = [P_j \in \mathbb{Z}_p : 1 \le j \le i], i$ is even, $\alpha \in \mathbb{Z}_p$, shift $s \in [1, p-2]$ and r which defines the Kronecker substitution K_r.

Output: $\bar{F} \in \mathbb{Z}_p[y_1, y_2, \cdots, y_m]$ or **FAIL**.

1 Run the Berlekamp-Massey algorithm on P to obtain the polynomial $\lambda(z)$.
2 **if** $\deg(\lambda, z) = \frac{i}{2}$ **then** return **FAIL end** // More images are needed
3 Compute the roots of $\lambda(z)$ in $\mathbb{Z}_p[z]$ to obtain the monomial evaluations $\hat{m}_i$.
4 Let $\hat{m} \subset \mathbb{Z}_p$ be the set of monomial evaluations $\hat{m}_i$ and let $t = |\hat{m}|$.
5 **if** $t \ne \deg(\lambda, z)$ **then** return **FAIL end** // $\lambda(z)$ is wrong.
6 Solve $\alpha^{e_i} = \hat{m}_i$ for e_i with $e_i \in [0, p-2]$ // The exponents are found here.
7 Let $\hat{M} = [y^{e_i} : i = 1, 2 \cdots, t]$ // These are the monomials
8 $F \leftarrow$ VanderSolver $(\hat{m}, [P_1, \cdots P_t], s, \hat{M})$ // $F \in \mathbb{Z}_p[y]$.
9 $\bar{F} \leftarrow K_r^{-1}(F) \in \mathbb{Z}_p[y_1, \cdots, y_m]$.// Invert the Kronecker map K_r.
10 **return** $\bar{F}$

We use the Berlekamp-Massey algorithm [1] to find the term bounds for the leading term polynomials in $f_k(y_1, \cdots, y_m)$ and $g_k(y_1, \cdots, y_m)$ by computing the corresponding feedback polynomial $\lambda(z)$ after $i = 2, 4, 6, \cdots, \cdots$ points and we wait until $\deg(\lambda, z) < \frac{i}{2}$. The condition $\deg(\lambda, z) < \frac{i}{2}$ ensures that $\lambda(z)$ is correct with high probability. This process of determining these term bounds is done by the two calls to Subroutine BMStep in Lines 24 and 25 of Algorithm DixonRes. If Subroutine BMStep succeeds in getting the correct term bound with high probability then the output F_k or G_k is not equal to FAIL. By design it follows that the polynomials F_k or G_k are the highest degree terms in the numerator and denominator of $\frac{f_k(y_1, \cdots, y_m)}{g_k(y_1, \cdots, y_m)}$.

Next, Algorithm DixonRes sends the leading term polynomials F_k and G_k to Subroutine Remove-Shift in Lines 27 and 28 to interpolate other lower degree polynomial terms in $f_k(y_1, \cdots, y_m)$ and $g_k(y_1, \cdots, y_m)$. However, the term bound that was sufficient for interpolating the leading term polynomials might be too small for other lower degree polynomial terms in $f_k(y_1, \cdots, y_m)$ and $g_k(y_1, \cdots, y_m)$. If this happens then Subroutine Remove-Shift will output FAIL. Thus more univariate images and auxiliary rational functions are computed in Algorithm DixonRes and a new term bound is found.

We need to solve shifted transposed Vandermonde systems using Subroutine VanderSolver [9] because Algorithms DixonRes and SparseKron randomized their evaluation points with a shift $s \in [1, p-2]$. To solve the shifted transposed Vandermonde system

$$
Va = \begin{bmatrix} \hat{m}_1^s & \hat{m}_2^s & \cdots & \hat{m}_t^s \\ \hat{m}_1^{s+1} & \hat{m}_2^{s+1} & \cdots & \hat{m}_t^{s+1} \\ \vdots & \vdots & \vdots & \vdots \\ \hat{m}_1^{s+t-1} & \hat{m}_2^{s+t-1} & \cdots & \hat{m}_t^{s+t-1} \end{bmatrix} \begin{bmatrix} a_1 \\ a_2 \\ \vdots \\ a_t \end{bmatrix} = \begin{bmatrix} v_0 \\ v_1 \\ \vdots \\ v_{t-1} \end{bmatrix} = B,
$$

where $\hat{m}_i$ are the monomial evaluations, we use Zippel's $O(t^2)$ algorithm [25] to first solve the transposed Vandermonde system

$$
Wc = \begin{bmatrix} 1 & 1 & \cdots & 1 \\ \hat{m}_1 & \hat{m}_2 & \cdots & \hat{m}_t \\ \vdots & \vdots & \vdots & \vdots \\ \hat{m}_1^{t-1} & \hat{m}_2^{t-1} & \cdots & \hat{m}_t^{t-1} \end{bmatrix} \begin{bmatrix} c_1 \\ c_2 \\ \vdots \\ c_t \end{bmatrix} = \begin{bmatrix} v_0 \\ v_1 \\ \vdots \\ v_{t-1} \end{bmatrix} = B,
$$

which yields $c = W^{-1}B$. Notice that $V = WD$ where D is a $t \times t$ diagonal matrix with entries $D_{ii} = \hat{m}_i^s$. Thus we obtain the unknown coefficients a_i using $a_i = \hat{m}_i^{-s}c_i$ since $Va = B \implies (WD)a = B \implies (Da) = W^{-1}B = c \implies a = D^{-1}c$.

Subroutine 4: RatFun : Rational function interpolation using MQRFR [13]

Input: $H = [H_j \in \mathbb{Z}_p[x_1] : 1 \le j \le e_{\max}], \theta \in \mathbb{Z}_p^{e_{\max}}$ and degrees d_k, e_k.

Output: $A(z) = \frac{N(z)}{\hat{N}(z)} \in \mathbb{Z}_p(z)$ such that $\hat{N}(z) = 1 + \sum_{j=1}^{\deg(\hat{N}, z)} a_j z^j \in \mathbb{Z}_p[z]$.

1   $m(z) \leftarrow \prod_{i=1}^{e_k}(z - \theta_i) \in \mathbb{Z}_p[z]$.
2   Interpolate $U \in \mathbb{Z}_p[z]$ using points $(\theta_i, \text{coeff}(H_i, x_1^{d_k}) : 1 \le i \le e_k)$.
3   $A(z) \leftarrow \text{MQRFR}(m, U, p)$
4   **return** $A(z)$.

Algorithm 5: SparseKron

Comment: If Algorithm DixonRes does not succeed in getting the square free factor, then SparseKron gets more images using the support from Algorithm DixonRes with new primes, performs Chinese remaindering + rational number reconstruction.

Input: $\hat{S} = x_1^{d_T} + \sum_{k=0}^{T-1} \frac{f_k(y_1,y_2,\cdots,y_m)}{g_k(y_1,y_2,\cdots,y_m)} x_1^{d_k} \in \mathbb{Z}_{p_1}(y_1,\cdots,y_m)[x_1]$

Input: $\overline{M} \in \mathbb{Z}_{p_1}[x_1, y_1, \cdots, y_m]$ where prime p_1 is from Algorithm DixonRes

Input: Degree bounds $\{(\deg(f_k), \deg(g_k)) : 0 \le k \le T-1\}$ and $e_{\max}$.

Input: $\hat{D} = \deg(\det M, x_1) + 1$, and $d_T = \deg(S, x_1)$ as defined in 4.

Input: Black box **BB** : $\mathbb{Z}_q^{m+1} \to \mathbb{Z}_q$ where $q \ne p_1$.

Output: Square-free factor $\bar{F} \in \mathbb{Q}[x_1, y_1, \cdots, y_m]$ or **FAIL**.

1 Let $N = \max_{k=0}^{T-1}\{\#f_k, \#g_k\}$ and set $(p, P) \leftarrow (p_1, p_1)$.

2 **do**

3   Get a new 62 bit prime $q > P$.

4   Pick $\beta, \alpha \in (\mathbb{Z}_q \setminus \{0\})^m$, $\delta \in \mathbb{Z}_q^{\hat{D}}, \theta \in \mathbb{Z}_q^{e_{\max}}$ and $s \in [1, q-2]$ at random.

5   **for** $i = 0, 1, \cdots, N-1$ **do**

6   Set $\alpha_i^* = (\alpha_1^{s+i}, \alpha_2^{s+i} \cdots, \alpha_m^{s+i})$.

7   Let $Z_i = [\beta\theta_j + \alpha_i^* \in \mathbb{Z}_q^m : 1 \le j \le e_{\max}]$ be the evaluation points.

8      $H_i \leftarrow \text{PolyInterp}(\mathbf{BB}, Z_i, \delta, d_T, e_{\max})$

9   **if** $H_i = $ **FAIL** then return **FAIL end**

10   **end**

11   **for** $k = 0, 1, \cdots, T-1$ **do**

12     $\hat{m} \leftarrow [\hat{M}_i(\alpha) : 1 \le i \le \hat{n}]$ where $\hat{n} = \#f_k$ and $\hat{M} = \text{supp}(f_k)$.

13     $\bar{m} \leftarrow [\bar{M}_i(\alpha) : 1 \le i \le \bar{n}]$ where $\bar{n} = \#g_k$ and $\bar{M} = \text{supp}(g_k)$.

14   **if** *the monomial evaluations $\hat{m}_i$ or $\bar{m}_i$ are not distinct* **then**

15 | return **FAIL.**

16   **end**

17   **for** $j = 0, 1, 2, \cdots, N-1$ **do**

18 | $B_j \leftarrow \text{RatFun}(H_j, d_k, \theta, e_k, q)$ // $B_j = N_j(z)/\hat{N}_j(z) \in \mathbb{Z}_q(z)$.

19 | **if** $\deg(N_j, z) \ne \deg(f_k)$ *or* $\deg(\hat{N}_j, z) \ne \deg(g_k)$ **then**

20 | | return **FAIL.**

21 | **end**

22 | $(U_j(z), V_j(z)) \leftarrow (N_j(z) \times \text{LC}(\hat{N}_j, z), \hat{N}_j(z) \times \text{LC}(\hat{N}_j, z))$

23 | $(a_j, b_j) \leftarrow (U_j(0), V_j(0))$ // $a_j, b_j \in \mathbb{Z}_q$

24 | $F_k \leftarrow \text{VanderSolver}(\hat{m}, [a_1, \cdots, a_{\hat{n}}], s, \hat{M})$.

25 | $G_k \leftarrow \text{VanderSolver}(\bar{m}, [b_1, \cdots, b_{\bar{n}}], s, \bar{M})$.

26   **end**

27   **end**

28   $\hat{S} \leftarrow x_1^{d_T} + \sum_{k=0}^{T-1} \frac{F_k(y_1,y_2,\cdots,y_m)}{G_k(y_1,y_2,\cdots,y_m)} x_1^{d_k} \in \mathbb{Z}_q(y_1,\cdots,y_m)[x_1]$

29   $L \leftarrow \text{LCM}\{G_k \in \mathbb{Z}_q[y_1, y_2, \cdots, y_m] : 0 \le k \le T-1\}$

30   $\underline{M} \leftarrow \hat{S} \times L \in \mathbb{Z}_q[x_1, y_1, y_2, \cdots, y_m]$. // Clear the denominators.

31   Solve $\hat{F} \equiv \overline{M}$ mod p and $\hat{F} \equiv \underline{M}$ mod q using the Chinese remainder algorithm and set $p = p \times q$.

32   Apply rational number reconstruction on coefficients of $\hat{F}$ mod p to get $\overline{F}$

33   **if** $\overline{F} \ne $ **FAIL** then return $\overline{F}$ **end**

34   $(\overline{M}, P) \leftarrow (\hat{F}, q)$

35 **end**

Subroutine 6: VanderSolver

> **Input:** Vectors $\hat{m}, b \in \mathbb{Z}_p^t$, shift $s \in [1, p-2]$ and monomials $[M_1, \cdots, M_t]$
> **Output:** $F \in \mathbb{Z}_p[y_1, \cdots, y_m]$
> 1 Let $V_{ij} = \hat{m}_i^{s+j-1}$ for $1 \le i, j \le t$.
> 2 Solve $Va = b$ for the coefficients a_i using Zippel's $O(t^2)$ algorithm [26].
> 3 **return** $F = \sum_{i=1}^t a_i M_i$

Subroutine 7: PolyInterp

> **Input:** Black box **BB** : $\mathbb{Z}_p^{m+1} \to \mathbb{Z}_p$.
> **Input:** $Z = [Z_j \in \mathbb{Z}_p^m : 1 \le j \le e_{\max}], \delta \in \mathbb{Z}_p^{\hat{D}+1}$, degree $d_T = \deg(\det S, x_1)$.
> **Output:** $H = [\text{monic}(H_j) \in \mathbb{Z}_p[x_1] : 1 \le j \le e_{\max}]$ or **FAIL**.
> 1 **for** $j = 1, 2, \ldots, e_{max}$ **do**
> 2   Compute $G_j = (\mathbf{BB}(\delta_i, Z_j) : 1 \le i \le \hat{D} + 1)$.
> 3   Interpolate $B_j \in \mathbb{Z}_p[x_1]$ using points $(\delta_i, G_{j,i} : 1 \le i \le \hat{D} + 1)$.
> 4   Compute the square-free part $H_j = B_j / \gcd(B_j, B_j')$.
> 5   **if** $\deg(H_j, x_1) \ne d_T$ **then** return **FAIL end**
> 6 **end**
> 7 **return** $[\text{monic}(H_1), \cdots, \text{monic}(H_{e_{\max}})]$.

5 Implementation Notes and Benchmarks

We have implemented our new Dixon resultant algorithm in Maple. To improve the overall efficiency, we have implemented in C major subroutines such as evaluating a Dixon matrix at integer points modulo prime p, computing the determinant of an integer matrix over $\mathbb{Z}_p$ and performing dense rational function interpolation using the MQRFR algorithm modulo a prime [13]. Thus each probe to the black box is computed using C code. Our C code supports primes up to 63 bits in length.

5.1 Speeding Up Evaluation of the Dixon Matrix

In our experiments, the most expensive step in our algorithm was, and still is, evaluating the Dixon matrix M modulo a prime. Let p be a prime and let M be a $t \times t$ matrix of polynomials in $\mathbb{Z}[z_1, \ldots, z_n]$. We need to compute $\det(M(\alpha))$ mod p for many $\alpha \in \mathbb{Z}_p^n$. Often, over 80% of the time is spent computing $M(\alpha)$ mod p. The Maple command

```
> Eval(M,{seq(z[i]=alpha[i]}) mod p;
```

does what we want, however, because we want our implementation to handle many variables and fail with low probability, we want to use the largest primes the hardware can support which are 63 bit primes if we use signed 64 bit integers. Unfortunately, Eval uses hardware arithmetic for $p < 2^{31}$, otherwise, it uses software arithmetic which is relatively very slow. Also, Eval evaluates each polynomial in M independently, that is, if $M_{1,1} = 2z_1^3 z_2$ and $M_{2,2} = z_1^3 + 5z_3$

say, `Eval` computes α_1^3 twice. To speed up evaluations we have written a C program to compute $M(\alpha)$ for $p < 2^{63}$ using hardware arithmetic. In Maple, we first precompute a vector of degrees

$$D = \left[\max_{1 \leq i,j \leq t} \deg(M_{ij}, z_k) \ : \ 1 \leq k \leq n \right].$$

For each $\alpha \in \mathbb{Z}_p^n$ we call our C program from Maple with inputs M, α, D, p. To save multiplications our C program first computes power arrays

$$P_k = \left[\alpha_k^i \ : \ 0 \leq i \leq D_k \right] \quad \text{for} \ 1 \leq k \leq n$$

then uses these P_k to evaluate $M_{i,j}(\alpha)$ for $1 \leq i,j \leq t$. Maple uses two data structures for polynomials, the SUM-OF-PROD data structure and the POLY data structure. POLY was added to Maple in 2013 by Monagan and Pearce [19] to speed up polynomial arithmetic. Figure 1 shows the POLY data structure for the polynomial $f = 9\,xy^3z - 4\,y^3z^2 - 6\,xy^2z - 8\,x^3 - 5$. Figure 2 shows how the same polynomial is represented in the SUM-OF-PROD data structure. All boxes in Figs. 1 and 2 represent arrays. The first entry in each box is a header word; it encodes the object type and the array length.

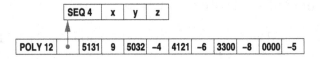

Fig. 1. Maple's POLY representation for $f = 9\,xy^3z - 4\,y^3z^2 - 6\,xy^2z - 8\,x^3 - 5$

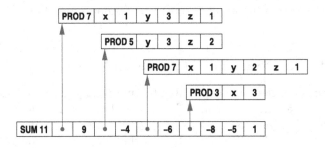

Fig. 2. Maple's SUM-OF-PROD representation for $f = 9\,xy^3z - 4\,y^3z^2 - 6\,xy^2z - 8\,x^3 - 5$

In POLY, if $M = z_1^{d_1} z_2^{d_2} \cdots z_n^{d_n}$ is a monomial in f, then M is encoded as the integer $d2^{nb} + \sum_{i=0}^{n-1} 2^{ib} d_i$ where $d = \sum_{i=1}^{n} d_i$ and $b = \lfloor 64/(n+1) \rfloor$. For example, the monomial xy^3z with $d = 5, b = 16, n = 3$ is encoded as the integer $5 \cdot 2^{48} + 2^{32} + 3 \cdot 2^{16} + 1$. This is depicted as 5131 in Fig. 1. This encoding allows Maple to compare two monomials in the graded monomial ordering using a single 64 bit integer comparison. Also, provided overflow does not occur, Maple can multiply two monomials using a single 64 bit integer addition.

When does Maple use POLY instead of SUM-OF-PROD? If a polynomial f has (i) all integer coefficients, (ii) more than one term, (iii) is not linear, and (iv) all monomials in f can be encoded in a 64 bit integer using B bits for d_i and $64 - nB$ bits for d, then it is encoded using POLY otherwise the SUM-OF-PROD representation is used. In a typical Dixon matrix both representations are used so we have to handle both and we need to know the details of both representations.

Also important for efficiency is how to multiply in $\mathbb{Z}_p$. We do not use the hardware division instruction which is very slow. Instead we use Roman Pearce's assembler implementation of Möller and Granlund [20] which replaces division with two multiplications and other cheap operations.

Table 2. Timings showing improvements for Heron5d and Tot systems

System	Eval	Determinant	Total	C-Eval	New Total
Heron5d	70.17s (66.2%)	9.74s (9.18%)	106.07s	18.02s (3.89x)	42.82s (2.48x)
Tot	635.75s (83.3%)	37.66s (4.9%)	763.2s	32.36s (19.64x)	150s (5.08x)

Table 2 shows the improvement obtained using our C code for evaluating a Dixon matrix M at integer points modulo a prime for both Tot and Heron5d systems. Column `Eval` contains the timings using `Eval` command and column C-Eval is the timings for the case when our C code was used. Column Determinant is the amount of time spent computing the determinant of integer matrices modulo a prime. Column Total contains the total CPU timings using `Eval` and column New Total is the new total CPU timings for both polynomial systems when the C code for matrix evaluation was used.

5.2 Timings

We present two tables for our Dixon resultant algorithm. Table 3 contains basic information about the polynomial systems that is stored on the web at www.cecm.sfu.ca/~mmonagan/code/DixonRes. This web address also contain our Maple and C codes and they are freely available for use. Table 3 includes timings comparing three methods. Columns 1–4 contain names of the polynomial systems, the number of equations in each system, the dimension of the Dixon matrix D and the rank of its maximal minor M respectively. The number of terms in the product of all the monic square-free factors in expanded form when the denominators are cleared is denoted by $\#S$, the number of terms in R labelled $\#R$ is in column 7 and column 6 labelled $t_{\max} = \max(\#f_{jk}, \#g_{jk})$. In column 8 named as DRes, we report the timings of our Dixon resultant algorithm. Column 9 contain timings of an efficient Maple implementation of the Gentleman & Johnson minor expansion method. The timings of a hybrid implementation of Zippel's sparse algorithm in Maple + C are given in column 11. All our experiments were performed on an Intel Xeon E5-2680 v2 processor using 1 core. The first prime used in our code is the 62 bit prime $p = (2^{50})(61)(67) + 1$.

Table 3. DixonRes versus Minor Expansion and Zippel's Interpolation

System	#Eq	n/m	dim D/Rank	#S	$t_{\max}$	#R	DRes	Minor	Cleaned	Zippel
Robot-t_1	4	4/7	$(32 \times 48)/20$	450	14	6924715	7.34s	2562.6s	188.4s	$> 10^5$s
Robot-t_2	4	4/7	$(32 \times 48)/20$	13016	691	16963876	316.99s	!	2559.6s	$> 10^5$s
Robot-b_1	4	4/7	$(32 \times 48)/20$	334	85	6385205	27.78s	182.4s	15.15s	$> 10^5$s
Robot-b_2	4	4/7	$(32 \times 48)/20$	11737	624	16801877	241.61s	!	2452.8s	$> 10^5$s
Heron5d	15	14/16	$(707 \times 514)/399$	823	822	12167689	23.12s	!	!	$> 10^5$s
Flex-v1	3	3/15	$(8 \times 8)/8$	5685	2481	45773	201s	5.09s	NA	308684.76s
Flex-v2	3	3/15	$(8 \times 8)/8$	12101	2517	45773	461.4s	5.02s	NA	308684.76s
Perimeter	6	6/4	$(16 \times 16)/16$	1980	303	9698	49.97s	18.23	NA	2360.27s
Pose	4	4/8	$(13 \times 13)/12$	24068	8800	24068	461.4s	4.48s	NA	21996.25s
Pendulum	3	2/3	$(40 \times 40)/33$	4667	243	19899	45.46s	1721.50s	NA	2105.321s
Tot	4	4/5	$(85 \times 94)/56$	8930	348	52982	82.11s	!	!	17370.07s
Image3d	10	10/9	$(178 \times 152)/130$	130	84	1456	2.34s	1.04s	NA	53.68s
Heron3d	6	5/7	$(16 \times 14)/13$	23	22	90	0.411s	0.014s	NA	0.738s
Nachtwey	6	6/5	$(11 \times 18)/11$	244	106	244	7.23s	0.424s	NA	5.36s
Storti	6	5/2	$(24 \times 113)/20$	12	4	32	0.177s	229.945s	NA	0.053s

! = ran out of memory, NA= Not Attempted

The Gentleman & Johnson minor expansion algorithm uses a lot of space. To reduce space and speed it up, we first divide each row i of the Dixon matrix M by the gcd of the entries in row i. Then we permute the Dixon matrix M by putting the sparsest columns at the left of the matrix. We call this method the cleaned version of the Gentleman & Johnson method. The timings for it are presented in column 10 labelled as Cleaned.

Our DixonRes algorithm outperforms Zippel's sparse interpolation. This was expected because #R is much larger than $t_{\max}$. Another reason is because more primes are needed to recover integer coefficients in R compared to the R_j's. Our algorithm is not always faster than the Gentleman & Johnson algorithm. The evaluation cost of the Dixon matrix is still the bottleneck of our algorithm while the determinant computation takes roughly 10% of the total time.

Some Dixon matrices have a block diagonal form and often, the determinant of all the blocks produce the same Dixon resultant R. For the timings recorded in Tables 2 and 3, we always compute the determinant of the smallest block after confirming that all blocks produce the same Dixon resultant. So, both #S and #R are the number of terms due to the determinant of the smallest block obtained. However for the Tot system, the 25×25 block matrix did not produce all the monic square-free factors R_j so we had to compute the determinant of the 31×31 block matrix. Details about the block structure of the Dixon matrices are provided in Table 4.

In Table 4, we provide details about block sizes of each Dixon matrix M and the number of black box probes required by our Dixon resultant algorithm to successfully interpolate the R_j's. The quantity Q in Table 4 is the number of black box probes done to obtain all degree bounds needed by Algorithm DixonRes. In Table 4, the quantity p_1 is the number of probes needed to get the

Table 4. Block structure and # of probes used by Algorithm DixonRes and Zippel's interpolation

System	Block Structure	Q	p_1	p_2	Zippel-probes
Robot-t_1	[8, 8]	3641	13000	-	-
Robot-t_2	[12]	5685	705796	-	-
Robot-b_1	[8, 8]	3901	91000	-	-
Robot-b_2	[12]	5489	529984	-	-
Heron5d	[49, 52, 48, 50, 49, 53, 50, 48]	307	62928	-	-
Flex-v1	[8]	1693	588060	-	3310871
Flex-v2	[8]	5017	2664948	-	3310871
Perimeter	[16]	1243	225828	-	230773
Pose	[12]	1072	525636	-	569513
Pendulum	[17, 16]	8971	114920	-	128322
Tot	[31, 25]	4261	420000	-	742099
Image3d	[13, 14, 14, 15, 18, 19, 18, 19]	401	12320	-	29415
Heron3d	[6, 7]	133	1392	-	3071
Nachtwey	[11]	576	39780	18020	12983
Storti	[20]	273	816	-	343

first image of the R_j's. If the rational number reconstruction process fails on the first image, then more primes are needed. The number of black box probes used for the second prime is p_2. One prime is typically enough to interpolate the R_j's. Zippel-probes represents the number of probes used by Zippel's algorithm to interpolate R. Note that the block structure depends on the variable elimination order. For example, we record that the block structure for Robot-b_1 is [8, 8]. For a different variable elimination order, we get [8, 4, 4].

6   Conclusion

We have designed and implemented a new Dixon resultant algorithm that computes the monic square-free factors of the Dixon resultant R of a parametric polynomial system using sparse interpolation tools. We have shown that there is a huge reduction in the number of terms when the monic square-free factors of R are interpolated instead of interpolating R. We have also shown that a Kronecker substitution can be used to reduce the problem of interpolating a multivariate rational function using Cuyt and Lee's method to a univariate rational function interpolation.

We implemented our algorithm in Maple and implemented several subroutines in C including the evaluation of the Dixon matrix modulo a prime. Our benchmarks showed that our algorithm is much faster than Zippel's sparse interpolation. We are currently working on the complexity analysis and the failure probability of our new Dixon resultant algorithm.

References

1. Atti, N.B., Lombardi, H., Diaz-Toca, G.M.: The Berlekamp-Massey algorithm revisited. Appli. Alebra Eng. Commun. **17**(4), 75–82 (2006)
2. Ben-Or, M., Tiwari, P.: A deterministic algorithm for sparse multivariate polynomial interpolation.In: Proceedings of STOC 2020, pp. 301–309, ACM (1988)
3. Chtcherba, A.D., Kapur, D.: On the efficiency and optimality of dixon-based resultant methods. In: Proceedings of ISSAC 2002, pp. 29–36, ACM (2002)
4. Cuyt, A., Lee, W.-S.: Sparse interpolation of multivariate rational functions. J. Theoretical Comp. Sci. **412**, 1445–1456 (2011)
5. Dixon, A.: On a form of the Eliminant of two quantics. Proc. Lond. Math. Soc. **2**, 468–478 (1908)
6. Dixon, A.: The eliminant of three quantics in two independent variables. Proc. Lond. Math. Soc. **2**, 49–69 (1909)
7. Gentleman, W.M., Johnson, S.C.: The evaluation of determinants by expansion by minors and the general problem of substitution. Math. Comput. **28**(126), 543–548 (1974)
8. Gerhard, J., Von zur Gathen, J.: Modern Computer Algebra. Cambridge University Press, New York (2013)
9. Hu, J., Monagan, M.: A fast parallel sparse polynomial GCD algorithm. In: Proceedings of ISSAC 2016, pp. 271–278, ACM (2016)
10. Kapur, D., Saxena, T.: Extraneous factors in the dixon resultant formulation. In: Proceedings of ISSAC 1997 pp. 141–148, ACM (1997)
11. Kapur, D., Saxena, T., Yang, L.: Algebraic and geometric reasoning using dixon resultants. In: Proceedings of ISSAC 1994, pp. 99–107, ACM (1994)
12. Kapur, D., Saxena, T.: Comparison of various multivariate resultant formulations. In: Proceedings of ISSAC 1995, pp. 187–194, ACM (1995)
13. Khodadad, S., Monagan, M.: Fast rational function reconstruction. In: Proceedings of ISSAC 2006, pp. 184–190, ACM (2006)
14. Lewis, R.H.: Dixon-EDF: the premier method for solution of parametric polynomial systems. In: Kotsireas, I.S., Martínez-Moro, E. (eds.) ACA 2015. SPMS, vol. 198, pp. 237–256. Springer, Cham (2017). https://doi.org/10.1007/978-3-319-56932-1_16
15. Lewis, R.H.: Resultants, implicit parameterizations, and intersections of surfaces. In: Davenport, J.H., Kauers, M., Labahn, G., Urban, J. (eds.) ICMS 2018. LNCS, vol. 10931, pp. 310–318. Springer, Cham (2018). https://doi.org/10.1007/978-3-319-96418-8_37
16. Lewis, R.: Private Communication
17. Lewis, R.: New Heuristics and Extensions of the Dixon Resultant for Solving Polynomial Systems, pp. 16–20. ACA, Montreal (2019)
18. Monagan, M.: Maximal quotient rational reconstruction: an almost optimal algorithm for rational reconstruction. In: Proceedings of ISSAC 2004, pp. 243–249, ACM (2004)
19. Monagan, M., Pearce, R.: The design of Maple's sum-of-products and POLY data structures for representing mathematical objects. Commun. Comput. Algebra **48**(4), 166–186, ACM (2014)
20. Möller, N., Grandlund, T.: Improved division by invariant integers. Trans. Comput. **60**(2), 165–175, IEEE (2011)
21. Storti, D.: Algebraic skeleton transform: a symbolic computation challenge, Submitted to Faculty Papers and Data, Mech. Eng. Res. Works Arch. http://hdl.handle.net/1773/48587

22. Tot, J.: Private Communication
23. Xiaolin, Q., Dingxiong, W., Lin, T., Zhenyi, J.: Complexity of constructing Dixon resultant matrix. Int. J. Comput. Math. **94**, 2074–2088 (2017)
24. Zhao, S., Fu, H.: An extended fast algorithm for constructing the Dixon resultant matrix. Sci. China Ser A Math. **48**, 131–143 (2005)
25. Zippel, R.: Probabilistic algorithms for sparse polynomials. In: Ng, E.W. (ed.) Symbolic and Algebraic Computation. LNCS, vol. 72, pp. 216–226. Springer, Heidelberg (1979). https://doi.org/10.1007/3-540-09519-5_73
26. Zippel, R.: Interpolating Polynomials from their values. J. Symbol. Comput. **9**, 375–403 (1990)

Distance Evaluation to the Set of Matrices with Multiple Eigenvalues

Elizaveta Kalinina$^{(\boxtimes)}$ [iD] and Alexei Uteshev [iD]

Faculty of Applied Mathematics, St. Petersburg State University,
7–9 Universitetskaya nab., St. Petersburg 199034, Russia
{e.kalinina,a.uteshev}@spbu.ru
http://www.apmath.spbu.ru

Abstract. The problem of finding the Frobenius distance in the $\mathbb{R}^{n \times n}$ matrix space from a given matrix to the set of matrices possessing multiple eigenvalues is considered. Two approaches are discussed: the one is reducing the problem to a constrained optimization problem in $\mathbb{R}^n$ with a quartic objective function, and the other one is connected with the singular value analysis for an appropriate matrix in $\mathbb{R}^{2n \times 2n}$. Several examples are presented including classes of matrices where the distance in question can be explicitly expressed via the matrix eigenvalues.

Keywords: Wilkinson's problem · Real perturbations · Frobenius norm · 2-norm

1 Introduction

Given a matrix $A \in \mathbb{R}^{n \times n}$ with distinct eigenvalues, we intend to find the distance from A to the set $\mathbb{D}$ of real matrices with multiple eigenvalues as well as the corresponding minimal perturbation, i.e., a matrix $E_* \in \mathbb{R}^{n \times n}$ of the minimal norm such that $B_* = A + E_* \in \mathbb{D}$.

The problem under consideration is known as *Wilkinson's problem* [21] and the desired distance, further denoted as $d(A, \mathbb{D})$, is called the *Wilkinson distance* of A [2,15]. Wilkinson's problem is closely related to ill-conditioning of eigenvalue problems. The ill-conditioning of a linear system is determined by the distance of the coefficient matrix from the set of singular matrices. For eigenvalue problems, the set of matrices with multiple eigenvalues plays the role of singularity [23]. The Wilkinson distance can be considered as a measure of sensitivity of the worst-conditioned eigenvalue of A. By eigenvalue perturbation theory, a matrix that is close to a defective matrix has an eigenvalue with large condition number. Conversely, any matrix with an ill-conditioned eigenvalue is close to a defective matrix [18,22].

For the spectral and the Frobenius norms, the problem has been studied intensively by Wilkinson [22–24] as well as by other researchers [2,4,5,10,14,18]. In the works [1,3,13,15], generalizations of Wilkinson's problem for the cases of

F. Boulier et al. (Eds.): CASC 2022, LNCS 13366, pp. 206–224, 2022.
https://doi.org/10.1007/978-3-031-14788-3_12

prescribed eigenvalues or multiplicities and matrix pencils are studied. However, several aspects of the problem still need further clarification.

The present paper is devoted to the stated problem for the case of Frobenius norm. It is organized as follows.

In Sect. 2, we start with algebraic background for the stated problem. We first detail the structure of the set $\mathbb{D}$ in the matrix space. The cornerstone notion here is the **discriminant** of a characteristic polynomial of a matrix. Being a polynomial function in the entries of the matrix, the discriminant permits one to translate the problem of evaluation of $d(A, \mathbb{D})$ to that of finding the distance from a point to an algebraic manifold in the matrix space. This makes it possible to attack the problem within the framework of the approach already exploited by the present authors in the preceding studies [11,12] on the distance to instability in the matrix space. The approach is aimed at the construction of the so-called **distance equation**, i.e., the univariate equation whose zero set contains all the critical values of the squared distance function. Its construction is theoretically feasible via application of symbolic methods for elimination of variables in an appropriate multivariate algebraic system. Unfortunately, the practical realization faces the variable flood difficulty, where the number of variables grows rapidly with the order of the matrix.

To bypass this, we reformulate the problem in terms of the minimal perturbation matrix. In Sect. 3, we prove that this matrix is a rank 1 matrix. Then we reduce the problem of its finding to that of a constrained optimization n-variate problem with an objective function of order 4. Some examples are presented illuminating the applicability of the developed algorithm.

The discovered property of the perturbation matrix makes it possible to look at the problem from the other side. Generically, the 2-norm of a matrix does not equal its Frobenius norm. However, for the rank 1 matrix (and this is exactly the case of the minimal perturbation matrix), these norms coincide. This allows one to verify the results obtained in the framework of symbolic approach with the counterpart obtained for the 2-norm case [14]. This issue is discussed in Sect. 4 while in Sect. 5, both approaches are illustrated for three classes of matrices where the distance $d(A, \mathbb{D})$ can be explicitly expressed via the eigenvalues of A. These happen to be symmetric, skew-symmetric and orthogonal matrices. Quite unexpected for the authors became the fact that, for some classes, each of their representative had a continuum of nearest matrices in $\mathbb{D}$.

Notation. For a matrix $A \in \mathbb{R}^{n \times n}$, $f_A(\lambda)$ denotes its characteristic polynomial, $\mathrm{adj}(A)$ stands for its adjoint matrix, $d(A, \mathbb{D})$ denotes the distance from A to the set $\mathbb{D}$ of matrices possessing a multiple eigenvalue. E_* and $B_* = A + E_*$ stand for, correspondingly, the (minimal) perturbation matrix and the nearest to A matrix in $\mathbb{D}$ (i.e., $d(A, \mathbb{D}) = \|A - B_*\|$); we then term by λ_* the multiple eigenvalue of B_*. I (or I_n) denotes the identity matrix (of the corresponding order). $\mathcal{D}$ (or $\mathcal{D}_\lambda$) denotes the discriminant of a polynomial (with subscript indicating the variable).

Remark. All the computations were performed in CAS Maple 15.0. (**Linear Algebra** package and functions **discrim**, and **resultant**). Although all the approximate computations have been performed within the accuracy 10^{-40}, the final results are rounded to 10^{-6}.

2   Algebraic Preliminaries

It is well-known that in the $(n + 1)$-dimensional space of the polynomial $f(\lambda) = a_0\lambda^n + a_1\lambda^{n-1} + \cdots + a_n, n \geq 2$ coefficients, the manifold of polynomials with multiple zeros is defined by the equation

$$D(a_0, a_1, \ldots, a_n) = 0 \quad where \; D := \mathcal{D}_\lambda(f(\lambda)) \tag{1}$$

denotes the discriminant of the polynomial. Discriminant can be represented in different ways, for instance, as the Sylvester determinant

$$\mathcal{D}_\lambda(a_0\lambda^4 + a_1\lambda^3 + a_2\lambda^2 + a_3\lambda + a_4) = \frac{1}{4^2} \begin{vmatrix} a_1 & 2a_2 & 3a_3 & 4a_4 & 0 & 0 \\ 0 & a_1 & 2a_2 & 3a_3 & 4a_4 & 0 \\ 0 & 0 & a_1 & 2a_2 & 3a_3 & 4a_4 \\ 0 & 0 & 4a_0 & 3a_1 & 2a_2 & a_3 \\ 0 & 4a_0 & 3a_1 & 2a_2 & a_3 & 0 \\ 4a_0 & 3a_1 & 2a_2 & a_3 & 0 & 0 \end{vmatrix} .$$

The discriminant $D(a_0, a_1, \ldots, a_n)$ is a homogeneous polynomial over $\mathbb{Z}$ of order $2n - 2$ in its variables, and it is irreducible over $\mathbb{Z}$.

The following result [16] is much less known.

Theorem 1 (Jacobi). *If $f(\lambda)$ possesses a unique multiple zero λ_* and its multiplicity equals 2, then the following ratio is valid*

$$1 : \lambda : \lambda^2 : \cdots : \lambda^n = \frac{\partial D}{\partial a_n} : \frac{\partial D}{\partial a_{n-1}} : \frac{\partial D}{\partial a_{n-2}} : \cdots : \frac{\partial D}{\partial a_0} . \tag{2}$$

To solve the problem stated in Introduction, one needs to transfer the discriminant manifold (1) into the matrix space. The corresponding manifold is then defined by a homogeneous polynomial of order $n(n - 1)$ in the matrix entries:

$$\mathfrak{D}(B) := \mathcal{D}_\lambda(f_B(\lambda)) = 0 . \tag{3}$$

We will further denote this manifold in $\mathbb{R}^{n^2}$ as $\mathbb{D}$. The problem of distance evaluation between a given matrix A and $\mathbb{D}$ can be viewed as a constrained optimization problem:

$$d^2(A, \mathbb{D}) = \min_{B \in \mathbb{R}^{n \times n}} \|B - A\|^2 \text{ subject to } (3). \tag{4}$$

Consider the Lagrange function for this problem

$$F(B, \mu) := \|B - A\|^2 - \mu\mathfrak{D}(B) .$$

Evidently, $\partial F/\partial\mu = 0$ is equivalent to (3). Differentiation with respect to the entries of B yields

$$2(b_{jk} - a_{jk}) - \mu\partial\mathfrak{D}(B)/\partial b_{jk} = 0 \quad \text{for} \quad \{j,k\} \subset \{1,\ldots,n\}. \tag{5}$$

Since the system (3)–(5) is an algebraic one, it admits application of symbolic methods of elimination of variables. We attach to the considered system an extra equation

$$z = \|B - A\|^2 \tag{6}$$

and then aim at finding the so-called distance equation

$$\mathcal{F}(z) = 0$$

resulting from the elimination of all the variables but z from this system. Positive zeros of this equation are the critical values of the squared distance function for the problem (4).

Example 1. For the matrix $A = [a_{jk}]_{j,k=1}^2$ with the characteristic polynomial $f_A(\lambda)$, the system (5) is linear with respect to $\{b_{jk}\}_{j,k=1}^2$ and the distance equation is easily computed as

$$\mathcal{F}(z) := 4096(a_{12} - a_{21})^2 \left[(a_{11} - a_{22})^2 + (a_{12} + a_{21})^2\right]$$

$$\times \left\{[4z - \mathcal{D}_\lambda(f_A(\lambda))]^2 - 16(a_{12} - a_{21})^2 z\right\} = 0. \tag{7}$$

It turns out that for any matrix A such that $\mathcal{D}_\lambda(f_A(\lambda)) \neq 0$, the distance equation is the quadratic one (7) where $d^2(A, \mathbb{D})$ equals its minimal zero.

For the matrix

$$A = \begin{bmatrix} s & t \\ -t & s \end{bmatrix} \quad \text{where } t > 0,$$

polynomial $\mathcal{F}(z)$ vanishes identically. Equation (7) possesses a multiple zero, namely $z = t^2$, and $d(A, \mathbb{D}) = t$. Surprisingly, this distance is provided by a continuum of perturbation (and thus nearest in $\mathbb{D}$) matrices, namely

$$E_* = \frac{t}{2}\begin{bmatrix} \sin\varphi & -1 + \cos\varphi \\ 1 + \cos\varphi & -\sin\varphi \end{bmatrix}, \quad \text{where } \varphi \in [0, 2\pi).$$

This example causes an anxious expectation of difficulties to appear while solving the stated distance evaluation problem for the case of orthogonal or skew-symmetric matrices A. □

For a general case, computation of the distance equation via the solution of the system (3)–(5)–(6) is a hardly executable task due to a drastic increase in the number of variables (i.e., the entries of matrix B) to be eliminated. To overcome this difficulty, let us reformulate the problem in terms of the entries of the perturbation matrix.

3   Distance Equation and Perturbation Matrix

Theorem 2. *Matrices E_* are B_* are linked by the equality*

$$E_* = \varkappa\left[f_*(B_*)\right]^\top , \tag{8}$$

were

$$f_*(\lambda) := \frac{f_{B_*}(\lambda)}{\lambda - \lambda_*},$$

and $\varkappa \in \mathbb{R}$ is some scalar.

Proof. We start with system (5) resulting from application of the Lagrange method to problem (4). Compute $\partial\mathfrak{D}(B)/\partial b_{jk}$ as a composite function with the coefficients of characteristic polynomial $f_B(\lambda) = \lambda^n + p_1\lambda^{n-1} + \cdots + p_n$ treated as intermediate variables:

$$\frac{\partial\mathfrak{D}(B)}{\partial b_{jk}} = \frac{\partial\mathfrak{D}(B)}{\partial p_0}\frac{\partial p_0}{\partial b_{jk}} + \frac{\partial\mathfrak{D}(B)}{\partial p_1}\frac{\partial p_1}{\partial b_{jk}} + \cdots + \frac{\partial\mathfrak{D}(B)}{\partial p_n}\frac{\partial p_n}{\partial b_{jk}}.$$

(We set here $p_0 := 1$ and thus the first term in the right-hand side is just 0). Under the condition $\mathfrak{D}(B) = 0$ (i.e., the matrix $B = B_*$ possesses a multiple eigenvalue λ_*), the Jacobi ratio (2) is fulfilled

$$1 : \lambda_* : \lambda_*^2 : \cdots : \lambda_*^n = \frac{\partial\mathfrak{D}}{\partial p_n} : \frac{\partial\mathfrak{D}}{\partial p_{n-1}} : \frac{\partial\mathfrak{D}}{\partial p_{n-2}} : \cdots : \frac{\partial\mathfrak{D}}{\partial p_0}.$$

Therefore,

$$\frac{\partial\mathfrak{D}}{\partial p_\ell} = \kappa\lambda_*^{n-\ell} \text{ for } \ell \in \{1,\ldots,n\}$$

and for some constant $\kappa \in \mathbb{R}$. Consequently

$$\frac{\partial\mathfrak{D}(B)}{\partial b_{jk}} = \kappa\left(\lambda_*^n\frac{\partial p_0}{\partial b_{jk}} + \lambda_*^{n-1}\frac{\partial p_1}{\partial b_{jk}} + \cdots + \frac{\partial p_n}{\partial b_{jk}}\right) = \kappa\frac{\partial f_B(\lambda_*)}{\partial b_{jk}}.$$

The preceding considerations lead to a conclusion that the system of Eqs. (5) is equivalent to the matrix equation

$$2(B_* - A) = \mu\kappa\,\partial f_B(\lambda_*)/\partial B\big|_{B=B_*}.$$

Next utilize the formula of differentiation of characteristic polynomial with respect to the matrix [19]:

$$\partial f_B(\lambda)/\partial B = \left[\operatorname{adj}(\lambda I - B)\right]^\top .$$

Equality (8) then follows from the representation of the adjoint matrix for $\lambda_* I - B_*$ as $f_*(B_*)$ with $f_*(\lambda)$ standing for the quotient on division of $f_B(\lambda)$ by $\lambda - \lambda_*$.

Corollary 1. *Matrices $E_*^\top$ and B_* commute and*

$$(\lambda_* I - B_*)E_*^\top = \mathbb{O}_{n\times n}.$$

Corollary 2. *If A does not have a multiple eigenvalue, then E_* is the rank 1 matrix with only zero eigenvalues.*

Proof. Matrix $f_*(B_*) = \operatorname{adj}(\lambda_* I - B_*)$ is the rank 1 matrix, since its columns are the eigenvectors of the matrix B_* corresponding to λ_* (Cayley–Hamilton theorem).

We next prove that $\operatorname{tr}(\operatorname{adj}(\lambda_* I - B_*)) = 0$. For any matrix B with spectrum $\{\lambda_j\}_{j=1}^n$, matrix $\operatorname{adj}(\lambda I - B)$ has the spectrum [17] (part VII, problem 48):

$$\left\{ \frac{f_B(\lambda)}{\lambda - \lambda_j} \right\}_{j=1}^n .$$

Thus,

$$\operatorname{tr}(E_*) = \operatorname{tr}(\operatorname{adj}(\lambda_* I - B_*)) = f'_{B_*}(\lambda_*) = 0.$$

Corollary 3. *Matrix E_* is normal to B_*, i.e., $\operatorname{tr}(B_*^\top E_*) = 0$.*

Corollary 4. $\operatorname{tr}(B_*) = \operatorname{tr}(A)$.

Theorem 3. *The value $d^2(A, \mathbb{D})$ is contained in the set of critical values of the function*

$$G(U) := U^\top A A^\top U - \left(U^\top A U\right)^2 \quad \text{subject to } U^\top U = 1, U \in \mathbb{R}^n \qquad (9)$$

If U_* is the vector providing $d^2(A, \mathbb{D})$, then the perturbation matrix can be computed by the formula

$$E_* = U_* U_*^\top (\kappa I - A) \quad \text{where } \kappa := U_*^\top A U_*. \qquad (10)$$

Proof. Due to Corollary 2, the singular value decomposition for the perturbation matrix E is represented as

$$E = \sigma U \cdot V^\top \qquad (11)$$

under restrictions

$$U^\top U = 1, \ V^\top V = 1, \ U^\top V = 0. \qquad (12)$$

From the condition $\operatorname{tr}((A + E)E^\top) = 0$ we deduce that $\sigma = -\operatorname{tr}(AVU^\top) = -U^\top AV$. Formulate the constrained optimization problem

$$\min(-U^\top AV) \quad \text{subject to (12)}. \qquad (13)$$

The derivatives of the corresponding Lagrange function

$$L(U, V, \mu_1, \mu_2, \mu_3) := -U^\top AV - \mu_1(U^\top U - 1) - \mu_2(V^\top V - 1) - \mu_3 U^\top V$$

result in the system of linear equations

$$\partial L / \partial U = -AV - 2\mu_1 U - \mu_3 V = \mathbb{O}_{n \times 1}, \qquad (14)$$

$$\partial L / \partial V = -A^\top U - 2\mu_2 V - \mu_3 U = \mathbb{O}_{n \times 1} \qquad (15)$$

with respect to U and V. Multiplication of (14) by $U^\top$ and (15) by $V^\top$ results (in accordance with (12)) in

$$\mu_3 = -V^\top AV = -U^\top AU. \tag{16}$$

Multiplication of (14) by $U^\top$ while (15) by $V^\top$ yields

$$-2\mu_1 = -2\mu_2 = U^\top AV = -\sigma \tag{17}$$

and, provided this value is not 0,

$$V = -\frac{1}{2\mu_2}(A^\top + \mu_3 I)U. \tag{18}$$

Substituting (18) in (11) and taking into account (16), we arrive at (10).

If $\mu_1 = \mu_2 = 0$ then system (14)–(15) is reduced to $AV = -\mu_3 V, A^\top U = -\mu_3 U$. This implies that the matrix A should possess a real eigenvalue κ_1 with the corresponding right and left eigenvectors V_1 and U_1 satisfying the condition $U_1^\top V_1 = 0$. We claim that, in this case, matrix A has a multiple eigenvalue. For the sake of simplicity, we prove this statement under an extra assumption that all the eigenvalues $\kappa_1, \ldots, \kappa_n$ of A are real. Suppose, by contradiction, that they are distinct. One has then

$$\kappa_1 U_1^\top V_j = U_1^\top AV_j = \kappa_j U_1^\top V_j \quad \Rightarrow \quad U_1^\top V_j = 0 \quad for\ j \in \{2, \ldots, n\}$$

and for V_j standing for the right eigenvector corresponding to κ_j. Therefore, U_1 is normal to all the vectors $V_1, V_2, \ldots, V_n$ composing a basis of $\mathbb{R}^n$. The contradiction proves the assertion. The statement of the theorem remains valid with the corresponding critical value of (9) equal to 0.   □

To find the critical values of the function (9), the Lagrange multipliers method is to be applied with the objective function $G(U) - \mu(U^\top U - 1)$. This results into the system

$$AA^\top U - (U^\top AU)(A + A^\top)U - \mu U = \mathbb{O}_{n \times 1} \tag{19}$$

where every equation is now just cubic with respect to the entries of U. This is an essential progress compared to the system (3)–(5)–(6), and makes it feasible to manage the procedure of elimination of variables from the system (19) accomplished with $z - G(U) = 0$ and $U^\top U = 1$ (at least for the matrices of the order $n \leq 8$).

Unfortunately, the new system possesses some extraneous solutions, i.e., those not corresponding to the critical values of the distance function.

Example 2. For the matrix

$$A = \begin{bmatrix} 0 & 1 \\ 13 & -6 \end{bmatrix},$$

the system

$$u_1^2 + u_2^2 = 1, \quad u_2 \partial G/\partial u_1 - u_1 \partial G/\partial u_2 = 0$$

possesses solutions

$$u_1 = \pm\frac{1}{58}\sqrt{2900 + 82\sqrt{22}}, \ u_2 = \pm\frac{1}{58}\sqrt{464 - 87\sqrt{22}}$$

that yield the value $z = 0$. The true distance equation is given by (7), and $d^2(A, \mathbb{D}) = -12\sqrt{58} + 94$ is provided by another solution of the system, namely

$$u_1 = \pm\frac{1}{58}\sqrt{1682 + 203\sqrt{58}}, \ u_2 = \pm\frac{1}{58}\sqrt{1682 - 203\sqrt{58}}.$$

□

The appearance of such extraneous solutions is caused by the non-equivalence of the passage from the original stated problem to that from Theorem 3. For instance, representation (10) is deduced under an extra condition of non-vanishing of value (17).

Example 3. For the Frobenius matrix

$$A = \begin{bmatrix} 0 & 1 & 0 \\ 0 & 0 & 1 \\ -91 & -55 & -13 \end{bmatrix},$$

the distance equation

$$\mathcal{F}(z) := 330760907004023420582465447\,z^6 - 3770391988613062890801451788647\,z^5$$

$$+ 9378649027038813210344501839167\,z^4 - 7718682760987209701497925039997\,z^3$$

$$+ 21107097878782151768402265062447\,z^2 - 51058410014045251854039449677\,z$$

$$+ 319295875259784560640000 = 0$$

possesses the following real zeros

$$z_1 \approx 0.739335, \ z_2 \approx 0.765571, \ z_3 \approx 0.980467, \ z_4 \approx 11396.658548.$$

One has $d(A, \mathbb{D}) = \sqrt{z_1} \approx 0.859846$ and

$$E_* \approx \begin{bmatrix} 0.198499 & -0.195124 & -0.530440 \\ 0.204398 & -0.200922 & -0.546202 \\ -0.000907 & 0.000891 & 0.002424 \end{bmatrix},$$

$$B_* = A + E_* \approx \begin{bmatrix} 0.198499 & 0.804875 & -0.530440 \\ 0.204398 & -0.200923 & 0.453797 \\ -91.000907 & -54.999108 & -12.997576 \end{bmatrix}.$$

The latter matrix possesses the double eigenvalue $\lambda_* \approx 0.824777$. □

Example 4. For the matrix

$$
A = \begin{bmatrix}
5 & -36 & -57 & 85 \\
80 & 90 & 74 & 27 \\
9 & -91 & 81 & 65 \\
-12 & 78 & 5 & -63
\end{bmatrix},
$$

the distance equation is represented by the order 12 irreducible over $\mathbb{Z}$ polynomial $\mathcal{F}(z)$ with the absolute value of coefficients up to 10^{100}. Its real zeros are

$$z_1 \approx 87.614714, \; z_2 \approx 2588.509661, \; z_3 \approx 17853.256334, \; z_4 \approx 32194.078324.$$

One has $d(A, \mathbb{D}) = \sqrt{z_1} \approx 9.360273$ and

$$
E_* \approx \begin{bmatrix}
3.350324 & -0.177130 & -3.704042 & -0.328216 \\
2.489713 & -0.131630 & -2.752569 & -0.243906 \\
2.565863 & -0.135656 & -2.836760 & 0.251366 \\
3.898666 & -0.206121 & -4.310276 & 0.381935
\end{bmatrix},
$$

with the matrix

$$
B_* = A + E_* \approx \begin{bmatrix}
8.350324 & -36.177130 & -60.704042 & 84.671784 \\
82.489713 & 89.868370 & 71.247430 & 26.756094 \\
11.565863 & -91.135656 & 78.163240 & 64.748634 \\
-8.101333 & 77.793879 & 0.689724 & -63.381935
\end{bmatrix}
$$

possessing the double eigenvalue $\lambda_* \approx 69.081077$. □

Some empirical conclusions resulting from about 30 generated matrices of the orders up to $n = 20$. Generically,

(a) The extraneous factor equals z^n, and on its exclusion one has
(b) the order of the distance equation $\mathcal{F}(z) = 0$ equals $n(n-1)$, and, if computed symbolically w.r.t. the entries of A, $\mathcal{F}(0)$ has a factor $[\mathcal{D}_\lambda(f_A(\lambda))]^2$;
(c) $d^2(A, \mathbb{D})$ equals the minimal positive zero of this equation.

Complete computational results for some examples are presented in [20]. For the matrices A with integer entries within $[-99, +99]$ (generated by Maple 15.0. **RandomMatrix** package) we point out some complexity estimates for the distance equation computation (PC AMD FX-6300 6 core 3.5 GHz)

n	deg $\mathcal{F}(z)$	coefficient size	number of real zeros	timing (s)
5	20	$\sim 10^{170}$	10	0.03
10	90	$\sim 10^{780}$	28	0.13
20	380	$\sim 10^{3500}$	36	1940

The adequacy of the results has been extra checked via the nearest matrix B_* computation. This matrix should

(a) possess a double eigenvalue;
(b) have the value $\|B_* - A\|$ equal to the square root of the least positive zero of $\mathcal{F}(z)$;
(c) satisfy the system of equations (3)–(5) (this property has been tested only for the orders $n \le 8$);
(d) have the number of real eigenvalues which differs from that of the matrix A at most by 2.

4   Singular Values

Let $A \in \mathbb{R}^{n \times n}$ be a nonsingular matrix with the singular value decomposition as follows

$$A = WD_nV^\top, \tag{20}$$

where $D_n = \operatorname{diag}\{\sigma_1, \sigma_2, \ldots, \sigma_n\}$, with singular values $\sigma_1 \ge \sigma_2 \ge \ldots \ge \sigma_n > 0$.

The following result [6,8] gives us the distance to the nearest matrix with rank $k < n$.

Theorem 4. *One has*

$$\min_{\operatorname{rank} B = k} \|A - B\| = \|A - A_k\| = \begin{cases} \sigma_{k+1}, & \text{for the 2-norm,} \\ \left[\displaystyle\sum_{i=k+1}^{n} \sigma_i^2 \right]^{1/2} & \text{for the Frobenius norm.} \end{cases}$$

Here

$$A_k = WD_kV^\top, \quad D_k = \operatorname{diag}\{\sigma_1, \sigma_2, \ldots, \sigma_k, 0, \ldots, 0\}.$$

According to this theorem, the Frobenius distance from the nonsingular A to the set of matrices with multiple eigenvalues satisfies the following inequality:

$$d(A, \mathbb{D}) \le \sqrt{\sigma_{n-1}^2 + \sigma_n^2}.$$

As for the distance $d(A, \mathbb{D})$ in the 2-norm, the following result [14] is known:

Theorem 5. *Let the singular values of the matrix*

$$M = \begin{bmatrix} A - \lambda I_n & \gamma I_n \\ \mathbb{O}_{n \times n} & A - \lambda I_n \end{bmatrix} \tag{21}$$

be ordered like $\sigma_1(\lambda, \gamma) \ge \sigma_2(\lambda, \gamma) \ge \ldots \ge \sigma_{2n}(\lambda, \gamma) \ge 0$. Then one has

$$d(A, \mathbb{D}) = \min_{\lambda \in \mathbb{C}} \max_{\gamma \ge 0} \sigma_{2n-1}(\lambda, \gamma).$$

It is well-known that for the matrix $A \in \mathbb{R}^{n \times n}, n \ge 2$, Frobenius norm and the 2-norm are related by the inequality [7]

$$\|A\|_2 \le \|A\|_F \le \sqrt{n}\|A\|_2.$$

It is also known, that $||A||_2 = ||A||_F$ iff $\mathrm{rank}(A) = 1$. According to Corollary 2, both norms coincide for the minimal perturbation E_*. This results in an algorithm for $d(A, \mathbb{D})$ computation that is an alternative to that treated in Sect. 3.

To find singular values of the matrix (21), i.e., zeros of the polynomial

$$\det(MM^\top - \mu I_{2n}) \tag{22}$$

$$= \det \begin{bmatrix} (A - \lambda I_n)(A - \lambda I_n)^\top + \gamma^2 I_n - \mu I_n & \gamma(A - \lambda I_n)^\top \\ \gamma(A - \lambda I_n) & (A - \lambda I_n)(A - \lambda I_n)^\top - \mu I_n \end{bmatrix}$$

treated with respect to μ, is a nontrivial task. We will restrict our consideration to the classes of matrices A where application of Schur formula for the determinant of the block matrix (22) is possible, i.e., transforming it into

$$\det(\mu^2 I_n - \mu[2(A - \lambda I_n)(A - \lambda I_n)^\top + \gamma^2 I_n] + [(A - \lambda I_n)(A - \lambda I_n)^\top]^2). \tag{23}$$

These happen to be symmetric, skew-symmetric, and orthogonal matrices. Singular values of the matrix (21) can be expressed explicitly via the eigenvalues of this matrix.

5   Distance via Matrix Eigenvalues

5.1   Symmetric Matrix

Theorem 6. *Let A be a symmetric matrix with distinct eigenvalues $\lambda_1, \lambda_2, \ldots, \lambda_n$. Then*

$$d(A, \mathbb{D}) = \frac{1}{2} \min_{1 \le k < \ell \le n} |\lambda_k - \lambda_\ell|.$$

If this minimum is attained at the eigenvalues λ_2 and $\lambda_1, \lambda_2 > \lambda_1$, then the perturbation can be found as

$$E_* = \frac{1}{4}(\lambda_2 - \lambda_1)(P_1 + P_2)(P_1 - P_2)^\top, \tag{24}$$

where P_1 and P_2 are the eigenvectors of A corresponding to λ_1 and λ_2 respectively with $||P_1|| = ||P_2|| = 1$.

Remark. Generically, matrices E_* and $B_* = A + E_*$ are not the symmetric ones.

Proof. For $j \in \{1, \ldots, m\}$, denote P_j the eigenvector of A corresponding to λ_j with $||P_j|| = 1$. Then $P = (P_1, P_2, \ldots, P_n)$ is the orthogonal matrix such that

$$P^\top A P = \Lambda \quad \text{where } \Lambda = \mathrm{diag}\,\{\lambda_1, \lambda_2, \ldots, \lambda_n\}.$$

Since the orthogonal transformation does not influence the Frobenius distance, we reduce $d(A, \mathbb{D})$ to $d(\Lambda, \mathbb{D})$.

In this case, $\Lambda - \lambda I = (\Lambda - \lambda I)^\top$ and these matrices commute. Hence, the expression (23) is valid. Therefore, the singular values of the matrix (21) are the zeros of the polynomials

$$\mu^2 - \mu(2(\lambda_j - \lambda)^2 + \gamma^2) + (\lambda_j - \lambda)^4, \ j \in \{1, 2, \ldots, n\},$$

namely

$$\mu_{1,2}^{(j)} = \frac{2(\lambda_j - \lambda)^2 + \gamma^2 \pm \gamma\sqrt{\gamma^2 + 4(\lambda_j - \lambda)^2}}{2}.$$

Differentiating w.r.t. γ, we get the single stationary point $\gamma = 0$. According to [14], to find the 2-norm distance from $A - \lambda I$ to the manifold of matrices with multiple zero eigenvalue, one should find the singular values σ_n and σ_{n-1} for the matrix $(A - \lambda I)$. They are $|\lambda_k - \lambda|$ and $|\lambda_\ell - \lambda|$ for some k, ℓ. The minimal w.r.t. λ value of σ_{n-1} comes up to $|\lambda_k - \lambda_\ell|/2$ where $\lambda_k - \lambda = \lambda - \lambda_\ell$.

Assume that

$$\min_{1 \le k < \ell \le n} |\lambda_k - \lambda_\ell| = |\lambda_1 - \lambda_2|.$$

Denote

$$Q := \begin{bmatrix} \frac{1}{\sqrt{2}} & -\frac{1}{\sqrt{2}} & 0 & \ldots & 0 \\ \frac{1}{\sqrt{2}} & \frac{1}{\sqrt{2}} & 0 & \ldots & 0 \\ 0 & 0 & 1 & \ldots & 0 \\ \ldots & \ldots & \ldots\ldots & & \\ 0 & 0 & 0 & \ldots & 1 \end{bmatrix}, \text{ then } Q^\top \Lambda Q = \begin{bmatrix} \frac{\lambda_1+\lambda_2}{2} & \frac{\lambda_2-\lambda_1}{2} & 0 & \ldots & 0 \\ \frac{\lambda_2-\lambda_1}{2} & \frac{\lambda_1+\lambda_2}{2} & 0 & \ldots & 0 \\ 0 & 0 & 1 & \ldots & 0 \\ \ldots & & \ldots & & \\ 0 & 0 & 0 & \ldots & 1 \end{bmatrix}.$$

For this matrix, $\tilde{E}_* = \begin{bmatrix} 0 & \frac{\lambda_1-\lambda_2}{2} & 0 & \ldots & 0 \\ 0 & 0 & 0 & \ldots & 0 \\ \ldots & \ldots & \ldots\ldots & & \\ 0 & 0 & 0 & \ldots & 0 \end{bmatrix}$. Obviously, we get

$$E_* = QP\tilde{E}_* P^\top Q^\top = \frac{\lambda_1 - \lambda_2}{4}(P_1 + P_2)(P_1 - P_2)^\top.$$

$\square$

Example 5. For the matrix

$$A = \frac{1}{9} \begin{bmatrix} -269 & -98 & 76 \\ -98 & -296 & 22 \\ 76 & 22 & -209 \end{bmatrix},$$

one has

$$\lambda_1 = -45, \lambda_2 = -25, \lambda_3 = -16, P_1 = [2/3, 2/3, -1]^\top, P_2 = [-1/3, 2/3, 2/3]^\top.$$

$$d(A, \mathbb{D}) = \frac{|-25 + 16|}{2} = \frac{9}{2} \text{ and } E_* = \begin{bmatrix} -3/4 & 3/4 & 0 \\ -3/4 & 3/4 & 0 \\ -3 & 3 & 0 \end{bmatrix}.$$

5.2 Skew-Symmetric Matrix

Theorem 7. *Let the nonzero eigenvalues of a skew-symmetric matrix A be*

$$\pm b_1 i, \pm b_2 i, \ldots, \pm b_m i \ where \ 0 < b_1 < b_2 < \ldots < b_m.$$

Then

$$d(A, \mathbb{D}) = b_1$$

and the minimal perturbation can be found as

$$E_* = -b_1 \Re(P_1)\Im(P_1)^\top, \tag{25}$$

where P_1 is the eigenvector of A corresponding to the eigenvalue $b_1 i$ with $\|\Re(P_1)\| = \|\Im(P_1)\| = 1$.

Proof. For $j \in \{1, \ldots, m\}$, denote P_j the eigenvector of A corresponding to $b_j i$ with $\|\Re(P_j)\| = \|\Im(P_j)\| = 1$. If A possesses the zero eigenvalue, denote by P_0 the corresponding eigenvector with $\|P_0\| = 1$. Then the orthogonal matrix

$$P = (\Re(P_1), \Im(P_1), \Re(P_2), \Im(P_2), \ldots, \Re(P_m), \Im(P_m), \{P_0\})$$

is such that

$$P^\top A P = \Upsilon \quad \text{where } \Upsilon := \operatorname{diag}\{\Upsilon_1, \Upsilon_2, \ldots, \Upsilon_m, \{0\}\},$$

$$\Upsilon_k := \begin{bmatrix} 0 & b_k \\ -b_k & 0 \end{bmatrix}, k \in \{1, 2, \ldots, m\}$$

(we set in braces the entries of the matrices corresponding to the case of existence of zero eigenvalue for A).

Since an orthogonal transformation does not influence the Frobenius distance, we reduce $d(A, \mathbb{D})$ to $d(\Upsilon, \mathbb{D})$. In this case,

$$(\Upsilon - \lambda I)(\Upsilon - \lambda I)^\top = \operatorname{diag}\{\tilde{\Upsilon}_1, \tilde{\Upsilon}_2, \ldots, \tilde{\Upsilon}_m, \{0\}\},$$

where

$$\tilde{\Upsilon}_k := \begin{bmatrix} b_k + \lambda^2 & 0 \\ 0 & b_k^2 + \lambda^2 \end{bmatrix} \ for \ k \in \{1, \ldots, m\}.$$

It is evident that

$$(\Upsilon - \lambda I)(\Upsilon - \lambda I)^\top (\Upsilon - \lambda I) = (\Upsilon - \lambda I)^2 (\Upsilon - \lambda I)^\top.$$

Hence, the expression (23) is valid.

Therefore, the singular values of matrix (21) are the zeros of the polynomials

$$\mu^2 - \mu(2(\lambda + b_k)^2 + \gamma^2) + (\lambda^2 + b_k^2)^2, \ k \in \{1, 2, \ldots, m\},$$

namely

$$\mu_{1,2}^{(k)} = \frac{1}{2}\left[2(\lambda + b_k)^2 + \gamma^2 \pm \gamma\sqrt{\gamma^2 + 4(\lambda^2 + b_k^2)} \right].$$

Differentiating w.r.t. γ, we get a single stationary point $\gamma = 0$. According to [14], to find the 2-norm distance from $\Upsilon - \lambda I$ to the manifold of matrices with multiple zero eigenvalue, it is sufficient to compute the singular values σ_n and σ_{n-1} of this matrix. They are

$$\text{either } \sigma_n = \sigma_{n-1} = \sqrt{b_k^2 + \lambda^2} \text{ for some } k, \text{ or } \sigma_{n-1} = \sqrt{b_k^2 + \lambda^2}, \sigma_n = |\lambda|.$$

The minimal w.r.t. λ value of σ_{n-1} comes up to b_1 when $\lambda = 0$.

The corresponding perturbation

$$E_* = P \begin{bmatrix} 0 & -b_1 & 0 & \dots & 0 \\ 0 & 0 & 0 & \dots & 0 \\ \dots & \dots & \dots & \dots & \dots \\ 0 & 0 & 0 & \dots & 0 \end{bmatrix} P^\top = -b_1 \Re(P_1) \Im(P_1)^\top.$$

$\square$

Corollary 5. *In the notation of Theorem 7, the distance $d(A, \mathbb{D})$ is provided by a continuum of perturbations E_* contained in the set*

$$\left\{ -b_1 (\eta \Re(P_1) + \theta \Im(P_1))(-\eta \Im(P_1) + \theta \Re(P_1))^\top \mid \{\eta, \theta\} \subset \mathbb{R}, \eta^2 + \theta^2 = 1 \right\}.$$

5.3   Orthogonal Matrix

Theorem 8. *Let $n \geq 3$, and the eigenvalues of an orthogonal matrix A, other than ± 1, be*

$$\cos\alpha_1 \pm i \sin\alpha_1, \cos\alpha_2 \pm i \sin\alpha_2, \dots, \cos\alpha_m \pm i \sin\alpha_m, \tag{26}$$

where $0 < \sin\alpha_1 \leq \sin\alpha_2 \leq \dots \leq \sin\alpha_m$. Then

$$d(A, \mathbb{D}) = \sin\alpha_1, \tag{27}$$

and the minimal perturbation can be found as

$$E_* = -(\sin\alpha_1) \Re(P_1) \Im(P_1)^\top, \tag{28}$$

where P_1 is the eigenvector of A corresponding to the eigenvalue $\cos\alpha_1 + i \sin\alpha_1$ with $\|\Re(P_1)\| = \|\Im(P_1)\| = 1$.

We present two independent proofs for this result: the first one following from Theorem 3 while the second one exploiting the considerations of Sect. 4.

Proof. I. Since $AA^\top = I$, the objective function (9) can be transformed into

$$G(U) = 1 - \left(U^\top A U\right)^2,$$

and system (19) is then replaced by

$$\left(U^\top A U\right)\left(A^\top + A\right) U - \mu U = \mathbb{O}. \tag{29}$$

Multiply it by $U^\top A^\top$:

$$(U^\top AU)\,[U^\top (A^\top)^2 U + 1 - \mu] = 0,$$

and we get two alternatives:

$$\text{either } U^\top AU = 0 \quad \text{or } \mu = 1 + U^\top A^2 U.$$

If the second alternative takes place, substitute the expression for μ into (29):

$$(U^\top AU)\,(A^\top + A)U - (1 + U^\top A^2 U)U = \mathbb{O}.$$

Wherefrom it follows that

$$(A^\top + A)U = \frac{1 + U^\top A^2 U}{U^\top AU} U. \tag{30}$$

If there exists a solution $U = U_* \neq \mathbb{O}$ for this equation, then U_* is necessarily an eigenvector of $A^\top + A$ corresponding to the eigenvalue

$$\nu_* = (1 + U_*^\top A^2 U_*)/(U_*^\top AU_*).$$

Matrix $A^\top + A$ is a symmetric one with the eigenvalues $2\cos\alpha_1,\dots,2\cos\alpha_m$ of the multiplicity 2 and, probably, ± 2. Substitution $U = U_*$ into (30) and multiplication by $U_*^\top$ yields

$$\nu_* = 2U_*^\top AU_* = 2\cos\alpha_j \quad \text{for some } j.$$

Therefore, the critical values of the function $G(U)$ are in the set $\{1 - \cos^2\alpha_j\}_{j=1}^m$. This results in (27).

The alternative $U_*^\top AU_* = 0$ for $U_*^\top U_* = 1$ corresponds to the case where A possesses eigenvalues $\pm\mathbf{i}$. The result (27) remains valid. □

Proof. II. For $j \in \{1,\dots,m\}$, denote by P_j the eigenvectors of A corresponding to the eigenvalue $\cos\alpha_j \pm \mathbf{i}\sin\alpha_j$ with $\|\Re(P_j)\| = \|\Im(P_j)\| = 1$. Denote $P_{[1]}$ and $P_{[-1]}$ the eigenvectors corresponding to the eigenvalues 1 and -1 correspondingly (if any) with $\|P_{[1]}\| = \|P_{[-1]}\| = 1$. Then the orthogonal matrix

$$P = (\Re(P_1), \Im(P_1), \Re(P_2), \Im(P_2),\dots, \Re(P_m), \Im(P_m), \{P_{[1]}, P_{[-1]}\})$$

is such that

$$P^\top AP = \Omega \quad \text{where } \Omega = \text{diag}\,\{\Omega_1, \Omega_2,\dots,\Omega_m, \{1, -1\}\},$$

where

$$\Omega_k := \begin{bmatrix} \cos\alpha_k & \sin\alpha_k \\ -\sin\alpha_k & \cos\alpha_k \end{bmatrix} \quad \text{for } k \in \{1, 2,\dots, m\}$$

(we set in braces the entries of the matrices corresponding to the case of existence of either of eigenvalues 1 or -1 or both for A).

Since the orthogonal transformation does not influence the Frobenius distance, we reduce $d(A, \mathbb{D})$ to $d(\Omega, \mathbb{D})$. In this case,

$$(\Omega - \lambda I)(\Omega - \lambda I)^\top = \operatorname{diag}\{\tilde{\Omega}_1 \tilde{\Omega}_2, \ldots, \tilde{\Omega}_m, \{1,1\}\},$$

where

$$\tilde{\Omega}_k := \begin{bmatrix} (\cos\alpha_k - \lambda)^2 & 0 \\ 0 & (\cos\alpha_k - \lambda)^2 \end{bmatrix} \quad \text{for } k \in \{1, 2, \ldots, m\}.$$

It is evident that

$$(\Omega - \lambda I)(\Omega - \lambda I)^\top(\Omega - \lambda I) = (\Omega - \lambda I)^2(\Omega - \lambda I)^\top.$$

Hence, expression (23) is valid. In this case, the singular values of the matrix (21) are the zeros of the polynomials

$$\mu^2 - \mu(2((\cos\alpha_k - \lambda)^2 + \sin^2\alpha_k) + \gamma^2) + (\cos\alpha_k - \lambda)^2 + \sin^2\alpha_k,$$

namely:

$$\mu_{1,2}^{(k)} = \frac{2((\cos\alpha_k - \lambda)^2 + \sin^2\alpha_k) + \gamma^2 \pm \gamma\sqrt{\gamma^2 + 4((\cos\alpha_k - \lambda)^2 + \sin^2\alpha_k)}}{2}.$$

Differentiating w.r.t. γ, we get a single stationary point $\gamma = 0$. According to [14], to find the 2-norm distance from $\Omega - \lambda I$ to the manifold of matrices with multiple zero eigenvalue, one should find the singular values σ_n and σ_{n-1} of this matrix. They are either

$$\sigma_n = \sigma_{n-1} = \sqrt{(\cos\alpha_k - \lambda)^2 + \sin^2\alpha_k}$$

for some k or

$$\sigma_{n-1} = \sqrt{(\cos\alpha_k - \lambda)^2 + \sin^2\alpha_k}, \sigma_n = |1 - \lambda|.$$

The minimal value of σ_{n-1} w.r.t. λ comes up to $\sin\alpha_1$ in both cases. The minimal perturbation

$$E_* = P \begin{bmatrix} 0 & -\sin\alpha_1 & 0 \ldots 0 \\ 0 & 0 & 0 \ldots 0 \\ \vdots & & & \vdots \\ 0 & 0 & 0 \ldots 0 \end{bmatrix} P^\top = -(\sin\alpha_1)\Re(P_1)\Im(P_1)^\top.$$

$\square$

Corollary 6. *In the notation of Theorem 8, the distance $d(A, \mathbb{D})$ is provided by a continuum of perturbations E_* contained in the set*

$$\left\{ (-\sin\alpha_1)(\eta\Re(P_1) + \theta\Im(P_1))(-\eta\Im(P_1) + \theta\Re(P_1))^\top \mid \{\eta, \theta\} \subset \mathbb{R}, \eta^2 + \theta^2 = 1 \right\}.$$

Example 6. For the matrix

$$A = \frac{1}{3} \begin{bmatrix} -2 & -2 & 1 \\ 1 & -2 & -2 \\ -2 & 1 & -2 \end{bmatrix},$$

one has

$$\lambda_{1,2} = -\frac{1}{2} \pm i\frac{\sqrt{3}}{2}, \lambda_3 = -1, \ P_1 = \left[-\frac{2}{\sqrt{6}}, \frac{1}{\sqrt{6}}, \frac{1}{\sqrt{6}} \right]^{\mathsf{T}} + i \left[-\frac{1}{\sqrt{2}}, \frac{1}{\sqrt{2}}, 0 \right]^{\mathsf{T}}.$$

Here $d(A, \mathbb{D}) = \sqrt{3}/2 \approx 0.866025$ and there are infinite number corresponding perturbation matrices (10) generated by columns U_* chosen from the span of $\Re(P_1)$ and $\Im(P_1)$. For instance:

$$U_* := \Re(P_1) \qquad\qquad U_* := \Im(P_1)$$
$$\Downarrow \qquad\qquad\qquad\qquad \Downarrow$$
$$E_* = \begin{bmatrix} 0 & 1/2 & -1/2 \\ 0 & -1/4 & 1/4 \\ 0 & -1/4 & 1/4 \end{bmatrix}; \quad E_* = \begin{bmatrix} 1/4 & 1/4 & -1/2 \\ -1/4 & -1/4 & 1/2 \\ 0 & 0 & 0 \end{bmatrix}.$$

In the both cases, spectrum of matrix B_* is $\{-1, -1/2, -1/2\}$.   □

Remark. In all the cases, where the distance $d(A, \mathbb{D})$ is achieved at $\gamma = 0$ and two minimal singular values of the matrix (21) coincide, i.e., $\sigma_{2n-1}(\lambda, 0) = \sigma_{2n}(\lambda, 0)$, we have found the rank 1 minimal perturbation whilst in the work [14] it is described as a rank 2 matrix.

6   Conclusions

We have investigated Wilkinson's problem for the distance evaluation from a given matrix to the set of matrices possessing multiple eigenvalues. The structure of the perturbation matrix is clarified that gives us an opportunity to compute symbolically the distance equation with the zero set containing the critical values of the squared distance function.

Computational complexity of the proposed solution is (traditionally to analytical approach) high. Although this payment should be agreed with regard to the reliability of the computation results, we still hope to reduce it in further investigations.

There exists a definite similarity of the considered problem to that of Routh–Hurwitz distance to instability computation. For instance, the approach suggested in Sect. 3 has its counterpart in the one developed by Ch. Van Loan for the distance to instability problem [11, 12]. This is also a subject of subsequent discussions.

Acknowledgments. The authors are grateful to Prof. Evgenii V. Vorozhtsov and to the anonymous referees for valuable suggestions that helped to improve the quality of the paper.

References

1. Ahmad, S.S., Alam, R.: On Wilkinson's problem for matrix pencils. ELA **30**, 632–648 (2015)
2. Alam, R., Bora, S.: On sensitivity of eigenvalues and eigendecompositions of matrices. Linear Algebra Appl. **396**, 273–301 (2005)
3. Armentia, G., Gracia, J.-M., Velasco, F.-E.: Nearest matrix with a prescribed eigenvalue of bounded multiplicities. Linear Algebra Appl. **592**, 188–209 (2020)
4. Demmel, J.W.: Computing stable eigendecompositions of matrices. Linear Algebra Appl. **79**, 163–193 (1986)
5. Demmel, J.W.: On condition numbers and the distance to the nearest ill-posed problem. Numer. Math. **51**, 251–289 (1987)
6. Eckart, C., Young, G.: The approximation of one matrix by another of lower rank. Psychometrika **1**, 211–218 (1936)
7. Golub, G., Van Loan, Ch.: Matrix Computations, 3rd edn. The Johns Hopkins University Press, Baltimore (1996)
8. Higham, N.G.: Matrix nearness problems and applications. In: Applications of matrix theory, pp. 1–27. Oxford University Press, New York (1989)
9. Horn, R.A., Johnson, Ch.: Matrix Analysis, 2nd edn. Cambridge University Press, New York (2013)
10. Lippert, R.A., Edelman, A.: The computation and sensitivity of double eigenvalues. In: Chen, Z., Li, Y., Micchelli, C.A., Xu, Y. (eds.) Advances in Computational Mathematics: Proceedings, pp. 353–393. Gaungzhou International Symposium, Dekker, New York (1999)
11. Kalinina, E.A., Smol'kin, Y.A., Uteshev, A.Y.: Routh – Hurwitz stability of a polynomial matrix family. Real perturbations. In: Boulier, F., England, M., Sadykov, T.M., Vorozhtsov, E.V. (eds.) CASC 2020. LNCS, vol. 12291, pp. 316–334. Springer, Cham (2020). https://doi.org/10.1007/978-3-030-60026-6_18
12. Kalinina, E., Uteshev, A.: On the real stability radius for some classes of matrices. In: Boulier, F., England, M., Sadykov, T.M., Vorozhtsov, E.V. (eds.) CASC 2021. LNCS, vol. 12865, pp. 192–208. Springer, Cham (2021). https://doi.org/10.1007/978-3-030-85165-1_12
13. Kokabifar, E., Loghmani, G.B., Karbassi, S.M.: Nearest matrix with prescribed eigenvalues and its applications. J. Comput. Appl. Math. **298**, 53–63 (2016)
14. Malyshev, A.: A formula for the 2-norm distance from a matrix to the set of matrices with multiple eigenvalues. Numer. Math. **83**, 443–454 (1999)
15. Mengi, E.: Locating a nearest matrix with an eigenvalue of prespecified algebraic multiplicity. Numer. Math. **118**, 109–135 (2011)
16. Netto, E.: Rationale Funktionen einer Veränderlichen; ihre Nullstellen. In: Meyer, W.F. (Ed.) Encyklopadie der Mathematischen Wissenschaften mit Einschluss ihrer Anwendungen, Teubner, Leipzig, Germany, 1898–1904, vol. 1, pp. 227–254 (1898). https://doi.org/10.1007/978-3-663-16017-5_7
17. Pólya, G., Szegö, G.: Problems and Theorems in Analysis II. Springer, Berlin (1976). https://doi.org/10.1007/978-3-642-61983-0
18. Ruhe, A.: Properties of a matrix with a very ill-conditioned eigenproblem. Numer. Math. **15**, 57–60 (1970)
19. Turnbull, H.W.: Matrix differentiation of the characteristic function. Proc. Edinb. Math. Soc. Second Ser. **II**, 256–264 (1931)
20. Uteshev, A.: Notebook (2022). http://vmath.ru/vf5/matricese/optimize/distancee/casc2022ex. Accessed 21 June 2022

21. Wilkinson, J.H.: The Algebraic Eigenvalue Problem. Oxford University Press, New York (1965)
22. Wilkinson, J.H.: Note on matrices with a very ill-conditioned eigenproblem. Numer. Math. **19**, 176–178 (1972)
23. Wilkinson, J.H.: On neighbouring matrices with quadratic elementary divisors. Numer. Math. **44**, 1–21 (1984)
24. Wilkinson, J.H.: Sensitivity of eigenvalues. Util. Math. **25**, 5–76 (1984)

On Boundary Conditions Parametrized by Analytic Functions

Markus Lange-Hegermann[1]($\boxtimes$) and Daniel Robertz[2]

[1] Technische Hochschule Ostwestfalen-Lippe, inIT (Institute Industrial IT), Lemgo, Germany
markus.lange-hegermann@th-owl.de
[2] Lehrstuhl für Algebra und Zahlentheorie, RWTH Aachen University, Aachen, Germany
daniel.robertz@rwth-aachen.de

Abstract. Computer algebra can answer various questions about partial differential equations using symbolic algorithms. However, the inclusion of data into equations is rare in computer algebra. Therefore, recently, computer algebra models have been combined with Gaussian processes, a regression model in machine learning, to describe the behavior of certain differential equations under data. While it was possible to describe polynomial boundary conditions in this context, we extend these models to analytic boundary conditions. Additionally, we describe the necessary algorithms for Gröbner and Janet bases of Weyl algebras with certain analytic coefficients. Using these algorithms, we provide examples of divergence-free flow in domains bounded by analytic functions and adapted to observations.

Keywords: Gaussian processes · Boundary conditions · Gröbner bases · Partial differential equations

1 Introduction

Differential algebra is concerned with structural properties of systems of ordinary and partial differential equations (ODEs and PDEs) and provides algorithms for their analysis [1,31]. The properties unveiled by these algorithms correspond to intrinsic properties of the solutions of the system. At the same time these algorithms isolate equations of interest via elimination, transform systems into normal forms [8], describe singularities [24], allow to investigate control-theoretic properties [22,23], or detect the size of solution sets [17,18,20].

Usually, PDEs come with additional information on the evaluation of functions. For example in inverse problems, parameters in differential equations are being estimated from data points. Or in theoretical and numerical methods for PDEs, boundary conditions, i.e., evaluations of functions on manifolds, ensure well-posedness. Data points and boundary conditions have rarely been addressed

© The Author(s), under exclusive license to Springer Nature Switzerland AG 2022
F. Boulier et al. (Eds.): CASC 2022, LNCS 13366, pp. 225–245, 2022.
https://doi.org/10.1007/978-3-031-14788-3_13

by algebraic means, with the exception of modeling of boundary conditions by integro-differential operators [35, 38].

Seemingly disconnected from these algebraic algorithms are Gaussian Processes (GPs) [34], a general regression technique, which arise as limit of large neural networks [29] and generalize linear (ridge) regression, Kriging, and many spline models. GPs describe probability distributions on function spaces. As such,

(1) they can be conditioned on observations given as data points using Bayes' rule in closed form, which avoids overfitting,
(2) they admit an extensive dictionary between their mathematical properties and their covariance function, which allows to prescribe intended behavior,
(3) form the maximum entropy prior distribution under the assumption of a finite mean and variance in the unknown behavior, and
(4) the class of GPs is closed under various operations like conditioning, marginalization, and linear operators.

They are typically used in applications when data is rare or expensive to produce, e.g., in active learning [50], biology [11], anomaly detection [3] or engineering [45]. The mean function of the posterior is used for regression and the variance quantifies uncertainty. In that sense, they allow to deal with data, noise, and uncertainty in a way algebraic algorithms usually cannot.

The inclusion of algebraic methods for differential equations into covariance functions of GPs began by divergence-free and curl-free vector fields [25, 40], extended to electromagnetic fields [43, 47] and strain fields [14]. These approaches were formalized in [15], building on [39]. Then, [19] used Gröbner bases and worked out the necessity of systems being controllable. Boundary conditions were added to the setup in [21], restricted to simple polynomial boundaries.

In this paper, we develop algebraic algorithms suitable for this framework to deal with analytic boundary conditions. These algorithms might take

(i) parametrizable linear systems of differential equations,
(ii) assumptions on the solutions of the differential equations, e.g. smoothness,
(iii) various forms of boundary conditions specified by analytic functions, and
(iv) (noisy or noiseless) evaluations of functions at finitely many points

as inputs. They yield a probability distribution on the solution space of the differential equation given by a GP, which has the above properties (1)–(4).

Our approach is as follows. We construct a first parametrization of the solution set of the system of differential equations by finding a matrix whose row nullspace is generated by the equations of the given system. We take a second parametrization of the boundary condition. Then, we construct a parametrization of the intersection of the images of these two parametrizations. Algorithmically, this requires Gröbner bases over a Weyl algebra enlarged by various analytic functions, for which we develop the necessary theory and algorithms. After this symbolic approach, numeric algorithms incorporate measurement data into the GP.

In this setup, ODEs are trivial, both algebraically, as parametrizable linear systems of ODEs with constant or variable coefficients are isomorphic to free systems due to the Jacobson form [12], and also from the stochastic point of view, as boundary conditions in ODEs can be modelled by conditioning on data points [16]. Hence, we focus on PDEs.

From the point of view of machine learning, the results of this paper allow to incorporate information into the covariance structure of a GP prior. This prior is supported by solutions of the differential equation and the boundary conditions. In particular, rare measurement data can refine and improve this prior knowledge, instead of being necessary to learn this prior knowledge.

The contributions of this paper can be summarized as follows:

(a) we develop Gröbner basis algorithms for Weyl algebras over certain rings of analytic functions (cf. Sects. 5 and 6),
(b) we study boundary conditions parametrized by analytic functions, in particular how they constrain GPs (cf. Sect. 7), and
(c) we construct GP priors for solution sets of PDEs including boundary conditions (cf. Sect. 8).

2 Gaussian Processes

A *Gaussian Process (GP)* $g = \mathcal{GP}(\mu, k)$ defines a probability distribution on the evaluations of functions $D \to \mathbb{R}^\ell$ where $D \subseteq \mathbb{R}^d \equiv \mathbb{R}^{1 \times d}$ such that function values $g(x_1), \ldots, g(x_n)$ at points $x_1, \ldots, x_n \in D$ are jointly (multivariate) Gaussian. A GP g is specified by a *mean function* $\mu : D \to \mathbb{R}^\ell : x \mapsto E(g(x))$ and a positive semidefinite[1] *covariance function*

$$k : D \times D \longrightarrow \mathbb{R}^{\ell \times \ell}_{\succeq 0} : (x, x') \longmapsto E\left((g(x) - \mu(x))(g(x') - \mu(x'))^T\right).$$

Any finite set of evaluations of g follows the multivariate Gaussian distribution

$$\begin{bmatrix} g(x_1) \\ \vdots \\ g(x_n) \end{bmatrix} \sim \mathcal{N}\left(\begin{bmatrix} \mu(x_1) \\ \vdots \\ \mu(x_n) \end{bmatrix}, \begin{bmatrix} k(x_1, x_1) & \ldots & k(x_1, x_n) \\ \vdots & \ddots & \vdots \\ k(x_n, x_1) & \ldots & k(x_n, x_n) \end{bmatrix} \right).$$

Now, one knows where a function value $g(x)$ is supposed to be (mean $\mu(x)$), which ignorance we have about $g(x)$ (variance $k(x, x)$), and how two function values $g(x_1)$ and $g(x_2)$ are related (covariance $k(x_1, x_2)$). GPs are popular functional priors in Bayesian inference due to their maximum entropy property [13].

Assume the probabilistic regression model $y = g(x)$ for a GP $g = \mathcal{GP}(0, k)$. Normalizing the data to mean zero justifies assuming a prior mean function zero.

---

[1] The function k is positive (semi)definite if and only if for any $x_1, \ldots, x_n \in D$ the matrix $K = (k(x_i, x_j))_{i,j} \in \mathbb{R}^{n\ell \times n\ell}$ is positive (semi)definite, i.e., $K \succeq 0$.

Conditioning the GP on training data points $(x_i, y_i) \in D \times \mathbb{R}^{1 \times \ell}$ for $i = 1, \dots, n$ by Bayes' theorem yields the posterior

$$p(\, g(x) = y \mid g(x_i) = y_i \,) = \frac{p(\, g(x_i) = y_i \mid g(x) = y \,)}{p(\, g(x_i) = y_i \,)} \cdot p(\, g(x) = y \,),$$

where i always runs from 1 to n. All of these distributions are multivariate Gaussian. Hence, the posterior $p(\, g(x) = y \mid g(x_i) = y_i \,)$ is again a GP and can be computed in closed form via linear algebra:

$$\mathcal{GP} \Big(\quad x \mapsto y k(X, X)^{-1} k(X, x),$$
$$(x, x') \mapsto k(x, x') - k(x, X) k(X, X)^{-1} k(X, x') \quad \Big), \tag{1}$$

where $y \in \mathbb{R}^{1 \times \ell n}$ denotes the row vector obtained by concatenating the y_i and $k(x, X) \in \mathbb{R}^{\ell \times \ell n}$ resp. $k(X, x) \in \mathbb{R}^{\ell n \times \ell}$ resp. $k(X, X) \in \mathbb{R}^{\ell n \times \ell n}_{\succeq 0}$ denote the (covariance) matrices obtained by concatenating the blocks $k(x, x_j)$ resp. $k(x_j, x)$ resp. $k(x_i, x_j)$ to a matrix. In case of noisy data $(y_i)_j$, one adds the *noise variance* $var((y_i)_j)$ to the $((i-1)\ell + j)$-th diagonal entry of $k(X, X)$. The Cholesky decomposition improves numerical stability regarding the inversion of the positive definite matrix $k(X, X)$ [34]. In the posterior (1), the mean function can be used as regression model and its variance as model uncertainty.

The class of GPs is closed under linear operators once mild assumptions hold, e.g. the derivative of a GP with differentiable realizations is again a GP.

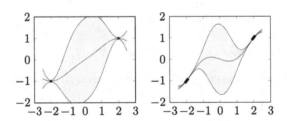

Fig. 1. Left: a regression plot (mean and the 2σ confidence bands) of a GP with mean zero and squared exponential covariance function conditioned on the points $(-2, -1)$ and $(2, 1)$ with noise variance of 0.1^2. Right: the GP is additionally conditioned on derivative 1 with noise 0.1^2 at both data points.

Given a set of functions $G \subseteq Y^D$ and $b : Y \to Z$, then the *pushforward* is $b_* G = \{b \circ f \mid f \in G\} \subseteq Z^D$. The *pushforward* of a stochastic process $g : D \to Y$ by $b : Y \to Z$ is defined as

$$b_* g : D \longrightarrow Z : d \longmapsto (b \circ g)(d).$$

Lemma 1 ([21, Lemma 2.2]). *Let $\mathcal{F}$ and $\mathcal{G}$ be spaces of functions defined on a set D with product σ-algebra of the function evaluations. Let $g = \mathcal{GP}(\mu(x), k(x, x'))$ with realizations in $\mathcal{F}$ and $B : \mathcal{F} \to \mathcal{G}$ a linear, measurable operator which commutes with expectation w.r.t. the measure induced by g on $\mathcal{F}$ and by B_*g on $\mathcal{G}$. Then, the* Gaussian Process (GP) B_*g *of g under B is a GP with*

$$B_*g = \mathcal{GP}(B\mu(x), Bk(x, x')(B')^T) ,$$

where B' denotes the operation of B on functions with argument x'.

Example 1. Let $g = \mathcal{GP}(0, k(x, x'))$ be a GP with realizations (a.s.) in the set $C^1(\mathbb{R}, \mathbb{R})$ of differentiable functions. The pushforward GP

$$\left[\tfrac{\partial}{\partial x}\right]_* g := \mathcal{GP}\left(0, \frac{\partial^2}{\partial x \partial x'} k(x, x')\right)$$

describes derivatives of the GP g [6, §5.2]. The one-argument derivative $\frac{\partial}{\partial x} k(x, x')$ yields the cross-covariance between on the one hand a function evaluation $g(x')$ of g at $x' \in \mathbb{R}$ and on the other hand its derivative $([\tfrac{\partial}{\partial x}]_* g)(x)$ evaluated at $x \in \mathbb{R}$. We use this to include data of derivatives into a model in Fig. 1.

3 Solution Sets of Operator Equations

This section discusses how GPs describe the real vector space $\mathcal{F} = C^\infty(D, \mathbb{R})$, a candidate set of solutions for the linear differential equations, and how such GPs interplay with linear operators. Assume that $D \subset \mathbb{R}^d$ is compact and $\mathcal{F}$ is endowed with the usual Fréchet topology generated by the separating family

$$\|f\|_a := \sup_{\substack{i \in \mathbb{Z}_{\geq 0}^d \\ |i| \leq a}} \sup_{x \in D} \left| \frac{\partial^{|i|}}{\partial x^i} f(x) \right| \tag{2}$$

of seminorms for all $a \in \mathbb{Z}_{\geq 0}$, where $i = (i_1, \ldots, i_d) \in \mathbb{Z}_{\geq 0}^d$ is a multi-index with $|i| = i_1 + \ldots + i_d$. The *squared exponential covariance function*

$$k_\mathcal{F} : \mathbb{R}^d \times \mathbb{R}^d \longrightarrow \mathbb{R} : (x_i, x_j) \longmapsto \exp\left(-\frac{1}{2}\sum_{a=1}^d (x_{i,a} - x_{j,a})^2\right) \tag{3}$$

induces an adapted GP prior in $\mathcal{F} = C^\infty(D, \mathbb{R})$.

Proposition 1. *The scalar GP $g_\mathcal{F} = \mathcal{GP}(0, k_\mathcal{F})$ has realizations dense (a.s.) in $\mathcal{F}$ with respect to the Fréchet topology defined by Eq. (2).*

Proof. We show that the realizations of $g_{\mathcal{F}}$ are densely contained in $\mathcal{F}$ in three steps: first, the realizations are contained in $\mathcal{F}$, i.e., smooth; second, the elements of the reproducing kernel Hilbert space (RKHS)[2] $\mathcal{H}(g_{\mathcal{F}})$ of the GP $g_{\mathcal{F}}$, are realizations; and third, the RKHS $\mathcal{H}(g_{\mathcal{F}})$ is dense in $\mathcal{F}$.

First, show that the realizations of $g_{\mathcal{F}}$ lie in $\mathcal{F}$. They are continuously differentiable, as $k_{\mathcal{F}}$ is twice continuously differentiable [6, (9.2.2)]. Continue inductively, as the covariance $\frac{\partial^2}{\partial x \partial x'} k_{\mathcal{F}}(x, x')$ of the derivative of $g_{\mathcal{F}}$ is again smooth.

For the second step, we note that $C^\infty(D, \mathbb{R})$ is Radon as D is compact, hence $g_{\mathcal{F}}$ induces a Radon measure on $\mathcal{F}$. For any Radon measure, $\mathcal{H}(g_{\mathcal{F}})$ is contained in the topological support of the measure induced by $g_{\mathcal{F}}$ by [4, Thm. 3.6.1]. For this, $\mathcal{F} = C^\infty(D, \mathbb{R})$ needs to be locally convex, which it is being Fréchet.

For the third step, by [41, Prop. 4], $\mathcal{H}(g_{\mathcal{F}})$ is continuously contained in $\mathcal{F}$ and dense by [41, Thm. 12, Prop. 42] or [41, after proof of Cor. 38]. $\square$

The following three $\mathbb{R}$-algebras R model linear operator equations by making $\mathcal{F}$ a left R-module. Sections 5 and 6 introduce Gröbner bases for such rings.

Example 2. The polynomial ring $R = \mathbb{R}[\partial_{x_1}, \ldots, \partial_{x_d}]$ models linear PDEs with constant coefficients, where ∂_{x_i} acts on $\mathcal{F} = C^\infty(D, \mathbb{R})$ via partial derivative with respect to x_i.

Example 3. Let $f_1, \ldots, f_n \in \mathcal{F}$ be functions. The ring $R = \mathbb{R}[f_1, \ldots, f_n]$ is commutative and models boundary conditions by multiplication, see Sect. 7.

Example 4. Let $F \subseteq \mathcal{F}$ be an $\mathbb{R}$-algebra closed under partial derivatives. To combine linear differential equations with boundary conditions, consider the Weyl algebra $R = \mathbb{R}[F]\langle \partial_{x_1}, \ldots, \partial_{x_d} \rangle$. The non-commutative relation $\partial_{x_i} f = f \partial_{x_i} + \frac{\partial f}{\partial x_i}$ represents the product rule of differentiation for $f \in F$ and $1 \leq i \leq d$.

Operators defined over these three rings satisfy the assumptions of Lemma 1: multiplication commutes with expectations and the dominated convergence theorem implies that expectation commutes with derivatives, as realizations of $g_{\mathcal{F}}$ are continuously differentiable. Furthermore, these rings act continuously on $\mathcal{F}$: the Fréchet topology makes derivation continuous by construction, and multiplication by elements in $\mathcal{F}$ is bounded as D is compact, which implies continuity in the Fréchet space $\mathcal{F}$. In particular, we have the following:

Corollary 1. *Let $\mathcal{F} = C^\infty(D, \mathbb{R})$ be the space of smooth functions defined on a compact set $D \subset \mathbb{R}^d$. Let $g = \mathcal{GP}(\mu(x), k(x, x'))$ with realizations in $\mathcal{F}^{\ell''}$ and $B : \mathcal{F}^{\ell''} \to \mathcal{F}^\ell$ a linear operator over one of the operator rings in Examples 2, 3, or 4. Then, the pushforward GP $B_* g$ is again Gaussian with*

$$B_* g = \mathcal{GP}(B\mu(x), Bk(x, x')(B')^T),$$

where B' denotes the operation of B on functions with argument x'.

---

[2] For $\mathcal{GP}(0, k)$, the set $\mathcal{H}^0(g)$ generated as a vector space by the $x \mapsto k(x_i, x)$ for $x_i \in D$ with scalar product $\langle k(x_i, -), k(x_j, -) \rangle := k(x_i, x_j)$ is a pre-Hilbert space. Its closure $\mathcal{H}(g)$ is the *reproducing kernel Hilbert space* of the GP g [2].

4 Parametrizations

We consider solution sets of linear differential equations, how to parametrize them by a suitable matrix B and thereby describe them by a GP B_*g. Let R be one of the rings from the previous section, $\mathcal{F}$ the left R-module $C^\infty(D, \mathbb{R})$ and $A \in R^{\ell' \times \ell}$. Define the *solution set* $\mathrm{sol}_\mathcal{F}(A) := \{f \in \mathcal{F}^{\ell \times 1} \mid Af = 0\}$ of A. We say that a GP is *in* a function space, if its realizations are a.s. contained in said space. We first describe the interplay of GPs and solution sets of operators.

Lemma 2 ([19, Lemma 2.2]). *Let $g = \mathcal{GP}(\mu, k)$ be a GP in $\mathcal{F}^{\ell \times 1}$. Then g is a GP in the solution set $\mathrm{sol}_\mathcal{F}(A)$ of $A \in R^{\ell' \times \ell}$ if and only if both μ is contained in $\mathrm{sol}_\mathcal{F}(A)$ and $A_*(g - \mu)$ is the constant zero process.*

This lemma motivates how to construct GPs with realizations in $\mathrm{sol}_\mathcal{F}(A)$: find a $B \in R^{\ell \times \ell''}$ with $AB = 0$ [15]. Then, taking any GP $g = \mathcal{GP}(0, k)$ in $\mathcal{F}^{\ell'' \times 1}$, the realizations of B_*g are (possibly strictly) contained in $\mathrm{sol}_\mathcal{F}(A)$, as $A_*(B_*g) = (AB)_*g = 0_*g = 0$. One prefers to enlarge B to approximate *all* solutions in $\mathrm{sol}_\mathcal{F}(A)$ by B_*g, i.e., the realizations of B_*g should be dense in $\mathrm{sol}_\mathcal{F}(A)$. Call $B \in R^{\ell \times \ell''}$ a *parametrization* of $\mathrm{sol}_\mathcal{F}(A)$ if $\mathrm{sol}_\mathcal{F}(A) = B\mathcal{F}^{\ell'' \times 1}$. Such a parametrization does not always exist, e.g., for the matrix $A = [\partial_{x_1}]$.

Proposition 2 ([21, Proposition 3.5]). *Let $B \in R^{\ell \times \ell''}$ be a parametrization of $\mathrm{sol}_\mathcal{F}(A)$. Let $g_\mathcal{F}^{\ell'' \times 1}$ be the GP of ℓ'' i.i.d. copies of $g_\mathcal{F}$, the GP with squared exponential covariance $k_\mathcal{F}$ (3). Then, $B_*g_\mathcal{F}^{\ell'' \times 1}$ has realizations dense in $\mathrm{sol}_\mathcal{F}(A)$.*

We summarize how to algorithmically decide whether a parametrization exists and how to compute it in the positive case. Computations directly over the space of functions $\mathcal{F}$ are infeasible. Hence, we compute over R instead. Inferring results over $\mathcal{F}$ is possible once $\mathcal{F}$ is an injective[3] R-module, i.e., $\mathrm{Hom}_R(-, \mathcal{F})$ is exact. Luckily, for PDEs with constant coefficients we have the following:

Theorem 1 ([7,26] [31, §(54)]). *Let $R = \mathbb{R}[\partial_{x_1}, \ldots, \partial_{x_d}]$ be as in Example 2 and $D \subset \mathbb{R}^d$ convex. Then, $\mathcal{F} = C^\infty(D, \mathbb{R})$ is an injective R-module.*

With this in mind, we recall the construction of parametrizations.

Theorem 2 ([49, Thm. 2] [31, §7.(24)] [5,32,33,37]). *Let R be a ring and $\mathcal{F}$ an injective left R-module. Let $A \in R^{\ell' \times \ell}$. Let B be the right nullspace of A and A' the left nullspace of B. Then $\mathrm{sol}_\mathcal{F}(A')$ is the largest subset of $\mathrm{sol}_\mathcal{F}(A)$ that is parametrizable, B parametrizes $\mathrm{sol}_\mathcal{F}(A')$, and $\mathrm{sol}_\mathcal{F}(A)$ is parametrizable if and only if the rows of A and A' generate the same row module, i.e., if all rows of A' are contained in the row module generated by A.*

---

[3] In algebraic system theory, one usually works with injective cogenerators $\mathcal{F}$ [32]. Injective cogenerators allow to infer back from analysis in $\mathcal{F}$ to algebra over R. In our setting, this step back is superfluous, as the algebra cannot encode data points.

Gröbner bases turn Theorem 2 effective, as they allow to compute the right nullspace B of A, the left nullspace A' of B and decide whether the rows of A' are contained in the row space of A over R. We have the following criterion.

Theorem 3 ([31, §7.(21)]). *A system* $\mathrm{sol}_{\mathcal{F}}(A)$ *is parametrizable if and only if it is controllable. If A is not parametrizable, then the solution set* $\mathrm{sol}_{\mathcal{F}}(A')$ *is the subset of controllable behaviors in* $\mathrm{sol}_{\mathcal{F}}(A)$*, where A' is defined as in Theorem 2.*

Solution sets of differential equations and polynomial boundary conditions can be intersected [21].

Theorem 4 ([21, Theorem 5.2]). *Let $B_1 \in R^{\ell \times \ell_1'}$ and $B_2 \in R^{\ell \times \ell_2'}$. Denote by $C := \begin{bmatrix} C_1 \\ C_2 \end{bmatrix} \in R^{(\ell_1' + \ell_2') \times m}$ the right-nullspace of the matrix $B := \begin{bmatrix} B_1 & B_2 \end{bmatrix} \in R^{\ell \times (\ell_1' + \ell_2')}$. Then $B_1 C_1 = -B_2 C_2$ parametrizes solutions of $B_1 \mathcal{F}^{\ell_1'} \cap B_2 \mathcal{F}^{\ell_2'}$.*

Here, B_1 might be a matrix of differential operators and B_2 a matrix of polynomial functions, and we consider both matrices over a common ring R.

5   Rings of Differential Operators over Differential Algebras

We have considered parametrizations by differential operators and in Sect. 7 we consider parametrizations of boundary conditions by analytic functions. For their combination in Sect. 8, we now extend classical Gröbner and Janet bases.

Let $D \subset \mathbb{R}^d$ be connected and denote by $\delta_1, \ldots, \delta_d$ the commuting derivations in the coordinate directions of $\mathbb{R}^d$. Let K be a differential algebra over the real numbers[4] $\mathbb{R}$ generated by analytic functions $f_1, \ldots, f_r \colon D \to \mathbb{R}$. For algorithmic reasons assume that K is finitely presented as a differential algebra over $\mathbb{R}$ as

$$K = \mathbb{R}\{f_1, \ldots, f_r\} \cong \mathbb{R}\{F_1, \ldots, F_r\}/P,$$

where P is a prime differential ideal of $\mathbb{R}\{F_1, \ldots, F_r\}$, generated by

$$\{\, \delta_j F_i - g_{i,j} \mid i = 1, \ldots, r, \; j = 1, \ldots, d \,\}, \tag{4}$$

where $g_{i,j} \in \mathbb{R}[F_1, \ldots, F_r]$ are (non-differential) polynomials in $F_1, \ldots, F_r$, and the above isomorphism is given by $f_i \mapsto F_i + P$. In particular, the generators $f_1, \ldots, f_r$ of K are algebraically independent over $\mathbb{R}$. Then K is isomorphic to $\mathbb{R}[f_1, \ldots, f_r]$ as an $\mathbb{R}$-algebra, and K is Noetherian, factorial, and a GCD domain.

Example 5. For the differential algebra $K = \mathbb{Q}\{x, y, \exp(x^2 + y^2 - 1)\}$ with derivations $\delta_1 = \partial/\partial x$, $\delta_2 = \partial/\partial y$ we have $K \cong \mathbb{Q}\{F_1, F_2, F_3\}/P$, where

$$\delta_1 F_1 - 1, \quad \delta_2 F_1, \quad \delta_1 F_2, \quad \delta_2 F_2 - 1, \quad \delta_1 F_3 - 2F_1 F_3, \quad \delta_2 F_3 - 2F_2 F_3$$

---

[4] Of course, the constructions in this and the following section work over any sufficiently algorithmic differential field of characteristic zero, not only $\mathbb{R}$. In practice, we assume to work over a computable subfield of $\mathbb{R}$.

generate the prime differential ideal P such that

$$\mathbb{Q}\{F_1, F_2, F_3\} \longrightarrow K : F_1 \longmapsto x, \; F_2 \longmapsto y, \; F_3 \longmapsto \exp(x^2 + y^2 - 1)$$

is an epimorphism of differential algebras over $\mathbb{Q}$ mapping precisely P to zero.

Definition 1. *Let the* ring of differential operators $R = K\langle \partial_1, \ldots, \partial_d \rangle$ *be the iterated Ore extension of K defined by*

$$\begin{aligned} \partial_i \, a &= a \, \partial_i + \delta_i(a), && a \in K, && i = 1, \ldots, d, \\ \partial_i \, \partial_j &= \partial_j \, \partial_i, && && i, j = 1, \ldots, d. \end{aligned}$$

Remark 1. The ring R is (left) Noetherian, because K is Noetherian (cf., e.g., [28, Thm. 1.2.9 (iv)]). Moreover, R has the left Ore property, i.e., every pair of non-zero elements of R has a non-zero common left multiple [28, Thm. 2.1.15], which, in particular, implies the existence of a skew field of fractions of R.

We define the set of *monomials* of R as

$$\mathrm{Mon}(R) = \{\, f_1^{\alpha_1} \ldots f_r^{\alpha_r} \, \partial_1^{\beta_1} \ldots \partial_d^{\beta_d} \mid \alpha_1, \ldots, \alpha_r, \beta_1, \ldots, \beta_d \in \mathbb{Z}_{\geq 0} \,\}.$$

It is a basis of R as an $\mathbb{R}$-vector space: every $p \in R$ has a unique representation

$$p = \sum_{m \in \mathrm{Mon}(R)} c_m \, m, \tag{$*$}$$

where $c_m \in \mathbb{R}$ and only finitely many c_m are non-zero.

A *monomial ordering* $<$ on R is a total ordering on $\mathrm{Mon}(R)$ satisfying

$$\begin{aligned} f_1^0 \ldots f_r^0 \, \partial_1^0 \ldots \partial_d^0 = 1 < m && \text{for all } 1 \neq m \in \mathrm{Mon}(R), \\ m_1 < m_2 \; \Rightarrow \; f_i \, m_1 < f_i \, m_2 && \text{for all } m_1, m_2 \in \mathrm{Mon}(R), \, i = 1, \ldots, r, \\ m_1 < m_2 \; \Rightarrow \; m_1 \, \partial_j < m_2 \, \partial_j && \text{for all } m_1, m_2 \in \mathrm{Mon}(R), \, j = 1, \ldots, d. \end{aligned}$$

For every $0 \neq p \in R$ the $<$-greatest monomial m occurring with non-zero coefficient c_m in the representation ($*$) of p is called the *leading monomial* of p and is denoted by $\mathrm{lm}(p)$. Its coefficient c_m is called the *leading coefficient* of p and is denoted by $\mathrm{lc}(p)$. For a subset S of R we let $\mathrm{lm}(S) = \{\, \mathrm{lm}(s) \mid 0 \neq s \in S \,\}$.

Example 6. The *weighted degree-reverse-lexicographical ordering* $<$ with weights $w = (w_1, \ldots, w_{r+d}) \in \mathbb{Q}_{>0}^{r+d}$ (weighted deg-rev-lex) is defined by

$$f_1^{\alpha_1} \ldots f_r^{\alpha_r} \partial_1^{\alpha_{r+1}} \ldots \partial_d^{\alpha_{r+d}} < f_1^{\alpha_1'} \ldots f_r^{\alpha_r'} \partial_1^{\alpha_{r+1}'} \ldots \partial_d^{\alpha_{r+d}'}$$

$$\Longleftrightarrow \left(-\sum_{j=1}^{r+d} w_i \alpha_i, \alpha_{r+d}, \ldots, \alpha_1 \right) >_{\mathrm{lex}} \left(-\sum_{j=1}^{r+d} w_i \alpha_i', \alpha_{r+d}', \ldots, \alpha_1' \right),$$

where $\alpha_i, \alpha_i' \in \mathbb{Z}_{\geq 0}$ and $>_{\mathrm{lex}}$ compares tuples lexicographically.

Example 7. We let the *elimination ordering* $<$ on R (eliminating $\partial_1, \ldots, \partial_d$) be

$$f_1^{\alpha_1} \ldots f_r^{\alpha_r} \partial_1^{\beta_1} \ldots \partial_d^{\beta_d} < f_1^{\alpha_1'} \ldots f_r^{\alpha_r'} \partial_1^{\beta_1'} \ldots \partial_d^{\beta_d'}$$

$$\Longleftrightarrow \left(\partial_1^{\beta_1} \ldots \partial_d^{\beta_d} \prec_\partial \partial_1^{\beta_1'} \ldots \partial_d^{\beta_d'} \quad \text{or} \right.$$

$$\left. \partial_1^{\beta_1} \ldots \partial_d^{\beta_d} = \partial_1^{\beta_1'} \ldots \partial_d^{\beta_d'} \quad \text{and} \quad f_1^{\alpha_1} \ldots f_r^{\alpha_r} \prec_f f_1^{\alpha_1'} \ldots f_r^{\alpha_r'} \right),$$

where $\alpha_i, \alpha_i', \beta_j, \beta_j' \in \mathbb{Z}_{\geq 0}$ and where $\prec_\partial$ and $\prec_f$ are the deg-rev-lex ordering on the polynomial algebras $\mathbb{Q}[\partial_1, \ldots, \partial_d]$ and $\mathbb{Q}[f_1, \ldots, f_r]$, respectively.

Assumption 1. *The monomial ordering* $<$ *on* R *is chosen such that the leading monomial of*

$$\partial_j f_i = f_i \partial_j + \delta_j(f_i) = f_i \partial_j + g_{i,j}$$

with respect to $>$ *is* $f_i \partial_j$, *for all* $i = 1, \ldots, r$ *and* $j = 1, \ldots, d$. *(Recall that* $f_i \partial_j + g_{i,j}$ *is the representation (*) of* $\partial_j f_i$ *taking the generators (4) of the prime differential ideal* P *into account.)*

In what follows, we make Assumption 1, which is met if $>$ is a degree-reverse-lexicographical ordering with weights $(v_1, \ldots, v_r, w_1, \ldots, w_d)$ satisfying

$$w_j \geq \max_{i=1,\ldots,r} \left(\sum_{k=1}^r v_k \deg_{f_k}(g_{i,j}) - v_i \right) \qquad \text{for all } j = 1, \ldots, d,$$

or if $>$ is an elimination ordering as in Example 7.

Before introducing Janet bases for left ideals of R we recall the concept of *Janet division*, which we formulate for ideals of the free commutative semigroup $(\mathbb{Z}_{\geq 0})^{r+d}$ in our context. Note that if I is a non-zero left ideal of R, then the exponent vectors $(\alpha_1, \ldots, \alpha_r, \beta_1, \ldots, \beta_d)$ of all elements of $\mathrm{lm}(I)$ form an ideal of $(\mathbb{Z}_{\geq 0})^{r+d}$ due to the definition of a monomial ordering and Assumption 1. The bijection between $\mathrm{Mon}(R)$ and $(\mathbb{Z}_{\geq 0})^{r+d}$ may as well be chosen to be, e.g.,

$$\varepsilon \colon \mathrm{Mon}(R) \longrightarrow (\mathbb{Z}_{\geq 0})^{r+d} : f_1^{\alpha_1} \ldots f_r^{\alpha_r} \partial_1^{\beta_1} \ldots \partial_d^{\beta_d} \longmapsto (\beta_1, \ldots, \beta_d, \alpha_1, \ldots, \alpha_r),$$

which is the bijection we usually work with.

Recall that every ideal of $(\mathbb{Z}_{\geq 0})^{r+d}$ is finitely generated; moreover, it has a unique minimal generating set. For $k \in \{1, \ldots, r+d\}$ we denote by $\mathbf{1}_k$ the multi-index with 1 in position k and 0 elsewhere. Following M. Janet (cf., e.g., [36]) we make the following definition in terms of exponent vectors.

Definition 2. *Let* $A \subset (\mathbb{Z}_{\geq 0})^{r+d}$ *be finite and* $\alpha = (\alpha_1, \ldots, \alpha_{r+d}) \in A$. *Then* $\varepsilon^{-1}(\mathbf{1}_k)$ *is said to be* multiplicative *for the monomial* $\varepsilon^{-1}(\alpha)$ *if and only if*

$$\alpha_k = \max\{\alpha_k' \mid (\alpha_1', \ldots, \alpha_{r+d}') \in A \text{ with } \alpha_1' = \alpha_1, \ldots, \alpha_{k-1}' = \alpha_{k-1}\}.$$

Let $M \subset \mathrm{Mon}(R)$ *be finite. Then for every* $m \in M$ *we obtain a partition* $\mu(m, M) \uplus \overline{\mu}(m, M)$ *of* $\{f_1, \ldots, f_r, \partial_1, \ldots, \partial_d\}$, *where each element of* $\mu(m, M)$ *is multiplicative for* m *and each element of* $\overline{\mu}(m, M)$ *is non-multiplicative for* m.

Example 8. Let $r = 2$, $n = 1$, $M = \{ f_1 f_2^2, f_1^2 f_2, f_2 \partial_1^2, f_1 \partial_1^2 \}$. Using the above bijection ε we obtain

$$\mu(f_1 f_2^2, M) = \{ f_2 \}, \qquad \mu(f_1^2 f_2, M) = \{ f_1, f_2 \},$$
$$\mu(f_2 \partial_1^2, M) = \{ \partial_1, f_2 \}, \qquad \mu(f_1 \partial_1^2, M) = \{ \partial_1, f_1, f_2 \}.$$

Definition 3. *Let $M \subset \mathrm{Mon}(R)$ be finite. We define two supersets of M in $\mathrm{Mon}(R)$ as follows:*

$$\langle M \rangle = \bigcup_{m \in M} \{ f_1^{\phi_1} \dots f_r^{\phi_r} m \, \partial_1^{\psi_1} \dots \partial_d^{\psi_d} \mid \phi_i, \psi_j \in \mathbb{Z}_{\geq 0} \},$$

$$[M] = \biguplus_{m \in M} \{ f_1^{\phi_1} \dots f_r^{\phi_r} m \, \partial_1^{\psi_1} \dots \partial_d^{\psi_d} \mid \phi_i, \psi_j \in \mathbb{Z}_{\geq 0},$$

$$\phi_i = 0 \text{ if } f_i \notin \mu(m, M) \text{ and } \psi_j = 0 \text{ if } \partial_j \notin \mu(m, M) \},$$

where the latter union is disjoint by construction of Janet division. The set M of monomials is said to be Janet complete *if* $[M] = \langle M \rangle$.

Any finite subset M of $\mathrm{Mon}(R)$ has a unique smallest (finite) Janet complete superset of M, which we call the *Janet completion* of M [36, Subsect. 2.1.1].

Definition 4. *Let I be a non-zero left ideal of R. Using the notation of Definition 3, a finite generating set $G \subset R \setminus \{0\}$ for I is called a* Gröbner basis *for I with respect to the monomial ordering $<$ if $\langle \mathrm{lm}(G) \rangle = \mathrm{lm}(I)$. If moreover, $\mathrm{lm}(G)$ is Janet complete, i.e., $[\mathrm{lm}(G)] = \langle \mathrm{lm}(G) \rangle = \mathrm{lm}(I)$, then G is called a* Janet basis *for I with respect to $<$.*

Assumption 1 facilitates a multivariate polynomial division in R.

Remark 2. Suppose $L \subset R \setminus \{0\}$ is finite and $\mathrm{lm}(L)$ is Janet complete. Let $p_1 \in R \setminus \{0\}$. If $\mathrm{lm}(p_1) \in [\mathrm{lm}(L)]$, then there exists a unique $p_2 \in L$ such that

$$\mathrm{lm}(p_1) = f_1^{\phi_1} \dots f_r^{\phi_r} \, \mathrm{lm}(p_2) \, \partial_1^{\psi_1} \dots \partial_d^{\psi_d}$$

for certain $\phi_i, \psi_j \in \mathbb{Z}_{\geq 0}$, where $\phi_i = 0$ if $f_i \notin \mu(\mathrm{lm}(p_2), \mathrm{lm}(L))$ and $\psi_j = 0$ if $\partial_j \notin \mu(\mathrm{lm}(p_2), \mathrm{lm}(L))$. Therefore, subtracting $\mathrm{lc}(p_1) f_1^{\phi_1} \dots f_r^{\phi_r} \partial_1^{\psi_1} \dots \partial_d^{\psi_d} p_2$ from $\mathrm{lc}(p_2) p_1$ yields either zero or an element of R whose leading monomial is less than $\mathrm{lm}(p_1)$. Since a monomial ordering $<$ does not admit infinitely descending chains of monomials, this reduction procedure always terminates.

Iterated reduction, as just defined, modulo a Gröbner basis or a Janet basis for the left ideal I allows to decide membership to I.

Proposition 3. *Let G be a Gröbner basis or a Janet basis for the left ideal I of R with respect to any monomial ordering $<$, and let $p \in R$. Then we have $p \in I$ if and only if the remainder of reduction of p modulo G is zero.*

Remark 3. Given a finite generating set L for a non-zero left ideal I of R and given a monomial ordering $<$ as above, a Janet basis for I with respect to $<$ can be computed in finitely many steps [36]. After a preliminary pairwise reduction of elements of L ensuring that the leading monomials of elements of L are pairwise different and that $\varepsilon(\mathrm{lm}(L))$ is the unique minimal generating set of the ideal of $(\mathbb{Z}_{\geq 0})^{r+d}$ it generates, multiplicative and non-multiplicative variables are determined for each leading monomial (with respect to $\mathrm{lm}(L)$) and L is replaced by its Janet completion. Reduction of left multiples of elements of L by non-multiplicative variables may yield non-zero remainders in I. Augmenting L by such elements results in a larger ideal $\varepsilon(\mathrm{lm}(L))$ of $(\mathbb{Z}_{\geq 0})^{r+d}$ than previously. Since every ascending chain of such ideals becomes stationary after finitely many steps, by iteration of these steps, one obtains a generating set G for I whose left multiples by non-multiplicative variables reduce to zero modulo G, which is a Janet basis for I with respect to $<$.

6   Module-Theoretic Constructions

The techniques of Sect. 5 can be extended to effectively deal with finitely presented left (and right) R-modules and module homomorphisms between them.

Let R be as in the previous section and $q \in \mathbb{N}$. We choose the standard basis $e_1, \ldots, e_q$ of the free left R-module $R^{1 \times q}$ and define the set of monomials

$$\mathrm{Mon}(R^{1 \times q}) = \{\, f_1^{\alpha_1} \ldots f_r^{\alpha_r}\, \partial_1^{\beta_1} \ldots \partial_d^{\beta_d}\, e_k \mid \alpha_i, \beta_j \in \mathbb{Z}_{\geq 0},\, k = 1, \ldots, q \,\}.$$

Then every element of $R^{1 \times q}$ has a unique representation as in $(*)$, where $\mathrm{Mon}(R)$ is replaced by $\mathrm{Mon}(R^{1 \times q})$. By generalizing the notion of monomial ordering defined in Sect. 5 to total orderings on $\mathrm{Mon}(R^{1 \times q})$, one can extend the reduction procedure described in Remark 2 and indeed any algorithm computing Gröbner or Janet bases for left ideals of R to one that computes Gröbner or Janet bases for submodules $R^{1 \times p}A$ of $R^{1 \times q}$, where $A \in R^{p \times q}$. In particular, membership to such a submodule can be decided by reduction, and therefore, computations with residue classes in $R^{1 \times q}/R^{1 \times p}A$ can be performed effectively.

We recall some relevant monomial orderings on $R^{1 \times q}$.

Example 9. A monomial ordering $\prec$ on R can be extended to monomial orderings $<$ on $R^{1 \times q}$ in different ways, for example, by defining

$$m_1 e_k < m_2 e_l \quad \Longleftrightarrow \quad \left(m_1 \prec m_2 \quad \text{or} \quad \left(m_1 = m_2 \quad \text{and} \quad k > l \right) \right)$$

("term-over-position"), or by defining

$$m_1 e_k < m_2 e_l \quad \Longleftrightarrow \quad \left(k > l \quad \text{or} \quad \left(k = l \quad \text{and} \quad m_1 \prec m_2 \right) \right)$$

("position-over-term"), where $m_1, m_2 \in \mathrm{Mon}(R)$ and $k, l \in \{1, \ldots, q\}$.

Example 10. Let $s \in \{1, \ldots, q-1\}$ and $\prec_1$, $\prec_2$ be monomial orderings on $R^{1 \times s}$ and $R^{1 \times (q-s)}$, with standard bases $e_1, \ldots, e_s$ and $e_{s+1}, \ldots, e_q$, respectively. A *monomial ordering $<$ on $R^{1 \times q}$ eliminating $e_1, \ldots, e_s$* is defined by

$$m_1\, e_k < m_2\, e_l \iff \Big(l \leq s < k \quad \text{or}$$

$$\Big(k \leq s \quad \text{and} \quad l \leq s \quad \text{and} \quad m_1\, e_k \prec_1 m_2\, e_l \Big) \quad \text{or}$$

$$\Big(k > s \quad \text{and} \quad l > s \quad \text{and} \quad m_1\, e_k \prec_2 m_2\, e_l \Big) \Big),$$

where m_1, $m_2 \in \mathrm{Mon}(R)$ and k, $l \in \{1, \ldots, q\}$.

Remark 4. Let $\varphi \colon R^{1 \times a} \to R^{1 \times b}$ be a homomorphism of left R-modules, represented by a matrix $A \in R^{a \times b}$. A Janet basis for the nullspace of φ can be computed as follows. Join the two standard bases of $R^{1 \times a}$ and $R^{1 \times b}$ to obtain the basis $e_1, \ldots, e_a, e_{a+1}, \ldots, e_{a+b}$ of $R^{1 \times a} \oplus R^{1 \times b} \cong R^{1 \times (a+b)}$. Let $<$ be a monomial ordering on $R^{1 \times (a+b)}$ as defined in Example 10 for $q = a + b$, $s = a$ and certain $\prec_1$ and $\prec_2$, i.e., eliminating $e_1, \ldots, e_a$. Then let J_0 be a Janet basis, with respect to $<$, for the submodule of $R^{1 \times (a+b)}$ generated by the rows of the matrix $(A \quad I_a) \in R^{a \times (b+a)}$, where I_a is the identity matrix. Now $J := \{ w \in R^{1 \times a} \mid (0, w) \in J_0 \}$ is a Janet basis for the nullspace of φ with respect to $\prec_2$ (cf. also [37, Ex. 3.10], [36, Ex. 3.1.27]).

Remark 5. An involution $\theta : R \to R$ of R allows to reduce computations with right R-modules to computations with left R-modules. More precisely, if we have $\theta(r_1 + r_2) = \theta(r_1) + \theta(r_2)$ and $\theta(r_1 r_2) = \theta(r_2)\,\theta(r_1)$ and $\theta(\theta(r)) = r$ for all r_1, r_2, $r \in R$, then any right R-module M is turned into a left R-module $\widetilde{M} := M$ (as abelian groups) via $r\, m := m\, \theta(r)$, where $r \in R$, $m \in \widetilde{M}$, and vice versa. The involution θ is extended to matrices by (cf. also [37, Rem. 3.11])

$$\theta(A) := (\theta((A^{tr})_{i,j}))_{1 \leq i \leq q, 1 \leq j \leq p} \in R^{q \times p}, \qquad A \in R^{p \times q}.$$

Since for $A \in R^{p \times q}$, $B \in R^{q \times r}$ we have $A B = 0$ if and only if $\theta(B)\,\theta(A) = 0$, the computation of nullspaces of homomorphisms of right R-modules is reduced to the situation described in Remark 4. For R introduced in Definition 1 we choose

$$\theta : R \to R, \quad \theta|_K := \mathrm{id}_K, \quad \theta(\partial_j) := -\partial_j, \quad j = 1, \ldots, d.$$

7 Parametrizing Boundary Conditions

This section constructs parametrizations of functions satisfying certain boundary conditions, independent of the parametrization of differential equations.

We restrict ourselves to boundary conditions parametrized by analytic functions for two reasons. First, this allows algebraic algorithms. Second, due to the limiting behaviour of GPs when conditioning on more data points, closed sets of functions are preferable, see Theorem 6. For approximate resp. asymptotic resp.

partially unknown boundary conditions for GPs see [42] resp. [44] resp. [10]. For a theoretic approach to endow RKHS with boundary information see [30].

Let again $\mathcal{F} = C^\infty(D, \mathbb{R})$ with Fréchet topology from (2) be the set of smooth functions on $D \subset \mathbb{R}^d$ compact, $K = \mathbb{R}\{f_1, \ldots, f_r\}$ with analytic functions $f_i : D \to \mathbb{R}$, and let $R \supseteq K$ be the Ore extension of K.

This section is based on two theorems. The first one describes closed modules satisfying a Nullstellensatz via their Taylor expansion. Denote by T_p the Taylor series of a (vector or matrix of) smooth function(s) around a point $p \in D$.

Theorem 5 (Whitney's Spectral Theorem; [48], [46, V Theorem 1.3]). *An $\mathcal{F}$-module $M \leq C^\infty(D, \mathbb{R})^\ell$ has topological closure $\overline{M} = \bigcap_{p \in D} T_p^{-1}(T_p(M))$.*

The second theorem specifies that analytic functions generate closed modules.

Theorem 6 ([27, Theorem 4], [46, VI Theorem 1.1]). *Let C be an $m \times n$-matrix of analytic functions on $D \subset \mathbb{R}^d$ and $\phi \in (C^\infty(D, \mathbb{R}))^m$. Then there is a $\psi \in (C^\infty(D, \mathbb{R}))^n$ with $\phi = C \cdot \psi$ if and only if for all $p \in D$ the $T_p(\phi)$ are an $\mathbb{R}[[x_1 - p_1, \ldots, x_d - p_d]]$-linear combination of the columns of $T(C)$.*

7.1 Boundary Conditions for Function Values of Single Functions

We begin parametrizing functions which are zero on an analytic set M, e.g. Dirichlet boundary conditions which prescribe values at the boundary ∂D.

We define boundaries $M \subseteq D$ implicitly via

$$M = \mathcal{V}(I) := \{m \in D \mid b(m) = 0 \text{ for all } b \in I\} \subseteq D,$$

where $I \trianglelefteq K$ is an ideal of equations. For any analytic set $M \subseteq D$ we have $M = \mathcal{V}(\mathcal{I}(M))$, where $\mathcal{I}(M) = \{b \in \mathcal{F}^\ell \mid b(m) = 0 \text{ for all } m \in M\} \subseteq \mathcal{F}$ is the (closed and radical) ideal of functions vanishing at M. If I is radical (it is automatically closed by Theorem 6, as generated by analytic functions), then $\mathcal{I}(\mathcal{V}(I)) = I$. Hence, any set of analytic function defined on D which generates a radical ideal parametrizes functions vanishing at its zero set. More formally:

Proposition 4. *Let $B' \in K^{1 \times \ell}$ be a row of analytic functions whose entries generate a radical $\mathcal{F}$-ideal $I = B'\mathcal{F}^\ell \leq \mathcal{F}$ of smooth functions. Then, I is the set $\{f \in \mathcal{F} \mid f_{|\mathcal{V}(I)} = 0\}$ of smooth functions vanishing at $\mathcal{V}(I)$.*

Proof. The condition $f_{|\mathcal{V}(I)} = 0$ restricts the zeroth order Taylor coefficients by homogeneous equations. All functions satisfying such restrictions are contained in the closure $\overline{I}$ of I by Whitney's Spectral Theorem 5. The $\mathcal{F}$-module parametrization $I = B'\mathcal{F}^\ell$ uses analytic functions as generators, which ensures that the ideal I is already equal to its closure $\overline{I}$ by Theorem 6. $\square$

We now compare constructions of rows B' of functions in Proposition 4.

Example 11. Functions $\mathcal{F} = C^\infty([0,1]^d, \mathbb{R})$ with Dirichlet boundary conditions $f(\partial D) = 0$ at the boundary of the domain $D = [0,1]^d$ are parametrized by

$$B_1' = \left[\prod_{i=1}^d x_i(x_i - 1) \right] \tag{5}$$

over $K = \mathbb{R}\{x_1, \ldots, x_n\} = \mathbb{R}[x_1, \ldots, x_n]$, by

$$B_2' = \left[1 - \exp\left((-1)^{d+1} \cdot \frac{\prod_{i=1}^d x_i(x_i-1)}{\delta} \right) \right] \tag{6}$$

over $K = \mathbb{R}\{\exp(x_1^2), \exp(x_1), x_1, \ldots, \exp(x_d^2), \exp(x_d), x_d\}$, or by[5]

$$B_3' = \left[\sqrt{ \prod_{i=1}^d \left(1 + \frac{\exp\left(-\frac{x_i^2}{\delta}\right) - 2\exp\left(-\frac{x_i^2 - x_i + 1}{\delta}\right) + \exp\left(-\frac{(x_i-1)^2}{\delta}\right)}{\exp\left(-\frac{1}{\delta}\right) - 1} \right) } \right] \tag{7}$$

for any $\delta > 0$. See [9, Section 3] for the special case $d = 2$ in (5). For practical differences of these formalizations of boundary conditions see Remark 6.

Block diagonal matrices parametrize boundaries of a vector of $\ell > 1$ functions. Also, restrictions on sets with higher codimension can be defined.

Example 12. The following three matrices $\left[1 - \exp\left(-\frac{|x|}{\delta}\right) \quad 1 - \exp\left(-\frac{|y|}{\delta}\right) \right]$, $\left[1 - \exp\left(-\frac{\sqrt{x^2+y^2}}{\delta}\right) \right]$, and $\left[1 - \exp\left(-\frac{x^2+y^2}{\delta}\right) \right]$ parametrize functions $f \in \mathcal{F} = C^\infty(\mathbb{R}^3, \mathbb{R})$ with $f(0,0,z) = 0$. The last parametrization is analytic.

7.2   Boundary Conditions for Derivatives and Vectors

Boundary conditions with vanishing derivatives can be constructed using multiplicities in the (no longer radical) ideal. The proof of the following proposition again follows from Theorems 5 and 6, in a similar way to Proposition 4.

Proposition 5. *Let $B' \in K^{\ell \times \ell'}$ be a matrix of analytic functions whose columns generate an $\mathcal{F}$-module $M = B'\mathcal{F}^{\ell'} \leq \mathcal{F}^\ell$ of smooth functions. Then,*

$$\left\{ f \in \mathcal{F}^\ell \, \middle| \, \forall p \in D \; \exists a_i \in \mathbb{R}[[x_1 - p_i, \ldots, x_d - p_d]] \right.$$

$$\left. \forall 1 \leq i \leq \ell : T_p(f_i) = \sum_{j=1}^{\ell'} T_p(b_{ij}) a_j \right\}$$

is the closed set of smooth functions sharing the same vanishing lower order Taylor coefficients as the columns of B'.

---

[5] B_3' is obtained as the product of the standard deviations obtained by conditioning d one-dimensional squared exponential covariances to the data points $(0,0)$ and $(1,0)$.

Example 13. Functions $\mathcal{F} = C^{\infty}([0,1]^d, \mathbb{R})$ with Dirichlet boundary conditions $f(\partial D) = 0$ and Neumann boundary condition $\frac{\partial f}{\partial n}(\partial D) = 0$ for n the normal to the boundary ∂D of the domain $D = [0,1]^d$ are parametrized by

$$B' = \left[1 - \exp\left((-1)^{d+1} \cdot \tfrac{\prod_{i=1}^{d} x_i^2 (x_i - 1)^2}{\delta} \right) \right], \tag{8}$$

constructed by squaring the exponent from the parametrization in (6), or

$$B' = \left[\prod_{i=1}^{d} \left(1 + \frac{\exp\left(-\frac{x_i^2}{\delta}\right) - 2\exp\left(-\frac{x_i^2 - x_i + 1}{\delta}\right) + \exp\left(-\frac{(x_i-1)^2}{\delta}\right)}{\exp\left(-\frac{1}{\delta}\right) - 1} \right) \right], \tag{9}$$

constructed by the squaring of the parametrization (7) for any $\delta > 0$.

Remark 6. In applications, the non-polynomial parametrizations from Examples 11 and 13 are more suitable. We demonstrate the effect by pushforward GPs obtained from these parametrizations in Fig. 2.

The polynomial pushforward from Example 11 yields the variance $x^2 \cdot (x-1)^2$, which strongly varies in the input interval $[0,1]$. The analytic pushforwards from Example 11 also set the variance to zero at the boundary, but quickly return to the original variance, and never exceed it. Even the speed of returning to the original variance can be controlled by changing the parameter δ.

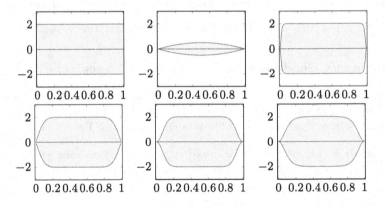

Fig. 2. GPs, represented by their mean function and two standard deviations. Upper left: a GP g with mean zero and square exponential covariance function. Upper middle: pushforward of the GP g by $x \cdot (x-1)$ has a strong global influence. Upper right resp. lower left: pushforward of the GP g by (6) resp. (7). Lower middle resp. right: pushforward of the GP g by (8) resp. (9) set the function and its derivative to zero at the boundary. Set $\delta := \frac{1}{100}$.

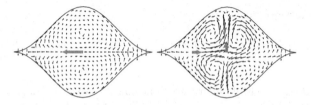

Fig. 3. The mean fields of the GP for divergence-free fields in the interior of $y^2 = \sin(x)^4$ from Example 14, which are conditioned on the data $(-1,0)$ at $(\frac{\pi}{2},0)$ (left) and on $(1,0)$ resp. $(0,1)$ at $(\frac{\pi}{4},0)$ resp. $(\frac{\pi}{2},0)$. The data is plotted artificially larger in gray. The flow at the analytic boundary is zero.

8   Examples

Now, we intersect (Theorem 4) solution sets of differential equations and analytic boundary conditions (Sect. 7) using the algorithms from Sects. 5 and 6.

Example 14. Consider divergence-free fields in the region in $\mathbb{R}^2$ bounded by $f := y^2 - \sin(x)^4$ for $x \in [0,\pi]$. Hence, consider

$$A = \begin{bmatrix} \partial_x & \partial_y \end{bmatrix}, B_1 = \begin{bmatrix} \partial_y \\ -\partial_x \end{bmatrix}, B_2 = \begin{bmatrix} f & 0 \\ 0 & f \end{bmatrix}.$$

The Matrix $C = \begin{bmatrix} f^2 \\ \partial_y f \\ -\partial_x f \end{bmatrix}$ from Theorem 4 yields the parametrization

$$\begin{bmatrix} \partial_y f^2 \\ -\partial_x f^2 \end{bmatrix} = \begin{bmatrix} f^2 \partial_y + 4 \cdot f \cdot y \\ -f^2 \partial_x + 8 \cdot f \cdot \sin(x)^3 \cos(x) \end{bmatrix}$$

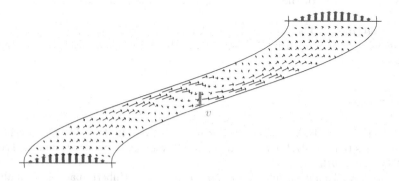

Fig. 4. The mean fields of the GP for divergence-free fields from Example 15, which are conditioned on $v = (0,-1)$ at $(0,1)$. The flow at the left and right boundary is zero, at the bottom resp. top there is flow into resp. out of the region. Both the data point resp. the inhomogeneous boundary conditions are plotted artificially larger in gray resp. dark gray.

and the push forward covariance

$$k \cdot \overline{f} \cdot \begin{bmatrix} 16y_1y_2 + 4\delta_y \cdot (f_1y_2 - f_2y_1) - (\delta_y^2 - 1) \cdot \overline{f} & (f_1\delta_y + 4y_1) \cdot \text{fsc}_2 \\ (f_2\delta_y - 4y_2) \cdot \text{fsc}_1 & \text{fsc}_1 \cdot \text{fsc}_2 - \overline{f} \cdot (2\delta_x^2 - 1) \end{bmatrix}$$

of the squared exponential covariance function $k = \exp(-\frac{1}{2}((x_1 - x_2)^2 + (y_1 - y_2)^2))$, where $f_1 = f(x_1, y_1)$, $f_2 = f(x_2, y_2)$, $\overline{f} = f_1 \cdot f_2$, $\delta_x = x_1 - x_2$, $\delta_y = y_1 - y_2$, $\text{sc}(x) = 8\sin(x)^3\cos(x)$, $\text{fsc}_1 = \text{sc}(x_1) - \delta_x \cdot f_1$, and $\text{fsc}_2 = \text{sc}(x_2) + \delta_x \cdot f_2$. For an illustration of this covariance see Fig. 3.

Example 15. Consider divergence-free fields in the compact domain D bounded by $-\frac{\pi}{2} \le y \le \frac{\pi}{2}$ and $3\sin(y) \le x \le 3\sin(y) + 2$. Hence, consider

$$A = \begin{bmatrix} \partial_x & \partial_y \end{bmatrix}, B_1 = \begin{bmatrix} \partial_y \\ -\partial_x \end{bmatrix}, B_2 = \begin{bmatrix} f & 0 \\ 0 & f \end{bmatrix}$$

for $f = (y - \frac{\pi}{2}) \cdot (y + \frac{\pi}{2}) \cdot (x - 3\sin(y)) \cdot (x - 3\sin(y) - 2)$. As the first entry in the column C is f^2, such fields can be parametrized by $\begin{bmatrix} \partial_y f^2 \\ -\partial_x f^2 \end{bmatrix}$. Pushing forward the squared exponential covariance function yields a covariance too big to display.

To encode non-zero boundary conditions we use a non-zero mean. Using the potential $p := -\frac{1}{4} \cdot (3\sin(y) - x + 3) \cdot (3\sin(y) - x)^2$ yields the divergence-free

$$\mu := \begin{bmatrix} -\frac{9}{4} \cdot \cos(y) \cdot (3\sin(y) - x + 2) \cdot (3\sin(y) - x) \\ -\frac{3}{4} \cdot (3\sin(y) - x + 2) \cdot (3\sin(y) - x) \end{bmatrix} = \begin{bmatrix} \partial_y \\ -\partial_x \end{bmatrix} p$$

satisfying the left and right boundaries and non-zero flow through at top and bottom. The GP $\mathcal{GP}(\mu, k)$ hence models of divergence-free fields in D with no flow on the sinoidal boundary left or right, but flow into D from the bottom and out of D at the top of the region. See Fig. 4 for a demonstration.

Acknowledgment. The authors thank Andreas Besginow for discussions and the anonymous reviewers for helpful comments, both of which improved the contents of this paper.

References

1. Bächler, T., Gerdt, V., Lange-Hegermann, M., Robertz, D.: Algorithmic Thomas decomposition of algebraic and differential systems. J. Symb. Comput. **47**(10), 1233–1266 (2012)
2. Berlinet, A., Thomas-Agnan, C.: Reproducing Kernel Hilbert Spaces in Probability and Statistics. Kluwer, Boston (2004)
3. Berns, F., Lange-Hegermann, M., Beecks, C.: Towards Gaussian processes for automatic and interpretable anomaly detection in industry 4.0. In: Proceedings of the International Conference on Innovative Intelligent Industrial Production and Logistics - IN4PL (2020)

4. Bogachev, V.I.: Gaussian Measures. Mathematical Surveys and Monographs, vol. 62. American Mathematical Society, Providence (1998)
5. Chyzak, F., Quadrat, A., Robertz, D.: Effective algorithms for parametrizing linear control systems over Ore algebras. Appl. Algebra Eng. Commun. Comput. **16**(5), 319–376 (2005)
6. Cramér, H., Leadbetter, M.R.: Stationary and Related Stochastic Processes. Dover Publications Inc., Mineola (2004)
7. Ehrenpreis, L.: Solution of some problems of division. I. Division by a polynomial of derivation. Am. J. Math. **76**, 883–903 (1954)
8. Gerdt, V.P., Lange-Hegermann, M., Robertz, D.: The MAPLE package TDDS for computing Thomas decompositions of systems of nonlinear PDEs. Comput. Phys. Commun. **234**, 202–215 (2019)
9. Graepel, T.: Solving noisy linear operator equations by Gaussian processes: application to ordinary and partial differential equations. In: Proceedings of the Twentieth International Conference on International Conference on Machine Learning, pp. 234–241 (2003)
10. Gulian, M., Frankel, A., Swiler, L.: Gaussian process regression constrained by boundary value problems. Comput. Methods Appl. Mech. Eng. **388**, Paper No. 114117, 18 (2022)
11. Honkela, A., et al.: Genome-wide modeling of transcription kinetics reveals patterns of RNA production delays. In: Proceedings of the National Academy of Sciences (2015)
12. Jacobson, N.: The Theory of Rings. American Mathematical Society Mathematical Surveys, vol. II. American Mathematical Society (1943)
13. Jaynes, E.T.: Probability Theory. Cambridge University Press, Cambridge (2003)
14. Jidling, C., et al.: Probabilistic modelling and reconstruction of strain. Nucl. Instrum. Methods Phys. Res. Sect. B Beam Interact. Mater. Atoms **436**, 141–155 (2018)
15. Jidling, C., Wahlström, N., Wills, A., Schön, T.B.: Linearly constrained gaussian processes. In: Advances in Neural Information Processing Systems (2017)
16. John, D., Heuveline, V., Schober, M.: GOODE: a Gaussian off-the-shelf ordinary differential equation solver. In: Proceedings of the 36th International Conference on Machine Learning (2019)
17. Lange-Hegermann, M.: Counting solutions of differential equations. Ph.D. thesis, RWTH Aachen (2014)
18. Lange-Hegermann, M.: The differential dimension polynomial for characterizable differential ideals. In: Böckle, G., Decker, W., Malle, G. (eds.) Algorithmic and Experimental Methods in Algebra, Geometry, and Number Theory, pp. 443–453. Springer, Cham (2017). https://doi.org/10.1007/978-3-319-70566-8_18
19. Lange-Hegermann, M.: Algorithmic linearly constrained Gaussian processes. In: Advances in Neural Information Processing Systems (2018)
20. Lange-Hegermann, M.: The differential counting polynomial. Found. Comput. Math. **18**(2), 291–308 (2018)
21. Lange-Hegermann, M.: Linearly constrained Gaussian processes with boundary conditions. In: International Conference on Artificial Intelligence and Statistics. PMLR (2021)
22. Lange-Hegermann, M., Robertz, D.: Thomas decompositions of parametric nonlinear control systems. IFAC Proc. Vol. **46**(2), 296–301 (2013)

23. Lange-Hegermann, M., Robertz, D.: Thomas decomposition and nonlinear control systems. In: Quadrat, A., Zerz, E. (eds.) Algebraic and Symbolic Computation Methods in Dynamical Systems. ADD, vol. 9, pp. 117–146. Springer, Cham (2020). https://doi.org/10.1007/978-3-030-38356-5_4

24. Lange-Hegermann, M., Robertz, D., Seiler, W.M., Seiß, M.: Singularities of algebraic differential equations. Adv. Appl. Math. **131**, Paper No. 102266, 56 (2021)

25. Macêdo, I., Castro, R.: Learning divergence-free and curl-free vector fields with matrix-valued kernels. Instituto Nacional de Matematica Pura e Aplicada, Brasil, Technical report (2008)

26. Malgrange, B.: Existence et approximation des solutions des équations aux dérivées partielles et des équations de convolution (1955/56). http://aif.cedram.org/item?id=AIF_1955_6_271_0

27. Malgrange, B.: Division des distributions. Séminaire Schwartz (1959)

28. McConnell, J.C., Robson, J.C.: Noncommutative Noetherian Rings. Graduate Studies in Mathematics, vol. 30. American Mathematical Society, Providence (2001)

29. Neal, R.M.: Priors for infinite networks. In: Neal, R.M. (ed.) Bayesian Learning for Neural Networks, pp. 29–53. Springer, New York (1996). https://doi.org/10.1007/978-1-4612-0745-0_2

30. Nicholson, J., Kiessler, P., Brown, D.A.: A kernel-based approach for modelling Gaussian processes with functional information. arXiv preprint arXiv:2201.11023 (2022)

31. Oberst, U.: Multidimensional constant linear systems. Acta Appl. Math. **20**(1–2), 1–175 (1990)

32. Quadrat, A.: Systèmes et Structures - Une approche de la théorie mathématique des systèmes par l'analyse algébrique constructive. Habilitation thesis, Université de Nice (Sophia Antipolis), France, April 2010

33. Quadrat, A.: Grade filtration of linear functional systems. Acta Appl. Math. **127**, 27–86 (2013)

34. Rasmussen, C.E., Williams, C.K.I.: Gaussian Processes for Machine Learning. MIT Press, Cambridge (2006)

35. Regensburger, G., Rosenkranz, M., Middeke, J.: A skew polynomial approach to integro-differential operators. In: ISSAC 2009–Proceedings of the 2009 International Symposium on Symbolic and Algebraic Computation, pp. 287–294. ACM, New York (2009)

36. Robertz, D.: Formal Algorithmic Elimination for PDEs. Springer, Cham (2014). https://doi.org/10.1007/978-3-319-11445-3

37. Robertz, D.: Recent progress in an algebraic analysis approach to linear systems. Multidimension. Syst. Signal Process. **26**(2), 349–388 (2014). https://doi.org/10.1007/s11045-014-0280-9

38. Rosenkranz, M., Regensburger, G.: Solving and factoring boundary problems for linear ordinary differential equations in differential algebras. J. Symb. Comput. **43**(8), 515–544 (2008)

39. Särkkä, S.: Linear operators and stochastic partial differential equations in Gaussian process regression. In: Honkela, T., Duch, W., Girolami, M., Kaski, S. (eds.) ICANN 2011. LNCS, vol. 6792, pp. 151–158. Springer, Heidelberg (2011). https://doi.org/10.1007/978-3-642-21738-8_20

40. Scheuerer, M., Schlather, M.: Covariance models for divergence-free and curl-free random vector fields. Stoch. Models **28**(3), 433–451 (2012)

41. Simon-Gabriel, C.J., Schölkopf, B.: Kernel distribution embeddings: universal kernels, characteristic kernels and kernel metrics on distributions. J. Mach. Learn. Res. **19**, Paper No. 44, 29 (2018)
42. Solin, A., Kok, M.: Know your boundaries: constraining Gaussian processes by variational harmonic features. In: Proceedings of the 22nd International Conference on Artificial Intelligence and Statistics (2019)
43. Solin, A., Kok, M., Wahlström, N., Schön, T.B., Särkkä, S.: Modeling and interpolation of the ambient magnetic field by Gaussian processes. IEEE Trans. Robot. **34**(4), 1112–1127 (2018)
44. Tan, M.H.Y.: Gaussian process modeling with boundary information. Statist. Sinica **28**(2), 621–648 (2018)
45. Thewes, S., Lange-Hegermann, M., Reuber, C., Beck, R.: Advanced Gaussian process modeling techniques. In: Design of Experiments (DoE) in Powertrain Development (2015)
46. Tougeron, J.C.: Idéaux de fonctions différentiables. Ergebnisse der Mathematik und ihrer Grenzgebiete, Springer, Heidelberg (1972). https://doi.org/10.1007/978-3-662-59320-2
47. Wahlström, N., Kok, M., Schön, T.B., Gustafsson, F.: Modeling magnetic fields using Gaussian processes. In: Proceedings of the 38th International Conference on Acoustics, Speech, and Signal Processing (ICASSP) (2013)
48. Whitney, H.: On ideals of differentiable functions. Am. J. Math. **70**, 635–658 (1948)
49. Zerz, E., Seiler, W.M., Hausdorf, M.: On the inverse syzygy problem. Commun. Algebra **38**(6), 2037–2047 (2010)
50. Zimmer, C., Meister, M., Nguyen-Tuong, D.: Safe active learning for time-series modeling with Gaussian processes. In: Advances in Neural Information Processing Systems (2018)

Computing the Integer Hull of Convex Polyhedral Sets

Marc Moreno Maza and Linxiao Wang[⊠]

University of Western Ontario, London, ON, Canada
{mmorenom,lwang739}@uwo.ca

Abstract. In this paper, we discuss a new algorithm for computing the integer hull P_I of a rational polyhedral set P, together with its implementation in Maple and in the C programming language. Our implementation focuses on the two-dimensional and three-dimensional cases. We show that our algorithm computes the integer hull efficiently and can deal with polyhedral sets with large numbers of integer points.

Keywords: Polyhedral set · Integer hull · Parametric polyhedron

1 Introduction

The integer points of rational polyhedral sets are of great interest in various areas of scientific computing. Two such areas are *combinatorial optimization* (in particular integer linear programming) and *compiler optimization* (in particular, the analysis, transformation and scheduling of for-loop nests in computer programs), where a variety of algorithms solve questions related to the points with integer coordinates belonging to a given polyhedron. Another area is at the crossroads of computer algebra and polyhedral geometry, with topics like toric ideals and Hilbert bases, see for instance [24] by Thomas.

One can ask different questions about the integer points of a polyhedral set, ranging from "whether or not a given rational polyhedron has integer points" to "describing all such points". Answers to that latter question can take various forms, depending on the targeted application. For plotting purposes, one may want to enumerate all the integer points of a 2D or 3D polytope. Meanwhile, in the context of combinatorial optimization or compiler optimization, more concise descriptions are sufficient and more effective.

For a rational convex polyhedron $P \subseteq \mathbb{Q}^d$, defined either by the set of its facets or by that of its vertices, one such description is the *integer hull* P_I of P, that is, the convex hull of $P \cap \mathbb{Z}^d$. The set P_I is itself polyhedral and can be described either by its facets, or its vertices. One important family of algorithms for computing the vertex set of P_I relies on the so-called *cutting plane method*, originally introduced by Gomory in [10] to solve integer linear programs (ILP) and mixed-integer programming (MILP) problems. This method is based on finding a sequence of linear inequalities (cuts) to reduce the feasible region to the

F. Boulier et al. (Eds.): CASC 2022, LNCS 13366, pp. 246–267, 2022.
https://doi.org/10.1007/978-3-031-14788-3_14

original ILP problem. Chvátal [6] and Schrijver [22] gave a geometrical description of the cutting plane method and developed a procedure to compute P_I based on it. Schrijver gave a full proof and a complexity study of this method in [20]. Another approach for computing P_I uses the *branch and bound method*, introduced by Land and Doig in the early 1960s in [15]. This method recursively divides P into sub-polyhedra, then the vertices of the integer hull of each part of the partition are computed.

There are also authors studying the relations between the vertices of P_I and the vertices of P. The authors of [11] provided an algorithm for finding the vertices of a polytope associated to the Knapsack integer programming problem. This algorithm computes boxes covering the input polyhedron and such that each box contains at most one vertex of P_I. Following that same approach, the authors [4] could give an upper bound on the number of those boxes, as well as a running estimate for enumerating the integer vertices of a polytope.

Since an integer hull is the convex hull of all the integer points within a polyhedral set, a straightforward way of computing the integer hull is enumerating all its integer points, followed by a convex hull computation. There is a family of studies focusing on enumerating or counting the lattice points of a given polyhedral set. A well-known theory on that latter subject was proposed by Pick [18]. In particular, the celebrated Pick's theorem provides a formula for the area of a simple polygon P with integer vertex coordinates, in terms of the number of integer points within P and on its boundary. In the 1990s, Barvinok [3] created an algorithm for counting the integer points inside a polyhedron, which runs in polynomial time, for a fixed dimension of the ambient space. Later studies such as [27] gave a simpler approach for lattice point counting, which divides a polygon into *right-angle triangles* and calculates the number of lattice points within each such triangle.

Verdoolaege et al. present in [25] a novel method for lattice point counting, based on Barvinok's decomposition for counting the number of integer points in a non-parametric polytope. In [23], Seghir, Loechner and Meister deal with the more general problem of counting the number of images by an affine integer transformation of the lattice points contained in a parametric polytope. In 2004, the software package LattE presented in [16] for lattice point enumeration offers the first implementation of Barvinok's algorithm. Other algorithms, such as [12] by Jing and Moreno Maza, compute an irredundant representation of the integer points of P in terms of "simpler" polyhedral sets, each of them given by a triangular-by-block system of linear inequalities.

Normaliz [5] is a software library for the computation of Hilbert bases of rational cones and the normalizations of affine monoids. The Hilbert basis of a convex cone C is a minimal set of integer vectors such that every integer vector in C is a conical combination of the vectors in the Hilbert basis with integer coefficients. The computation of a Hilbert basis of a simplicial cone can be done by enumerating all lattice points of paralleltopes. From there, Normaliz provides a command for computing the integer hull of a given polyhedral set based on enumeration and convex hull computation.

Polymake [1] is a software system that includes several algorithms for convex hull computation and lattice points enumeration (including those of LattE and `Normaliz`). `Polymake` uses these algorithms to compute the integer hulls of various kinds of input polyhedral sets.

Since the integer hull P_I of P is completely determined by its vertices, it is natural to ask for the number of vertices in an integer hull of a polyhedron. The earliest study by Cook, Hartmann, Kannan and McDiarmid, in [8], shows that the number of vertices of P_I is related to the *size* (as defined in [20]) of the coefficients of the inequalities that describe P. Let $x = p/q$ be a rational number, where p and q are coprime integers, the size of x is defined as

$$size(x) = 1 + \lceil (\log(|p| + 1)) \rceil + \lceil (\log(|q| + 1)) \rceil.$$

For a linear inequality $a_n x_n + \cdots + a_1 x_1 + a_0 \le 0$, its size is $\sum size(a_i)$. For a polyhedron $P = \{\mathbf{x} \mid A\mathbf{x} \le \vec{b}\}$ where $A \in \mathbb{Q}^{m \times n}$ and $\vec{b} \in \mathbb{Q}^m$. Cook, Hartmann, Kannan and McDiarmid showed that the number of vertices of the integer hull of P is bounded over by $2m^n(6n^2\varphi)^{n-1}$ where φ is the maximum size of any of the m inequalities. More recent studies such as [26] and [4] use different approaches to reach similar or slightly improved estimates. We also discussed this question in our CASC 2021 paper [17].

In this paper, we present our algorithm for computing P_I and we report on the performance of its implementation as a new command of MAPLE's library `PolyhedralSets` [19] as well as in the C programming language. We present benchmarks for both implementations in Sect. 6. Our results show that our algorithm is very efficient comparing to the well known library Normaliz [5] especially when the input polyhedral set is large in volume.

Our algorithm has three main steps:

Normalization: during this step, we construct a new polyhedral set Q from P as follows. Consider in turn each facet F of P:
1. if the hyperplane H supporting F contains an integer point, then H is a hyperplane supporting a facet of Q,
2. otherwise one slides H towards the center of P along the normal vector of F, stopping as soon as one hits a hyperplane H' containing an integer point, then making H' a hyperplane supporting a facet of Q.

The resulting polyhedral set Q clearly has the same integer hull as P; computing Q is a preparation phase for the following step.

Partitioning: during this step, we search for integer points inside Q so as to partition P into smaller polyhedral sets, the integer hulls of which can easily be computed. We observe that every vertex of Q which is an integer point is also a vertex of Q_I. Now, for every vertex V of Q which is not an integer point we look, on each facet F to which V belongs, for an integer point $C_{V,F}$ that is "close" to V (ideally as close as possible to V). All the points $C_{V,F}$ together with the vertices of Q are used to build that partition of Q. Each part of the partition is a polyhedron R which:
1. either has integer points as vertices (making the computation of the integer hull R_I trivial),
2. or has a small volume so that any algorithm (including exhaustive search) can be applied to compute R_I.

Merging: Once the integer hull of each part of the partition is computed and given by the list of its vertices, an algorithm for computing the convex hull of a set points, such as QuickHull [2], can be applied to deduce P_I.

The paper is organized as follows. Section 2 is a brief review of polyhedral geometry. Sections 4 and 5 present our algorithms in the 2D and 3D cases, respectively. Section 3.2 gathers key arguments supporting our algorithm, essentially based on the concept of the Hermite Normal Form of a matrix. Section 6 reports on our experimentation with the proposed algorithms.

2   Preliminaries

In this review of polyhedral geometry, we follow the concepts and notations of Schrijver's book [20], As usual, we denote by $\mathbb{Z}$, $\mathbb{Q}$ and $\mathbb{R}$ the ring of integers, the field of rational numbers and the field of real numbers. Unless specified otherwise, all matrices and vectors have their coefficients in $\mathbb{Z}$. A subset $P \subseteq \mathbb{Q}^d$ is called a *convex polyhedron* (or simply a *polyhedron*) if $P = \{\mathbf{x} \in \mathbb{Q}^d \mid A\mathbf{x} \leq \vec{b}\}$ holds, for a matrix $A \in \mathbb{Q}^{m \times d}$ and a vector $\vec{b} \in \mathbb{Q}^m$, where m and d are positive integers; we call the linear system $\{A\mathbf{x} \leq \vec{b}\}$ an *H-representation* of P. Hence, a polyhedron is the intersection of finitely many affine half-spaces. Here an affine half-space is a set of the form $\{\mathbf{x} \in \mathbb{Q}^d \mid \vec{w}^t\mathbf{x} \leq \delta\}$ for some nonzero vector $\vec{w} \in \mathbb{Z}^d$ and an integer number δ.

A non-empty subset $F \subseteq P$ is a *face* of P if $F = \{\mathbf{x} \in P \mid A'\mathbf{x} = \vec{b}'\}$ for some subsystem $A'\mathbf{x} \leq \vec{b}'$ of $A\mathbf{x} \leq \vec{b}$. A face of P, distinct from P, and with maximum dimension is a *facet* of P. The *lineality space* of P is $\{\mathbf{x} \in \mathbb{Q}^d \mid A\mathbf{x} = \vec{0}\}$ and P is said *pointed* if its lineality space has dimension zero. Note that, in this paper, we only consider pointed polyhedra. For a pointed polyhedron P, the inclusion-minimal faces are the *vertices* of P.

We are interested in computing P_I the *integer hull* of P, that is, the smallest convex polyhedron containing the integer points of P. In other words, P_I is the intersection of all convex polyhedra containing $P \cap \mathbb{Z}^d$. Assume that P is pointed. Then, $P = P_I$ if and only if every vertex of P is integral, see [21]. Thus, the convex hull of all the vertices of P_I is P_I itself.

3   Two Core Constructions of our Algorithm

In this section, we emphasize two constructions supporting respectively the *normalization* and *partitioning* steps of our algorithm. Both constructions deal with "algebraic aspects", that is, with the fact that we are solving for the integer solutions of a system of linear inequalities. These two constructions are inspired respectively by [17] and [12].

3.1   Normalization

Considering the rational polyhedron $P = \{\mathbf{x} \in \mathbb{Q}^d \mid A\mathbf{x} \leq \vec{b}\}$, with the notations of Sect. 2, we observe that one can compute a vector $\vec{e} \in \mathbb{Z}^m$ so that the rational polyhedron $Q = \{\mathbf{x} \in \mathbb{Q}^d \mid A\mathbf{x} \leq \vec{e}\}$ satisfies:

1. $P_I = Q_I$, and
2. the supporting hyperplane of every facet of Q has at least one integer point. Notice that this does not necessarily means that the new facet has an integer point.

In the introduction, the construction of Q is referred as the *normalization* step. We construct Q from P as follows:

1. consider each facet F of P in turn; if the hyperplane H supporting F does not contain an integer point, then one "slides" H towards the center of P along the normal vector of F, stopping as soon as a hyperplane H' containing an integer point is reached, otherwise keep H unchanged;
2. the resulting polyhedron is Q, for which rational consistency must be checked, which can be done efficiently using a method based on linear programming.

The "sliding process" described above informally is performed as follows. Let the equation below define the hyperplane H supporting F:

$$a_1 x_1 + \cdots + a_d x_d = b, \tag{1}$$

where $a_1, \ldots, a_d, b$ can be assumed to be integers. The fact that $\mathbb{Z}$ is an Euclidean domain (and thus a principal ideal domain) implies that H has integer points if and only if we have:

$$\gcd(a_1, \ldots, a_d) \mid b. \tag{2}$$

If the hyperplane H supporting F does not have integer points and P is included in the half-space $a_1 x_1 + \cdots + a_d x_d \leq b$, then H' is given by:

$$a_1 x_1 + \cdots + a_d x_d \leq g \left\lfloor \frac{b}{g} \right\rfloor, \tag{3}$$

with $g := \gcd(a_1, \ldots, a_d)$.

Summing things up, we denote by `Normalization`(P) a function call returning the polyhedron Q.

3.2   Partitioning

The other algebraic construction in our algorithm supports the *partition* step briefly explained in the introduction. The underlying question is the following: given a vertex V of P which is not an integer point and given a facet F of P to which V belongs, find on F an integer point $C_{V,F}$, if any, which is "close" to V (ideally as close to V as possible).

If P is two-dimensional, thus, if F is one-dimensional, then the question is easily answered by elementary arguments, see our previous paper [17]. If P has dimension $d \geq 3$, thus, if F has dimension $d - 1$, then we take advantage of the *Hermite normal form* of a matrix. In the sequel of this section, we review this concept. and use it to compute the integer hull of a facet of a polyhedron. Finally, we solve the question of finding an integer point $C_{V,F}$ on F (if any) as close as possible to V.

Hermite Normal Form. Consider a positive integer $p \leq d$ and a linear system $C\mathbf{x} = \mathbf{s}$ where $C \in \mathbb{Z}^{p \times d}$ is a full row-rank matrix and $\mathbf{s} \in \mathbb{Z}^p$ is a vector. There exists a uni-modular matrix $U \in \mathbb{Z}^{d \times d}$ so that $CU = [\mathbf{0}H]$ where $\mathbf{0} \in \mathbb{Z}^{p \times (d-p)}$ is the null matrix and H is the column-style Hermite normal form of C. We write $U = [U_L U_R]$ where $U_L \in \mathbb{Z}^{d \times (d-p)}$ and $U_R \in \mathbb{Z}^{d \times p}$. Therefore, the matrix $H \in \mathbb{Z}^{p \times p}$ is non-singular and the following properties hold:

1. $C\mathbf{x} = \mathbf{s}$ has integer solutions if and only if $H^{-1}\mathbf{s}$ is an integer vector,
2. every integer solution of $C\mathbf{x} = \mathbf{s}$ has the form $U_R H^{-1}\mathbf{s} + U_L\mathbf{z}$, where $\mathbf{z} \in \mathbb{Z}^{d-p}$ is arbitrary.

Determining the Integer Hull of a Facet. Let $\vec{c}^t\mathbf{x} = s$ be the equation of the hyper-plane supporting F, thus with $\vec{c} \in \mathbb{Z}^d$ and $s \in \mathbb{Z}$. Let $U \in \mathbb{Z}^{d \times d}$ be a uni-modular matrix so that $\vec{c}^t U = [\mathbf{0}H]$ where $\mathbf{0} \in \mathbb{Z}^{1 \times (d-1)}$ is the null matrix and H is the column-style Hermite normal form of $\vec{c}^t$ regarded as a matrix of $\mathbb{Z}^{1 \times d}$. We write $U = [U_L U_R]$ where $U_L \in \mathbb{Z}^{d \times (d-1)}$ and $U_R \in \mathbb{Z}^{d \times 1}$. Let $\mathbf{v} := U_R H^{-1}s$. Then, from the above paragraph on Hermite Normal Form, we know that the integer points of the hyper-plane supporting F are of the form $\mathbf{x} = \mathbf{v} + U_L\mathbf{z}$ where $\mathbf{z} \in \mathbb{Z}^{d-1}$ is arbitrary. The facet F is described by a system of linear inequalities in $\mathbb{Q}^d$ with $\mathbf{x}$ as unknown vector. Substituting $\mathbf{v} + U_L\mathbf{z}$ for $\mathbf{x}$ yields a system of linear inequalities in $\mathbb{Q}^{d-1}$ (with $\mathbf{z}$ as unknown vector) representing a rational polyhedron $G \subseteq \mathbb{Q}^{d-1}$. With these notations and hypotheses, we have the following.

Theorem 1. *The vertices of the integer hull G_I of G are in one-to-one correspondence with the vertices of the integer hull F_I of F via the map*

$$R_F : \begin{cases} \mathbb{Q}^{d-1} & \to & \mathbb{Q}^d \\ \mathbf{z} & \longmapsto & \mathbf{x} = \mathbf{v} + U_L\mathbf{z}. \end{cases} \tag{4}$$

In particular, we have $R_F(G_I) = F_I$.

PROOF ▷ The proof follows from seven claims.

Claim 1. R_F is injective. Indeed, the matrix U is uni-modular, thus the columns of U are linearly independent, and the map $\mathbf{z} \longmapsto U_L\mathbf{z}$ is injective.

Claim 2. The image of R_F is F. Since R_F is an injective affine map from $\mathbb{Q}^{d-1}$ to $\mathbb{Q}^d$, it follows that the image of R_F is an affine space of dimension $d - 1$. Therefore, in order to prove the claim, it suffices to prove that for every $\mathbf{z} \in \mathbb{Z}^{d-1}$ we have $R_F(\mathbf{z}) \in F$. Since $F \cap \mathbb{Z}^d \neq \emptyset$ (as a consequence of the normalization step of our algorithm) there exists $\mathbf{z}_0 \in \mathbb{Z}^{d-1}$ so that $\mathbf{x}_0 := \mathbf{v} + U_L \mathbf{z}_0 \in F \cap \mathbb{Z}^d$ holds. Let $\mathbf{z} \in \mathbb{Z}^{d-1}$. Define $\mathbf{x} := R_F(\mathbf{z})$. We have:

$$\mathbf{x} = \mathbf{v} + U_L \mathbf{z}_0 + U_L(\mathbf{z} - \mathbf{z}_0) = \mathbf{x}_0 + U_L(\mathbf{z} - \mathbf{z}_0).$$

We deduce:

$$\vec{c}^t \mathbf{x} = \vec{c}^t \mathbf{x}_0 + \vec{c}^t U_L(\mathbf{z} - \mathbf{z}_0) = s + 0 = s,$$

which proves that $R_F(\mathbf{z}) \in F$ holds.

Claim 3. $R_F^{-1}(H)$ is a half-space of $\mathbb{Q}^{d-1}$ for any half-space H of $\mathbb{Q}^d$. Indeed, for any $\mathbf{x} \in \mathbb{Q}^d$ of the form $\mathbf{v} + U_L \mathbf{z}$, with $\mathbf{z} \in \mathbb{Q}^{d-1}$, we have

$$\vec{a}^t \mathbf{x} \geq b \quad \Longleftrightarrow \quad \vec{a}^t U_L \mathbf{z} \geq b - \vec{a}^t \mathbf{v},$$

where $H : \vec{a}^t \mathbf{x} \geq b$ is an arbitrary half-space of $\mathbb{Q}^d$.

Claim 4. The integer points of F are in one-to-one correspondence with the integer points of $\mathbb{Z}^{d-1}$. This claim follows directly from the properties of the Hermite Normal Form.

Claim 5. $R_F^{-1}(S)$ is a polyhedron of $\mathbb{Q}^{d-1}$ for any polyhedron S of $\mathbb{Q}^d$. Indeed, let $S := \cap_i H_i$ be a polyhedron of $\mathbb{Q}^{d-1}$ given as the intersection of finitely many half-spaces of $\mathbb{Q}^{d-1}$. We have

$$R_F^{-1}(S) = R_F^{-1}(\cap_i H_i) = \cap_i R_F^{-1}(H_i).$$

The conclusion follows with Claim 3.

Claim 6. $R_F(T)$ is a polyhedron of $\mathbb{Q}^d$ for any polyhedron T of $\mathbb{Q}^{d-1}$. The proof is similar to that of Claim 5.

Claim 7. We have: $R_F(G_I) = F_I$. Let S be the set of all polyhedra of $\mathbb{Q}^d$ containing $F \cap \mathbb{Z}^d$. Let T be the set of all polyhedra of $\mathbb{Q}^{d-1}$ containing $G \cap \mathbb{Z}^d$, where $G = R_F^{-1}(F)$. Then, by definition of F_I and G_I, we have:

$$F_I = \bigcap_{S \in \mathcal{S}} S \quad \text{and} \quad G_I = \bigcap_{T \in \mathcal{T}} T.$$

From Claim 5, we have:

$$R_F^{-1}(F_I) = \bigcap_{S \in \mathcal{S}} R_F^{-1}(S) \supseteq \bigcap_{T \in \mathcal{T}} T = G_I.$$

From Claim 6, and since R_F is injective, we have:

$$R_F(G_I) = \bigcap_{T \in \mathcal{T}} R_F(T) \supseteq \bigcap_{S \in \mathcal{S}} S = F_I.$$

Therefore, we have $R_F(G_I) = F_I$. Now we can prove the theorem. Since R_F is a bijective affine map from $\mathbb{Q}^{d-1}$ to F, it maps affine subspaces of dimension $0 \leq d' < d$ of $\mathbb{Q}^{d-1}$ to affine subspaces of dimension d' of F. Combined with Claims 5 and 6, this latter observation implies that faces of dimension $0 \leq d' < d$ of G_I are mapped to faces of dimension d' of F_I. Therefore, the vertices of G_I are in one-to-one correspondence with the vertices of F_I.

Theorem 1 shows that one can reduce the computation of the vertices of F_I to computing the vertices of G_I.

Based on that observation, we denote by $\texttt{HNFProjection}(F, d)$ a function call returning the ordered pair (G, R_F).

Finding an Integer Point $C_{V,F}$ on F (If Any) Close to V. Let us return now to the question of finding an integer point $C_{V,F}$ on F (if any) as close as possible to V. A second consequence of Theorem 1 is that we can compute an integer point $C_{V,F}$ simply by choosing a point $R_F(W)$ at minimum Euclidean distance to V, where W ranges in the set of the vertices of G_I. As mentioned, such a point may not be an integer point of F at minimum Euclidean distance to V, but if F is large enough (that is, if its area is large enough) then $C_{V,F}$ is a good approximate solution to this optimization problem.

4   Integer Hull of a 2D Polyhedral Set

In this section, we present our algorithm for computing the integer hull of a 2D polyhedral set. We first give a high-level introduction of the algorithm, then we present its sub-routines, a more precise presentation of the general algorithm together with the implementation details.

As introduced in Sect. 1, our main idea is to partition the input 2D-polyhedral set into several smaller areas, compute the integer hulls of each area and find a convex hull of all these integer hulls.

In Sect. 2, we explained that an integer hull is a convex polyhedral set whose vertices are all integer points. Therefore, given a polyhedral set that is not an integer hull, if we can replace each fractional vertex with some integer ones, we will obtain the integer hull of the input polyhedral set. Of course, during this replacement process, we should not exclude any integer points, otherwise the result would not be valid.

To replace the fractional vertices, we need to look at the areas around those vertices that are the corners of the input polyhedral set. We do that by partitioning the input such that each fractional vertex is included in a "small" triangle, for which the integer hull is computed by a straightforward method.

Other than these corners, there is the central part of the input, ideally this should be the part that covers most of the area of the input. To make the computation of the central part easier, we construct the partition by ensuring the central area is already an integer hull. In the final step, we combine the corner parts and the central part using a convex hull algorithm to compute the final output.

To meet all the requirements above, we propose the following method to partition the input. First, we normalize the input using procedure $\texttt{Normalization(P)}$. For each fractional vertex, we find the closest integer point to it on each of its adjacent facets. For a 2D polyhedral set, each vertex has exactly two adjacent facets, therefore, two "closest integer points". We partition the input by connecting each of these closest integer point pairs. Thus, in most cases a corner part would be a triangle with vertices of a fractional vertex and its two closest integer points. In some special cases when some facets contain no integer point, we combine the adjacent vertices and their closest integer points to form a polyhedral set has two and only two integer vertices. The central part is an integer hull with vertices of all these closest integer points and all the integer vertices of the input.

The details of the sub-routines as well as the general algorithm are given in the following sections.

4.1   Algorithm

In this section, we consider an input polyhedral set P defined by a system of linear inequalities

$$\begin{cases} a_{11}x_1 + a_{21}x_2 \leq b_1, \\ a_{12}x_1 + a_{22}x_2 \leq b_2, \\ \quad \cdots \\ a_{1n}x_1 + a_{2n}x_2 \leq b_n, \end{cases}$$

where $\gcd(a_{1i}, a_{2i}, b_i) = 1$ for $i \in \{1, \ldots, n\}$. We assume that this representation of P is irredudant, that is, the defining linear inequalities of P are in one-to-one correspondence with the facets of P. In this paper, we follow the convention of MAPLE's `PolyhedralSets` library and refer to these inequalities as the `relations` of P.

Following the informal description of the algorithm above, for each fractional vertex, we need to find the closest integer points on the facets adjacent to this vertex. But we first notice that it is possible that the supporting hyperplane of a facet, and therefore the facet itself, do not have any integer points. Therefore, the first step of our algorithm is to normalize the `relations` of the input using the "sliding process" described in Sect. 3.1.

In the next step, `closestIntegerPoints` (Algorithm 1), we find the closest integer point to each fractional vertex on its adjacent facets. From the proof of Lemma 1 in [17] we know that, on a line $a_1x + a_2y = b$, a point is an integer point if and only if it has x value of $x \equiv \frac{b}{a_1} \mod a_2$. We can use this observation to find the closest integer point on a line to a given point. We also deal with the case where a facet does not contain any integer point.

Next, we need to construct the corner polyhedral sets and compute their integer hulls. Then, we find the convex hull of all these integer hulls (see Algorithm 2). Lemma 4 in [17] shows that the vertices of this final convex hull are the vertices of P_I.

For a fractional vertex $V[i]$, if neither $V_C[i][1]$ nor $V_C[i][2]$ is NULL, then the corner is a triangle with vertices $[V[i], V_C[i][1], V_C[i][2]]$. If one or both of $V_C[i][1]$ and $V_C[i][2]$ are NULL, which means there is no integer point on one or both adjacent facets of $V[i]$, we construct the corner as follow.

Algorithm 1: Compute the closest integer points to each fractional vertex on its adjacent facets

---

1 **Function** closestIntegerPoints(V)

    Input: V, a list of the vertices of P

    Output: V_C, a list of pairs where $V_C[i][1]$ and $V_C[i][2]$ store the closest integer points of vertex $V[i]$ on its two adjacent facets.

2   **for** $i = 1, \ldots, n$ **do**

3   Let $V[i_1]$ and $V[i_2]$ be the vertices adjacent to $V[i]$

4   **for** $j = 1, 2$ **do**

5   **if** *there are integer points between* $[V[i], V[i_j]]$ **then**

6             $V_C[i][j] \leftarrow$ closest integer point to $V[i]$ on $[V[i], V[i_j]]$

7   **else**

8             $V_C[i][j] \leftarrow NULL$

9   **return** V_C

---

1. Let V_P be an empty set.
2. Let's say facet f is adjacent to $V[i]$ and does not contain integer point, we add both vertices of f, $V[i]$ and $V[j]$, to V_P.
3. Check the adjacent facets of all the vertices in V_P, if some of them does not contain integer point go to step 2, until no new fractional vertex can be added to V_P.
4. For every vertex in V_P add any existing "closest integer point" to V_P.
5. In the end, V_P contains several fractional points and at most two integer points and we construct a polyhedral set with V_P as the vertex set.

To compute the integer hull of a corner, we use a brute-force method that searches for all the integer points within the corner polyhedral set and then compute the convex hull of all these points. [8] has showed that, the size and shape of the corner polyhedral set only depends on the coefficients, a_{ij}, of the relations of the input but not the constant terms b_i. This implies that the size of the area that we need to do exhaustive search on is not related to the size of the input polyhedral set P so that the time complexity of our algorithm is not related to the volume of the input polyhedral set.

With all the sub-routines introduced above, we present our integer hull algorithm (Algorithm 3) for 2D polyhedral sets. We discuss some of the implementation details in Sect. 6.

4.2   An Example

In this section, we use the following example to show how our 2D algorithm works. The input is a polyhedral set defined by

$$\begin{cases} 2x + 5y \leq 64, \\ -7x - 5y \leq -20, \\ 3x - 6y \leq -7. \end{cases}$$

Algorithm 2: Construct and compute the integer hulls of the corner polyhedral sets

---

1 **Function** cornerIntegerHulls(V)
    Input:
        – V, the list of the vertices of the input polyhedral set
        – V_C, the output from Algorithm 1

    Output: A list of the vertices of the integer hull of P
2     $V_I \leftarrow \{\}$
3   **for** $i = 1, \ldots, n$ **do**
4   **if** $V[i]$ *is an integer point* **then**
5             $V_I \leftarrow V_I \cup \{V[i]\}$
6   **else**
7             $T \leftarrow$ ConstructCorner($V[i], V_C$)
                /* create a corner polyhedral set as we described above */
8             $A \leftarrow$ AllIntegerPoints(T)   /* find all the integer points in T */
9             $V_{\text{tmp}} \leftarrow$ ConvexHull(A)
                /* compute the vertices of the convex hull of A */
10             $V_I \leftarrow V_I \cup \{V_{\text{tmp}}\}$
11   **return** ConvexHull(V_I)

The first step we need to do is to normalize the facets. In this example, there is
only one facet which is given by the relation $3x - 6y \leq -7$. We replace it with
$3x - 6y \leq -9$ (see Fig. 1).

Next we need to find the closest integer points to each fractional vertex on
its adjacent facets. In our case, all three vertices are fractional, so we need to
find two integer points for each (see Fig. 2a). And as we discussed in Sect. 4, the
center part of the input is already an integer hull, so no action needed for this
area. As we can see in Fig. 2b, the center part takes most of the volume of the
input, by doing so we cut down the size of the problem.

Then we just need to compute the integer hulls of the small corner triangles
and use the results to compute the final output (see Fig. 3).

5 Integer Hull of a 3D Polyhedral Set

With the 2D algorithm in place, we can move on to a higher dimension. In this
section, we present our integer hull algorithm for 3D polyhedral sets. The general
idea behind the algorithm is the same as that of the 2D algorithm. We want to
partition the input into smaller polyhedral sets and separate the parts into two
categories, the ones with fractional vertices for which we need to compute the

---

Algorithm 3: Compute the integer hull of a given 2D polyhedral set

---

1 Function IntegerHull2D(P)

    Input: P, a 2D PolyhedralSet object

    Output: I, a list of the vertices of the integer hull of P

2   Process corner cases

3 $Q \leftarrow$ Normalization(P)

4 $V \leftarrow$ Vertices(Q)

5 $V_C \leftarrow$ closestIntegerPoints(V)

6 $I \leftarrow$ cornerIntegerHulls(V, V_C)

7   **return** I

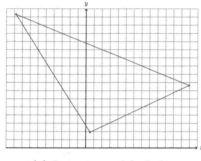

(a) Input is a polyhedral set

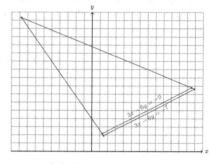

(b) Normalize the input

Fig. 1. Input and replaceNonIntegerFacets

integer hulls as sub-problems and the other ones that are already integer hulls themselves. After processing all the sub-problems, we combine the results of all the parts together and compute the final result.

5.1 Algorithm

The first step of the 3D algorithm is the same as that in Sect. 4.1. We normalize the facets as is in Sect. 3.1. Similarly, we want to find the closest integer points to the fractional vertices on their adjacent facets. Every fractional vertex and its closest integer points would form a small polyhedral set. For example, Fig. 4a is an example input and the green areas in Fig. 4b are the polyhedral sets formed by fractional vertices and their closest integer points.

Figure 5a shows the center part of the input, this is a polyhedral set with vertices that are all the closest integer points. In the 2D problem, the corner polyhedral sets and the center part formed a partition of the input. But in the 3D case, there are areas that are not covered by these parts, to be precise, these are the areas near the edges (see Fig, 5b).

In order to form a complete partition, we need another set of sub-polyhedral sets. As is shown in Fig. 6a, for an edge that has at least one fractional vertex, the

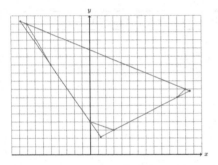
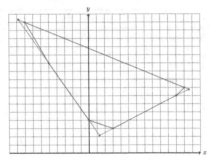

(a) For a fractional vertex, find the integer point on each adjacent facet that is closest to it and construct a triangle with the three points

(b) The center part is already an integer hull so we don't need to do anything

Fig. 2. Partition the input

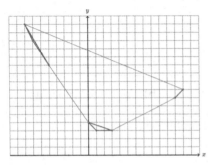
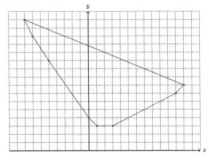

(a) Apply the previous two steps to each fractional vertex

(b) Find the convex hull of all the result vertices from the previous steps

Fig. 3. Compute the integer hulls of the parts and the final result

two vertices of the edge and the closest integer points to the fractional vertices (or vertex) form a polyhedral set. If we construct one such polyhedral set for each edge, we can cover all the missing areas in Fig. 5b.

For the parts that are not integer hulls already, we use a brute-force method to compute their integer hulls, that is, we use exhaustive search to find all the integer points within the part and compute the convex hull of the points. To cut down the area that needs exhaustive search, we further partition the edge polyhedral sets if possible. If there are integer points on an edge, we find the closest one to each fractional vertex and partition the polyhedral set into three parts, see Fig. 6b for an example.

Finding, on a given segment S, the integer point closest to a given vertex of S is relatively simple in the 2D problem, but in the 3D case, we need to address the following, more complicated, question: finding, on a given bounded

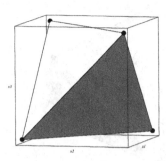

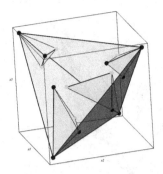

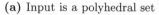

(a) Input is a polyhedral set

(b) Every fractional vertex and its closest integer points form a small polyhedral set

Fig. 4. Input and fractional vertices

3D polyhedron F, an integer point closest to a given vertex of F. A natural step towards answering this question is to represent all the integer points of F, which, itself, is an integer hull problem. Since the 3D polyhedron F is "flat", we can project it to a 2D ambient space and use our algorithm from Sect. 4.

Here we use the procedure HNFProjection(F, d) which is introduced in detail in Sect. 3.2. Recall that this procedure will return an ordered pair (G, R_F) where R_F gives the map between a 3D point to a 2D point.

Having a 2D polyhedral set F_P, we use our Algorithm 3 to compute the vertices of the integer hull of F_P. Although the HNF method keeps the integer points in the projection, it can not keep the distance among the points in general, so we must find the original image of the vertices of the integer hull of F_P.

Now that we have the integer hull of a facet, we can search for the closest integer points to each of its vertices. Here we decide to use the closest vertex of the integer hull instead of the actual closest integer point. Using the closest integer vertex might slow down the later steps but only by a very small amount. Searching for the actual point would be another optimization problem and this would be less efficient looking at the whole picture.

As mentioned above, in order to form a complete partition of the input polyhedral set, we need to carefully consider every edge that has at least one fractional vertex. To this end, we use Algorithm 1 to find the closest integer points to a fractional vertex on its adjacent edges. Now that we have all the "closest integer points" we need, we can construct the parts that are the "blue", "red" and "green" regions in Figs. 4, 5 and 6. Since all the vertices of the "red" polyhedral set are integer points, work remains to be done only in the "green" and "blue" polyhedral sets.

Before we present the complete algorithm, there are some corner cases that need to be considered. Similar to our 2D problem, the input polyhedral set could be not fully dimensional. Again we use Hermite Normal Form (HNF) to project

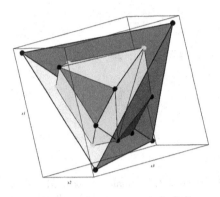 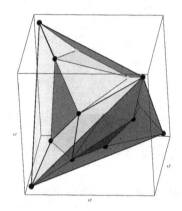

(a) The center part is already an integer hull so we don't need to do anything

(b) There are areas that are not covered by any part

Fig. 5. The center part and the corners

the input to 2D space, and deal with it as a 2D problem. Another corner case would be after applying `Normalization`: no facets have integer points, in this case we use the brute-force approach for the whole input.

With all the sub-routines in order, here is our algorithm, Algorithm 5, for computing the integer hull of a bounded 3D polyhedral set.

6 Implementation and Experimentation

We have implemented both the 2D and 3D algorithms in both MAPLE and the C/C++ programming language. The MAPLE version is available in 2022 release of MAPLE as the `IntegerHull` command of the `PolyhedralSets` library. In this section, we discuss implementation details and the experimentation with our implementations. All the benchmarks are done on an Intel i5-8300H CPU at 2.30 GHz with 16 GB of memory. As we discussed in Sect. 1, there are studies (such as [11] and [4]) developing approaches to enumerate the vertices of P_I using their relations with the vertices of P but to our knowledge no implementation of such methods exist. So in the following sections we compare our implementation with the existing implementation of the naive method (enumeration of all integer points, followed by the computation of their convex hull) for verification and proof of concept.

6.1 The MAPLE Implementation

For the MAPLE version, we use the functions provided by the `PolyhedralSets` library for polyhedral set manipulation such as construction, getting the vertices and faces. To obtain the adjacency information among the faces we need to compute the face lattice of the input polyhedral set; the `PolyhedralSets` library

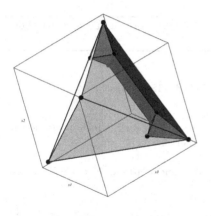 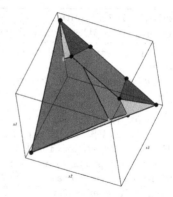

(a) The area around an edge   (b) When there are integer points on the edge

Fig. 6. Polyhedral sets that cover the edge areas

provides the command Graph for that task. We compare our MAPLE implementation with another MAPLE package. In the 2019 Maple Conference, Jing and Moreno-Maza introduced the ZPolyhedralSets package, presented in [13]. A ZPolyhedralSet is the intersection of a polyhedral set and a lattice. The integer hull of a polyhedral set is equal to a ZPolyhedralSet when the ZPolyhedralSet is defined using the standard integer lattice (which represents all the points with integer coordinates).

The ZPolyhedralSet package provides the EnumerateIntegerPoints command, which finds and outputs all the integer points within a ZPolyhedralSet object. Given a polyhedral set, to obtain the same result that our algorithm computes, which is the list of the vertices of the integer hull, we use the command EnumerateIntegerPoints to find all the integer points within the input, then we the command use ConvexHull from ComputationalGeometry to compute the vertices.

Table 1 shows the time spent in our algorithm (IntegerHull) and the above two-step method (EIP+CH) to obtain the same result. The inputs are triangles with different volumes. As we discussed in Sect. 2, the cost for finding all the integer points is related to the volume of the input and we can see the trend in the "EIP+CH" columns. Time spent by our algorithm does not seem to depend on the volume of the input.

From Algorithm 3, we can see that the complexity of our algorithm depends on the number of facets and the number of fractional vertices in the input. Table 2 shows the running time of both algorithms (IntegerHull and EIP+CH) when the inputs are hexagons. The running time for IntegerHull is roughly double the time for triangle inputs.

Algorithm 4: Compute the closest integer points on a facet F to the vertices on it in a 3D polyhedral set

---

1 **Function** `closestIntegerPoints3D`(F, V)

    Input:
- F, a facet of P in the form of a `PolyhedralSet` object
- V, a list of the vertices of P

    Output: V_C, a list where $V_C[i]$ is the integer point on F which is the closest to $V[i]$, if $V[i]$ is in F, and [] otherwise

2     $F_P, R_F \leftarrow$ `HNFProjection`$(F, 3)$
    `/* Make a projection` F_P `of the 3D bounded plane` F `onto 2D space using Hermite Normal Form */`

3     $V_{tmp} \leftarrow$ `IntegerHull2D`(F_P)
                    `/* Find the vertices of the integer hull of` F_P `*/`

4     $V_F \leftarrow$ `IntegerPointIn3D`(V_{tmp}, R_F)
    `/* Find the orignal image of the points in` V_{tmp} `*/ using the map` R_F

5     $n \leftarrow |V|$

6   **for** $i = 1, \ldots, n$ **do**

7   **if** $V[i]$ in F and $V_F \neq []$ **then**

8             $V_C[i] \leftarrow$ closest point to $V[i]$ in V_F

9   **else**

10             $V_C[i] \leftarrow []$

11   **return** V_C

Tables 3 and 4 show the running times of the same two algorithms when the input is a tetrahedron and a bipyramid respectively. The result is similar to that of the 2D algorithm where the running time increases if there are more facets and vertices. One thing that we need to notice is that the running time of our algorithm grows as the volume increases, this is due to the way we deal with the parts that are around the edges. As we discussed in Sect. 5.1, if there is no integer point on an edge, the sub-polyhedral set would include the whole edge and its volume depends on the length of the edge. Recall that we use exhaustive search for the sub-polyhedral sets thus the running time depends on the volume of the input polyhedral set.

6.2 The C/C++ Implementation

For the C/C++ implementation, we follow the representations in the C library `cddlib` by Komei Fukuda [9] for the polyhedral set computations. GMP rational arithmetic is used until the integer coordinates are obtained to ensure correctness. Our implementation can take polyhedral sets in either the V-representation or the H-representation as input; `cddlib` is used for representation conversion and some redundancy removal.

Algorithm 5: Compute the integer hull of a given 3D polyhedral set

---

1 **Function** IntegerHull3D(P)
   | **Input**: P, a 3D PolyhedralSet object
   | **Output**: I, a list of the vertices of the integer hull of P
2 | Process corner case: P is not fully dimensional
3 | $Q \leftarrow$ Normalization(P)
4 | $V \leftarrow$ Vertices(Q)
5 | $P \leftarrow Q$
6 | $F \leftarrow$ Facets(P)
7 | **for** *each $F[i]$ in F* **do**
8 | | $V_C[i] \leftarrow$ closestIntegerPoints3D(F[i], V)
   | | /* V_C is a 2D list where $V_C[i][j]$ contains the closest integer
   | | point to $V[j]$ on $F[i]$ */
9 | $E \leftarrow$ Edges(P)
10 | **for** *each $E[i]$ in E* **do**
11 | | $V_E \leftarrow$ closestIntegerPointsOnEdge($E[i], V$)
12 | $V_{\text{list}} \leftarrow$ PartitionP(V, V_C, V_E)
   | /* $V_{\text{list}} = [V_1, \ldots, V_n]$ where V_i contains the vertices of one part */
13 | $I \leftarrow \{\}$
14 | **for** *each $V_{list}[i]$ in V_{list}* **do**
15 | | $P_{\text{list}} \leftarrow$ PolyhedralSet($V_{\text{list}}[i]$)
16 | | $A_I \leftarrow$ AllIntegerPoints(P_{list})
17 | | $I \leftarrow I \bigcup$ ConvexHull(A_I)
18 | **return** ConvexHull(I)

As we have discussed in Sect. 3.2 we use part of the algorithm in [12] to partition the polyhedral sets and we follow that same article for the enumeration of the integer points in the corners. We implemented Algorithm 2.4.10 in [7] and Algorithm 3 in [12] for the procedure HNFProjection. We also implemented the algorithm introduced by Kaibel and Pfetsch in [14] for the computations of the face lattice.

To verify our implementation, we compare our results with that of the Normaliz library [5]. We also implemented a naive procedure based on enumeration and convex hull computation to obtain the integer hull. Note that Algorithm 3 in [12] only enumerate the integer points inside the given polyhedral set while for Normaliz, if the input is not homogeneous Normaliz homogenizes it by raising the input to a higher dimension, therefore, Normaliz enumerates more points than we do for the same input.

Tables 5 and 6 show the time spent in these three different approaches for computing the integer hulls of the same inputs. Since the I/O formats are different for Normaliz and cddlib, we only measured the timings for the integer hull computation part but not the I/O parts of the programs. Especially, for Normaliz we only timed the function call "MyCone.compute(ConeProperty::IntegerHull)".

Table 1. Integer hulls of triangles

Volume	27.95		111.79		11179.32	
Algorithm	IntegerHull	EIP+CH	IntegerHull	EIP+CH	IntegerHull	EIP+CH
Time(s)	0.172	0.410	0.244	0.890	0.159	58.083

Table 2. Integer hulls of hexagons

Volume	58.21		5820.95		23283.82	
Algorithm	IntegerHull	EIP+CH	IntegerHull	EIP+CH	IntegerHull	EIP+CH
Time(s)	0.303	0.752	0.275	31.357	0.304	123.159

The examples are named as xdy_z, where x is the dimension of the input (all the examples are full dimensional). Each y represents a set of examples that are of the same shape which means these polyhedral sets $A\mathbf{x} <= \mathbf{b}$ share the same coefficient matrix A while the vector $\mathbf{b}$ varies. xdy_0 is the smallest (volume wise) example in a set, for $z = 1, 2, 3$, vector $\mathbf{b}$ get multiplied by $2, 5, 10$ respectively. For the 2D examples, $2d1$ has 6 vertices, $2d2$ has 4 vertices and $2d3$ has 3 vertices. And for the 3D examples, $3d1$ has 12 facets, 8 vertices and 18 edges, $3d2$ has 4 facets, 4 vertices and 6 edges and $3d3$ has 6 facets, 5 vertices and 9 edges.

The result is consistent with our observation in [17]. For the same family of input, the time spent by our algorithm is relatively stable while for both our naive implementation and Normaliz, the larger the volume of the input is, the more time they need to do the computation since often time larger polyhedral sets contain more integer points for enumeration.

Table 3. Integer hulls of tetrahedrons (4 facets, 4 vertices and 6 edges)

Volume	447.48		6991.89		55935.2	
Algorithm	IntegerHull	EIP+CH	IntegerHull	EIP+CH	IntegerHull	EIP+CH
Time(s)	1.202	6.892	1.498	67.814	1.517	453.577

Table 4. Integer hulls of triangular bipyramids (6 facets, 5 vertices and 9 edges)

Volume	412.58		7050.81		60417.63	
Algorithm	IntegerHull	EIP+CH	IntegerHull	EIP+CH	IntegerHull	EIP+CH
Time(s)	1.476	5.711	1.573	60.233	1.728	512.101

Table 5. Timing (ms) for computing integer hull of 2D examples

example	IntegerHull	Naive	Normaliz
2d1_0	0.451	0.565	2.837
2d1_1	0.478	0.657	1216.238
2d1_2	0.396	0.682	740.559
2d1_3	0.443	1.134	472.447
2d2_0	0.413	1.128	1258.422
2d2_1	0.411	2.714	1242.081
2d2_2	0.393	16.079	2622.995
2d2_3	0.449	47.145	10218.368
2d3_0	0.284	0.768	835.730
2d3_1	0.339	1.676	462.116
2d3_2	0.286	6.883	1559.401
2d3_3	0.324	25.637	5072.894

Table 6. Timing (ms) for computing integer hull of 3D examples

example	IntegerHull	Naive	Normaliz
3d1_0	51.727	11.396	274.364
3d1_1	52.034	13.483	1018.449
3d1_2	60.821	21.106	2330.534
3d1_3	54.350	79.219	15346.996
3d2_0	4.488	0.826	851.495
3d2_1	4.615	0.923	956.666
3d2_2	4.624	1.527	793.192
3d2_3	5.522	4.394	1318.150
3d3_0	11.049	21.235	7862.109
3d3_1	16.001	145.068	N/A
3d3_2	23.822	2082.559	N/A
3d3_3	24.162	N/A	N/A

7   Conclusion and Future Work

In this paper, we introduced a new algorithm for computing the integer hull
of a convex polyhedral set. Our algorithm takes into consideration geometric
properties of the input polyhedral set in order to make the computation more
efficient. We implemented the proposed algorithm for two-dimensional and three-
dimensional input in both MAPLE and C/C++. The efficiency of this algorithm
depends mainly on the shape of the input while the size of the input has little

impact. We show in Sect. 6 that our algorithm can deal with inputs of very large volumes that algorithms depending on enumeration can not process.

The main steps of our algorithm are normalization, partition and merging. Our algorithm can be stated for polyhedral sets of arbitrary dimension and a MAPLE implementation is work in progress. Another on-going development is an algebraic complexity analysis of our algorithm.

We sketch below our algorithm for computing the integer hull of a d-dimensional convex polyhedral set P:

1. normalize the input using the procedure introduced in Sect. 3.1,
2. for each vertex, find the closest integer points to it on each of its adjacent faces,
3. for each face of dimension from 0 to $d - 2$, construct a "corner polyhedral set" using the integer points we obtained from step 2,
4. compute the integer hull of each corner polyhedral set,
5. compute the convex hull of all the integer hulls from step 4,
6. this convex hull is the integer hull of P.

References

1. Assarf, B., Gawrilow, E., Herr, K., Joswig, M., Lorenz, B., Paffenholz, A., Rehn, T.: Computing convex hulls and counting integer points with polymake. Math. Program. Comput. **9**(1), 1–38 (2017)
2. Barber, C.B., Dobkin, D.P., Huhdanpaa, H.: The quickhull algorithm for convex hulls. ACM Trans. Math. Softw. **22**(4), 469–483 (1996)
3. Barvinok, A.I.: A polynomial time algorithm for counting integral points in polyhedra when the dimension is fixed. Math. Oper. Res. **19**(4), 769–779 (1994)
4. Berndt, S., Jansen, K., Klein, K.: New bounds for the vertices of the integer hull. In: Le, H.V., King, V. (eds.) 4th Symposium on Simplicity in Algorithms, SOSA 2021, Virtual Conference, January 11–12, 2021. pp. 25–36. SIAM (2021)
5. Bruns, W., Ichim, B., Römer, T., Sieg, R., Söger, C.: Normaliz: algorithms for rational cones and affine monoids . J. Algebra **324** (2010)
6. Chvátal, V.: Edmonds polytopes and a hierarchy of combinatorial problems. Discret. Math. **4**(4), 305–337 (1973)
7. Cohen, H.: A Course in Computational Algebraic Number Theory, vol. 8. Springer-Verlag, Berlin (1993). https://doi.org/10.1007/978-3-662-02945-9
8. Cook, W.J., Hartmann, M., Kannan, R., McDiarmid, C.: On integer points in polyhedra. Combinatorica **12**(1), 27–37 (1992)
9. Fukuda, K.: cdd. c: C-implementation of the double description method for computing all vertices and extremal rays of a convex polyhedron given by a system of linear inequalities. Department of Mathematics, Swiss Federal Institute of Technology, Lausanne, Switzerland (1993)
10. Gomory, Ralph E..: Outline of an algorithm for integer solutions to linear programs *and* an algorithm for the mixed integer problem. In: Jünger, M., et al. (eds.) 50 Years of Integer Programming 1958-2008, pp. 77–103. Springer, Heidelberg (2010). https://doi.org/10.1007/978-3-540-68279-0_4
11. Hayes, A.C., Larman, D.G.: The vertices of the knapsack polytope. Discret. Appl. Math. **6**(2), 135–138 (1983)

12. Jing, R.-J., Moreno Maza, M.: Computing the integer points of a polyhedron, I: algorithm. In: Gerdt, V.P., Koepf, W., Seiler, W.M., Vorozhtsov, E.V. (eds.) CASC 2017. LNCS, vol. 10490, pp. 225–241. Springer, Cham (2017). https://doi.org/10.1007/978-3-319-66320-3_17

13. Jing, R., Moreno Maza, M.: The z_polyhedra library in maple. In: Gerhard, J., Kotsireas, I.S. (eds.) Maple in Mathematics Education and Research - Third Maple Conference, MC 2019, Waterloo, Ontario, Canada, October 15–17, 2019, Proceedings of the Communications in Computer and Information Science, vol. 1125, pp. 132–144. Springer, Cham (2019). https://doi.org/10.1007/978-3-030-81698-8

14. Kaibel, V., Pfetsch, M.E.: Computing the face lattice of a polytope from its vertex-facet incidences. Comput. Geom. **23**(3), 281–290 (2002)

15. Land, A., Doig, A.: An automatic method of solving discrete programming problems. Econometric **28**, 497–520 (1960)

16. Loera, J.A.D., Hemmecke, R., Tauzer, J., Yoshida, R.: Effective lattice point counting in rational convex polytopes. J. Symb. Comput. **38**(4), 1273–1302 (2004)

17. Moreno Maza, M., Wang, L.: On the pseudo-periodicity of the integer hull of parametric convex polygons. In: Boulier, F., England, M., Sadykov, T.M., Vorozhtsov, E.V. (eds.) CASC 2021. LNCS, vol. 12865, pp. 252–271. Springer, Cham (2021). https://doi.org/10.1007/978-3-030-85165-1_15

18. Pick, G.: Geometrisches zur zahlenlehre. Sitzenber. Lotos (Prague) **19**, 311–319 (1899)

19. Maple polyhedralsets package (2021), https://www.maplesoft.com/support/help/maple/view.aspx?path=PolyhedralSets

20. Schrijver, A. (Ed.): Theory of Linear and Integer Programming. Wiley, New York (1986)

21. Schrijver, A.: Theory of Linear and Integer Programming. Wiley, New York (1999)

22. Schrijver, A., et al.: On cutting planes. Combinatorics **79**, 291–296 (1980)

23. Seghir, R., Loechner, V., Meister, B.: Integer affine transformations of parametric Z-polytopes and applications to loop nest optimization. ACM Trans. Archit. Code Optim. **9**(2), 8:1–8:27 (2012)

24. Thomas, R.R.: Integer programming: Algebraic methods. In: Floudas, C.A., Pardalos, P.M. (eds.) Encyclopedia of Optimization, 2nd edn., pp. 1624–1634. Springer, Boston (2009). https://doi.org/10.1007/978-0-387-74759-0

25. Verdoolaege, S., Seghir, R., Beyls, K., Loechner, V., Bruynooghe, M.: Counting integer points in parametric polytopes using Barvinok's rational functions. Algorithmica **48**(1), 37–66 (2007)

26. Veselov, S., Chirkov, A.Y.: Some estimates for the number of vertices of integer polyhedra. J. Appl. Ind. Math. **2**(4), 591–604 (2008)

27. Yanagisawa, H.: A simple algorithm for lattice point counting in rational polygons (2005)

A Comparison of Algorithms for Proving Positivity of Linearly Recurrent Sequences

Philipp Nuspl[1]($\boxtimes$) and Veronika Pillwein[2]

[1] Doctoral Program Computational Mathematics, Johannes Kepler University,
Linz, Austria
philipp.nuspl@jku.at
[2] Research Institute for Symbolic Computation, Johannes Kepler University,
Linz, Austria
veronika.pillwein@risc.jku.at

Abstract. Deciding positivity for recursively defined sequences based on only the recursive description as input is usually a non-trivial task. Even in the case of C-finite sequences, i.e., sequences satisfying a linear recurrence with constant coefficients, this is only known to be decidable for orders up to five. In this paper, we discuss several methods for proving positivity of C-finite sequences and compare their effectiveness on input from the Online Encyclopedia of Integer Sequences (OEIS).

Keywords: Difference equations · Inequalities · Holonomic sequences

1 Introduction

A sequence is called D-finite (or P-recursive or holonomic), if it satisfies a linear recurrence with polynomial coefficients. These sequences appear in many applications, e.g., in combinatorics or as coefficient sequences of special functions [7,32]. They are interesting from the symbolic computation point of view, as they can be represented by a finite amount of data – the recurrence coefficients and sufficiently many initial values. Several closure properties hold for holonomic sequences and there exist summation algorithms that work with this representation for input and output. These methods are used to automatically prove and derive identities for holonomic sequences. When it comes to automatic proving of inequalities on holonomic sequences, there are not many algorithms available. Gerhold and Kauers [10] introduced a method in 2005 that can be used for sequences satisfying (a system of) recurrences including in particular holonomic sequences. This method (together with variations of it) has been applied successfully on several examples [17,29,31]. Still, a priori it is not known in general whether the procedure terminates [22].

The research was funded by the Austrian Science Fund (FWF) under the grant W1214-N15, project DK15.

In this paper, we restrict our study to C-finite sequences, i.e., holonomic sequences with constant coefficients, and the problem of proving positivity. This is known to be decidable for integer linear recurrences of order 2 [12], order 3 [23], order at most 5 and is related to difficult number theoretic problems for higher order [28]. We give an overview on some methods which can be used to prove the positivity of C-finite sequences, including the Gerhold–Kauers method and the most used variation (Algorithms 1 and 2 below). Other methods are based on theoretical results that, as far as we know, have not yet been implemented and tested on practical input on a larger scale. For testing the effectiveness of these different algorithms, we use input from the Online Encyclopedia of Integer Sequences (OEIS) [27] that are likely candidates for positive sequences. Our implementations are done both in SageMath and Mathematica and the source files as well as testing data are made available online (see links in Sect. 4).

2   Preliminaries

We introduce some notations and definitions that will be used throughout the paper. We always assume that $\mathbb{Q} \subseteq \mathbb{K} \subsetneq \mathbb{R}$ is some number field. We denote the field of algebraic numbers by $\overline{\mathbb{Q}}$ and the field of real algebraic numbers by $\mathbb{A} := \overline{\mathbb{Q}} \cap \mathbb{R}$. We denote the $\mathbb{K}$-vector space of sequences by $\mathbb{K}^{\mathbb{N}}$ and let σ denote the shift operator $\sigma((c(n))_{n \in \mathbb{N}}) := (c(n+1))_{n \in \mathbb{N}}$.

2.1   Linear Recurrence Sequences

We denote the Ore algebra of shift operators by $\mathbb{K}[x]\langle\sigma\rangle$. Let $\mathcal{A} = \sum_{i=0}^{r} p_i(x)\sigma^i \in \mathbb{K}[x]\langle\sigma\rangle$. If $p_r \neq 0$, then r is called the *order* of $\mathcal{A}$ and $\max_{i=0,\ldots,r} \deg(p_i)$ is called the *degree* of $\mathcal{A}$. The operator $\mathcal{A}$ acts on a sequence $c \in \mathbb{K}^{\mathbb{N}}$ in the natural way as

$$\mathcal{A}c = (p_0(n)c(n) + \cdots + p_r(n)c(n+r))_{n \in \mathbb{N}}.$$

A sequence $c \in \mathbb{K}^{\mathbb{N}}$ is called *D-finite* (or P-recursive or holonomic) if there is a non-zero operator $\mathcal{A} \in \mathbb{K}[x]\langle\sigma\rangle$ with $\mathcal{A}c = 0$, i.e., the sequence satisfies a linear recurrence with polynomial coefficients. We call $\mathcal{A}$ an *annihilating operator* of c. It is well known that D-finite sequences form a computable difference ring [21]. The minimal possible order r of an annihilating operator is also called the *order* of the sequence c. The *degree* of c is then just defined as the degree of this operator.

A D-finite sequence c is called *C-finite* if it satisfies a linear recurrence with constant coefficients, i.e., if there are $\gamma_0, \ldots, \gamma_r \in \mathbb{K}$ with $\gamma_r \neq 0$ such that

$$\gamma_0 c(n) + \cdots + \gamma_r c(n+r) = 0, \quad \text{for all } n \in \mathbb{N}. \tag{1}$$

Again, the order of c as a C-finite sequence is the minimal r (note that the order of c considered as a C-finite sequence can be different from the order considered as a D-finite sequence, cf. Lemma 3). The set of C-finite sequences is again a computable difference ring. Every such sequence can be uniquely described by the coefficients of the recurrence $\gamma_0, \ldots, \gamma_r$ and sufficiently many initial values $c(0), \ldots, c(r-1)$.

2.2   Characteristic Polynomial

For an operator $\mathcal{A} = \sum_{i=0}^{r} p_i(x)\sigma^i \in \mathbb{K}[x]\langle\sigma\rangle$ the *characteristic polynomial* is defined as

$$\chi(\mathcal{A}) := \mathrm{lc}_x \left(\sum_{i=0}^{r} p_i(x)y^i \right) \in \mathbb{K}[y].$$

The roots of $\chi(\mathcal{A})$ are called *eigenvalues* and usually govern the asymptotic behavior of sequences which are annihilated by $\mathcal{A}$ [22].

We can extend the notion of the characteristic polynomial to the left-Noetherian ring $\mathbb{K}(x)\langle\sigma\rangle$. For a univariate polynomial $p \in \mathbb{K}[x]$ we denote by $\mathrm{coeff}\,(p, i) \in \mathbb{K}$ the coefficient of x^i in p. For a rational function $\frac{p(x)}{q(x)}$ with coprime $p, q \in \mathbb{K}[x]$ we define the *degree* as $\deg(p/q) = \deg(p) - \deg(q)$ and call

$$\mathrm{lc}(p/q) := \mathrm{coeff}\,(p/q, \deg(p/q)) := \mathrm{lc}(p)/\mathrm{lc}(q)$$

the *leading coefficient* of p/q. Now, for an operator $\mathcal{A} = \sum_{i=0}^{r} \frac{p_i(x)}{q_i(x)}\sigma^i \in \mathbb{K}(x)\langle\sigma\rangle$ with $\deg(\mathcal{A}) := \max_{i=0,\dots,r} \deg(p_i/q_i)$ we define the characteristic polynomial as

$$\chi(\mathcal{A}) := \sum_{\substack{i=0 \\ \deg(p_i/q_i)=\deg(\mathcal{A})}}^{r} \mathrm{lc}(p_i/q_i)y^i \in \mathbb{K}[y].$$

Next, in Lemma 1 and Lemma 2, we state some basic properties of the characteristic polynomial. Since we could not find references for those, we add the proofs for sake of completeness.

Lemma 1. *Let $\mathcal{A}, \mathcal{B} \in \mathbb{K}(x)\langle\sigma\rangle$. Then $\chi(\mathcal{AB}) = \chi(\mathcal{A})\chi(\mathcal{B})$.*

Proof. Let $\mathcal{A} := \sum_{i=0}^{r} p_i(x)\sigma^i \in \mathbb{K}(x)\langle\sigma\rangle$ and $\mathcal{B} := \sum_{j=0}^{s} q_j(x)\sigma^j \in \mathbb{K}(x)\langle\sigma\rangle$ and $d_{\mathcal{A}} := \max_{i=0,\dots,r} \deg p_i, d_{\mathcal{B}} := \max_{j=0,\dots,s} \deg q_j \in \mathbb{Z}$ their respective degrees. We show that $\mathcal{AB}$ has degree $d_{\mathcal{A}} + d_{\mathcal{B}}$. By the definition of multiplication in $\mathbb{K}(x)\langle\sigma\rangle$ and the properties of the degree of a rational function, the degree of $\mathcal{AB}$ is certainly bounded by $d_{\mathcal{A}}+d_{\mathcal{B}}$. Let i', j' be maximal such that $\deg p_{i'} = d_{\mathcal{A}}$ and $\deg q_{j'} = d_{\mathcal{B}}$. We show that the coefficient of $\sigma^{i'+j'}$ of $\mathcal{AB}$ has degree $d_{\mathcal{A}} + d_{\mathcal{B}}$. This coefficient is given by $\sum_{l=0}^{i'+j'} p_l(x)q_{i'+j'-l}(x + l)$. Because of the choice of i', j' we have

$$\deg(p_l(x)q_{i'+j'-l}(x)) = \deg(p_l(x)) + \deg(q_{i'+j'-l}(x + l)) < d_{\mathcal{A}} + d_{\mathcal{B}}$$

for all $l \neq i'$. For $l = i'$, we have $\deg(p_l(x)q_{i'+j'-l}(x)) = d_{\mathcal{A}} + d_{\mathcal{B}}$, so by the properties of the degree we have

$$\deg\left(\sum_{l=0}^{i'+j'} p_l(x)q_{i'+j'-l}(x + l) \right) = \max_{l=0,\dots,i'+j'} (\deg\,(p_l(x)) + \deg\,(q_{i'+j'-l}(x + l)))$$

$$= d_{\mathcal{A}} + d_{\mathcal{B}}.$$

Next, we show that all coefficients of $\chi(\mathcal{A})\chi(\mathcal{B})$ and $\chi(\mathcal{AB})$ agree. Let $i \in \{0, \ldots, r+s\}$. Then,

$$\mathrm{coeff}\,(\chi(\mathcal{A}), i) = \mathrm{coeff}\,(p_i(x), d_{\mathcal{A}}), \quad \mathrm{coeff}\,(\chi(\mathcal{B}), i) = \mathrm{coeff}\,(q_i(x), d_{\mathcal{B}})$$

and therefore

$$\mathrm{coeff}\,(\chi(\mathcal{A})\chi(\mathcal{B}), i) = \sum_{j=0}^{i} \mathrm{coeff}\,(p_j(x), d_{\mathcal{A}})\,\mathrm{coeff}\,(q_{i-j}(x), d_{\mathcal{B}}).$$

In the first part of the proof, we have shown that $\mathcal{AB}$ has degree $d_{\mathcal{A}} + d_{\mathcal{B}}$. Therefore,

$$\mathrm{coeff}\,(\chi(\mathcal{AB}), i) = \mathrm{coeff}\,\left(\sum_{j=0}^{i} p_j(x)q_{i-j}(x+j), d_{\mathcal{A}} + d_{\mathcal{B}}\right)$$

$$= \sum_{j=0}^{i} \mathrm{coeff}\,(p_j(x)q_{i-j}(x+j), d_{\mathcal{A}} + d_{\mathcal{B}})$$

$$= \sum_{j=0}^{i} \mathrm{coeff}\,(p_j(x), d_{\mathcal{A}})\,\mathrm{coeff}\,(q_{i-j}(x+j), d_{\mathcal{B}})$$

$$= \sum_{j=0}^{i} \mathrm{coeff}\,(p_j(x), d_{\mathcal{A}})\,\mathrm{coeff}\,(q_{i-j}(x), d_{\mathcal{B}}).$$

$\square$

Suppose $\mathcal{A}$ is an annihilator of a and $\mathcal{B}$ an annihilator of b. Then, the *least common left multiple* $\mathrm{lclm}(\mathcal{A}, \mathcal{B})$ is an annihilator of $a + b$ [19].

Lemma 2. *Let $\mathcal{A}, \mathcal{B} \in \mathbb{K}[x]\langle\sigma\rangle$. Then*

$$\chi(\mathcal{A}) \mid \chi(\mathrm{lclm}(\mathcal{A}, \mathcal{B})) \text{ and } \chi(\mathcal{B}) \mid \chi(\mathrm{lclm}(\mathcal{A}, \mathcal{B})).$$

In particular, we have

$$\mathrm{lcm}(\chi(\mathcal{A}), \chi(\mathcal{B})) \mid \chi(\mathrm{lclm}(\mathcal{A}, \mathcal{B})).$$

Proof. Let $\mathcal{C} \in \mathbb{K}(x)\langle\sigma\rangle$ be such that $\mathcal{CA} = \mathrm{lclm}(\mathcal{A}, \mathcal{B})$. Then, with Lemma 1 we have

$$\chi(\mathrm{lclm}(\mathcal{A}, \mathcal{B})) = \chi(\mathcal{CA}) = \chi(\mathcal{C})\chi(\mathcal{A}).$$

$\square$

Example 1. In Lemma 2, divisibility cannot be replaced with equality. Consider $\mathcal{A} := 1 + \sigma$ and $\mathcal{B} := x + (x+1)\sigma$. Then,

$$\chi(\mathcal{A}) = \chi(\mathcal{B}) = 1 + y,$$

but

$$\chi(\mathrm{lclm}(\mathcal{A}, \mathcal{B})) = \chi(x + (2x+2)\sigma + (x+2)\sigma^2) = 1 + 2y + y^2.$$

An operator $\mathcal{A} = \sum_{i=0}^{r} p_i \sigma^i \in \mathbb{K}[x]\langle \sigma \rangle$ is called *balanced* if

$$\deg p_0 = \deg p_r = \max_{i=0,\ldots,r} \deg p_i.$$

Equivalently, $\mathcal{A}$ is balanced if and only if the degree of $\chi(\mathcal{A}) \in \mathbb{K}[y]$ equals the order of $\mathcal{A}$ and the trailing coefficient of $\chi(\mathcal{A})$ is non-zero, i.e., $y \nmid \chi(\mathcal{A})$.

2.3   Positivity

Suppose we are given a C-finite sequence c in terms of a recurrence and sufficiently many initial values. Our goal is to prove $c(n) > 0$ for all $n \in \mathbb{N}$ (i.e., show that c is positive) or to find an index $n_0 \in \mathbb{N}$ such that $c(n_0) \leq 0$. The very same methods can always be applied to show non-negativity instead of strict positivity of a sequence.

If b, c are C-finite sequences, then the inequality $b > c$ (or $b \geq c$) can easily be reduced to the positivity problem. The sequence $b - c$ is again C-finite. Hence, proving the equivalent positivity problem $b - c > 0$ (or $b - c \geq 0$) shows the original inequality.

Suppose c is C-finite satisfying the recurrence (1). Let $k \in \mathbb{N}$ be minimal such that $\gamma_k \neq 0$. Now, define $\tilde{c} := \sigma^k c$. Then, $\tilde{c}$ is again C-finite satisfying the recurrence

$$\gamma_k \tilde{c}(n) + \cdots + \gamma_r \tilde{c}(n + r - k) = 0, \quad \text{for all } n \in \mathbb{N}.$$

The sequence c is positive if and only if the sequence $\tilde{c}$ and the initial values $c(0), \ldots, c(k-1)$ are positive. Therefore, we can (and will) always assume that a C-finite sequence c is given by a recurrence with coefficients $\gamma_0, \ldots, \gamma_r$ with $\gamma_0, \gamma_r \neq 0$. Such a sequence c can then always be written as a polynomial-linear combination of exponential sequences. One can compute polynomials $p_1, \ldots, p_m \in \overline{\mathbb{Q}}[x]$ and pairwise distinct non-zero constants $\lambda_1, \ldots, \lambda_m \in \overline{\mathbb{Q}}$ such that

$$c(n) = \sum_{i=1}^{m} p_i(n) \lambda_i^n, \quad \text{for all } n \in \mathbb{N}. \tag{2}$$

These λ_i are called the *eigenvalues* of c and they are the roots of the characteristic polynomial $\sum_{i=0}^{r} \gamma_i y^i \in \mathbb{K}[y]$ of the minimal order recurrence of c. More precisely, if λ_i is a root of multiplicity d_i, then $\deg(p_i) = d_i - 1$. Hence, $r = \sum_{i=1}^{m} d_i$ [21].

Two sequences b, c which are non-zero from some term on are called asymptotically equivalent if $\lim_{n \to \infty} \frac{b(n)}{c(n)} = 1$. In this case, we write $b \sim c$. The asymptotic behavior of c is governed by the k eigenvalues of maximal modulus, we call them the *dominant eigenvalues*. We assume $|\lambda_1| = \cdots = |\lambda_k| > |\lambda_{k+1}| \geq \cdots \geq |\lambda_m|$. Let $d := \max_{i=1,\ldots,k} \deg p_i$. Then, $c(n) \sim n^d \sum_{i=1}^{k} \text{coeff}(p_i, d) \lambda_i^n$ [21].

In the special case that we have a unique dominant eigenvalue (i.e., $k = 1$) we have $c(n) \sim \gamma n^d \lambda_1^n$ for some γ [21]. Hence, c can only be a positive sequence if $\gamma, \lambda_1 \in \mathbb{A}$ and $\gamma > 0, \lambda_1 > 0$. Then, c is positive if and only if $c(n)/\lambda_1^n$ is positive. Therefore, it is sufficient to show positivity of a sequence

$$p(n) + \sum_{i=1}^{s} \left(o_i(n)\xi_i^n + \overline{o_i}(n)\overline{\xi_i}^n \right) + \sum_{i=1}^{l} q_i(n)\rho_i^n \tag{3}$$

with $p \in \mathbb{A}[x], o_1, \ldots, o_s \in \overline{\mathbb{Q}}[x], q_1, \ldots, q_l \in \mathbb{A}[x]$ and constants $\xi_1, \ldots, \xi_s \in \overline{\mathbb{Q}}, \rho_1, \ldots, \rho_l \in \mathbb{A}$ where the leading coefficient of p is positive [28].

3   Algorithms

In this section we give an overview over some methods which can be used to prove positivity of a C-finite sequence. Algorithms 1 and 2 introduced below in Sects. 3.1, 3.2 can be applied to D-finite sequences. As such they can be used to prove positivity of C-finite sequences. However, sometimes C-finite sequences satisfy a D-finite recurrence of lower order, which is better suited as input for these methods. In Sect. 3.3, we discuss when such a D-finite recurrence exists. A method based on the combination of Algorithms 1 and 2 as well as on the closed form of a C-finite sequence is introduced in Sect. 3.5. The methods described in Sects. 3.4 and 3.6 also make use of the closed form of C-finite sequences. They are based on known results, but we believe that they had not been implemented so far.

3.1   Algorithm 1

In 2008 [10], a method based on cylindrical algebraic decomposition [1,3,5,6] (CAD) was introduced which can be used to show positivity of sequences that can be defined recursively along some discrete parameter. This procedure, however, is not guaranteed to terminate. For D-finite sequences of small order conditions which guarantee the termination of the algorithm were found [22,30].

We give a short description of Algorithm 1 from [22]. For a D-finite sequence c of order r, the $\mathbb{Q}(x)$-vector space which is generated by the shifts of c is finitely generated [21]. In fact, it is generated by $c, \ldots, \sigma^{r-1}c$, i.e.,

$$\langle \sigma^i c \mid i \in \mathbb{N} \rangle_{\mathbb{Q}(x)} = \langle c, \ldots, \sigma^{r-1}c \rangle_{\mathbb{Q}(x)}.$$

Hence, for all $\rho \in \mathbb{N}$ there are rational functions $q_{\rho,0}(x), \ldots, q_{\rho,r-1}(x) \in \mathbb{K}(x)$ with $c(n + \rho) = \sum_{i=0}^{r-1} q_{\rho,i}(n)c(n + i)$ for all $n \in \mathbb{N}$. The idea now is to check with CAD whether $c(n), \ldots, c(n+r-1) > 0$ implies $c(n+r) > 0$ where $c(n+r)$ can be written in terms of the $c(n), \ldots, c(n + r - 1)$. If this is true, then by induction it would be sufficient to check finitely many initial values to deduce positivity of the entire sequence. If, however, this cannot be shown, then we can add $c(n + r) > 0$ to the hypothesis and show $c(n + r + 1) > 0$. This process is iterated. In the iteration step $\rho \geq r$ we try to show positivity of the formula

$$\Phi(\rho, c) := \forall y_0, \ldots, y_{r-1}, x \in \mathbb{R}: \left(x \geq 0 \wedge \bigwedge_{j=0}^{\rho-1} \sum_{i=0}^{r-1} q_{j,i}(x) y_i > 0 \right)$$

$$\implies \sum_{i=0}^{r-1} q_{\rho,i}(x) y_i > 0.$$

Formula $\Phi(\rho, c)$ is a generalized induction formula over the reals. It is certainly sufficient to prove the initial induction step and has the advantage of being a valid input for CAD. Here, we give a slightly adjusted version which searches for an index n_0 such that the sequence $\sigma^{n_0} c$ is positive, i.e., it checks whether the sequence is *eventually* positive (hence, we denote the algorithm by Algorithm 1e). If such an n_0 can be found by the algorithm, then it is sufficient to check the initial values $c(0), \ldots, c(n_0 - 1)$ of the sequence to prove positivity of c.

Algorithm 1e. Adjusted version of Algorithm 1 from [22]

Input : D-finite sequence c of order r
output: n_0 such that $\sigma^{n_0} c$ is positive
$n, n_0 \leftarrow 0$
$d \leftarrow c$
while $n < r$ *or* $\neg\Phi(n, d)$ **do**
  if $d(n) > 0$ **then**
  | $n \leftarrow n + 1$
  else
      $n_0 \leftarrow n_0 + n + 1$
      $d \leftarrow \sigma^{n+1} d$
      $n \leftarrow 0$
return n_0

Clearly, Algorithm 1e is not guaranteed to terminate. E.g., if the input sequence c is negative, then the algorithm never terminates. Suppose the sequence c is eventually positive, i.e., there exists an $n_0 \in \mathbb{N}$ such that $\sigma^{n_0} c$ is positive. Since $\chi(c) = \chi(\sigma^{n_0} c)$, the same termination conditions for Algorithm 1 in [22] now also apply to Algorithm 1e.

Example 2. The sequence A001903 is C-finite of order 3 satisfying

$$c(n) - c(n+1) + c(n+2) - c(n+3) = 0$$

with initial values $c(0) = 1, c(1) = 7, c(2) = 9$. Algorithm 1e terminates for this sequence for $n = 4$ showing that c is positive.

3.2   Algorithm 2

Algorithm 2 in [22] again uses CAD to prove positivity of a D-finite sequence. The idea is to check whether there is a $\mu > 0$ such that $c(n + 1) \geq \mu c(n)$ for all $n \in \mathbb{N}$. By induction, if there is a $\mu > 0$ such that $c(n + 1) \geq \mu c(n), \ldots, c(n + r - 1) \geq \mu c(n + r - 2)$ implies $c(n + r) \geq \mu c(n + r - 1)$, then it is again sufficient to check finitely many initial values to prove positivity of c. Hence, the important step in the algorithm is to use CAD to check whether there exists a $\mu > 0$ such that the formula

$$\Psi(\xi, \mu, c) := \forall y_0, \ldots, y_{r-1} \in \mathbb{R} \ \forall x \in \mathbb{R}_{\geq \xi}: \left(y_0 > 0 \wedge \bigwedge_{i=0}^{r-2} y_{i+1} \geq \mu y_i \right)$$

$$\implies \sum_{i=0}^{r-1} q_i(x) y_i \geq \mu y_{r-1}$$

is valid where $q_i \in \mathbb{K}(x)$ are such that $c(n + r) = \sum_{i=0}^{r-1} q_i(n) c(n + i)$ for all $n \in \mathbb{N}$.

Again, we give a slightly adjusted version which searches for an index n_0 such that the sequence $\sigma^{n_0} c$ is positive. If the input sequence c is eventually positive, then the same termination conditions as for Algorithm 2 in [22] apply in this adjusted version.

Algorithm 2e. Adjusted version of Algorithm 2 from [22]

Input : D-finite sequence c of order r
output: n_0 such that $\sigma^{n_0} c$ is positive
$n, n_0 \leftarrow 0$
$d \leftarrow c$
$\Psi(\xi, \mu) \leftarrow$ quantifier free formula equivalent to $\Psi(\xi, \mu, d)$
for $n = 0, 1, \ldots$ **do**
   if $d(n) \leq 0$ **then**
      $n_0 \leftarrow n_0 + n + 1$
      $d \leftarrow \sigma^{n+1} d$
      $\Psi(\xi, \mu) \leftarrow$ quantifier free formula equivalent to $\Psi(\xi, \mu, d)$
      $n \leftarrow 0$
   else if $\exists \mu > 0: \bigwedge_{i=0}^{r-2} d(n + i + 1) \geq \mu d(n + i) \wedge \Psi(n, \mu)$ **then**
      return n_0

Example 3. The sequence A005682 is C-finite of order 6 satisfying

$$c(n) + c(n + 2) - 2c(n + 5) + c(n + 6) = 0$$

with initial values $c = \langle 1, 2, 4, 8, 15, 28, \ldots \rangle$. Algorithm 2e terminates for this sequence at $n = 0$ showing that c is positive. Algorithm 1e cannot show positivity of c in 60 s.

3.3   D-finite Reduction

Clearly, every C-finite sequence is also D-finite. Sometimes, C-finite sequences satisfy shorter D-finite recurrences. In these cases, it can be helpful to use this shorter D-finite recurrence as the next example shows.

Example 4. Let c be the sequence defined by $c(n) = n^2 + 1$ for all $n \in \mathbb{N}$ (A002522). If c is considered as a C-finite sequence of order 3, then neither Algorithm 1e nor Algorithm 2e terminate in 60 s. If c is, however, considered as a D-finite sequence of order 1 and degree 2, then both algorithms terminate and show that c is indeed positive.

The next lemma shows that we can find a shorter D-finite recurrence of a C-finite sequence c if and only if c has eigenvalues of higher multiplicities or equivalently the characteristic polynomial of c is not squarefree.

Lemma 3. *Let c be a C-finite sequence of order r with $y \nmid \chi(c)$. Then, c is D-finite of order $m < r$ if and only if $\chi(c)$ is not squarefree.*

Proof. Suppose c is given as in (2).

$\Longleftarrow$: The sequences $p_i(n)\lambda_i^n$ are D-finite of order 1 and degree d_i over $\overline{\mathbb{Q}}$. Hence, by the bounds for closure properties of D-finite sequences, $c(n)$ is D-finite of order at most m over $\overline{\mathbb{Q}}$ [21]. [9, Lemma 2] shows that the sequence is then also D-finite over $\mathbb{K}$ with the same order and degree. In particular, if $\chi(c)$ is not squarefree, then $r = \sum_{i=1}^{m} d_i > m$.

$\Longrightarrow$: Suppose c satisfies a D-finite recurrence of order $m < r$ and degree d

$$\sum_{i=0}^{m} p_i(n)c(n+i) = 0 \quad \text{for all } n \in \mathbb{N} \tag{4}$$

with $p_i(n) = \sum_{k=0}^{d} p_{i,k} n^k$ where not all $p_{i,k}$ are zero. Furthermore, suppose that c is C-finite of order r with pairwise distinct eigenvalues $\lambda_1, \ldots, \lambda_r \in \overline{\mathbb{Q}}$, i.e., $c(n)$ can be written as $c(n) = \sum_{j=1}^{r} \gamma_j \lambda_j^n$ for some $\gamma_j \in \overline{\mathbb{Q}}$. Using this closed form in (4) yields

$$\sum_{k=0}^{d} \left(\sum_{i=0}^{m} \sum_{j=1}^{r} p_{i,k} \gamma_j \lambda_j^{n+i} \right) n^k = 0. \tag{5}$$

Let $\gamma_{k,j} := \sum_{i=0}^{m} p_{i,k} \gamma_j \lambda_j^i$, then (5) is equivalent to $\sum_{k=0}^{d} \left(\sum_{j=1}^{r} \gamma_{k,j} \lambda_j^n \right) n^k = 0$. For $n = 0, \ldots, r(d+1) - 1$ we get a homogeneous linear system for the $\gamma_{k,j}$. The corresponding matrix is regular [24, Theorem 2.2.1],[13, Proposition 2.11], so $\gamma_{k,j} = 0$ for all k, j. Let k be such that $p_{i,k} \neq 0$ for some i. Then,

$$0 = \sum_{j=1}^{r} \lambda_j^n \sum_{i=0}^{m} p_{i,k} \gamma_j \lambda_j^i = \sum_{i=0}^{m} \sum_{j=1}^{r} p_{i,k} \gamma_j \lambda_j^{n+i} = \sum_{i=0}^{m} p_{i,k} c(n+i).$$

Hence, c satisfies a C-finite recurrence of order $m < r$, a contradiction to c being C-finite of order r. $\qquad\square$

The proof of Lemma 3 shows that precisely the polynomial factors can be reduced in the D-finite recurrence, i.e., the m in the statement of Lemma 3 is the number of distinct eigenvalues of the sequence, which is also denoted by m in Eq. (2). The degree of the D-finite recurrence can be bounded by

$$(m(m+1) - m) \max_{i=1,\ldots,m} d_i = m^2 \max_{i=1,\ldots,m} d_i \leq r^3$$

using [18, Theorem 2].

In practice, we can easily check whether $\chi(c)$ is squarefree by checking whether $\chi(c)$ and its derivative are coprime. The shorter D-finite recurrence can then be either found by guessing or by computing it explicitly from the closed form of c.

3.4   Classical Algorithm for Sequences with Unique Dominant Eigenvalue

If a C-finite sequence has a unique dominant eigenvalue, checking positivity of the sequence is known to be decidable [28]. In this section, we give a full description of such an algorithm based on that result.

As discussed in Sect. 2.3 we can assume that a C-finite sequence c is given in its closed form representation, i.e., as

$$c(n) = p(n) + r(n), \tag{6}$$

where $p \in \mathbb{A}[x]$ with $\mathrm{lc}(p) > 0$ and $r(n) = \sum_{i=1}^{m} p_i(n)\lambda_i^n$ with $p_i \in \overline{\mathbb{Q}}[x], \lambda_i \in \overline{\mathbb{Q}}$ and $1 > |\lambda_1| \geq |\lambda_2| \geq \cdots \geq |\lambda_m|$. The idea is now to compute an $\varepsilon \in (0,1)$ and $n_0, n_1 \in \mathbb{N}$ such that $|r(n)| < (1 - \varepsilon)^n$ for $n \geq n_0$ and $p(n) \geq (1 - \varepsilon)^n$ for $n \geq n_1$. Then, clearly $c(n)$ is positive from $\max(n_0, n_1)$ on. The initial values can be checked separately again.

Algorithm C. Positivity for sequences with dominant eigenvalues [28]

Input : C-finite sequence c of the form (6)
output: true if $c(n) > 0$ for all $n \in \mathbb{N}$ and false otherwise
$\varepsilon \leftarrow \frac{1-|\lambda_1|}{2}$
compute n_0 such that $|r(n)| < (1-\varepsilon)^n$ for all $n \geq n_0$
compute n_1 such that $p(n) \geq (1-\varepsilon)^n$ for all $n \geq n_1$
if $c(n) > 0$ *for* $n = 0, \ldots, \max(n_0, n_1)$ **then**
  | **return** *true*
else
  └ **return** *false*

For a polynomial $p_i(x) = \sum_{j=0}^{d_i} \gamma_{i,j} x^j \in \mathbb{A}$ of degree d_i we can easily compute a constant $c_i \in \mathbb{A}$ such that $|p_i(n)| \leq c_i n^{d_i}$ for all $n \geq 1$. For example, we can choose $c_i := \sum_{j=0}^{d_i} |\gamma_{i,j}|$. Let $c := \sum_{i=1}^{m} c_i$ and $d := \max(d_1, \ldots, d_m)$, i.e., the maximal multiplicity of the eigenvalues $\lambda_1, \ldots, \lambda_m$. Furthermore, let $\varepsilon := \frac{1-|\lambda_1|}{2}$. Then, $1 - \varepsilon = |\lambda_1| + \varepsilon$.

First, we show how n_0 can be found such that $|r(n)| < (1 - \varepsilon)^n$ for $n \geq n_0$. Let $\mu := \frac{|\lambda_1| + \varepsilon}{|\lambda_1|}$. If $d = 0$, then

$$|r(n)| \leq c|\lambda_1|^n < (1 - \varepsilon)^n \iff \frac{\log(c)}{\log(\mu)} < n.$$

Hence, we can choose $n_0 := \lceil \frac{\log(c)}{\log(\mu)} \rceil$ in this case. If $d > 0$, then

$$|r(n)| \leq c\, n^d |\lambda_1|^n < (1 - \varepsilon)^n \iff \log(c^{1/d}) < \frac{n}{d} \log(\mu) - \log(n).$$

The derivative of the right-hand side of this inequality is positive if $n > \frac{d}{\log(\mu)}$, i.e., from $\lceil \frac{d}{\log(\mu)} \rceil$ on the sequence on the right-hand side is monotonously increasing. Hence, if the inequality is true for some $n_0 \geq \lceil \frac{d}{\log(\mu)} \rceil$, then it is true for all $n \geq n_0$. Checking these values one by one, we will find a suitable n_0 eventually.

If the polynomial $p(x) = p_0$ is just constant, then $p(n) \geq (1 - \varepsilon)^n$ if and only if $n \geq \frac{\log(p_0)}{\log(1-\varepsilon)}$. Otherwise, we can compute the largest real root x_1 of the derivative of $p(x)$. If $p(n_1) \geq (1 - \varepsilon)^{n_1}$ for any $n_1 \geq \lceil x_1 \rceil$, then the inequality holds for all $n \geq n_1$.

Example 5. The sequence A000126 is C-finite of order 4 satisfying

$$c(n) - c(n + 1) - 2c(n + 2) + 3c(n + 3) - c(n + 4) = 0$$

with initial values $c = \langle 1, 2, 4, 8, \ldots \rangle$. The sequence has the unique dominant root $\frac{1+\sqrt{5}}{2}$. Algorithm 1e and Algorithm 2e do not terminate in 60 s whereas Algorithm C terminates after checking the first 14 terms.

3.5   Combination of Algorithm 1 and Algorithm 2

In the case that the C-finite sequence has a unique dominant eigenvalue, we can combine the closed form representation of the sequence together with Algorithm 1e and Algorithm 2e. As we know that the polynomial term $p(n)$ in (3) certainly dominates the exponential terms, we can find indices n_i using Algorithm 1e and Algorithm 2e from which on the exponential sequences are dominated by the polynomial term. These input sequences have very low order (maximum order 3). Therefore, the termination criteria in [22] show that these algorithms terminate in most instances.

As Algorithm 2e terminates for essentially all sequences of order 2, the real algebraic part of Algorithm P certainly terminates.

Theorem 1. *Algorithm P terminates if $s = 0$, i.e., if all eigenvalues of c are real algebraic.*

Algorithm P. Positivity for sequences with dominant eigenvalues

---

Input : C-finite sequence c of the form (3)
output: true if $c(n) > 0$ for all $n \in \mathbb{N}$ and **false** otherwise
for $i \leftarrow 1$ **to** s **do**
$\quad \lfloor \; n_{i,\overline{\mathbb{Q}}} \leftarrow$ Algorithm 1e applied to $\frac{p(n)}{s+l} + o_i(n)\xi_i^n + \overline{o_i}(n)\overline{\xi_i}^n$
for $i \leftarrow 1$ **to** l **do**
$\quad \lfloor \; n_{i,\mathbb{A}} \leftarrow$ Algorithm 2e applied to $\frac{p(n)}{s+l} + q_i(n)\rho_i^n$
$n_0 \leftarrow \max(n_{1,\overline{\mathbb{Q}}}, \dots, n_{s,\overline{\mathbb{Q}}}, n_{1,\mathbb{A}}, \dots, n_{l,\mathbb{A}})$
if $c(n) > 0$ *for* $n = 0, \dots, n_0$ **then**
$\quad |$ **return** *true*
else
$\quad \lfloor$ **return** *false*

Proof. Each sequence $h(n) := \frac{p(n)}{s+l} + q_i(n)\rho_i^n$ is the sum of two balanced D-finite sequences g, f over $\mathbb{A}$ satisfying the recurrences

$$-p(n+1)g(n) + p(n)g(n+1) = 0, \quad -q_i(n+1)\rho_i f(n) + q_i(n)f(n+1) = 0$$

with characteristic polynomials

$$\chi(\mathcal{G}) = \mathrm{lc}(p)(y - 1), \quad \chi(\mathcal{F}) = \mathrm{lc}(q_i)(y - \rho_i),$$

where $\mathcal{G}, \mathcal{F}$ denote the annihilating operators of g, f, respectively. As these characteristic polynomials are coprime, Lemma 2 yields

$$\chi(\mathcal{H}) = \chi(\mathcal{G})\chi(\mathcal{F}) = \gamma(y - 1)(y - \rho_i)$$

for some constant γ where $\mathcal{H}$ denotes the annihilating operator of h. In particular, $\mathcal{H}$ is balanced. Furthermore, $h \sim p(n)$ by construction. With [22, Theorem 3], Algorithm 2e terminates with input h. □

It is conjectured that Algorithm 1e terminates for sequences of order 3 if the eigenvalues are complex. This is the case if we apply Algorithm 1e. Hence, if the conjecture is true, Algorithm P terminates for all C-finite sequences with a unique dominant eigenvalue.

Theorem 2. *Assume Conjecture 1 from [22] is true. Then, Algorithm P terminates.*

Proof. The proof of Theorem 1 already shows that the algorithm terminates for the real algebraic eigenvalues. Analogously, in the complex case, the sequences $h(n) := \frac{p(n)}{s+l} + o_i(n)\xi_i^n + \overline{o_i}(n)\overline{\xi_i}^n$ are D-finite of order 3 with a balanced annihilating operator $\mathcal{H}$ with characteristic polynomial

$$\chi(\mathcal{H}) = \gamma(y - 1)(y - \xi_i)(y - \overline{\xi_i})$$

for some constant γ. With Conjecture 1, Algorithm 1e terminates on this input. □

Example 6. The sequence A002248 is C-finite of order 4 satisfying the recurrence

$$4c(n) - 8c(n+1) + 7c(n+2) - 4c(n+3) + c(n+4) = 0$$

with initial values $c = \langle 2, 8, 14, 16, \ldots \rangle$. The sequence has the unique dominant eigenvalue 2. Neither Algorithm 1e nor Algorithm 2e terminate in 60 s. However, both Algorithm C and Algorithm P terminate in negligible time.

3.6   Decomposition into Non-degenerate Sequences

A C-finite sequence c is called *degenerate* if the ratio $\frac{\lambda_i}{\lambda_j}$ of two distinct eigenvalues λ_i, λ_j is a root of unity. Every C-finite sequence c can be written as the interlacing of non-degenerate and zero-sequences $c_1, \ldots, c_k$ [8, Theorem 1.2]. For proving inequalities for C-finite sequences this decomposition often turned out useful [26,28,35]. For proving positivity of c we can compute this decomposition and prove positivity for every subsequence $c_1, \ldots, c_k$.

One can explicitly compute the eigenvalues of a C-finite sequence and check whether the ratio of two eigenvalues is a root of unity [4]. Hence, a naive algorithm can decompose a sequence c into k subsequences

$$c_1(n) = c(kn), \ldots, c_k(n) = c(kn + k - 1)$$

and check whether all these subsequences are either zero or non-degenerate. Eventually, for large enough k, this is the case. This already works well in practice as we see in Sect. 4. A more efficient algorithm is given in [36].

If decomposition into subsequences is used together with Algorithm C or Algorithm P, then it is more efficient to check whether every subsequence has a unique dominant root (which can be done numerically with arbitrary-precision arithmetic) instead for checking degeneracy. The main bottleneck (cf. Example 8) is usually the computations of the subsequences. Hence, an efficient implementation should certainly aim to minimize the computations of these subsequences.

Example 7. The sequence A000115 is C-finite of order 8 and satisfies the recurrence

$$c(n) - c(n+1) - c(n+2) + c(n+3)$$
$$-c(n+5) + c(n+6) + c(n+7) - c(n+8) = 0.$$

with initial values $c = \langle 1, 1, 2, 2, 3, 4, 5, 6, \ldots \rangle$. It has 6 dominant eigenvalues and is degenerate. It can be decomposed into 10 non-degenerate sequences with unique dominant eigenvalues. For these subsequences, Algorithm C and Algorithm P both have no problem showing positivity.

4   Comparison

As far as we are aware the only implementations of the algorithms presented in Sect. 3 are implementations of the Gerhold–Kauers method for Mathematica in the package SumCracker [16] and for SageMath [34]. We have implemented the presented algorithms in SageMath (using QEPCAD-B) and in Mathematica and tested them on C-finite sequences which could be obtained from the OEIS by guessing.

4.1 Test Set

We used guessing on the terms given in the OEIS to check for each sequence whether it is C-finite. To have reasonable certainty that the guessed recurrence is indeed correct we make sure that the corresponding linear systems are overdetermined with at least 15 more equations than variables. We take the first 1000 of these sequences for which the first 500 terms are strictly positive and are therefore highly likely to be positive altogether[1].

The maximal order of these sequences is 42. The following table shows the number of sequences of each given order:

order	1	2	3	4	5	6	7	8	9	10	11	12	13	14	15	> 15
	73	134	117	139	120	80	87	36	47	27	31	14	17	10	10	58

More than half of these sequences, 567, have a unique dominant eigenvalue. There are 102, 40, 70, 32 sequences with 2, 3, 4, 5 distinct dominant eigenvalues, respectively. Hence, there are 139 sequences with more than 6 distinct dominant eigenvalues.

About half of the sequences, 513, have a characteristic polynomial which is not squarefree. By Lemma 3 these are the sequences which have a shorter D-finite recurrence.

4.2 SageMath Implementation

The methods for proving inequalities for C-finite sequences (and in a limited way for D-finite sequences) are part of the rec_sequences package which is itself based on the ore_algebra package [20]. SageMath provides an interface to QEPCAD-B which allows CAD computations [2,33]. This is used in the implementations of Algorithm 1 and Algorithm 2. For Algorithm C, we rely on fast arbitrary-precision arithmetic using the library Arb which is included in Sage-Math [15]. To decompose a sequence into subsequences with a unique dominant eigenvalue, we decompose the sequence into k subsequences and check, using arbitrary-precision arithmetic, whether all of these have a unique dominant eigenvalue. If they do not have a unique dominant eigenvalue, we increase k by one. The main bottleneck when decomposing is by far the computation of the subsequences. Checking whether a subsequence has a unique dominant eigenvalue or proving positivity of a sequence with a unique dominant eigenvalue using Algorithm C only takes negligible time in our examples.

---

[1] A table with these sequences and additional information is given on the website https://www3.risc.jku.at/people/pnuspl/PositivityCFinite. It also contains the detailed results of the SageMath and Mathematica tests.

The package is publicly available[2]. We give a list of the methods that can be used on C-finite sequences to show positivity. Every method has a parameter `strict` which is `True` by default and indicates whether strict positivity or non-negativity should be shown. The additional parameter `time` can be used to give an upper bound (in seconds) after which the algorithms should be terminated, the default value is -1, indicating that they should not stop prematurely.

- `is_positive_algo1` implements Algorithm 1 from [22]. As an additional parameter `bound` can be specified which gives an upper bound on the number of iterations.
- `is_positive_algo2` implements Algorithm 2 from [22]. Again, `bound` can be specified. This method is also implemented for general D-finite sequences and can be called using `is_positive` on D-finite sequences.
- `is_positive_dominant_root` implements Algorithm C for sequences with a unique dominant eigenvalue.
- `is_positive_dominant_root_decompose` first tries to decompose the sequence into sequences with a unique dominant eigenvalue and zero sequences and calls Algorithm C on each of those.
- `is_positive` is a combination of all these algorithms which additionally uses a reduction to D-finite sequences if possible. This method is also applied if the comparison operators `>, <, >=, <=` are used.

The following example session shows how the methods can be used.

```
sage: from rec_sequences.CFiniteSequenceRing import *
sage: C = CFiniteSequenceRing(QQ)
sage: f = C([1,1,-1], [0,1]) # Fibonacci numbers
sage: f.is_positive(strict=False)
True
sage: var("n")
sage: c1 = C(n^2+1) # A002522
sage: c1 >= 0 # use is_positive implicitly
True
sage: c2 = C([1, -1, -1, 1, 0, -1, 1, 1, -1],
sage:         [1, 1, 2, 2, 3, 4, 5, 6]) # A000115
sage: c2.is_positive_dominant_root_decompose()
True
sage: c = C(1/100 * (-3)^n + 100 * 2^n)
sage: c > 0
False
```

Using the above mentioned methods, 987 out of the 1000 sequences from the test set could be proven to be positive where each method was given 60 s. The following table gives an overview on the number of sequences which could be

[2] The package can be obtained from https://github.com/PhilippNuspl/rec_sequences. Extensive documentation and instructions for the installation can be found under the same link. The version used to run the experiments is available at https://github.com/PhilippNuspl/rec_sequences/tree/v0.1-exp.

proven to be positive by each method ("Comb." stands for a combination of the algorithms and a "D" indicates that decomposition of the sequence is used):

Algo. 1	Algo. 2	Algo. C	D, Algo. C	Comb.
384	327	566	984	986

It is clear that decomposing the sequences and using Algorithm C is the most powerful method. The implementation of Algorithm C is very fast and takes at most 0.3 s for every example we considered.

Example 8. The sequence A008628 is C-finite of order 13 satisfying

$$c(n) - c(n+1) - 2c(n+2) + c(n+3) + 2c(n+4) - c(n+6)$$
$$+c(n+7) - 2c(n+9) - c(n+10) + 2c(n+11) + c(n+12) - c(n+13) = 0$$

with initial values $= c = \langle 1,1,2,3,5,7,10,13,18,23,31,38,49 \rangle$. If the sequence is decomposed into 30 subsequences, then all of the subsequences have a unique dominant root and positivity of these subsequences can be shown easily with Algorithm C. It takes about 2 min to show positivity of the sequence c and 98% of the time is used to compute the subsequences in the decomposition.

Allowing more than 60 s for each sequence, all 1000 sequences can be shown to be positive using decomposition into subsequences with a unique dominant eigenvalue and Algorithm C for these subsequences.

4.3   Mathematica Implementation

The Mathematica package `Positivity` encompasses several of the algorithms described in Sect. 3. It is part of `RISCErgoSum` which is a collection of Mathematica packages developed at RISC[3]. The package `GeneratingFunctions` is used to compute closure properties of C-finite sequences [25]. Our package, therefore, uses the same syntax as Mallinger's package for defining sequences. For the quantifier elimination steps in Algorithm 1e and Algorithm 2e, we use the Mathematica method `Resolve`. It might be interesting to compare different quantifier elimination procedures for our concrete examples. Following, we give a list of the methods contained in the `Positivity` package. All methods can be used in a strict version to show strict positivity of a sequence (this is the default) or a non-strict version to show non-negativity of a sequence using the parameter `Strict` set to `False`. If the parameter `Verbose` is set to `True`, then more information about the different computation steps are printed.

---

[3] It can be obtained from https://www3.risc.jku.at/research/combinat/software/ergosum/RISC/PositiveSequence.html. A demo notebook can be found on the same webpage. The source code is available on GitHub. The version used to run the experiments is available at https://github.com/PhilippNuspl/PositiveSequence/tree/v0.1-exp.

- KPAlgorithm1 implements Algorithm 1e, i.e., for a C-finite or D-finite sequence an index n_0 is returned from which the sequence is guaranteed to be positive. If the parameter Eventual is set to False, then the traditional Algorithm 1 from [22] is executed which returns True if the sequence is positive or False if the sequence is not positive.
- KPAlgorithm2 implements Algorithm 2e and Algorithm 2 from [22], analogous to KPAlgorithm1.
- AlgorithmDominantRootClassic is an implementation of Algorithm C.
- AlgorithmDominantRootCAD provides an implementation of Algorithm P.
- AlgorithmClassic and AlgorithmCAD first decompose the sequence into nondegenerate and zero sequences and check positivity of these subsequences with AlgorithmDominantRootClassic and AlgorithmDominantRootCAD, respectively.
- PositiveSequence combines some of the previous algorithms.

The methods can be used in the following way:

In[1]:= $\ll$ **RISC`Positivity`**

In[2]:= $f = \mathbf{RE}[\{\{0, 1, 1, -1\}, \{0, 1\}\}, c[n]];$

In[3]:= **PositiveSequence**$[f, \mathbf{Strict} \rightarrow \mathbf{False}]$ (*Fibonacci*)

Out[3]= True

In[4]:= $\mathbf{c1} = \mathbf{SeqFromExpr}[n^2 + 1, c[n]];$

In[5]:= **PositiveSequence**$[\mathbf{c1}]$ (*A002522*)

Out[5]= True

In[6]:= $\mathbf{c2} = \mathbf{RE}[\{\{0, 1, -1, -1, 1, 0, -1, 1, 1, -1\}, \{1, 1, 2, 2, 3, 4, 5, 6\}\}, c[n]];$

In[7]:= **AlgoClassic**$[\mathbf{c2}]$ (*A000115*)

Out[7]= True

In[8]:= $\mathbf{c3} = \mathbf{SeqFromExpr}[1/100 * (-3)^n + 100 * 2^n, c[n]];$

In[9]:= **PositiveSequence**$[\mathbf{c3}]$

Out[9]= False

Comparing the different algorithms on the test set we see similar results as in the SageMath implementation. Every method was again aborted after 60 s. 980 out of the 1000 sequences could be shown to be positive by at least one of the methods. The following table shows the number of sequences which could be proven positive by each method:

Algo. 1	Algo. 2	Algo. C	Algo. P	D, Algo. C	D, Algo. P	Comb.
387	325	526	528	940	942	980

A more precise comparison of Algorithm C and Algorithm P shows that the two methods are not only equally powerful on the test set, but their runtime for the individual examples is also very similar. One can, however, expect that this is due to the specific implementation as the next example indicates. Hence, if provided by the computer algebra system, implementations based on numerical arbitrary-precision computations should be prefered over implementations based on algebraic number computations or quantifier elimination methods.

Example 9. The C-finite sequence A003520 is C-finite of order 5 satisfying

$$c(n) + c(n+4) - c(n+5) = 0$$

with initial values $c(0) = \cdots = c(4) = 1$. The sequence has a unique dominant root. The Mathematica implementations of Algorithm C and Algorithm P both take several seconds. The SageMath implementation based on arbitrary-precision ball arithmetic instead of computations with algebraic numbers takes less than 0.1 s.

Increasing the time shows that the combined algorithm can show the positivity of 996 sequences with a time limit of 12 hours per sequence.

5   Conclusions

Summarizing, we have investigated some well known and new methods for showing positivity of C-finite sequences. To our knowledge, most of these algorithms were never implemented and it was not clear how well they perform on practical examples. It turned out that the methods are already powerful enough to prove the positivity of most C-finite sequences from the OEIS in reasonable time.

The given algorithms already cover most of the sequences appearing in combinatorial examples. One can, however, construct examples of non-degenerate sequences which have multiple dominant eigenvalues. For sequences with up to 5 dominant eigenvalues, positivity is still known to be decidable [28]. Other algorithms for showing positivity are given for instance in [11] and [14]. It would certainly be interesting to check whether and how these methods can be applied and implemented in practice and how their runtime compares to the algorithms presented here.

Acknowledgments. We like to thank Ralf Hemmecke for providing helpful feedback on the Mathematica implementation. We thank the referees for their careful reading and their valuable suggestions that helped improve the quality of the paper. In particular, the suggestion of one of the reviewers to use arbitrary-precision arithmetic greatly improved the implementation of the classical method for showing positivity.

References

1. Basu, S., Roy, S., Pollack, R., Roy, M.F.: Algorithms in Real Algebraic Geometry. Algorithms and Computation in Mathematics, Springer, Heidelberg (2003). https://doi.org/10.1007/3-540-33099-2
2. Brown, C.W.: QEPCAD B: a program for computing with semi-algebraic sets using CADs. SIGSAM Bull. **37**(4), 97–108 (2003)
3. Caviness, B.F., Johnson, J.R.: Quantifier Elimination and Cylindrical Algebraic Decomposition. Texts & Monographs in Symbolic Computation, Springer, Vienna (1998). https://doi.org/10.1007/978-3-7091-9459-1
4. Cohen, H.: A Course in Computational Algebraic Number Theory. Graduate Texts in Mathematics, Springer, Heidelberg (2013). https://doi.org/10.1007/978-3-662-02945-9
5. Collins, G.E.: Quantifier elimination for real closed fields by cylindrical algebraic decompostion. In: Brakhage, H. (ed.) GI-Fachtagung 1975. LNCS, vol. 33, pp. 134–183. Springer, Heidelberg (1975). https://doi.org/10.1007/3-540-07407-4_17
6. Collins, G.E., Hong, H.: Partial cylindrical algebraic decomposition for quantifier elimination. J. Symb. Comput. **12**(3), 299–328 (1991)
7. Olver, F.W.J., et al.: NIST Digital Library of Mathematical Functions. Release 1.1.0 of 2020–12-15 (2021). http://dlmf.nist.gov
8. Everest, G., van der Poorten, A., Shparlinski, I., Ward, T.: Recurrence Sequences. Mathematical Surveys and Monographs, American Mathematical Society, Providence, USA (2015)
9. Gerhold, S.: Combinatorial Sequences: Non-Holonomicity and Inequalities. Ph.D. Thesis, Johannes Kepler University Linz (2005)
10. Gerhold, S., Kauers, M.: A Procedure for proving special function inequalities involving a discrete parameter. In: Proceedings of ISSAC 2005, Beijing, China, 24–27 July 2005. pp. 156–162 (2005)
11. Gourdon, X., Salvy, B.: Effective asymptotics of linear recurrences with rational coefficients. Discrete Math. **153**(1–3), 145–163 (1996)
12. Halava, V., Harju, T., Hirvensalo, M.: Positivity of second order linear recurrent sequences. Discrete Appl. Math. **154**(3), 447–451 (2006)
13. Halava, V., Harju, T., Hirvensalo, M., Karhumäki, J.: Skolem's Problem: On the Border Between Decidability and Undecidability. Technical Report (2005)
14. van der Hoeven, J.: Fuchsian holonomic sequences (2021). https://hal.archives-ouvertes.fr/hal-03291372/
15. Johansson, F.: Arb: efficient arbitrary-precision midpoint-radius interval arithmetic. IEEE Trans. Comput. **66**, 1281–1292 (2017)
16. Kauers, M.: SumCracker: a package for manipulating symbolic sums and related objects. J. Symb. Comput. **41**(9), 1039–1057 (2006)
17. Kauers, M.: Computer algebra and power series with positive coefficients. In: Proceedings of FPSAC 2007, pp. 1–7 (2007)
18. Kauers, M.: Bounds for D-finite closure properties. In: Proceedings of ISSAC 2014, Kobe, Japan, pp. 288–295. Association for Computing Machinery, New York, NY, USA (2014)
19. Kauers, M.: Algorithms for D-finite functions. In: JNCF 2015, Cluny, France (2015)
20. Kauers, M., Jaroschek, M., Johansson, F.: Ore polynomials in Sage. In: Computer Algebra and Polynomials: Applications of Algebra and Number Theory, pp. 105–125. Springer, Cham (2015). https://doi.org/10.1007/978-3-319-15081-9_6

21. Kauers, M., Paule, P.: The Concrete Tetrahedron. Texts and Monographs in Symbolic Computation, Springer, Vienna (2011). https://doi.org/10.1007/978-3-7091-0445-3

22. Kauers, M., Pillwein, V.: When can we detect that a P-finite sequence is positive? In: Proceedings of ISSAC 2010, Munich, Germany, pp. 195–201. Association for Computing Machinery, New York, NY, USA (2010)

23. Laohakosol, V., Tangsupphathawat, P.: Positivity of third order linear recurrence sequences. Discrete Appl. Math. **157**(15), 3239–3248 (2009)

24. Li, H.C.: Studies on Generalized Vandermonde Matrices: Their Determinants, Inverses, Explicit LU Factorizations, with Applications. Ph.D. Thesis, National Chengchi University (2006)

25. Mallinger, C.: Algorithmic Manipulations and Transformations of Univariate Holonomic Functions and Sequences. Diplomarbeit, Johannes Kepler University Linz (1996)

26. Mignotte, M., Shorey, T.N., Tijdeman, R.: The distance between terms of an algebraic recurrence sequence. J. für die reine und angewandte Mathematik **349**, 63–76 (1984)

27. OEIS Foundation Inc.: The On-Line Encyclopedia of Integer Sequences (2022). http://www.oeis.org

28. Ouaknine, J., Worrell, J.: Positivity problems for low-order linear recurrence sequences. In: SODA 2014: Proceedings of the Twenty-Fifth Annual ACM-SIAM Symposium on Discrete Algorithms, pp. 366–379 (2014)

29. Pillwein, V.: Positivity of certain sums over Jacobi kernel polynomials. Adv. Appl. Math. **41**(3), 365–377 (2008)

30. Pillwein, V.: Termination conditions for positivity proving procedures. In: Proceedings of ISSAC 2013, Boston, USA, 26–29 June 2013. pp. 315–322 (2013)

31. Pillwein, V.: On the positivity of the Gillis-Reznick-Zeilberger rational function. Adv. Appl. Math. **104**, 75–84 (2019)

32. Stanley, R.P.: Enumerative Combinatorics: Vol. 2, Cambridge Studies in Advanced Mathematics, Cambridge University Press (1999)

33. The Sage Developers: SageMath, the Sage Mathematics Software System (Version 9.4) (2022). https://www.sagemath.org

34. Uray, M.J.: On proving inequalities by cylindrical algebraic decomposition. Annales Univ. Sci. Budapest. Sect. Comp, pp. 231–252 (2020)

35. Vereshchagin, N.K.: Occurrence of zero in a linear recursive sequence. Mat. Zametki **38**(2), 609–615 (1985)

36. Yokoyama, K., Li, Z., Nemes, I.: Finding roots of unity among quotients of the roots of an integral polynomial. In: Proceedings of ISSAC 1995, Montreal, Quebec, Canada, 10–12 July 1995. pp. 85–89 (1995)

Stability Analysis of Periodic Motion of the Swinging Atwood Machine

Alexander Prokopenya[✉]

Institute of Information Technology, Warsaw University of Life Sciences – SGGW,
Nowoursynowska 159, 02-776 Warsaw, Poland
alexander_prokopenya@sggw.edu.pl

Abstract. The swinging Atwood machine is a conservative Hamiltonian system with two degrees of freedom that is essentially nonlinear. A general solution of its equations of motion cannot be written in symbolic form, only in some special case it is integrable. A very interesting peculiarity of the system is an existence of a state of dynamical equilibrium when the oscillating body of smaller mass balances a body of larger mass. This state is described by periodic solution of the equations of motion that is constructed in the form of power series in a small parameter. In this paper, we investigate the system dynamics in the neighbourhood of the periodic solution. Its perturbed motion is described in linear approximation by the fourth order system of differential equations with periodic coefficients. We computed a fundamental matrix for this system and found its characteristic exponents in the form of power series in a small parameter. We have shown that owing to oscillations the state of dynamical equilibrium of the swinging Atwood machine is stable in linear approximation. All the relevant symbolic calculations are performed with the aid of the computer algebra system Wolfram Mathematica.

Keywords: Swinging Atwood's machine · Periodic solution · Characteristic exponents · Stability · Computer algebra · Mathematica

1 Introduction

The swinging Atwood machine (SAM) is a well-known device that is obtained from a simple Atwood machine [1] when one body of mass m_1 is allowed to oscillate in a plane while the other body of mass $m_2 > m_1$ moves along a vertical (see [2] and Fig. 1). Owing to oscillations the system acquires two degrees of freedom and becomes essentially nonlinear; a general solution of its equations of motion cannot be written in symbolic form. As the system demonstrates very interesting dynamics, it has been a subject of many studies (see, for example, [3–10]). Detailed investigations have shown that only for the mass ratio m_2/m_1 being equal to three the system is integrable (see [5,7,9–11]). Numerical analysis of the equations of motion has shown that, depending on the mass ratio and initial conditions, the SAM can demonstrate different types of motion, namely, periodic, quasi-periodic, and chaotic (see [3,6,8,10]).

F. Boulier et al. (Eds.): CASC 2022, LNCS 13366, pp. 288–299, 2022.
https://doi.org/10.1007/978-3-031-14788-3_16

In [12] we studied numerically the equations of motion of the SAM and showed that a physical reason for such behaviour of the system is an increase of an averaged tension of the thread during oscillation. As this tension depends on the amplitude of oscillation one can choose initial conditions such that quasi-periodic motion of the system can take place. Although a simple Atwood's machine with two bodies of different mass cannot be in a state of equilibrium (see [1]), owing to oscillations the system has a dynamic equilibrium state described by a periodic solution of the equations of motion (see [13]). If both bodies are allowed to oscillate in a plane the system acquires additional degree of freedom and demonstrates a quasi-periodic motion even in the case of equal masses $m_2 = m_1$ (see [14,15]). Note that such unusual behaviour of the swinging Atwood machine is possible only due to oscillations of the bodies resulting in nonlinearity of the equations of motion.

In the present paper, we consider the SAM in case of small difference of masses of the bodies and planar oscillation of the mass m_1. Our main purpose is to study the stability of periodic motion of the SAM. It should be noted that the constructing and investigation of periodic solutions of the equations of motion often imply rather cumbersome symbolic computations, which are convenient to carry out using computer algebra systems (see, for example, [16–18]). In this work, all symbolic calculations are performed with the aid of the computer algebra system Wolfram Mathematica (see [19]).

The paper is organized as follows. In Sect. 2 we describe the model and derive the equations of motion in the form that is convenient for applying the perturbation theory. Then in Sect. 3 we demonstrate shortly an algorithm for constructing the periodic solution in the form of power series in a small parameter. Section 4 is devoted to the investigation of stability of periodic solution in linear approximation. Integrating the linearized system of four differential equations with periodic coefficients which describes the perturbed motion, we compute the fundamental matrix in the form of power series in a small parameter and find the characteristic exponents for the system. At last, we summarize the obtained results in Sect. 5.

2   Model Description

The swinging Atwood machine under consideration consists of two small massless pulleys and two bodies of masses $m_1 \leq m_2$ attached to opposite ends of a massless inextensible thread (see Fig. 1). The body m_1 is allowed to swing in vertical plane and it behaves like a pendulum of variable length while the body m_2 is constrained to move only along a vertical. Note that in case of a pulley of finite radius used in the simple Atwood machine a length of the pendulum changes not only due to rotation of the pulley but due to the thread winding on the pulley during oscillation, as well. The last effect was investigated theoretically and experimentally in [10] and it does not modify qualitatively the system motion. Replacing one pulley of finite radius by the two pulleys of negligible radius, we obtain the swinging Atwood machine, where the pendulum length varies only due to rotation of the pulleys. Placing the pulleys at some distance between each other enables to avoid collisions of the bodies during oscillations but does not change the physical properties of the system.

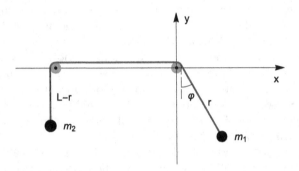

Fig. 1. The SAM with two small pulleys

The Lagrangian function of the system is (see [14])

$$\mathcal{L} = \frac{m_1 + m_2}{2}\dot{r}^2 + \frac{m_1}{2}r^2\dot{\varphi}^2 - m_2 gr + m_1 gr\cos\varphi, \tag{1}$$

where the dot over a symbol denotes the total derivative of the corresponding function with respect to time, g is a gravity acceleration, r is the distance between the pulley and the mass m_1, and the angle φ determines the deviation of the mass m_1 from the vertical.

To simplify analysis of the system it is expedient to introduce dimensionless variables. As we expect the body m_1 in the state of dynamic equilibrium behaves like a pendulum of a length R_0, the distance r can be made dimensionless by using R_0 as a characteristic distance, whereas the time t can be made dimensionless by using the inverse of the pendulum's natural frequency $\sqrt{g/R_0}$. Thus, making the substitutions $r \to rR_0$, $t \to t\sqrt{R_0/g}$, where r and t denote now the dimensionless variables, and dividing the Lagrangian by constant $m_1 gR_0$, we rewrite (1) in the form

$$\mathcal{L} = \frac{2+\varepsilon}{2}\dot{r}^2 + \frac{1}{2}r^2\dot{\varphi}^2 - (1+\varepsilon)r + r\cos\varphi, \tag{2}$$

where the parameter $\varepsilon = (m_2 - m_1)/m_1$ represents the ratio of the masses difference to the mass m_1. Note that the Lagrangian (2) depends on a single dimensionless parameter ε which we shall assume to be small ($0 \le \varepsilon \ll 1$).

Using (2), we obtain the equations of motion in the Lagrangian form (see [20])

$$(2+\varepsilon)\ddot{r} = -\varepsilon - (1 - \cos\varphi) + r\dot{\varphi}^2,$$
$$r\ddot{\varphi} = -\sin\varphi - 2\dot{r}\dot{\varphi}. \tag{3}$$

One can easily check that system (3) has an equilibrium solution $r = \text{const}$, $\varphi = 0$ only in the case of equal masses ($\varepsilon = 0$). This equilibrium state is unstable, and the system leaves it as soon as the mass m_1 gets even very small initial velocity (see [12]). On the other hand, in the case of different masses, the constant term $\varepsilon > 0$ in the right-hand side of the first Eq. (3) causes the uniformly accelerated motion of the Atwood machine in the absence of oscillations as it is in the classical Atwood's machine (see [1]). However, if the masses difference is sufficiently small one can expect that an averaged value of the oscillating functions in the right-hand side of the first Eq. (3) compensates the constant ε, and the smaller oscillating mass m_1 can balance the larger mass m_2. Our aim is to demonstrate that such a state of dynamical equilibrium of the system exists and it is described by the periodic solution of system (3).

3   Periodic Solution

To simplify the calculations we assume that the oscillations are small ($|\varphi| \ll 1$) and replace the sine and cosine functions by their expansions in power series accurate to the sixth order inclusive. As we will see later, such expansions are necessary to construct periodic solution accurate to the third order in ε. Then the system (3) takes the form

$$(2 + \varepsilon)\ddot{r} = -\varepsilon - \frac{1}{2}\varphi^2 + r\dot{\varphi}^2 + \frac{1}{24}\varphi^4 - \frac{1}{720}\varphi^6,$$

$$r\ddot{\varphi} = -\varphi - 2\dot{r}\dot{\varphi} + \frac{1}{6}\varphi^3 - \frac{1}{120}\varphi^5. \tag{4}$$

It is obvious that constant term ε in the right-hand side of the first Eq. (4) can vanish only if the amplitude of φ is proportional to $\sqrt{\varepsilon}$. In this case, the oscillating part of the distance r will be proportional to ε. Doing the substitution

$$r(t) \to 1 + \varepsilon r(t), \quad \varphi(t) \to \sqrt{\varepsilon}\varphi(t), \tag{5}$$

we reduce system (4) to the form

$$2\ddot{r} = -1 - \frac{1}{2}\varphi^2 + \dot{\varphi}^2 + \varepsilon(-\ddot{r} + r\dot{\varphi}^2 + \frac{1}{24}\varphi^4) - \frac{1}{720}\varepsilon^2\varphi^6, \tag{6}$$

$$\ddot{\varphi} + \varphi = -\varepsilon(r\ddot{\varphi} + 2\dot{r}\dot{\varphi} - \frac{1}{6}\varphi^3) - \frac{1}{120}\varepsilon^2\varphi^5. \tag{7}$$

One can readily check that a general solution to nonlinear system (6)–(7) cannot be found in symbolic form. As parameter ε is assumed to be small the Poincaré–Lindstedt perturbation technique for obtaining periodic solutions may be applied (see [21,22]). Note that in the case of $\varepsilon = 0$, Eq. (7) becomes independent of (6) and determines harmonic oscillations of the angle φ. Obviously, the amplitude of the corresponding function $\varphi(t)$ may be chosen in such a way that the constant part of the function in the right-hand side of (6) vanishes. Therefore, the corresponding solution $r(t)$ to (6) will be a bounded oscillating function. Taking into account the higher order terms in the right-hand sides of (6)–(7) for $\varepsilon > 0$ results in the appearance of corrections to zero-order solutions. Thus, we can look for an approximate solution to system (6)–(7) in the form of power series in ε:

$$r(t) = r_0(t) + \varepsilon r_1(t) + \varepsilon^2 r_2(t) + \varepsilon^3 r_3(t) + \dots, \tag{8}$$
$$\varphi(t) = \varphi_0(t) + \varepsilon \varphi_1(t) + \varepsilon^2 \varphi_2(t) + \varepsilon^3 \varphi_3(t) + \dots. \tag{9}$$

Computation of unknown functions $r_j(t), \varphi_j(t)$ in (8)–(9) is done in rather standard way but requires quite tedious symbolic computations (see [22]), which in this paper are performed using Wolfram Mathematica. Substituting (8)–(9) into (6)–(7) and collecting coefficients of equal powers of ε, we obtain the following system of linear differential equations:

$$\ddot{\varphi}_0 + \varphi_0 = 0, \tag{10}$$

$$2\ddot{r}_0 = -1 - \frac{1}{2}\varphi_0^2 + \dot{\varphi}_0^2, \tag{11}$$

$$\ddot{\varphi}_1 + \varphi_1 = r_0\varphi_0 - 2\dot{r}_0\dot{\varphi}_0 + \frac{1}{6}\varphi_0^3, \tag{12}$$

$$2\ddot{r}_1 = -\ddot{r}_0 + 2\dot{\varphi}_0\dot{\varphi}_1 + r_0\dot{\varphi}_0^2 - \varphi_0\varphi_1 + \frac{1}{24}\varphi_0^4, \tag{13}$$

$$\begin{aligned}\ddot{\varphi}_2 + \varphi_2 = {} & r_0\varphi_1 + r_1\varphi_0 - 2\dot{r}_0\dot{\varphi}_1 - 2\dot{r}_1\dot{\varphi}_0 + 2r_0\dot{r}_0\dot{\varphi}_0 \\ & - r_0^2\varphi_0 + \frac{1}{2}\varphi_0^2\varphi_1 - \frac{1}{6}r_0\varphi_0^3 - \frac{1}{120}\varphi_0^5,\end{aligned} \tag{14}$$

$$\begin{aligned}2\ddot{r}_2 = {} & -\ddot{r}_1 + 2\dot{\varphi}_0\dot{\varphi}_2 + \dot{\varphi}_1^2 + 2r_0\dot{\varphi}_0\dot{\varphi}_1 + r_1\dot{\varphi}_0^2 \\ & - \frac{1}{2}\varphi_1^2 - \varphi_0\varphi_2 + \frac{1}{6}\varphi_0^3\varphi_1 - \frac{1}{720}\varphi_0^6, \dots\end{aligned} \tag{15}$$

Obviously, Eqs. (10)–(15) may be solved in succession. Without loss of generality, we may assume that at the initial instant of time, the body m_1 is on the vertical ($\varphi(0) = 0$) and has some initial velocity $w_0 > 0$. The corresponding solution of Eq. (10) is

$$\varphi_0(t) = w_0 \sin t. \tag{16}$$

On substituting (16) into (11) we obtain

$$2\ddot{r}_1 = -1 + \frac{w_0^2}{4} + \frac{3}{4}w_0^2 \cos 2t. \tag{17}$$

As we are looking for an oscillating function $r_1(t)$ the amplitude w_0 is chosen from the condition that the constant term in the right-hand side of (17) vanishes. Due to this condition we set $w_0 = 2$ and solve Eq. (17) with initial condition $\dot{r}_1(0) = 0$. Then we obtain

$$r_1(t) = r_{10} - \frac{3}{8} \cos 2t, \tag{18}$$

where r_{10} is an arbitrary constant.

On substituting (16) and (18) with $w_0 = 2$ into (12) and reducing the trigonometric functions, we obtain

$$\ddot{\varphi}_1 + \varphi_1 = \left(2r_{10} - \frac{1}{8}\right)\sin t - \frac{53}{24}\sin 3t. \tag{19}$$

Equation (19) describes the forced oscillations of a pendulum, and to avoid an increase of the amplitude we need to eliminate a resonance term in the right-hand side. So putting $r_{10} = 1/16$ and solving differential equation (19) with initial condition $\varphi_1(0) = 0$, we find

$$\varphi_1(t) = \left(w_1 + \frac{53}{96}\right)\sin t + \frac{53}{192}\sin 3t, \tag{20}$$

where w_1 is an arbitrary constant.

On substituting (16), (18), and (20) into (13) and reducing the trigonometric functions, we derive the following differential equation

$$2\ddot{r}_2 = \frac{53}{96} + w_1 + \left(3w_1 + \frac{37}{64}\right)\cos 2t + \frac{105}{64}\cos 4t. \tag{21}$$

Again the unknown w_1 is chosen from the condition of vanishing constant terms in the right-hand side of (21), therefore, $w_1 = -53/96$. Then integrating (21) with the initial condition $\dot{r}_2(0) = 0$, we find

$$r_2(t) = r_{20} + \frac{69}{512}\cos 2t - \frac{105}{2048}\cos 4t, \tag{22}$$

where r_{20} is another arbitrary constant.

In order to find the solution more accurately we have to repeat such calculations step by step, solving successively linear differential equations (14), (15), and so on for the functions $\varphi_k(t)$ and $r_k(t)$ under the initial conditions $\varphi_k(0) = 0$, $\dot{r}_k(0) = 0$, $k = 1, 2, \ldots$. Each of the solutions $\varphi_k(t)$, $r_k(t)$ will contain an arbitrary constant which appears during integration and should be found from the condition that constant terms in the equation for $r_{k+1}(t)$ and resonance terms in the equation for $\varphi_{k+1}(t)$ vanish. We have done the calculations up to the third order in ε, and the corresponding periodic solutions are given by

$$r_p(t) = 1 + \frac{\varepsilon}{16}(1 - 6\cos 2t) - \frac{\varepsilon^2}{2048}(261 - 276\cos 2t + 105\cos 4t)$$
$$+ \frac{\varepsilon^3}{131072}(4275 - 8166\cos 2t + 5067\cos 4t - 1510\cos 6t), \tag{23}$$

$$\varphi_p(t) = \sqrt{\varepsilon}\left(2\sin t + \frac{53\varepsilon}{192}\sin 3t + \frac{\varepsilon^2}{16384}(2959\sin t\right.$$
$$\left. - 1699\sin 3t + \frac{5813}{5}\sin 5t)\right). \tag{24}$$

It follows from (23)–(24) that the initial length of the thread

$$r_p(0) = 1 - \frac{5\varepsilon}{16} - \frac{45}{1024}\varepsilon^2 - \frac{167}{65536}\varepsilon^3, \tag{25}$$

and the initial angular velocity

$$\dot{\varphi}_p(0) = \sqrt{\varepsilon}\left(2 + \frac{53\varepsilon}{64} + \frac{3675\varepsilon^2}{16384}\right), \tag{26}$$

corresponding to the periodic solution depend on parameter ε; for larger ε or larger masses difference, the initial velocity must increase to provide a larger amplitude of oscillations. Dependence of the initial length $r_p(0)$ on ε means that the frequency of oscillation depends on the amplitude; such dependence is typical of nonlinear oscillations (see [20, 22]).

4 Stability Analysis

The existence of periodic solution to equations of motion (4) means that for given value of parameter ε, one can choose initial conditions (25), (26), $\dot{r}_p(0) = 0$, and $\varphi_p(0) = 0$ such that the system is in the state of dynamical equilibrium when the bodies oscillate near some equilibrium positions. Note that for $\varepsilon > 0$, the system under consideration has no static equilibrium state when the coordinates $r(t)$, $\varphi(t)$ are some constants. So it is natural to investigate whether the system will remain in the neighborhood of the equilibrium if the initial conditions are perturbed or whether the periodic solution (23)–(24) is stable.

It should be noted that studying the stability of periodic solution is much more complicated in comparison to the case of equilibrium state stability and the relevant symbolic computations become much more cumbersome. First of all, we need to derive the equations of perturbed motion in the form of four first-order differential equations. Using (2) and doing the Legendre transformation (see [20]), we define the Hamiltonian in case of $|\varphi| \ll 1$

$$\mathcal{H} = \frac{p_r^2}{2(2 + \varepsilon)} + \frac{p_\varphi^2}{2r^2} + \varepsilon r + \frac{r}{2}\left(\varphi^2 - \frac{1}{12}\varphi^4 + \frac{1}{360}\varphi^6\right). \tag{27}$$

The equations of motion written in the Hamiltonian form are

$$\dot{r} = \frac{\partial \mathcal{H}}{\partial p_r} = \frac{p_r}{2 + \varepsilon}, \quad \dot{p}_r = -\frac{\partial \mathcal{H}}{\partial r} = -\varepsilon - \frac{1}{2}\varphi^2\left(1 - \frac{1}{12}\varphi^2 + \frac{1}{360}\varphi^4\right) + \frac{p_\varphi^2}{r^3},$$

$$\dot{\varphi} = \frac{\partial \mathcal{H}}{\partial p_\varphi} = \frac{p_\varphi}{r^2}, \quad \dot{p}_\varphi = -\frac{\partial \mathcal{H}}{\partial \varphi} = -r\varphi\left(1 - \frac{1}{6}\varphi^2 + \frac{1}{120}\varphi^4\right), \tag{28}$$

where p_r, p_φ are the conjugate momenta to r, φ, respectively.

One can readily check that periodic solution (23)–(24) satisfy Eqs. (28). To investigate its stability we define new canonical variables q_1, q_2, p_1, p_2 according to the rule

$$r \to r_p + q_1, \quad \varphi \to \varphi_p + q_2, \quad p_r \to p_{r0} + p_1, \quad p_\varphi \to p_{\varphi 0} + p_2, \tag{29}$$

where the momenta $p_{r0} = (2 + \varepsilon)\dot{r}_p, p_{\varphi 0} = r_p^2 \dot{\varphi}_p$ are obtained by substituting (23)–(24) into (28). Doing the canonical transformation (29) and expanding the Hamiltonian (27) into power series in terms of q_1, q_2, p_1, p_2 up to second order inclusive, we represent it in the form

$$\tilde{\mathcal{H}} = \mathcal{H}_0 + \mathcal{H}_1 + \mathcal{H}_2 + \ldots, \tag{30}$$

where $\mathcal{H}_k$ is the kth order homogeneous polynomial with respect to canonical variables q_1, q_2, p_1, p_2 which are considered as small perturbations of periodic solution (23)–(24). Note that zero-order term $\mathcal{H}_0$ in (30) can be omitted as a function of time which does not influence the equations of motion. The first-order term $\mathcal{H}_1$ is equal to zero because periodic solution (23)–(24) satisfy the unperturbed equations of motion (28). Therefore, the first non-zero term in the expansion (30) is a quadratic one that is

$$\mathcal{H}_2 = \frac{p_1^2}{2(2 + \varepsilon)} + \frac{3p_{\varphi 0}^2}{2r_p^4}q_1^2 + \frac{p_2^2}{2r_p^2} + \frac{r_p}{2}q_2^2 \left(1 - \frac{1}{2}\varphi_p^2 + \frac{1}{24}\varphi_p^4\right)$$
$$- \frac{2p_{\varphi 0}}{r_p^3}q_1 p_2 + q_1 q_2 \left(\varphi_p - \frac{1}{6}\varphi_p^3 + \frac{1}{120}\varphi_p^5\right). \tag{31}$$

The quadratic part $\mathcal{H}_2$ of the Hamiltonian determines the linearized equations of the perturbed motion which is convenient to write in the matrix form

$$\dot{x} = J \cdot H(t, \varepsilon)x, \tag{32}$$

where $x^T = (q_1, q_2, p_1, p_2)$ is a 4-dimensional vector, $J = \begin{pmatrix} 0 & E_2 \\ -E_2 & 0 \end{pmatrix}$, E_2 is the second-order identity matrix, and the fourth-order matrix-function $H(t, \varepsilon)$ is

$$H(t, \varepsilon) = \begin{pmatrix} \frac{3p_{\varphi 0}^2}{r_p^4} & \varphi_p & 0 & -\frac{2p_{\varphi 0}}{r_p^3} \\ \varphi_p & r_p & 0 & 0 \\ 0 & 0 & \frac{1}{2+\varepsilon} & 0 \\ -\frac{2p_{\varphi 0}}{r_p^3} & 0 & 0 & \frac{1}{r_p^2} \end{pmatrix}. \tag{33}$$

Note that the elements of matrix (33) are obtained by differentiation of $\mathcal{H}_2$:

$$H_{i,j} = \frac{\partial^2 \mathcal{H}_2}{\partial x_i \partial x_j}, \quad i, j = 1, 2, 3, 4.$$

It is clear that matrix $H(t, \varepsilon)$ is periodic function of time, and so the perturbed motion of the system is described by the linear system of four differential equations with periodic coefficients (32).

4.1   Computing the Monodromy Matrix

The systems of linear differential equations with periodic coefficients and their general properties have been studied quite well (see [23]). The behavior of solutions to system (32) is determined by its characteristic multipliers which are the eigenvalues of the monodromy matrix $X(2\pi, \varepsilon)$, where $X(t, \varepsilon)$ is a fundamental matrix for system (32) satisfying the initial condition $X(0) = E_4$. As periodic solution (23)–(24) is represented by power series in parameter ε, the matrix $H(t, \varepsilon)$ can also be represented in the form of power series

$$H(t, \varepsilon) = H_0(t) + \sqrt{\varepsilon}H_1(t) + \varepsilon H_2(t) + \varepsilon^{3/2}H_3(t) + \ldots, \tag{34}$$

where $H_k(t), k = 0, 1, 2, \ldots$, are continuous periodic fourth-order square matrices which are obtained by substitution of solution (23)–(24) into (33) and expanding each element of the matrix $H(t, \varepsilon)$ into power series in ε.

The fundamental matrix $X(t, \varepsilon)$ can be sought in the form of power series

$$X(t, \varepsilon) = X_0(t) + \sqrt{\varepsilon}X_1(t) + \varepsilon X_2(t) + \varepsilon^{3/2}X_3(t) + \ldots, \tag{35}$$

where $X_k(t), k = 0, 1, 2, \ldots$, are continuous matrix functions. On substituting (34) and (35) into (32) and collecting coefficients of equal powers of ε, we obtain the following sequence of differential equations:

$$\dot{X}_0 = JH_0X_0(t), \tag{36}$$

$$\dot{X}_k - JH_0X_k = \sum_{j=1}^{k} JH_j(t)X_{k-j}(t), \quad (k \geq 1). \tag{37}$$

The functions $X_k(t)$ must satisfy the following initial conditions:

$$X_0(0) = E_4, \quad X_k(0) = 0 \ (k \geq 1). \tag{38}$$

As H_0 is a constant matrix, Eq. (36) has a solution

$$X_0(t) = \exp(JH_0t). \tag{39}$$

Making a substitution

$$X_k(t) = \exp(JH_0t)Y_k(t), \tag{40}$$

we transform Eq. (37) to the form

$$\dot{Y}_k = \sum_{j=1}^{k} \exp(-JH_0t)JH_j(t)\exp(JH_0t)Y_{k-j}(t), \quad (k \geq 1), \tag{41}$$

where initial conditions for the functions $Y_k(t)$ are

$$Y_0(0) = E_4, \quad Y_k(0) = 0 \ (k \geq 1). \tag{42}$$

Now we can easily integrate Eq. (41) and its solution satisfying the initial conditions (42) is given by

$$Y_k(t) = \sum_{j=1}^{k} \int_0^t \exp(-JH_0\tau)JH_j(\tau)\exp(JH_0\tau)Y_{k-j}(\tau)d\tau, \quad (k \geq 1). \tag{43}$$

As the right-hand side of Eq. (43) determining $Y_k(t)$ depends only on $Y_0, Y_1, \ldots, Y_{k-1}$ the functions $Y_k(t)$ may be computed in succession. Such computations are performed with Wolfram Mathematica but the results are very bulky and we do not show them here. Finally, the monodromy matrix $X(2\pi, \varepsilon)$ of system (32) can be found in the form

$$X(2\pi, \varepsilon) = \exp(2\pi J H_0) \sum_{j=1}^{\infty} Y_k(2\pi)\varepsilon^{k/2}. \tag{44}$$

4.2   Characteristic Multipliers

Characteristic multipliers for system (32) are the eigenvalues of the monodromy matrix (44) and to find them we need to compute the monodromy matrix first. To find $X_0(t)$ it is not necessary to compute the exponential function of the matrix $J H_0 t$ according to (39). It is much easier to solve Eq. (36) with initial conditions (38) and

$$H_0 = \begin{pmatrix} 0 & 0 & 0 & 0 \\ 0 & 1 & 0 & 0 \\ 0 & 0 & 1/2 & 0 \\ 0 & 0 & 0 & 1 \end{pmatrix},$$

the corresponding solution is

$$X_0(t) = \begin{pmatrix} 1 & 0 & t/2 & 0 \\ 0 & \cos t & 0 & \sin t \\ 0 & 0 & 1 & 0 \\ 0 & -\sin t & 0 & \cos t \end{pmatrix}.$$

But the next steps require to multiply and integrate matrices as it follows from (43) and to do quite cumbersome symbolic calculations. So application of the computer algebra system Wolfram Mathematica turned out to be very helpful. We do not show here the intermediate results of calculations because they are quite bulky. Using the monodromy matrix which was computed up to the third order in parameter ϵ, we can write the characteristic equation determining the characteristic multipliers for system (32) in the form

$$\det(X(2\pi, \varepsilon) - \rho E_4) = (\rho - 1)^2(\rho^2 + 2B\rho + 1) = 0, \tag{45}$$

where

$$B = -2 + 3\pi^2\varepsilon - \frac{3\pi^2}{16}(17 + 4\pi^2)\varepsilon^2 + \frac{3\pi^2}{5120}(4845 + 2720\pi^2 + 128\pi^4)\varepsilon^3.$$

Solving (45), we obtain four characteristic multipliers

$$\rho_{1,2} = 1,$$

$$\rho_{3,4} = 1 \pm i\pi\sqrt{3\varepsilon} - \frac{3\pi^2}{2}\varepsilon \mp i\frac{\pi\sqrt{3}}{32}(17 + 16\pi^2)\varepsilon^{3/2} + \frac{3\pi^2}{32}(17 + 4\pi^2)\varepsilon^2.$$

Note that two characteristic multipliers $\rho_{1,2} = 1$ determine two independent periodic solutions to system (32). One can readily check that the absolute value of the second couple of the characteristic multipliers $\rho_{3,4}$ is equal to 1. They are complex conjugate and determine two purely imaginary characteristic exponents

$$\lambda_{3,4} = \frac{1}{2\pi} \log \rho = \pm i \frac{\sqrt{3\varepsilon}}{2} \left(1 - \frac{17}{32}\varepsilon + \frac{85}{256}\varepsilon^2 \right).$$

According to Floquet–Lyapunov theory (see [23]), four linearly independent solutions to system (32) with 2π-periodic matrix may be represented in the form

$$x_1(t) = f_1(t), \ x_2(t) = f_2(t), \ x_3(t) = \exp(\lambda_3 t)f_3(t), x_4(t) = \exp(\lambda_4 t)f_4(t), \ (46)$$

where $f_k(t), (k = 1, 2, 3, 4)$ are 2π-periodic functions. Therefore, in the case of $\varepsilon > 0$ solutions (46) describe the perturbed motion of the system in the bounded domain in the neighborhood of the periodic solution (23)–(24). It means this solution is stable in linear approximation, and so the SAM is an example of mechanical system in which the equilibrium state is stabilized by oscillations.

5   Conclusion

In the present paper, we have considered a swinging Atwood machine in the case when one body of smaller mass is permitted to oscillate in a vertical plane. Such a system has a state of equilibrium only in the case of equal masses but this state is unstable. Doing necessary symbolic computations, we have demonstrated that owing to oscillations the system has a dynamic equilibrium state described by a periodic solution of the equations of motion. It is a very interesting peculiarity of the system which takes place only due to the nonlinearity of the equations of motion.

We have found the initial conditions under which the equations of motion have periodic solution and proved its linear stability. Simulation of the system shows that this periodic motion is stable but its stability in Lyapunov sense still should be proved; so the problem requires further investigation. Note that the stability analysis of periodic solutions is a very complicated problem which involves quite tedious symbolic computations; so the application of computer algebra systems for doing such calculations is very helpful. In this work, we realized all the symbolic computations with the aid of the computer algebra systems Wolfram Mathematica.

References

1. Atwood, G.: A Treatise on the Rectilinear Motion and Rotation of Bodies. Cambridge University Press, Cambridge (1784)
2. Tufillaro, N.B., Abbott, T.A., Griffiths, D.J.: Swinging Atwood's machine. Am. J. Phys. **52**(10), 895–903 (1984)

3. Tufillaro, N.B.: Motions of a swinging Atwood's machine. J. Phys. **46**(9), 1495–1500 (1985)
4. Tufillaro, N.B.: Collision orbits of a swinging Atwood's machine. J. Phys. **46**(12), 2053–2056 (1985)
5. Tufillaro, N.B.: Integrable motion of a swinging Atwood's machine. Am. J. Phys. **54**(2), 142–143 (1986)
6. Tufillaro, N.B., Nunes, A., Casasayas, J.: Unbounded orbits of a integrable swinging Atwood's machine. Am. J. Phys. **56**(12), 1117–1119 (1988)
7. Casasayas, J., Nunes, A., Tufillaro, N.B.: Swinging Atwood's machine: integrability and dynamics. J. Phys. **51**(16), 1693–1702 (1990)
8. Nunes, A., Casasayas, J., Tufillaro, N.B.: Periodic orbits of the integrable swinging Atwood's machine. Am. J. Phys. **63**(2), 121–126 (1995)
9. Yehia, H.M.: On the integrability of the motion of a heavy particle on a tilted cone and the swinging Atwood's machine. Mech. Res. Commun. **33**(5), 711–716 (2006)
10. Pujol, O., Pérez, J.P., Ramis, J.P., Simo, C., Simon, S., Weil, J.A.: Swinging Atwood machine: Experimental and numerical results, and a theoretical study. Phys. D **239**(12), 1067–1081 (2010)
11. Elmandouh, A.A.: On the integrability of the motion of 3D-Swinging Atwood machine and related problems. Phys. Lett. A **380**(9), 989–991 (2016)
12. Prokopenya, A.N.: Motion of a swinging Atwood's machine: simulation and analysis with Mathematica. Math. Comput. Sci. **11**(3), 417–425 (2017). https://doi.org/10.1007/s11786-017-0301-9
13. Prokopenya, A.N.: Construction of a periodic solution to the equations of motion of generalized Atwood's machine using computer algebra. Program. Comput. Softw. **46**(2), 120–125 (2020). https://doi.org/10.1134/S0361768820020085
14. Prokopenya, A.N.: Modelling Atwood's machine with three degrees of freedom. Math. Comput. Sci. **13**(1–2), 247–257 (2019). https://doi.org/10.1007/s11786-018-0357-1
15. Prokopenya, A.N.: Searching for equilibrium states of Atwood's machine with two oscillating bodies by means of computer algebra. Program. Comput. Softw. **47**(1), 43–49 (2021). https://doi.org/10.1134/S0361768821010084
16. Prokopenya, A.N.: Determination of the stability boundaries for the Hamiltonian systems with periodic coefficients. Math. Model. Anal. **10**(2), 191–204 (2005). https://doi.org/10.1080/13926292.2005.9637281
17. Prokopenya, A.N.: Some symbolic computation algorithms in cosmic dynamics problems. Program. Comput. Softw. **32**(2), 71–76 (2006). https://doi.org/10.1134/S0361768806020034
18. Prokopenya, A.N.: Symbolic computation in studying stability of solutions of linear differential equations with periodic coefficients. Program. Comput. Softw. **33**(2), 60–66 (2007). https://doi.org/10.1134/S0361768807020028
19. Wolfram, S.: An Elementary Introduction to the Wolfram Language, 2nd edn. Wolfram Media, Champaign (2017)
20. Goldstein, H., Poole, C., Safko, J.: Classical Mechanics, 3rd edn. Addison Wesley, Boston (2000)
21. Grimshaw, R.: Nonlinear Ordinary Differential Equations. Blackwell Scientific Publications, Oxford (1990)
22. Nayfeh, A.H.: Introduction to Perturbation Techniques. Wiley, New York (1993)
23. Yakubovich, V.A., Starzhinskii, V.M.: Linear Differential Equations with Periodic Coefficients. Wiley, New York (1975)

New Heuristic to Choose a Cylindrical Algebraic Decomposition Variable Ordering Motivated by Complexity Analysis

Tereso del Río and Matthew England[✉]

Coventry University, Coventry, UK
delriot@uni.coventry.ac.uk, Matthew.England@coventry.ac.uk

Abstract. It is well known that the variable ordering can be critical to the efficiency or even tractability of the cylindrical algebraic decomposition (CAD) algorithm. We propose new heuristics inspired by complexity analysis of CAD to choose the variable ordering. These heuristics are evaluated against existing heuristics with experiments on the SMT-LIB benchmarks using both existing performance metrics and a new metric we propose for the problem at hand. The best of these new heuristics chooses orderings that lead to timings on average 17% slower than the virtual-best: an improvement compared to the prior state-of-the-art which achieved timings 25% slower.

1 Introduction

1.1 Cylindrical Algebraic Decomposition

A Cylindrical Algebraic Decomposition (CAD) of $\mathbb{R}^n$ is a decomposition of $\mathbb{R}^n$ into semi-algebraic cells that are cylindrically arranged. A cell being semi-algebraic means that it can be described by polynomial constraints. CADs are defined relative to a variable ordering, for example, $x_n \succ x_{n-1} \succ \cdots \succ x_1$. Then the cylindrical property means that the projections of any two cells in $\mathbb{R}^n$ onto a subspace $\mathbb{R}^i, i < n$ with respect to this variable ordering, are either equal or disjoint. I.e. the cells in $\mathbb{R}^n$ are arranged into cylinders above cells in $\mathbb{R}^{n-1}$, which are themselves arranged into cylinders above $\mathbb{R}^{n-2}$ and so on.

It can be very useful to find such decompositions satisfying a property such as sign-invariance for an input set of polynomials (i.e. each polynomial has constant sign in each cell). The principle is that given an infinite space, a sign-invariant decomposition gives a finite set of regions on each of which our system of study has invariant behaviour, and thus can be analyzed by testing a single sample point. When such a decomposition is also cylindrical and semi-algebraic we can use it to perform tasks like quantifier elimination.

Collins in 1975 [13] was the first to propose a feasible algorithm to build such sign-invariant decompositions for a given set of polynomials. This algorithm has two phases, projection and lifting, each of them consisting of n steps, where n is the number of variables in the given set of polynomials S_n.

In the first step of the projection phase the given set of polynomials, S_n is passed to a CAD projection operator to obtain a set of polynomials S_{n-1} in $n-1$ variables (without the biggest variable x_n). This process is iterated until a set of polynomials S_1 only in the variable x_1 is left: at this point the projection phase ends.

In the first step of the lifting phase a CAD of $\mathbb{R}^1$ is created by computing the t ordered roots of S_1, denoted $r_1, \ldots, r_t$, and building the CAD of $\mathbb{R}^1$ out of the cells $(-\infty, r_1), [r_1], (r_1, r_2), \ldots, (r_{n-1}, r_n), [r_n], (r_n, \infty)$. Note that a sample point can be taken from each of those cells.

For each of those cells, the sample point is substituted into the set of polynomials S_2 to obtain a set of polynomials in one variable. Then using this set a stack of cells is built on top of each cell by following the instructions in the previous paragraph. These stacks are combined later into a CAD of $\mathbb{R}^2$ and by iterating this process a CAD of $\mathbb{R}^n$ is eventually built, concluding the lifting phase and the algorithm.

The proof of correctness of CAD (allowing the conclusion of sign-invariance) relies on proving that the decompositions built over the sample point are representative of the behaviour over the entire cell: to conclude this the projection operator must produce polynomials whose zeros indicate where the behaviour would change. One representation of a CAD is as a tree of cells of increasing dimension; whose leaves are the cells in $\mathbb{R}^n$, nodes the cells in lower dimension, and branches representing the cylinders over projections.

Since the introduction of CAD by Collins, many improvements have been made to the algorithm. We do not detail them all here but refer the reader to the overview of the first 20 years in [14] and to e.g. the introduction of [3] for some of the more recent advances. We note in particular the recent developments in CAD projection [26] and the recent application of CAD technology within SMT and verification technology e.g. [24], [1] which has inspired new adaptations of the CAD algorithm such as [6]. The work of this paper is presented for traditional CAD, but we expect it would transfer easily to these recent contexts.

1.2   CAD Variable Ordering

It is well known that the variable ordering given can have a huge impact on the time and resources needed to build the CAD (see e.g. [15, 19, 23]). We demonstrate this for a very simple example in Fig. 1, where one choice leads to three times the number of cells than the other. In fact, [7] shows that the choice of variable ordering can even change the theoretical complexity for certain classes of problems.

Depending on the application the CAD is to be used for, we have a free or constrained choice of variable ordering. For example, to use a CAD for real quantifier elimination it is necessary to project variables in the order they are quantified, but there is freedom to swap the order of variables in quantifier blocks, and also to swap the order of the free (unquantified) variables and parameters. Making use of this

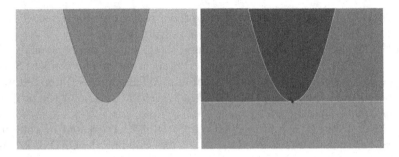

Fig. 1. CADs sign-invariant for $x^2 - y$ in the two possible orderings. The ordering $y \succ x$ generates the CAD on the left with only three cells (the coloured regions and the curve between). The ordering $x \succ y$ generates the CAD on the right with nine cells (the four coloured regions along with four line segments and the turning point of the curve).

freedom is an important optimisation. This paper aims to present a new heuristic to pick the variable ordering for the construction of a CAD.

1.3 Plan of the Paper

We continue in Sect. 2 by describing the previous heuristics developed for the problem. Then in Sect. 3 the proposed heuristics are presented. We then move onto our experimental evaluation: in Sect. 4 the methodology is detailed and in Secti. 5 the results obtained are analyzed. We finish with our conclusions and suggestions for future work in Sect. 6.

2 Previous Heuristics

Due to how critical the variable ordering can be, a variety of heuristics have already been proposed for making the choice. We will focus on two of these which are widely considered to constitute the current state-of-the-art: the Brown heuristic, presented in [5], and sotd, presented in [15]. We describe these in full over the coming subsections.

We acknowledge there are additional human-designed heuristics in the literature, but these are either even more expensive than sotd e.g. [4,27], or designed relative to a very specific CAD implementation e.g. [16].

We also acknowledge that there exists a family of machine learning methods to take this decisions e.g. [8,12,19,22,23] which have been shown to outperform the human-designed heuristics. We do not compare against these directly but note that the lessons learnt in this paper could inform another generation of these machine learnt heuristics.

2.1 The Brown Heuristic

The Brown heuristic was proposed by Brown in the notes to his ISSAC 2004 tutorial [5, Section 5.2]. It chooses to project the variable with:

1. lowest degree; breaking ties with
2. lowest value of the highest total degree term in which the variable appears; breaking ties with
3. lowest number of terms containing the variable.

Should there remain ties after the third measure then [5] does not specify what to do: we name the variables $x_1, x_2, \ldots$ according to the order in which they appear in the description of the problem and our implementation of the Brown heuristic chooses the variable with the lowest subindex first.

The text in [5] is also unclear on whether these measures are applied only to the input polynomials to produce the complete ordering, or whether they are applied to select an ordering one variable at a time, each time being applied to the polynomials obtained from projection with the last variable. Our implementation does the latter: this requires no additional projection computation above that required to build a single CAD, and matches the projection computation used by our new heuristic allowing for a fair comparison later.

For an example, let us consider the set $S_3 = \{x_3{}^3 + x_2{}^3 + x_2 - x_1{}^4, x_2{}^3 - x_1\}$. In S_3 the variable x_1 has degree 4, but x_3 and x_2 both have degree 3 so they tie in the first feature. However, x_3 reaches this maximum in only one term, while x_2 does so in two of them and hence x_3 will be the first CAD projected variable. The CAD projection of S_3 with respect to x_3 is $S_2 = \{x_2{}^3 + x_2 - x_1{}^4, x_2{}^3 - x_1\}$. In S_2 the variable x_1 has degree 4 while x_2 has degree 3, so x_2 is the second variable to be projected. This determines the variable ordering chosen with our implementation of the Brown heuristic: $x_3 \succ x_2 \succ x_1$.

The motivation of the Brown heuristic is to try to make the next projected set of polynomials as small as possible. The heuristic is "cheap" as the measures it uses require only easy to calculate characteristics of the input polynomials, and no algebraic computations such as projection.

Moreover, the Brown heuristic has been shown to achieve similar accuracy to more complicated heuristics [21, 23], like the one introduced in the next section. This means that when we include the cost of running the heuristics themselves, the Brown heuristic is actually superior.

Nevertheless, as will be seen later, it is possible to propose better heuristics by looking at the bigger picture rather than focusing on the next projected set.

2.2   The sotd Heuristics

The acronym *sotd* stands for 'sum of total degrees'. The heuristic sotd consists of computing the whole CAD projection of the input polynomials in every possible variable ordering and choosing the ordering whose projection has the smallest sum of total degrees throughout all the monomials in all the polynomials of the projection [15].

For example, given the set of polynomials S_3 defined above and following variable ordering $x_3 \succ x_1 \succ x_2$, then we find S_2 as defined above and them projection with respect to x_1 gives $S_1 = \{x_2, x_2{}^2 + 1, x_2{}^{11} - x_2{}^2 - 1\}$. The sum of all the degrees in the projected sets for this ordering is 43. It turns out that this is lowest such value possible from any of the six possible orderings. Hence the sotd heuristic would choose this variable ordering.

This heuristic is not very desirable at first glance, because $\mathcal{O}(n!)$ projection steps are needed to make the choice, far more than the n projections used to make a single CAD. This problem was spotted by the authors of [15] leading them to develop a "greedy" version of the heuristic in the same paper. This was greedy in the sense that instead of selecting the entire ordering at once using information from the whole projection phase for each possible ordering; it computed the ordering one variable at a time by comparing their metric on the output of a single projection step for each possible variable from the remaining ones. This version only requires $\frac{n(n+1)}{2}$ number of projections, still more than the amount of projection normally used to build a CAD. In our experiments, the choices it makes are much poorer than those made by sotd with the full projection information. In fact, some experiments show it even performs worse than the Brown heuristic, despite having more projection information available to make its choice. We see this later in our experiments (Table 1).

The metric *sotd*, i.e. summing all the degrees in a set of polynomials, was originally constructed as a measure of the overall size of a polynomial set. In [15] it was used along with other measures to demonstrate the effect of variable ordering on CAD computation. It was found to have a strong correlation with those other CAD complexity measures, but unlike them, it did not require the computation of the entire CAD. This led to its proposal for using it as a heuristic.

Thus, it seems the sotd measure was not designed primarily for use in a CAD variable ordering heuristics, allowing a gap for the new results of the present paper which presents a measure that is designed this way. This lack of tailoring to CAD can be seen in the way sotd and its greedy version give the same importance to all degrees found in the projected sets of polynomials regardless of the variable carrying that degree, whereas CAD works iteratively one variable at a time (treating the others as part of the coefficients when projection and substituting them for the sample point when lifting). Thus different variables carry different weights of effect on CAD computation.

In the next section, it will be shown how better heuristics can be proposed which take into account the potential growth in complexity of the polynomials we observe in CAD projection.

3   Our New Proposed Heuristics

It has been shown in [15] that the number of cells of a CAD is strongly correlated with the time taken to build that CAD. Hence the most recent CAD complexity analyses in e.g. [3,17,18,25] have studied a bound on the maximum number of cells that can be generated.

The idea of the proposed heuristic is to make a corresponding estimate on this maximum number of cells of the final CAD for each of the possible orderings and pick the ordering that minimizes this value.

We explain how this number can be computed or estimated if the whole CAD projection is known for each ordering. However, as CAD projection of polynomials can be an expensive operation, a greedy version of this heuristic that does not require any CAD projections to make the choice is also proposed.

3.1   Heuristic Motivated by a Complexity Analysis: mods

Define the *degree sum* of a variable x in a set of polynomials $S = \{p_1, \ldots, p_n\}$ as

$$D_x(S) = \sum_{i=1}^{n} d_x(p_i), \tag{1}$$

where $d_x(p)$ is the degree of x in the polynomial p. Thus the maximum number of unique real roots that the polynomials in S can have with respect to x, is $D_x(S)$.

To compute a CAD with the variable ordering $x_n \succ \cdots \succ x_1$ for the set of polynomials S_n, the set S_n must be projected with respect to x_n to obtain the set S_{n-1}; and in the same fashion the sets $S_{n-2}, S_{n-3}, \ldots, S_1$ are computed.

Thus when following the creation of a CAD as described in Sect. 1, in the first lifting step at most $2D_{x_1}(S_1) + 1$ cells can be created because x_1 will have at most $D_{x_1}(S_1)$ roots in S_1. Subsequently at most $(2D_{x_1}(S_1) + 1) \cdot (2D_{x_2}(S_2) + 1)$ cells will be built in the second lifting step (a similar limit applied for each stack above a cell from $\mathbb{R}^1$).

Hence, at the end of the lifting phase, when the CAD is completed, an upper bound on the number of cells is

$$\prod_{i=1}^{n} \left(2D_{x_i}(S_i) + 1\right). \tag{2}$$

As discussed earlier, the number of cells in a CAD is strongly correlated with the time needed to build such CAD. Therefore, choosing the ordering that minimizes the maximum number of cells in the final CAD sounds like a good idea if we want to choose a fast ordering. Hence, we want to choose the ordering that minimizes (2).

We note that the dominant term of (2) is

$$\prod_{i=1}^{n} D_{x_i}(S_i), \tag{3}$$

which we refer to as the **multiplication of degree sum (mods)**. By minimising (2) we are likely minimising this and so we refer to the heuristic that picks an ordering to minimise (2) as mods. As with our implementation of the Brown heuristic, we apply this to choose one variable at a time, projecting with respect to that variable after the choice and then applying the measure to the projection polynomials to make the next choice. In case where there is a tie on the measure then we pick the variable with the lowest subindex.

Consider our example set of polynomials $S_3 = \{x_3{}^3 + x_2{}^3 + x_2 - x_1{}^4, x_2{}^3 - x_1\}$. The degree sum of x_1 in S_3 is

$$D_{x_e}(S_3) = d_{x_e}(x_3{}^3 + x_2{}^3 + x_2 - x_1{}^4) + d_{x_e}(x_2{}^3 - x_1) = 3 + 0 = 3.$$

Suppose we built a CAD using the variable ordering $x_3 \succ x_1 \succ x_2$. Then as before we obtain $S_2 = \{-x_2{}^3 + x_1, x_1{}^4 - x_2{}^3 - x_2\}$ for which $D_{x_1}(S_2) = 5$, and $S_1 = \{x_2, x_2{}^2 + 1, x_2{}^{11} - x_2{}^2 - 1\}$ for which $D_{x_2}(S_1) = 14$. Therefore, for this example CAD the product (2) evaluates to 2233. It turns out that is the lowest value of (2) for all the possible orderings. Hence mods would have chosen this ordering.

3.2　Creating a Greedy Version of mods

As with sotd, the heuristic mods is relatively expensive, requiring the use of CAD projection operations in all different variable orderings. To reduce its cost we present a greedy version of this heuristic, that will simply choose to project the variable with the lowest degree sum (see (1)) in the set of polynomials[1].

This heuristic will be referred to as gmods. Note that unlike greedy-sotd, gmods does not use any projection information beyond that required to build a single CAD. The metric it is based on uses only easily extracted information from the polynomials. It is thus similar in cost to our implementation of the Brown heuristic.

For example, given the set of polynomials S_3 above we have $D_{x_1}(S_3) = 5$, $D_{x_2}(S_3) = 6$ and $D_{x_3}(S_3) = 3$. Thus gmods will select x_3 as the first variable for CAD projection. The CAD projection of S_3 with respect to x_3 gives S_2 as above. In S_2 the variable x_1 has degree sum 5 while x_2 has degree sum 6, so x_1 is the second variable to be projected, determining completely the variable ordering that gmods chooses: $x_3 \succ x_1 \succ x_2$.

3.3　Heuristic Motivated by Expected Number of Cells

Our mods heuristic is motivated to reduce the maximum number of cells that could be computed according to a complexity analysis. It is natural to ask whether we could be more accurate and seek take decisions according to an expected value of the number of cells rather than the maximum?

To calculate the maximum number of cells that can be generated, the degree of the polynomials has been used because it is the maximum number of real roots that a polynomial can have, so we may consider the expected number of roots of a polynomial. According to [20], the expected number of real roots for polynomials of small degree is proportional to the logarithm of its degree, at least for their definition of random polynomials. However, for a linear polynomial, this relation

---

[1] We note that this measure applied only to the original polynomials is one of the features that was generated algorithmically to train different machine learning classifiers to take decisions on sets of polynomials in [21].

would predict zero roots when it should be one, and so to address this we suggest a heuristic following this approach should add one before taking the logarithm.

Thus we hypothesise an expected number of cells in the final CAD as below (following the approach of Sect. 3.1):

$$\prod_{i=1}^{n} (2\log(D_{x_i}(S_i) + 1) + 1). \tag{4}$$

As before, we define a heuristic to pick the ordering that minimizes (4). Given the similarity to $\mathtt{mods}$ and the use of the logarithm we refer to this as $\mathtt{logmods}$.

For example, consider the set of polynomials S_3 as before and the variable ordering $x_3 \succ x_1 \succ x_2$ to produce S_2 and S_1 as before. We find $D_{x_3}(S_3) = 3$, $D_{x_1}(S_2) = 5$, and $D_{x_2}(S_1) = 14$. Therefore, (4) evaluates to 15.43, and it turns out that is the lowest value for all possible orderings, hence, $\mathtt{logmods}$ would choose this variable ordering.

4 Experiments and Benchmarking

4.1 Benchmarking

The three-variable problems in the QF_NRA category of the SMT-LIB [2] are used to build a dataset for comparing the different heuristics.

For each of those problems and all possible orderings, we timed (see Sect. 4.1) how long it takes to build a sign-invariant CAD for the polynomials involved, discarding the problems in which the creation of the CAD timed out for all orderings. After building all the possible CADs, a dataset of "unique" problems is created (see Sect. 4.1).

Of the 5942 original problems, in 343 of them, all the orderings timed out. And out of the remaining 5599 problems, only 1019 unique problems were found. These 1019 problems will be used as benchmarks to compare the heuristics presented in Sects. 2 and 3.

CAD Implementation. For our experiments we used the function $\mathtt{Cylindri}$ $\mathtt{calAlgebraicDecompose}$ in the MAPLE 2022 Library $\mathtt{RegularChains}$, whose implementation is described in [10]. This actually implements a somewhat different CAD algorithm to the classical approach described above. Instead of projecting and lifting it first decomposes complex space and then refines this to a CAD [11], with the current implementation doing the complex decomposition incrementally by polynomial [9]. As reported in these papers, this approach can avoid some superfluous cell divisions. However, there is still the same choice of variable ordering to be made which can be crucial [12] with the Brown heuristic observed previously to work similarly well for the regular chains based algorithms [23].

Timings. Timings are performed following the methodology of [19]. For each of the possible variable orderings, the polynomials defining the problems were given as input to the CAD in MAPLE with a time limit of 30 s. If none of the orderings finishes, all the orderings are attempted again with a time limit of 60 s.

Projection times are timed individually using our implementation in MAPLE of McCallum CAD projection (that returns the polynomials factorized) with a time limit of 10 s: these times are used to give a more meaningful comparison of heuristics that requires us to compute all the projections, with heuristics that do not need to do so.

Every CAD call was made in a separate MAPLE session launched from and timed in Python, to avoid MAPLE's caching of intermediate results from one benchmark or ordering that may help another. From each timing, 0.075 s were removed: the average time that MAPLE takes to open on the computer when called from Python. It was removed as this is not a cost that would normally be paid but as a consequence of the benchmarking.

Uniqueness. When studying the dataset it was observed that many examples were very similar to each other. Similar in the sense that they were described by very similar polynomials, resulting in CADs with equivalent tree structures for every variable ordering, making it likely that all aspects of the CAD generation were similar. It is well observed that there exist these families of very similar benchmarks in the SMT-LIB. Treating each of them as an independent benchmark could result in skewed experimental results. E.g. a heuristic that happens to perform well on a large family of almost identical benchmarks would receive a huge but unwarranted boost in the analysis if we do not take care.

To avoid this, the samples with the same number of cells in the CADs for all possible variable orderings are clustered and only one of them is included in the dataset. This ensures that there are no two problems with an equivalent CAD tree structure for each variable ordering.

4.2 Evaluation Metrics

Existing Evaluation Metrics. The most obvious metric to evaluate the choices of our heuristics is the total time taken to build CADs for all the problems with the orderings chosen by that heuristic: this metric will be referred to as total-time. Also, another metric that will be used to compare the different heuristics is the number of problems completed before timeout using the orderings chosen by the heuristic.

In previous studies such as [21] accuracy, i.e. the percentage of times that the fastest ordering is chosen, is used as one of the main metrics. However, as discussed in [22], for our context, accuracy is not the most meaningful metric. This is because it is well observed that the second-best ordering may only be very marginally worse than the best ordering and so picking that should also be considered accurate. Further, the timings may include small amounts of computational noise which change the ranking of orderings in such subsets and thus the accuracy score.

In [22] the authors proposed to address this by considering a heuristic as successful if it identifies any ordering that takes no more than 20% additional time than the optimal. This fitted their work on a machine learning classification problem, but this definition is not suitable for regression, or use to evaluate a continuous range of possibilities. It considers equally inaccurate an ordering that is 30% slower and an ordering that is three or four times slower, and even an ordering that timed out. We thus propose a new metric for use in the evaluation in place of accuracy.

Markup. We suggest measuring the amount of time that the chosen ordering takes above the time of the optimal ordering, as a percentage of the optimal ordering:

$$\frac{\text{heuristic\_time} - \text{optimal\_time}}{\text{optimal\_time}}.$$

This allows for problems of different sizes to be evaluated relative to their possible solutions. For example, suppose Problem A's optimal ordering took 10 s and Problem B's took 20 s. If the chosen ordering for Problem A took 2 s longer than optimal then the score would be 0.2; while if that happened for Problem B the score would be 0.1, recognizing that the excess 2 s is a less substantial markup for the larger problem.

However, this can lead to distortions for problems where the optimal ordering is really fast. For example, if the optimal ordering takes 0.02 s and the chosen ordering takes 4 s then the metric above would give that problem a very huge influence over the final score. To avoid that situation, and taking into consideration that anything below a second would likely be acceptable to use for constructing a CAD, we propose instead to add one to all the timings, i.e.

$$\text{Markup} = \frac{(\text{heuristic\_time} + 1) - (\text{optimal\_time} + 1)}{\text{optimal\_time} + 1}.$$

This measure still allows the evaluation of relative potential but reduces distortions from fast examples and computational noise. In the example above, the metric would evaluate to 3.9 instead of 199. We refer to this as *Markup*, i.e. a measure of how far from the optimal this choice was.

Markup combines the benefits of both accuracy and total-time. Like total-time does it can measure not only if a choice was worse than the virtual-best but also how worse it was. But it adapts better to the different sizes of examples, unlike total-time where performing slightly worse in a difficult problem can have more impact on the metric than performing really bad in an easy example. Like accuracy it gives the same relevance to all the instances, but unlike accuracy it does not define a choice as simply either right or wrong.

Timeouts. For computing markup and total-time we must decide how to deal with cases where the chosen variable ordering leads to a timeout in CAD computation. In this case, when an ordering does not finish within the time limit given it will be assumed that it would have taken twice the time limit given.

4.3   Metrics and Expensive Heuristics

Note that some of our heuristics are cheap, manipulating over data easily extracted from the polynomial, while others are expensive, requiring the use of CAD projection and thus algebraic computations. When analyzing an expensive heuristic we have the choice of ignoring the cost of the heuristic or taking it into account. It is clear that the latter is more realistic because without paying this cost it would not be possible to make the choice. However, the former way of analyzing the heuristic also brings some interesting insight. Therefore, when presenting the metrics for these heuristics (Table 1), the metric without including the cost of the heuristic will be shown between brackets.

For example, the number of examples marked as complete stands for the number of problems in which the CAD was constructed with the heuristic's choice of variable ordering before the timeout. To adjust this in expensive heuristics, we count as timeouts the problems in which the time taken to choose the ordering plus the time taken to build the CAD did not exceed the time limit. As the more realistic value, the latter is outside brackets and the former within.

5   Results and Analysis

The results given by the analysis of the different heuristics to choose the variable ordering for the 1019 benchmarks are summarized in Table 1, and a survival plot comparing the heuristics is presented in Fig. 2. To produce the survival plot, for each heuristic the times taken to solve the problems with the variable ordering chosen by the heuristic are sorted into increasing order to form a sequence (t_i), discarding the timed-out problems; and the points $(k, \sum_{i=1}^{k} t_i)$ are then plotted. This plot encapsulates visually a lot of information about the success of the heuristics on a given dataset (it does not say anything about heuristics relative performance on particular problem instances).

Table 1. Evaluation metrics for the different heuristics to choose the variable orderings for CAD. For the expensive heuristics, the metrics without taking into account the cost of the heuristic can be seen between brackets. In bold, the best measure of the metric out of all the heuristics.

Name	Accuracy	Total time	Markup	# Completed
sotd	0.43	11007(9656)	1.56(1.16)	931(946)
mods	**0.64**	8137(**6637**)	0.57(**0.13**)	979(**990**)
logmods	0.49	9085(7535)	0.91(0.47)	968(983)
greedy-sotd	0.4	15669(15533)	2.55(2.51)	840(841)
brown	0.56	7590	0.25	974
gmods	0.58	**6945**	**0.17**	**987**

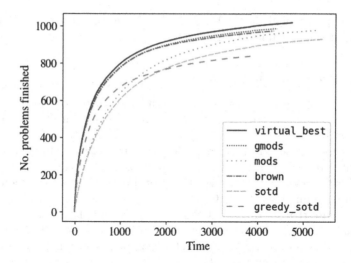

Fig. 2. Survival plot comparing heuristics on the benchmarks they can tackle before timeout

The first thing to note is that *any* heuristic is significantly better than making a random choice, giving further evidence on the critical need for attention to this decision. We give further analysis on the heuristics grouped by their relative costs.

5.1 Expensive Heuristics: sotd vs mods

As discussed earlier, one of the strategies that can be taken to choose an ordering is to compute all the projection phases for each variable ordering beforehand and base the decision on this information. Our new heuristic mods and the existing heuristic sotd follow this strategy. This approach requires a huge cost: it is possible to observe in Table 1 that one-fifth and one-seventh of the time taken to do a CAD by the ordering suggested by mods and sotd respectively is invested on deciding the ordering.

They both use the same information to take the decision, however, as can be seen in Table 1, mods outperforms sotd in all the presented metrics. The proposed heuristic picks the best ordering for almost two-thirds of the problems, while the existing one does so in less than half of them. The choices of mods reduce the total time by almost fifty minutes and solve almost 50 problems more with respect to the choices done by sotd.

Moreover, the ordering chosen by mods took on average 58% more time than the best ordering, while the choice of sotd took on average 160% more than it. Meaning that for a problem where the virtual-best ordering takes 10 s it is expected that the ordering proposed by mods takes 15.8 s while 26 s are expected from the ordering proposed by sotd.

When we compare mods to logmods we see that logmods is outperformed in all measured metrics. We thus conclude that the expected number of real roots for the polynomials in our benchmark set does not match well that of the random polynomials studied in [20].

5.2 Cheaper Heuristics: gmods vs brown

The heuristics presented in Sect. 5.1 exerted a large amount of effort to solve the problem of choosing the ordering, at odds with the behaviour of most algorithm optimization heuristics.

We now look at the cheaper heuristics: greedy-sotd which greatly reduces the amount of projection information used to make a decision (compared to sotd and mods) and brown and gmods which do not use any such information beyond that required to build the single CAD.

We first note that even without taking into account the higher cost of greedy-sotd, it is greatly outperformed by the two heuristics that do not make use of projection information at all. When comparing brown and gmods: both heuristics have similar accuracies, however, the choices of gmods reduce the total time by ten minutes, and solves 13 problems more with respect to the choices of brown. Moreover, the ordering chosen by gmods took on average 17% more time than the best ordering, while the choice of brown took on average 25% more than it. Meaning that for a problem where the optimal ordering takes 10 s it is expected that the ordering proposed by gmods takes 11.7 s while 12.5 s are expected from the ordering proposed by brown.

To further understand how these two heuristics compare, and if there are subsets of problems in which one of them performs better than the other, an adversarial plot is presented in Fig. 3. This plots for each benchmark the time taken by the two heuristics against each other. In that figure, it can be observed that most of the points are close to the diagonal line, implying that both heuristics perform similarly for most of the instances. However, it can be also observed that for some problems the ordering suggested by gmods timed out while brown proposed an ordering that completes and vice versa. This phenomenon is more common in favour of gmods but leaves open the possibility of a combination or meta-heuristic outperforming either.

5.3 Expensive vs Cheap Approach: mods vs gmods

It can be observed looking at the accuracies in Table 1 that mods picks better orderings than gmods and is superior in all other metrics if the cost of the heuristic is not taken into account (looking at the values between brackets in the same table). I.e. it has the strongest predictive power. However, as discussed in Sect. 4.2, accuracy is not the most interesting metric and gmods outperforms mods in all fair comparisons that take the cost of the heuristic into account.

In Fig. 2 we see that greedy-sotd outperforms sotd at first, solving many problems in a shorter time, but in the long run sotd ends up solving more in total. In fact, it is possible to observe that the greedy heuristics start ahead of the expensive heuristics. This implies that it is especially disadvantageous

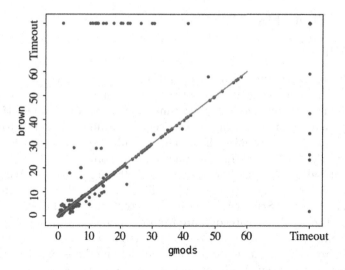

Fig. 3. Adversarial plot comparing gmods and brown.

to compute all projections when working on easy problems, and motivates a separate analysis excluding the easiest problems.

The results when we restrict to only the hardest 134 problems (those whose optimal time need more than 10 s) are plotted in Fig. 4. Now mods performs almost as well as brown. This further highlights the superiority of gmods over the rest in this particularly relevant slice of the problems where easy problems are excluded. Thus these expensive heuristics may still have a role as we expand our analysis to still harder problems.

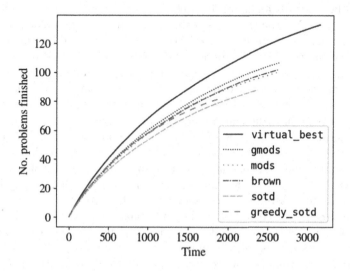

Fig. 4. Survival plot comparing heuristics on harder benchmarks only

6 Final Thoughts

6.1 Conclusions

The new heuristics motivated in this paper by the complexity analysis of CAD have clearly become the new state-of-the-art to choose the variable ordering for CAD. This leads to the most important conclusion of the paper: theoretical complexity analyses of an algorithm are a very powerful tool not just to compare algorithms but also to optimize them. These results also show that the benefits of a greedy or lazy approach in algebraic computation.

We would also highlight how out of the 5599 problems of three variables in the QF_NRA category of the SMTLIB library, only 1019 were found to be unique (in the sense of Sect. 4.1). I.e. three-quarters, at least of the three-variable problems in the QF_NRA category, are from the point of view of CAD copies of other problems in that category. Their differences may be important when using SAT solvers to study the logic, or other incomplete methods, but for an analysis of a complete solver they should be merged as in our methodology.

Finally, we note that `logmods` failed in our experiments, implying that for the polynomials found in the QF_NRA category of the SMTLIB library the expected number of roots is not proportional to the logarithm of their degree, and therefore they do not follow the same distribution as studied in [20]. This is not that surprising given it is well documented that polynomials from applications are often different to those generated from a simple random generator.

6.2 Future Work

An obvious future work is to experiment with the new heuristics on problems with more variables, higher degree, or data from other sources. We see a number of avenues beyond this for more research.

First we note that the results here should feed into work on machine learning methods to make such optimization decision. In fact, it would be interesting to study the extend to which the metrics presented here were used in the machine learning classifiers of [21] who created similar metrics among hundreds via an automated method. We are also interested in building simpler ML models using a restricted set of features. These could be more interpretable and thus offer further insights on how the features connect.

Next, we note that the actual CAD complexity analyses in e.g. [3,17,18,25] were performed not on the projection polynomials as a single set but an optimal arrangement of them. This is known as the (m,d)-property and stems from the PhD thesis of McCallum. It would be interesting to see if there is any heuristic that can be deduced from the analysis involving this property.

We are also interested to look if the additional information encoded in `mods` could be obtained more cheaply than using CAD projections. Especially since the heuristic does not use all the information in these projections, only degree information. This would allow us to make choices only 13% slower on average than the virtual best (based on the results between brackets – without heuristic cost – in Table 1). Alternatively, another greedy version of mods could be developed, in which, similarly to greedy-sotd, instead of computing the whole projection phase for each possible variable ordering, a projection step is done for each available variable.

Acknowledgments. The authors would like to thank AmirHosein Sadeghimanesh for his interesting conversations and his constructive criticism, and for sharing his Maple code to perform CAD projections. We also thank the anonymous reviewers whose comments helped us improve the paper.

The research of the first author is supported financially from a scholarship of Coventry University. The research of the second author is supported by EPSRC Grant EP/T015748/1, *Pushing Back the Doubly-Exponential Wall of Cylindrical Algebraic Decomposition* (the DEWCAD Project).

Research Data and Code Statement

Data and code necessary to generate the figures and results presented in this paper are available at: https://doi.org/10.5281/zenodo.6750528.

References

1. Ábrahám, E., Davenport, J.H., England, M., Kremer, G.: Deciding the consistency of non-linear real arithmetic constraints with a conflict driven search using cylindrical algebraic coverings. J. Log. Algebraic Methods Program. **119**, 100633 (2021). https://doi.org/10.1016/j.jlamp.2020.100633
2. Barrett, C., Fontaine, P., Tinelli, C.: The Satisfiability Modulo Theories Library (SMT-LIB) (2016), www.SMT-LIB.org
3. Bradford, R., Davenport, J.H., England, M., McCallum, S., Wilson, D.: Truth table invariant cylindrical algebraic decomposition. J. Symbol. Comput. **76**, 1–35, 100633 (2016). https://doi.org/10.1016/J.JSC.2015.11.002
4. Bradford, R., Davenport, J.H., England, M., Wilson, D.: Optimising Problem Formulation for Cylindrical Algebraic Decomposition. In: Carette, J., Aspinall, D., Lange, C., Sojka, P., Windsteiger, W. (eds.) CICM 2013. LNCS (LNAI), vol. 7961, pp. 19–34. Springer, Heidelberg (2013). https://doi.org/10.1007/978-3-642-39320-4_2
5. Brown, C.W.: Companion to the tutorial cylindrical algebraic decomposition. In: International Symposium on Symbolic and Algebraic Computation - ISSAC (2004). www.usna.edu/Users/cs/wcbrown/research/ISSAC04/handout.pdf
6. Brown, C.W.: Open non-uniform cylindrical algebraic decompositions. In: Proceedings of the International Symposium on Symbolic and Algebraic Computation, ISSAC, pp. 85–92. Association for Computing Machinery (2015). https://doi.org/10.1145/2755996.2756654
7. Brown, C.W., Davenport, J.H.: The complexity of quantifier elimination and cylindrical algebraic decomposition. In: Proceedings of the International Symposium on Symbolic and Algebraic Computation, ISSAC, pp. 54–60 (2007). https://doi.org/10.1145/1277548.1277557

8. Brown, C.W., Daves, G.C.: Applying machine learning to heuristics for real polynomial constraint solving. In: Bigatti, A.M., Carette, J., Davenport, J.H., Joswig, M., de Wolff, T. (eds.) ICMS 2020. LNCS, vol. 12097, pp. 292–301. Springer, Cham (2020). https://doi.org/10.1007/978-3-030-52200-1_29

9. Chen, C., Moreno Maza, M.: An incremental algorithm for computing cylindrical algebraic decompositions. In: Feng, R., Lee, W., Sato, Y. (eds.) Computer Mathematics, pp. 199–221. Springer, Heidelberg (2014). https://doi.org/10.1007/978-3-662-43799-5_17

10. Chen, C., Moreno Maza, M.: Cylindrical Algebraic decomposition in the RegularChains library. In: Hong, H., Yap, C. (eds.) ICMS 2014. LNCS, vol. 8592, pp. 425–433. Springer, Heidelberg (2014). https://doi.org/10.1007/978-3-662-44199-2_65

11. Chen, C., Moreno Maza, M., Xia, B., Yang, L.: Computing cylindrical algebraic decomposition via triangular decomposition. In: Proceedings of the International Symposium on Symbolic and Algebraic Computation, ISSAC, pp. 95–102 (2009). https://doi.org/10.1145/1576702.1576718

12. Chen, C., Zhu, Z., Chi, H.: Variable ordering selection for cylindrical algebraic decomposition with artificial neural networks. In: Bigatti, A.M., Carette, J., Davenport, J.H., Joswig, M., de Wolff, T. (eds.) ICMS 2020. LNCS, vol. 12097, pp. 281–291. Springer, Cham (2020). https://doi.org/10.1007/978-3-030-52200-1_28

13. Collins, G.E.: Quantifier elimination for real closed fields by cylindrical algebraic decompostion. In: Brakhage, H. (ed.) GI-Fachtagung 1975. LNCS, vol. 33, pp. 134–183. Springer, Heidelberg (1975). https://doi.org/10.1007/3-540-07407-4_17

14. Collins, G.E.: Quantifier Elimination by Cylindrical Algebraic Decomposition - Twenty Years of Progress. In: Caviness, B.F., Johnson, J.R. (eds.) Quantifier Elimination and Cylindrical Algebraic Decomposition. Texts and Monographs in Symbolic Computation, pp. 8–23. Springer, Vienna (1998). https://doi.org/10.1007/978-3-7091-9459-1_2

15. Dolzmann, A., Seidl, A., Sturm, T.: Efficient projection orders for CAD. In: Proceedings of the 2004 International Symposium on Symbolic and Algebraic Computation, ISSAC, pp. 111–118. ACM Press, New York, New York, USA (2004). https://doi.org/10.1145/1005285.1005303

16. England, M., Bradford, R., Chen, C., Davenport, J.H., Maza, M.M., Wilson, D.: Problem formulation for truth-table invariant cylindrical algebraic decomposition by incremental triangular decomposition. In: Watt, S.M., Davenport, J.H., Sexton, A.P., Sojka, P., Urban, J. (eds.) CICM 2014. LNCS (LNAI), vol. 8543, pp. 45–60. Springer, Cham (2014). https://doi.org/10.1007/978-3-319-08434-3_5

17. England, M., Bradford, R., Davenport, J.H.: Improving the use of equational constraints in cylindrical algebraic decomposition. In: Proceedings of the International Symposium on Symbolic and Algebraic Computation, ISSAC, pp. 165–172 (2015). https://doi.org/10.1145/2755996.2756678

18. England, M., Bradford, R., Davenport, J.H.: Cylindrical algebraic decomposition with equational constraints. J. Symbol. Comput. 100, 38–71 (2020). https://doi.org/10.1016/j.jsc.2019.07.019

19. England, M., Florescu, D.: Comparing machine learning models to choose the variable ordering for cylindrical algebraic decomposition. In: Kaliszyk, C., Brady, E., Kohlhase, A., Sacerdoti Coen, C. (eds.) CICM 2019. LNCS (LNAI), vol. 11617, pp. 93–108. Springer, Cham (2019). https://doi.org/10.1007/978-3-030-23250-4_7

20. Fairley, W.B.: The number of real roots of random polynomials of small degree. Indian J. Stat. Ser. B 38(2), 144–152 (1976), www.jstor.org/stable/25052004

21. Florescu, D., England, M.: Algorithmically generating new algebraic features of polynomial systems for machine learning. In: CEUR Workshop Proceedings 2460 (2019). https://doi.org/10.48550/1906.01455

22. Florescu, D., England, M.: Improved cross-validation for classifiers that make algorithmic choices to minimise runtime without compromising output correctness. In: Slamanig, D., Tsigaridas, E., Zafeirakopoulos, Z. (eds.) MACIS 2019. LNCS, vol. 11989, pp. 341–356. Springer, Cham (2020). https://doi.org/10.1007/978-3-030-43120-4_27

23. Huang, Z., England, M., Wilson, D.J., Bridge, J., Davenport, J.H., Paulson, L.C.: Using machine learning to improve cylindrical algebraic decomposition. Math. Comput. Sci. **13**(4), 461–488 (dec 2019). https://doi.org/10.1007/s11786-019-00394-8

24. Kremer, G., Ábrahám, E.: Fully incremental cylindrical algebraic decomposition. J. Symbol. Comput. **100**, 11–37 (2020). https://doi.org/10.1016/j.jsc.2019.07.018

25. Li, H., Xia, B., Zhang, H., Zheng, T.: Choosing the variable ordering for cylindrical algebraic decomposition via exploiting chordal structure. In: Proceedings of the International Symposium on Symbolic and Algebraic Computation, ISSAC, pp. 281–288 (2021). https://doi.org/10.1145/3452143.3465520

26. McCallum, S., Parusiński, A., Paunescu, L.: Validity proof of Lazard's method for CAD construction. J. Symbol. Comput. **92**, 52–69 (2019). https://doi.org/10.1016/j.jsc.2017.12.002

27. Wilson, D., England, M., Bradford, R., Davenport, J.H.: Using the distribution of cells by dimension in a cylindrical algebraic decomposition. In: Proceedings - 16th International Symposium on Symbolic and Numeric Algorithms for Scientific Computing, SYNASC 2014, pp. 53–60 (2015). https://doi.org/10.1109/SYNASC.2014.15

An Implementation of Parallel Number-Theoretic Transform Using Intel AVX-512 Instructions

Daisuke Takahashi[✉][iD]

Center for Computational Sciences, University of Tsukuba,
1-1-1 Tennodai, Tsukuba, Ibaraki 305-8577, Japan
daisuke@cs.tsukuba.ac.jp

Abstract. In this paper, we propose an implementation of the parallel number-theoretic transform (NTT) using Intel Advanced Vector Extensions 512 (AVX-512) instructions. The butterfly operation of the NTT can be performed using modular addition, subtraction, and multiplication. We show that a method known as the six-step fast Fourier transform algorithm can be applied to the NTT. We vectorized NTT kernels using the Intel AVX-512 instructions and parallelized the six-step NTT using OpenMP. We successfully achieved a performance of over 83 giga-operations per second on an Intel Xeon Platinum 8368 (2.4 GHz, 38 cores) for a 2^{20}-point NTT with a modulus of 51 bits.

Keywords: Number-theoretic transform · Modular multiplication · Intel AVX-512 instructions

1 Introduction

The fast Fourier transform (FFT) [5] is an algorithm that is widely used today in scientific and engineering computing. FFTs are often computed using complex or real numbers, but it is known that these transforms can also be computed in a ring and a finite field [14]. Such a transform is called the number-theoretic transform (NTT). The NTT is used for homomorphic encryption, polynomial multiplication, and multiple-precision multiplication.

Efficient arithmetic for NTTs has been proposed [7]. The number theory library (NTL) [15] is a C++ library for performing number-theoretic computations and implements NTT. Although the NTL is thread-safe, the parallel NTT is not supported. Spiral-generated modular FFTs have been proposed [11,12] and experiments were performed using 32-bit integers and 16-bit primes with Intel SSE4 instructions. An implementation of NTT using the Intel AVX-512IFMA (Integer Fused Multiply-Add) instructions has been proposed [2]. This implementation is available as the Intel Homomorphic Encryption (HE) Acceleration Library [3], an open-source C++ library that provides efficient implementations of integer arithmetic on finite fields. Intel HEXL targets the typical data size

F. Boulier et al. (Eds.): CASC 2022, LNCS 13366, pp. 318–332, 2022.
https://doi.org/10.1007/978-3-031-14788-3_18

$n = [2^{10}, 2^{17}]$ of NTTs used in homomorphic encryption [2] and is not parallelized by OpenMP.

In contrast, we consider accelerating NTT for larger data sizes by parallelization, targeting polynomial multiplication and multiple-precision multiplication. In this paper, we vectorize NTT kernels using the Intel AVX-512 instructions and parallelize NTT using OpenMP.

The remainder of this paper is organized as follows. Section 2 describes the number-theoretic transform (NTT). Section 3 presents the vectorization of the NTT kernels. Section 4 presents the proposed implementation of the parallel NTT. Section 5 presents the performance results. Finally, Sect. 6 presents concluding remarks.

2   Number-Theoretic Transform (NTT)

The discrete Fourier transform (DFT) is given by

$$y(k) = \sum_{j=0}^{n-1} x(j)\omega_n^{jk}, \quad 0 \le k \le n-1, \tag{1}$$

where $\omega_n = e^{-2\pi i/n}$ and $i = \sqrt{-1}$.

The DFT can be defined over rings and fields other than the complex field [14]. Equation (1) can be expressed in a field $\mathbb{F}_p = \mathbb{Z}/p\mathbb{Z}$, where p is a prime number:

$$y(k) = \sum_{j=0}^{n-1} x(j)\omega_n^{jk} \bmod p, \quad 0 \le k \le n-1, \tag{2}$$

in which ω_n is the primitive n-th root of unity. For example, if $\omega_n = 157$ for $n = 16$ and $p = 1297 (= 81 \times 16 + 1)$, then $\omega_n^{16} \equiv 1 \bmod 1297$ and $\omega_n^{16/2} = \omega_n^8 \equiv 1296 \not\equiv 1 \bmod 1297$.

The n-point NTT in Eq. (2) is directly computed by $O(n^2)$ arithmetic operations, but by applying an algorithm similar to the FFT, the number of arithmetic operations can be reduced to $O(n \log n)$. Applications of the Stockham algorithm [4,16], known as an out-of-place FFT algorithm, to radix-2, 4, and 8 NTTs are shown in Algorithms 1, 2, and 3, respectively. The multiplication by $\omega_n^{n/4}$ in line 13 of Algorithm 2 can be performed by a trivial multiplication by $-i$ in the radix-4 FFT. As a result, the total number of real arithmetic operations for the n-point FFT is reduced from $5n \log_2 n$ in the radix-2 FFT to $4.25n \log_2 n$ in the radix-4 FFT. However, the number of arithmetic operations cannot be reduced by increasing the radix because there is no such trivial multiplication in the NTT. Intel HEXL includes radix-2 and 4 NTT implementations.

When computing the NTT, modular multiplication takes up most of the computation time. Modular multiplication includes modulo operations, which are slow due to the integer division process. However, Montgomery multiplication [13] and Shoup's modular multiplication [7] are known to avoid this problem.

Algorithm 1. Stockham radix-2 NTT algorithm

Input: $n = 2^q$, $X_0(j) = x(j)$, $0 \leq j \leq n - 1$, and ω_n is the primitive n-th root of unity.

Output: $y(k) = X_q(k) = \sum_{j=0}^{n-1} x(j)\omega_n^{jk} \bmod p$, $0 \leq k \leq n - 1$

1: $l \leftarrow n/2$
2: $m \leftarrow 1$
3: **for** t **from** 1 **to** q **do**
4:   **for** j **from** 0 **to** $l - 1$ **do**
5:     **for** k **from** 0 **to** $m - 1$ **do**
6:       $c_0 \leftarrow X_{t-1}(k + jm)$
7:       $c_1 \leftarrow X_{t-1}(k + jm + lm)$
8:       $X_t(k + 2jm) \leftarrow (c_0 + c_1) \bmod p$
9:       $X_t(k + 2jm + m) \leftarrow \omega_n^{jm}(c_0 - c_1) \bmod p$
10:     **end for**
11:   **end for**
12:   $l \leftarrow l/2$
13:   $m \leftarrow 2m$
14: **end for**

Shoup's modular multiplication [7] is shown in Algorithm 4. In Algorithm 4, if β is a power-of-two integer, the truncated quotient of dividing AB' by β in line 1 can be calculated by right shifting, and the remainder of dividing $(AB - qN)$ by β in line 2 can be calculated by bit masking.

The decimation-in-frequency butterfly operation of the NTT is shown in the following expression:

$$\begin{cases} X = (x + y) \bmod p, \\ Y = \omega(x - y) \bmod p. \end{cases}$$

The value of Y in the butterfly operation can be calculated in Algorithm 4 using $A = x - y$, $B = \omega$, and $N = p$. Here, the value of $B' = \lfloor \omega\beta/p \rfloor$ can be calculated in advance.

When convolution is performed for polynomials using NTT, the modulus also needs to increase as the degree increases. If the modulus does not fit into the size of the machine word (e.g., 32 or 64 bits), it is known that the convolution can be performed by computing NTTs on multiple moduli and then reconstructing the moduli using the Chinese remainder theorem.

3 Vectorization of NTT Kernels

Intel Advanced Vector Extensions 512 (AVX-512) [8] is a 512-bit vector instruction set that consists of multiple extensions that can be implemented independently. All Intel AVX-512 implementations require only the core extension Intel AVX-512F (Foundation). The most direct way to use Intel AVX-512 instructions is to insert assembly-language instructions inline into the source code. However, this can be time-consuming and tedious. Intel thus provides API extension

---

Algorithm 2. Stockham radix-4 NTT algorithm

Input: $n = 4^q$, $X_0(j) = x(j)$, $0 \le j \le n - 1$, and ω_n is the primitive n-th root of unity.

Output: $y(k) = X_q(k) = \sum_{j=0}^{n-1} x(j)\omega_n^{jk} \bmod p$, $0 \le k \le n - 1$

1: $l \leftarrow n/4$
2: $m \leftarrow 1$
3: **for** t **from** 1 **to** q **do**
4:     **for** j **from** 0 **to** $l - 1$ **do**
5:         **for** k **from** 0 **to** $m - 1$ **do**
6:             $c_0 \leftarrow X_{t-1}(k + jm)$
7:             $c_1 \leftarrow X_{t-1}(k + jm + lm)$
8:             $c_2 \leftarrow X_{t-1}(k + jm + 2lm)$
9:             $c_3 \leftarrow X_{t-1}(k + jm + 3lm)$
10:           $d_0 \leftarrow (c_0 + c_2) \bmod p$
11:           $d_1 \leftarrow (c_0 - c_2) \bmod p$
12:           $d_2 \leftarrow (c_1 + c_3) \bmod p$
13:           $d_3 \leftarrow \omega_n^{n/4}(c_1 - c_3) \bmod p$
14:           $X_t(k + 4jm) \leftarrow (d_0 + d_2) \bmod p$
15:           $X_t(k + 4jm + m) \leftarrow \omega_n^{jm}(d_1 + d_3) \bmod p$
16:           $X_t(k + 4jm + 2m) \leftarrow \omega_n^{2jm}(d_0 - d_2) \bmod p$
17:           $X_t(k + 4jm + 3m) \leftarrow \omega_n^{3jm}(d_1 - d_3) \bmod p$
18:         **end for**
19:     **end for**
20:     $l \leftarrow l/4$
21:     $m \leftarrow 4m$
22: **end for**

---

sets, referred to as intrinsics [9], to facilitate implementation. The GCC [6], Clang [19], and Intel C compilers [9] support automatic vectorization using Intel AVX-512 instructions.

The NTT kernels include modular addition, subtraction, and multiplication. The modular addition $c = (a + b) \bmod N$ for $0 \le a, b < N$ can be replaced by the addition $c = a + b$ and the conditional subtraction $c - N$ when $c \ge N$. Such conditional subtraction involves a branch. However, the branch can be avoided by replacing it with the minimum operation $\min(c, c - N)$ for unsigned integer values c and N with wrap-around two's complement arithmetic [18]. The Intel AVX-512F instruction set supports the vpminuq instruction for the 64-bit unsigned integer minimum operation. Here, it is sufficient that each of a and b be less than 2^{63} in order for the calculation of $c = a + b$ not to overflow with the 64-bit unsigned integer addition. Similarly, the modular subtraction $c = (a - b) \bmod N$ for $0 \le a, b < N$ can be replaced by the subtraction $c = a - b$ and the minimum operation $c = \min(c, c + N)$ for unsigned integer values a, b, c, and N with wrap-around two's complement arithmetic.

Figures 1 and 2 show the modular additions and subtractions of packed 63-bit integers using Intel AVX-512 intrinsics. The intrinsics support the __m512i data type in Figs. 1 and 2. The __m512i data type can hold 64 8-bit integer

Algorithm 3. Stockham radix-8 NTT algorithm

Input: $n = 8^q$, $X_0(j) = x(j)$, $0 \leq j \leq n - 1$, and ω_n is the primitive n-th root of unity.

Output: $y(k) = X_q(k) = \sum_{j=0}^{n-1} x(j)\omega_n^{jk} \bmod p$, $0 \leq k \leq n - 1$

1: $l \leftarrow n/8$
2: $m \leftarrow 1$
3: **for** t from 1 to q **do**
4:     **for** j from 0 to $l - 1$ **do**
5:         **for** k from 0 to $m - 1$ **do**
6:             $c_0 \leftarrow X_{t-1}(k + jm)$
7:             $c_1 \leftarrow X_{t-1}(k + jm + lm)$
8:             $c_2 \leftarrow X_{t-1}(k + jm + 2lm)$
9:             $c_3 \leftarrow X_{t-1}(k + jm + 3lm)$
10:           $c_4 \leftarrow X_{t-1}(k + jm + 4lm)$
11:           $c_5 \leftarrow X_{t-1}(k + jm + 5lm)$
12:           $c_6 \leftarrow X_{t-1}(k + jm + 6lm)$
13:           $c_7 \leftarrow X_{t-1}(k + jm + 7lm)$
14:           $d_0 \leftarrow (c_0 + c_4) \bmod p$
15:           $d_1 \leftarrow (c_0 - c_4) \bmod p$
16:           $d_2 \leftarrow (c_2 + c_6) \bmod p$
17:           $d_3 \leftarrow \omega_n^{n/4}(c_2 - c_6) \bmod p$
18:           $d_4 \leftarrow (c_1 + c_5) \bmod p$
19:           $d_5 \leftarrow (c_1 - c_5) \bmod p$
20:           $d_6 \leftarrow (c_3 + c_7) \bmod p$
21:           $d_7 \leftarrow \omega_n^{n/4}(c_3 - c_7) \bmod p$
22:           $e_0 \leftarrow (d_0 + d_2) \bmod p$
23:           $e_1 \leftarrow (d_0 - d_2) \bmod p$
24:           $e_2 \leftarrow (d_4 + d_6) \bmod p$
25:           $e_3 \leftarrow \omega_n^{n/4}(d_4 - d_6) \bmod p$
26:           $e_4 \leftarrow (d_1 + d_3) \bmod p$
27:           $e_5 \leftarrow (d_1 - d_3) \bmod p$
28:           $e_6 \leftarrow \omega_n^{n/8}(d_5 + d_7) \bmod p$
29:           $e_7 \leftarrow \omega_n^{3n/8}(d_5 - d_7) \bmod p$
30:           $X_t(k + 8jm) \leftarrow (e_0 + e_2) \bmod p$
31:           $X_t(k + 8jm + m) \leftarrow \omega_n^{jm}(e_4 + e_6) \bmod p$
32:           $X_t(k + 8jm + 2m) \leftarrow \omega_n^{2jm}(e_1 + e_3) \bmod p$
33:           $X_t(k + 8jm + 3m) \leftarrow \omega_n^{3jm}(e_5 + e_7) \bmod p$
34:           $X_t(k + 8jm + 4m) \leftarrow \omega_n^{4jm}(e_0 - e_2) \bmod p$
35:           $X_t(k + 8jm + 5m) \leftarrow \omega_n^{5jm}(e_4 - e_6) \bmod p$
36:           $X_t(k + 8jm + 6m) \leftarrow \omega_n^{6jm}(e_1 - e_3) \bmod p$
37:           $X_t(k + 8jm + 7m) \leftarrow \omega_n^{7jm}(e_5 - e_7) \bmod p$
38:         **end for**
39:        **end for**
40:        $l \leftarrow l/8$
41:        $m \leftarrow 8m$
42: **end for**

Algorithm 4. Shoup's modular multiplication algorithm [7]

Input: A, B, N such that $0 \leq A, B < N$, $N < \beta/2$
        precomputed $B' = \lfloor B\beta/N \rfloor$
Output: $C = AB \bmod N$
1: $q \leftarrow \lfloor AB'/\beta \rfloor$
2: $C \leftarrow (AB - qN) \bmod \beta$
3: if $C \geq N$ then
4:     $C \leftarrow C - N$
5: return C.

```
__m512i _mm512_addmod_epu64(__m512i a, __m512i b, __m512i N)
/*  Compute (a + b) mod N. Requires 0 <= a, b < N < 2^63. */
{
  __m512i c;

  c = _mm512_add_epi64(a, b);
  c = _mm512_min_epu64(c, _mm512_sub_epi64(c, N));

  return c;
}
```

Fig. 1. Modular additions of packed 63-bit integers using Intel AVX-512 intrinsics

values, 32 16-bit integer values, 16 32-bit integer values, or 8 64-bit integer values. In addition, the intrinsics _mm512_add_epi64() and _mm512_sub_epi64() correspond to the vpaddq and vpsubq instructions, respectively.

We consider performing modular multiplication $c = ab \bmod N$ using Shoup's modular multiplication. If we set $\beta = 2^{64}$ in Algorithm 4, then the upper 64-bit half of the 64-bit $\times$ 64-bit $\rightarrow$ 128-bit unsigned integer multiplication is required. The Intel AVX-512DQ (Doubleword and Quadword) instruction set [8] supports the vpmullq instruction for the lower 64-bit half of the 64-bit $\times$ 64-bit $\rightarrow$ 128-bit integer multiplication, but does not support the upper 64-bit half of the 64-bit $\times$ 64-bit $\rightarrow$ 128-bit unsigned integer multiplication.

The Intel AVX-512F instruction set supports the vpmuludq instruction, which performs 32-bit $\times$ 32-bit $\rightarrow$ 64-bit unsigned integer multiplication. The upper 64-bit half of the 64-bit $\times$ 64-bit $\rightarrow$ 128-bit unsigned integer multiplication can be implemented by dividing the multiplicand and multiplier of a 64-bit unsigned integer into the upper and lower 32-bit unsigned integers, respectively, and using the vpmuludq instruction for 32-bit $\times$ 32-bit $\rightarrow$ 64-

```
__m512i _mm512_submod_epu64(__m512i a, __m512i b, __m512i N)
/*  Compute (a - b) mod N. Requires 0 <= a, b < N < 2^63. */
{
  __m512i c;

  c = _mm512_sub_epi64(a, b);
  c = _mm512_min_epu64(c, _mm512_add_epi64(c, N));

  return c;
}
```

Fig. 2. Modular subtractions of packed 63-bit integers using Intel AVX-512 intrinsics

```
__m512i _mm512_mulhi_epu64(__m512i a, __m512i b)
/*  Compute floor((a * b) / 2^64). Requires 0 <= a, b < 2^64. */
{
    __m512i a0, a1, b0, b1, c, t0, t1, t2, t3;

  a0 = _mm512_and_epi64(a, _mm512_set1_epi64(0xFFFFFFFF));
  a1 = _mm512_srli_epi64(a, 32);
  b0 = _mm512_and_epi64(b, _mm512_set1_epi64(0xFFFFFFFF));
  b1 = _mm512_srli_epi64(b, 32);
  t0 = _mm512_mul_epu32(a0, b0);
  t1 = _mm512_mul_epu32(a0, b1);
  t2 = _mm512_mul_epu32(a1, b0);
  t3 = _mm512_mul_epu32(a1, b1);
  t1 = _mm512_add_epi64(t1, _mm512_srli_epi64(t0, 32));
  t2 = _mm512_add_epi64(t2, _mm512_and_epi64(t1,
                              _mm512_set1_epi64(0xFFFFFFFF)));
  c = _mm512_add_epi64(_mm512_srli_epi64(t1, 32),
                       _mm512_add_epi64(_mm512_srli_epi64(t2, 32), t3));

  return c;
}
```

Fig. 3. The upper 64-bit half of the 64-bit $\times$ 64-bit $\to$ 128-bit unsigned integer multiplications of packed 64-bit integers using Intel AVX-512 intrinsics

bit unsigned integer multiplication. Figure 3 shows the upper 64-bit half of the 64-bit $\times$ 64-bit $\to$ 128-bit unsigned integer multiplications of packed 64-bit integers using Intel AVX-512 intrinsics. The intrinsics _mm512_and_epi64(), _mm512_set1_epi64(), _mm512_srli_epi64(), and _mm512_mul_epu32() correspond to the vpandq, vpbroadcastq, vpsrlq, and vpmuludq instructions, respectively.

Figure 4 shows Shoup's modular multiplications of packed 63-bit integers using Intel AVX-512 intrinsics, which correspond to $\beta = 2^{64}$ in Algorithm 4. In this program, the function _mm512_mulhi_epu64() shown in Fig. 3 is used. The intrinsic _mm512_mullo_epi64() corresponds to the vpmullq instruction. The conditional subtraction on lines 3 and 4 of Algorithm 4 is also performed

```
__m512i _mm512_mulmod_epu64(__m512i a, __m512i b, __m512i bb, __m512i N)
/*  Compute (a * b) mod N. Precomputed bb = floor((b * 2^64) / N).
    Requires 0 <= a, b < N < 2^63. */
{
    __m512i c, q;

  q = _mm512_mulhi_epu64(a, bb);
  c = _mm512_sub_epi64(_mm512_mullo_epi64(a, b),
                       _mm512_mullo_epi64(q, N));
  c = _mm512_min_epu64(c, _mm512_sub_epi64(c, N));

  return c;
}
```

Fig. 4. Shoup's modular multiplications of packed 63-bit integers using Intel AVX-512 intrinsics

```
__m512i _mm512_mulmod_epu64(__m512i a, __m512i b, __m512i bb, __m512i N)
/* Compute (a * b) mod N. Precomputed bb = floor((b * 2^52) / N).
   Requires 0 <= a, b < N < 2^51. */
{
  __m512i c, q;

  q = _mm512_madd52hi_epu64(_mm512_set1_epi64(0), a, bb);
  c = _mm512_sub_epi64(
      _mm512_madd52lo_epu64(_mm512_set1_epi64(0), a, b),
      _mm512_madd52lo_epu64(_mm512_set1_epi64(0), q, N));
  c = _mm512_and_epi64(c, _mm512_set1_epi64(0x000FFFFFFFFFFFFF));
  c = _mm512_min_epu64(c, _mm512_sub_epi64(c, N));

  return c;
}
```

Fig. 5. Shoup's modular multiplications of packed 51-bit integers using Intel AVX-512 intrinsics

Table 1. Inner-loop operations for radix-2, 4, and 8 NTT kernels

	Radix-2	Radix-4	Radix-8
Loads	2	4	8
Stores	2	4	8
Modular multiplications	1	4	12
Modular additions/subtractions	2	8	24
Total arithmetic operations	3	12	36
Byte/Operation ratio	10.667	5.333	3.556

using the vpsubq and vpminuq instructions in the same way as in the modular addition.

Intel AVX-512IFMA instructions [8] are supported by the Cannon Lake, Ice Lake, and Tiger Lake microarchitectures. The Intel AVX-512IFMA instruction set supports the vpmadd52luq and vpmadd52huq instructions, which multiply 52-bit unsigned integers and produce the low and high halves, respectively, of a 104-bit intermediate result. These halves are added to 64-bit accumulators. Since such operations are not supported in the C language, it is necessary to use Intel AVX-512 intrinsics or insert assembly-language instructions inline into the source code in order to use the Intel AVX-512IFMA instructions.

Figure 5 shows Shoup's modular multiplications of packed 51-bit integers using Intel AVX-512 intrinsics, which correspond to $\beta = 2^{52}$ in Algorithm 4. The intrinsics _mm512_madd52lo_epu64() and _mm512_madd52hi_epu64() correspond to the vpmadd52luq and vpmadd52huq instructions, respectively.

The Stockham radix-2, 4, and 8 NTTs are vectorized using the functions in Figs. 1, 2, 3, 4, and 5. Table 1 shows the inner-loop operations for radix-2, 4, and 8 NTT kernels. As mentioned in Sect. 2, the radix-4 or 8 NTT does not reduce the number of arithmetic operations compared to the radix-2 NTT. However, in view of the Byte/Operation ratio, the radix-8 NTT is preferable to the radix-2 and 4 NTTs. Although higher radix NTTs require more registers to hold intermediate results, processors that support the Intel AVX-512 instructions have 32 ZMM 512-bit registers.

A power-of-two point NTT (except for the 2-point NTT) can be performed by a combination of radix-8 and radix-4 steps containing at most two radix-4 steps. In other words, the power-of-two NTTs can be performed as a length $n = 2^p = 4^q 8^r$ ($p \geq 2$, $0 \leq q \leq 2$, $r \geq 0$).

4   Parallel Implementation of Number-Theoretic Transform

In Eq. (2), if n has factors n_1 and n_2 ($n = n_1 \times n_2$), then the indices j and k can be expressed as:

$$j = j_1 + j_2 n_1, \quad k = k_2 + k_1 n_2. \tag{3}$$

We can define x and y in Eq. (2) as two-dimensional arrays (in column-major order):

$$x(j) = x(j_1, j_2), \qquad 0 \leq j_1 \leq n_1 - 1, \qquad 0 \leq j_2 \leq n_2 - 1, \tag{4}$$
$$y(k) = y(k_2, k_1), \qquad 0 \leq k_1 \leq n_1 - 1, \qquad 0 \leq k_2 \leq n_2 - 1. \tag{5}$$

Substituting the indices j and k in Eq. (2) with the indices in Eq. (3), and using the relation of $n = n_1 \times n_2$, we can derive the following equation:

$$y(k_2, k_1) = \sum_{j_1=0}^{n_1-1} \sum_{j_2=0}^{n_2-1} x(j_1, j_2) \omega_n^{j_2 k_2 n_1} \omega_n^{j_1 k_2} \omega_n^{j_1 k_1 n_2} \bmod p. \tag{6}$$

In the same way as the six-step FFT algorithm [1,20], the following six-step NTT algorithm is derived from Eq. (6):

Step 1:   Transposition
$$x_1(j_2, j_1) = x(j_1, j_2).$$

Step 2: n_1 individual n_2-point multicolumn NTTs
$$x_2(k_2, j_1) = \sum_{j_2=0}^{n_2-1} x_1(j_2, j_1) \omega_n^{j_2 k_2 n_1} \bmod p.$$

Step 3:   Twiddle factor multiplication
$$x_3(k_2, j_1) = x_2(k_2, j_1) \omega_n^{j_1 k_2} \bmod p.$$

Step 4:   Transposition
$$x_4(j_1, k_2) = x_3(k_2, j_1).$$

Step 5: n_2 individual n_1-point multicolumn NTTs
$$x_5(k_1, k_2) = \sum_{j_1=0}^{n_1-1} x_4(j_1, k_2) \omega_n^{j_1 k_1 n_2} \bmod p.$$

Step 6:   Transposition
$$y(k_2, k_1) = x_5(k_1, k_2).$$

In the six-step NTT algorithm, two multicolumn NTTs are performed in steps 2 and 5. The locality of the memory reference in the multicolumn NTT is high. On the other hand, the three transpose steps (steps 1, 4, and 6) are typically the chief bottlenecks in cache-based processors. We can use cache blocking to reduce the number of cache misses in matrix transposition. An example of matrix transposition with cache blocking is shown in Fig. 6. Parameter NBLK is the blocking parameter. In Fig. 6, the outermost loop length may not have sufficient parallelism for manycore processors. A loop collapsing makes the length of a loop long by collapsing nested loops into a single-nested loop. By using the OpenMP collapse clause, the parallelism of the outermost loop can be expanded [17].

We parallelized the six-step NTT using OpenMP. Figure 7 shows a parallel implementation of the six-step NTT. In this program, transpose() is a function to transpose a matrix, mulmod() is a function to perform a modular multiplication, ntt2() is the Stockham NTT, and a variable omega is the primitive n-th root of unity.

5   Performance Results

For performance evaluation, we compared the performance of the following six implementations:

- Proposed implementation of the Stockham NTT (AVX-512DQ) with a modulus of 63 bits
- Proposed implementation of the six-step NTT (AVX-512DQ) with a modulus of 63 bits
- Proposed implementation of the Stockham NTT (AVX-512IFMA) with a modulus of 51 bits
- Proposed implementation of the six-step NTT (AVX-512IFMA) with a modulus of 51 bits

Table 2. Specifications of the platform

Platform	Intel Xeon Platinum processor
Number of cores	38
Number of threads	76
CPU type	Intel Xeon Platinum 8368 Ice Lake 2.4 GHz
L1 cache (per core)	I-cache: 32 KB D-cache: 48 KB
L2 cache (per core)	1.25 MB
L3 cache	57 MB
Main memory	DDR4-3200 256 GB
Theoretical peak performance	2.918 TFlops
OS	Linux 4.18.0-305.25.1.el8_4.x86_64

```
void transpose(uint64_t *a, uint64_t *b, uint64_t n1, uint64_t n2)
{
  uint64_t i, ii, j, jj;

#pragma omp parallel for collapse(2) private(i, j, jj)
  for (ii = 0; ii < n1; ii += NBLK)
    for (jj = 0; jj < n2; jj += NBLK)
      for (i = ii; i < min(ii + NBLK, n1); i++)
        for (j = jj; j < min(jj + NBLK, n2); j++)
          b[j + i * n2] = a[i + j * n1];
}
```

Fig. 6. Example of matrix transposition with cache blocking

```
uint64_t a[n1 * n2], b[n1 * n2], w[n1 * n2], ww[n1 * n2];
uint64_t i, j, omega, p;

    ...
/* Step 1: transpose n1*n2 to n2*n1 */
  transpose(a, b, n1, n2);

/* Step 2: n1 individual n2-point multicolumn NTTs */
#pragma omp parallel for
  for (j = 0; j < n1; j++)
    ntt2(&b[j * n2], n2, omega, p);

/* Step 3: twiddle factor multiplication modulo p */
#pragma omp parallel for
  for (i = 0; i < n1 * n2; i++)
    a[i] = mulmod(b[i], w[i], ww[i], p);

/* Step 4: transpose n2*n1 to n1*n2 */
  transpose(a, b, n2, n1);

/* Step 5: n2 individual n1-point multicolumn NTTs */
#pragma omp parallel for
  for (j = 0; j < n2; j++)
    ntt2(&b[j * n1], n1, omega, p);

/* Step 6: transpose n1*n2 to n2*n1 */
  transpose(b, a, n1, n2);
```

Fig. 7. Parallel implementation of the six-step NTT

– Intel HEXL 1.2.4 (AVX-512DQ) with a modulus of 62 bits
– Intel HEXL 1.2.4 (AVX-512IFMA) with a modulus of 50 bits

Intel HEXL uses a modified Shoup butterfly [7] that requires $p < \beta/4$ to reduce the number of conditional subtractions [2]. Therefore, the modulus sizes of Intel HEXL (AVX-512DQ) with $\beta = 2^{64}$ and Intel HEXL (AVX-512IFMA) with $\beta = 2^{52}$ are 62 bits and 50 bits, respectively.

The specifications of the platform are shown in Table 2. Note that Hyper-Threading [10] was enabled on the platform. The Intel C compiler (version 19.1.3.304) was used for the proposed implementations. The compiler options were icc -O3 -xICELAKE-SERVER -fno-alias -qopenmp. The compiler option -O3 enables optimizations for speed and more aggressive loop transformations. The compiler option -xICELAKE-SERVER specifies the generation of instructions for the Ice Lake microarchitecture. The compiler option -fno-alias specifies that aliasing is not assumed in a program. The compiler option -qopenmp specifies the enabling of the compiler to generate multi-threaded code based on the OpenMP directives. Intel HEXL could not be built successfully with the Intel C/C++ compiler, so the GNU C/C++ compiler (version 8.3.1) was used for Intel HEXL. The compiler option was gcc -O3.

Since the proposed implementation of the Stockham NTT is not parallelized, it was executed in a single thread. The proposed implementation of the six-step NTT was run with 1 to 76 threads. In the proposed implementations of the Stockham NTT and six-step NTT, the number of repetitions was doubled until the elapsed time of the forward NTT was greater than 1 second, and the average elapsed time was measured. The table for twiddle factors was prepared in advance. Since Intel HEXL does not support parallel execution, it was executed in a single thread. The performance of Intel HEXL was measured using the benchmark program included in the Intel HEXL source code.

On the Intel Xeon Platinum 8368, the environment variable KMP_AFFINITY= granularity=fine,compact was specified. The giga-operations per second (Gops) values are each based on $(3/2)n \log_2 n$ for a transform of size $n = 2^m$. This Gops value is calculated with modular addition, subtraction, and multiplication as one operation each, but several instructions are required to actually perform modular addition, subtraction, and multiplication.

Figure 8 shows the performance of NTTs using Intel AVX-512DQ instruction. As shown in Fig. 8, the proposed implementations of the Stockham NTT and six-step NTT (AVX-512DQ) are slower than Intel HEXL (AVX-512DQ) in a single-thread execution. One possible reason for this is that the modulus size of the proposed implementations of the Stockham NTT and six-step NTT (AVX-512DQ) is 63 bits, while the modulus size of Intel HEXL (AVX-512DQ) is 62 bits, reducing the number of instructions. While the six-step NTT is suitable for parallelization, it requires three matrix transpositions, and the overhead of these matrix transpositions is the reason why the proposed implementation of the six-step NTT is slower than the proposed implementation of the Stockham NTT in a single-thread execution. The Intel Xeon Platinum 8368 processor used in this performance evaluation has 57 MB of L3 cache, so up to 2^{20}-point NTT fits into the L3 cache. Although the six-step NTT and matrix transposition with cache blocking are effective when the data do not fit into the cache, Intel HEXL was only able to execute up to 2^{22}-point NTT, which may not have demonstrated the superiority of the proposed implementation of the six-step NTT. The proposed

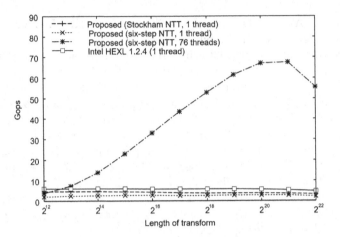

Fig. 8. Performance of NTTs using Intel AVX-512DQ instruction (Intel Xeon Platinum 8368, 38 cores)

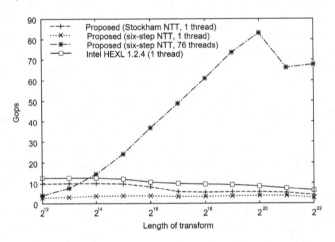

Fig. 9. Performance of NTTs using Intel AVX-512IFMA instruction (Intel Xeon Platinum 8368, 38 cores)

implementation of the six-step NTT (AVX-512DQ) is faster than Intel HEXL (AVX-512DQ) for $n \geq 2^{13}$ on 76 threads.

Figure 9 shows the performance of NTTs using Intel AVX-512IFMA instructions. The proposed implementations of the Stockham NTT and six-step NTT (AVX-512IFMA) are slower than Intel HEXL (AVX-512IFMA) in a single-thread execution, as shown in Fig. 9. One possible reason for this is that the modulus size of the proposed implementations of the Stockham NTT and six-step NTT (AVX-512IFMA) is 51 bits, while the modulus size of Intel HEXL (AVX-512IFMA) is 50 bits, reducing the number of instructions. The proposed implementation of the six-step NTT (AVX-512IFMA) is faster than Intel HEXL (AVX-512IFMA) for $n \geq 2^{14}$ on 76 threads.

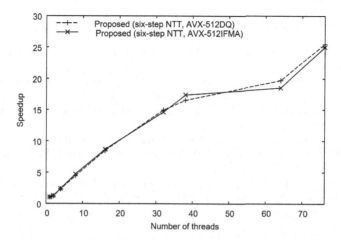

Fig. 10. Speedup for 2^{22}-point NTTs (Intel Xeon Platinum 8368, 38 cores)

Comparing Figs. 8 and 9, the proposed implementations of the Stockham NTT and six-step NTT (AVX-512IFMA) are faster than the proposed implementations of the Stockham NTT and six-step NTT (AVX-512DQ). The reason for this is that the proposed implementations of the Stockham NTT and six-step NTT (AVX-512IFMA) require fewer instructions to perform Shoup's modular multiplication using the Intel AVX-512DQ instruction. However, the modulus size is 63 bits for the proposed implementations of the Stockham NTT and six-step NTT (AVX-512DQ), while the modulus size is reduced to 51 bits for the proposed implementations of the Stockham NTT and six-step NTT (AVX-512IFMA).

Figure 10 shows the speedup for 2^{22}-point NTTs on the Intel Xeon Platinum 8368 when 1 to 76 threads are used. The results indicate that Hyper-Threading is effective for the proposed implementations of the six-step NTT (AVX-512DQ and AVX-512IFMA).

6   Conclusion

In this paper, we proposed the implementation of the parallel NTT using Intel AVX-512 instructions. The butterfly operation of the NTT can be performed using modular addition, subtraction, and multiplication. We showed that a method known as the six-step FFT algorithm could be applied to the NTT. We vectorized NTT kernels using the Intel AVX-512 instructions and parallelized the six-step NTT using OpenMP. We succeeded in obtaining a performance of over 83 Gops on an Intel Xeon Platinum 8368 (2.4 GHz, 38 cores) for a 2^{20}-point NTT with a modulus of 51 bits. These performance results demonstrate that the implemented parallel NTT uses cache memory effectively and exploits the Intel AVX-512 instructions.

Acknowledgments. This work was supported by JSPS KAKENHI Grant Number JP19K11989.

References

1. Bailey, D.H.: FFTs in external or hierarchical memory. J. Supercomput. **4**, 23–35 (1990)
2. Boemer, F., Kim, S., Seifu, G., de Souza, F.D.M., Gopal, V.: Intel HEXL: accelerating homomorphic encryption with Intel AVX512-IFMA52. In: Proceedings of 9th Workshop on Encrypted Computing & Applied Homomorphic Cryptography (WAHC 2021), pp. 57–62 (2021)
3. Boemer, F., et al.: Intel HEXL. https://github.com/intel/hexl
4. Cochran, W.T., et al.: What is the fast Fourier transform? IEEE Trans. Audio Electroacoust. **15**, 45–55 (1967)
5. Cooley, J.W., Tukey, J.W.: An algorithm for the machine calculation of complex Fourier series. Math. Comput. **19**, 297–301 (1965)
6. Free Software Foundation Inc: GCC, the GNU Compiler Collection. https://gcc.gnu.org/
7. Harvey, D.: Faster arithmetic for number-theoretic transforms. J. Symb. Comput. **60**, 113–119 (2014)
8. Intel Corporation: Intel 64 and IA-32 architectures software developer's manual, volume 1: Basic architecture. https://software.intel.com/content/dam/develop/public/us/en/documents/253665-sdm-vol-1.pdf (2020)
9. Intel Corporation: Intel C++ compiler 19.1 developer guide and reference (2020). https://software.intel.com/content/dam/develop/external/us/en/documents/19-1-cpp-compiler-devguide.pdf
10. Marr, D.T., et al.: Hyper-threading technology architecture and microarchitecture. Intel. Technol. J. **6**, 1–11 (2002)
11. Meng, L., Johnson, J.: Automatic parallel library generation for general-size modular FFT algorithms. In: Gerdt, V.P., Koepf, W., Mayr, E.W., Vorozhtsov, E.V. (eds.) CASC 2013. LNCS, vol. 8136, pp. 243–256. Springer, Cham (2013). https://doi.org/10.1007/978-3-319-02297-0_21
12. Meng, L., Johnson, J.R., Franchetti, F., Voronenko, Y., Maza, M.M., Xie, Y.: Spiral-generated modular FFT algorithms. In: Proceedings of 4th International Workshop on Parallel and Symbolic Computation (PASCO 2010), pp. 169–170 (2010)
13. Montgomery, P.L.: Modular multiplication without trial division. Math. Comput. **44**, 519–521 (1985)
14. Pollard, J.M.: The fast Fourier transform in a finite field. Math. Comput. **25**, 365–374 (1971)
15. Shoup, V.: NTL: a library for doing number theory. https://libntl.org
16. Swarztrauber, P.N.: FFT algorithms for vector computers. Parallel Comput. **1**, 45–63 (1984)
17. Takahashi, D.: An implementation of parallel 1-D real FFT on Intel Xeon phi processors. In: Gervasi, O., et al. (eds.) ICCSA 2017. LNCS, vol. 10404, pp. 401–410. Springer, Cham (2017). https://doi.org/10.1007/978-3-319-62392-4_29
18. Takahashi, D.: Computation of the 100 quadrillionth hexadecimal digit of π on a cluster of Intel Xeon phi processors. Parallel Comput. **75**, 1–10 (2018)
19. The Clang Team: clang: a C language family frontend for LLVM. https://clang.llvm.org/
20. Van Loan, C.: Computational Frameworks for the Fast Fourier Transform. SIAM Press, Philadelphia, PA (1992)

Locating the Closest Singularity in a Polynomial Homotopy

Jan Verschelde$^{(\boxtimes)}$ and Kylash Viswanathan

Department of Mathematics, Statistics, and Computer Science, University of Illinois at Chicago, 851 S. Morgan St. (m/c 249), Chicago, IL 60607-7045, USA
{janv,kviswa5}@uic.edu

Abstract. A polynomial homotopy is a family of polynomial systems, where the systems in the family depend on one parameter. If for one value of the parameter we know a regular solution, then what is the nearest value of the parameter for which the solution in the polynomial homotopy is singular? For this problem we apply the ratio theorem of Fabry. Richardson extrapolation is effective to accelerate the convergence of the ratios of the coefficients of the series expansions of the solution paths defined by the homotopy. For numerical stability, we recondition the homotopy. To compute the coefficients of the series we propose the quaternion Fourier transform. We locate the closest singularity computing at a regular solution, avoiding numerical difficulties near a singularity.

Keywords: Analytic continuation · Asymptotic expansion · Richardson extrapolation · Fabry · Fourier · Polynomial homotopy · Quaternion Taylor series · Singularity

1 Introduction

Polynomial homotopies define the deformation of polynomial systems, from systems with known solutions into systems that must be solved. We call a solution *regular* if the matrix of all partial derivatives evaluated at the solution has full rank, otherwise the solution is *singular*. We aim to locate the nearest singularity starting at a regular solution. Applying the ratio theorem of Fabry, we can detect singular points based on the coefficients of the Taylor series.

Theorem 1 (the ratio theorem of Fabry [11]). *If for the series* $x(t) = c_0 + c_1 t + c_2 t^2 + \cdots + c_n t^n + c_{n+1} t^{n+1} + \cdots$, *we have* $\lim_{n \to \infty} c_n/c_{n+1} = z$, *then*

- *z is a singular point of the series, and*
- *it lies on the boundary of the circle of convergence of the series.*

Then the radius of this circle is less than $|z|$.

Supported by the National Science Foundation under grant DMS 1854513.

While the proof of the theorem would take us deep into complex analysis [9, Chapter XI], one can immediately verify that the ratio c_n/c_{n+1} is the pole of Padé approximants ([1,38]) of degrees $[n/1]$, where n is the degree of the numerator, with linear denominator.

The ratio theorem of Fabry provides a radar to detect singularities in an adaptive step size control for continuation methods, as introduced in [39] (with a parallel implementation in [40]) and reproduced by [41]. Earlier applications of Padé approximants in deformation methods appeared in [20], in a symbolic context, and in [35] in a numerical setting. Empirically, in the plain application of this ratio theorem, already relatively few terms in the series appear to be sufficient to take nearby singularities into account.

The problem considered in this paper can be stated as follows. How many terms in the Taylor series do we need to locate the closest singularity with eight decimal places of accuracy? Answering this question exactly is not possible because of constants which differ for each series, but we can provide information about the order of the number of terms, e.g.: tens or hundreds.

We show that Richardson extrapolation (see [3] for a general formalism) effectively solves our problem. On monomial homotopies (defined in the next section), we can separate our problem from the required accuracy of the coefficients of the Taylor series. On examples, at 64 terms of the series, we obtain eight decimal places of accuracy in the location of the radius of convergence. In the third section, the justification for this successful application of Richardson extrapolation is proven. This is the first contribution of this paper.

The second contribution of this paper is the introduction of the quaternion Fourier transform [10,34] to compute the coefficients of the series. If we want to locate a singularity to full double precision, then, on examples, it appears that 512 terms in the series are needed. The Fast Fourier Transform scales well.

In the fifth section, we consider the application of Richardson extrapolation in an end game, when the path tracker approaches an isolated singular solution at the end of the path. Power series methods for singular solutions in [28] introduced the concept of the *end game operation range*. In this range, the continuation parameter has values for which the Puiseux series expansions are valid and where the numerical condition numbers still allow to compute sufficiently accurate approximations of the points on the path. In fixed precision, this range may be empty. Using multiple double precision for ill-conditioned problems is wasteful due to the slow convergence of Newton's method. For homotopies with a random complex gamma constant, we introduce the notion of the last pole. With this last pole, we *recondition* the homotopy with a shift and stretch transformation.

The new methods are illustrated in section six. Deflation restores the quadratic convergence of Newton's method for an isolated singular solution, of multiplicity μ. While [25] proves that μ is the upper bound on the number of deflation steps, the numerical decision to apply deflation is left to a singular value decomposition of the Jacobian matrix, which may not always be reliable enough. Although deflation has been addressed by many (e.g. [4,5,7,8,15–17,26,27,30]), the question on when to deflate is an open problem.

2   Monomial Homotopies

The examples of the homotopies in this section have only one singularity.

A *monomial homotopy* is defined by an exponent matrix $A \in \mathbb{Z}^{n \times n}$ and an n-dimensional coefficient vector $\mathbf{c}(t)$ of invertible power series:

$$\mathbf{h}(\mathbf{x}, t) = \mathbf{x}^A - \mathbf{c}(t) = \mathbf{0}, \tag{1}$$

with $\mathbf{x} = (x_1, x_2, \ldots, x_n)$, and the multi-index notation

$$\mathbf{a}_j = (a_{1,j}, a_{2,j}, \ldots, a_{n,j}), \quad \mathbf{x}^{\mathbf{a}_j} = x_1^{a_{1,j}} x_2^{a_{2,j}} \cdots x_n^{a_{n,j}}, \tag{2}$$

where $\mathbf{a}_j$ is the jth column of the matrix A.

For any specific value for t, the system $\mathbf{h}(\mathbf{x}, t) = \mathbf{0}$ reduces to a system with exactly two monomials in every equation. The solving of such a system happens via a unimodular coordinate transformation defined by the Hermite normal form of A. Singular solutions can occur only when $\mathbf{c}(t) = \mathbf{0}$, only for specific values of t. While monomial homotopies have thus no direct practical use, they provide good test cases to experiment with algorithms and to introduce new ideas.

2.1   A Square Root Homotopy

The simplest example of a monomial homotopy is

$$x^2 - 1 + t = 0, \quad \text{with solution} \quad x(t) = \pm\sqrt{1 - t}. \tag{3}$$

The two paths defined by this homotopy are shown in Fig. 1.

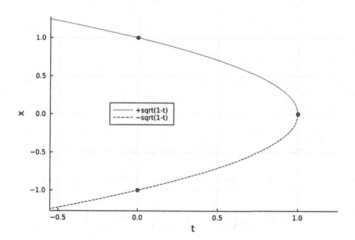

Fig. 1. Starting at $x = \pm 1$, the two paths converge to $x = 0$, as t moves from 0 to 1.

At $t = 1$, the two paths coincide at a double point. Our problem is to predict for which value of t this singularity happens *without computing* $x(t)$ *for* $t \approx 1$.

In the development of the solution $x(t) = \sqrt{1-t}$ in a Taylor series about $t = 0$, let c_n be the coefficient of t^n. Then the application of the ratio theorem of Fabry gives

$$\frac{c_n}{c_{n+1}} = \frac{2(n+1)}{2n-1} =: f(n), \quad \lim_{n \to \infty} f(n) = 1. \tag{4}$$

As the limit of the ratios equals one, we can predict the location of the singularity, already at the series development at $t = 0$. The main problem is the slow convergence of the series. Table 1 illustrates that in order to gain one extra bit of accuracy, we must double the value of n.

Table 1. $f(n) = \frac{2(n+1)}{2n-1}$ converges slowly to one. The error column lists $|f(n) - 1|$. The last column is the ratio of two consecutive errors. As n doubles, the error is cut in half.

n	$f(n)$	error	error ratio
2	2.00000000000000	1.00E+00	
4	1.42857142857143	4.29E−01	2.3333E+00
8	1.20000000000000	2.00E−01	2.1429E+00
16	1.09677419354839	9.68E−02	2.0667E+00
32	1.04761904761905	4.76E−02	2.0323E+00
64	1.02362204724409	2.36E−02	2.0159E+00
128	1.01176470588235	1.18E−02	2.0079E+00
256	1.00587084148728	5.87E−03	2.0039E+00
512	1.00293255131965	2.93E−03	2.0020E+00

Observe we can rewrite $f(n)$ of (4) as

$$f(n) = \frac{2(n+1)}{2n-1} = \frac{2n-1+3}{2n-1} = 1 + \frac{3}{2n-1} = 1 + \frac{3}{2n}\left(\frac{1}{1 - \frac{1}{2n}}\right) \tag{5}$$

$$= 1 + \frac{3}{2n}\left(1 + \frac{1}{2n} + \left(\frac{1}{2n}\right)^2 + \left(\frac{1}{2n}\right)^3 + \cdots\right). \tag{6}$$

As shown in Sect. 3, $f(n)$ has an asymptotic expansion of the form

$$f(n) = 1 + \gamma_1\left(\frac{1}{n}\right) + \gamma_2\left(\frac{1}{n}\right)^2 + \gamma_3\left(\frac{1}{n}\right)^3 + \cdots \tag{7}$$

for some coefficients $\gamma_1, \gamma_2, \gamma_3, \ldots$. If we double the value for n, we have

$$f(2n) = 1 + \gamma_1\left(\frac{1}{2n}\right) + \gamma_2\left(\frac{1}{2n}\right)^2 + \gamma_3\left(\frac{1}{2n}\right)^3 + \cdots \tag{8}$$

and then we eliminate γ_1 via a linear combination:

$$2f(2n) - f(n) = 1 + 2\gamma_2 \left(\frac{1}{2n}\right)^2 - \gamma_2 \left(\frac{1}{n}\right)^2 + 2\gamma_3 \left(\frac{1}{2n}\right)^3 - \gamma_3 \left(\frac{1}{n}\right)^3 + \cdots , \quad (9)$$

which results in an approximation with error $O(1/n^2)$.

This regular ratio of two consecutive errors allows for an effective application of Richardson extrapolation. The input to Richardson extrapolation are the values $f(2), f(4), f(8), \ldots, f(2^N)$. The output is $R_{i,j}$, the triangular table of extrapolated values. Then the extrapolation proceeds as follows:

1. The first column: $R_{i,1} = f(2^i)$, for $i = 1, 2, 3, \ldots, N$.
2. The next columns in the table are computed via

$$R_{i,j} = \frac{2^{i-j+1} R_{i,j-1} - R_{j-1,j-1}}{2^{i-j+1} - 1}, \quad (10)$$

for $i = i, i+1, \ldots, N$ and for $j = 2, 3, \ldots, N$.

Table 2 shows the errors $|R_{i,j} - 1|$ of the extrapolated values. Looking at the diagonal of Table 2, we see that we gain about two decimal places of accuracy at each doubling of n.

Table 2. Errors of Richardson extrapolation. The column E_0 is the error column of Table 1. The column E_j is the error obtained from extrapolating j times, applying formula (10). At $n = 64$ we have 8 correct decimal places and at $n = 512$, the full machine precision is attained.

n	E_0	E_1	E_2	E_3	E_4	E_5	E_6	E_7	E_8
2	1.0E+0								
4	4.3E−1	1.4E−1							
8	2.0E−1	6.7E−2	9.5E−3						
16	9.7E−2	3.2E−2	4.6E−3	3.1E−4					
32	4.8E−2	1.6E−2	2.3E−3	1.5E−4	4.9E−6				
64	2.4E−2	7.9E−3	1.1E−3	7.5E−5	2.4E−6	3.8E−8			
128	1.2E−2	3.9E−3	5.6E−4	3.7E−5	1.2E−6	1.9E−8	1.5E−10		
256	5.9E−3	2.0E−2	2.8E−4	1.9E−5	6.0E−7	9.5E−9	7.5E−11	2.9E−13	
512	2.9E−3	9.8E−4	1.4E−4	9.3E−6	3.0E−7	4.8E−9	3.8E−11	1.5E−13	4.4E−16

2.2   Two Paths Ending in a Cusp

Figure 2 is an example of a situation not covered by Theorem 1. Consider the homotopy

$$h(x, t) = x^2 - (t - 1)^4 = (x - (t - 1)^2)(x + (t - 1)^2) = 0, \quad (11)$$

which has the obvious two solutions $x(t) = \pm(t - 1)^2$.

In this case, the power series for both paths are polynomials of degree two, and there is no limit, as all coefficients $c_n = 0$, for $n > 2$. In [32], an algorithm to sweep an algebraic curve for singularities monitors the determinant of the Jacobian matrix along the curve. If the path of the determinant of the Jacobian matrix on the curve is concave up, then that is an indicator for undetected singularities.

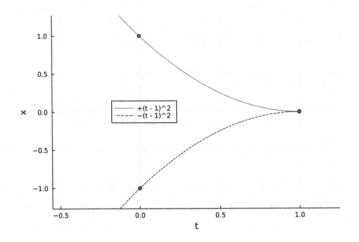

Fig. 2. Starting at $x = \pm 1$, the two paths converge to $x = 0$, as t moves from 0 to 1.

2.3   A Random 4-Dimensional Monomial Homotopy

In this section, we illustrate the need for multiple precision, even already in relatively low dimensions and degrees. Consider

$$\mathbf{h}(\mathbf{x}, t) = \begin{cases} x_1^7 x_2^7 x_3^7 x_4^7 = 1 - t, \\ x_1^7 x_2^3 x_3^2 = 1 - t, \\ x_2^5 x_3 x_4 = 1 - t, \\ x_2^7 x_3^2 x_4^2 = 1 - t. \end{cases} \qquad (12)$$

Storing the exponents of the monomials in the columns of $A = [\mathbf{a}_1, \mathbf{a}_2, \mathbf{a}_3, \mathbf{a}_4]$, $\mathbf{x}^A = (\mathbf{x}^{\mathbf{a}_1}, \mathbf{x}^{\mathbf{a}_2}, \mathbf{x}^{\mathbf{a}_3}, \mathbf{x}^{\mathbf{a}_4})$, the monomial homotopy $\mathbf{h}(\mathbf{x}, t)$ can be written as

$$\mathbf{x}^A = (1 - t)\mathbf{e}, \quad \mathbf{e} = \begin{bmatrix} 1 \\ 1 \\ 1 \\ 1 \end{bmatrix}, \quad A = \begin{bmatrix} 7 & 7 & 0 & 0 \\ 7 & 3 & 5 & 7 \\ 7 & 2 & 1 & 2 \\ 7 & 0 & 1 & 2 \end{bmatrix}, \quad \det(A) = -42. \qquad (13)$$

At $t = 0$, $(1,1,1,1)$ is one of the 42 solutions, as $42 = |\det(A)|$, computed via the Smith normal form of A.

In double precision, extrapolating on $x_1(t)$, the extrapolation does not get any more accurate than six decimal places. Working with coefficients computed with 32 decimal places running the algorithms of [2] (implemented in PHCpack [42]), the extrapolation gives eight decimal places of accuracy, similarly as in the square root homotopy.

For the examples in this section, Richardson extrapolation results in an accuracy of 8 decimal places when $n = 64$ and for $n = 512$, we can locate to singularity to the full double precision.

3   Asymptotic Expansions

Consider the coefficient c_n of t^n in the Taylor series. What happens if n grows:

$$\left| \frac{c_n}{c_{n+1}} \right| \rightarrow \begin{cases} |z| < 1 : \text{coefficients increase,} \\ |z| = 1 : \text{coefficients are constant,} \\ |z| > 1 : \text{coefficients decrease.} \end{cases} \tag{14}$$

Let $x(t)$ satisfy $h(x(t), t) = 0$, then in the series for $x(t)$, we may assume that for sufficiently large n, the magnitude of the nth coefficient is $|z|^n$. If we then set

$$t = |z|s \tag{15}$$

then the coefficient of s^n in the series $x(s)$ will have a magnitude close to one. By Lemma 1, the radius of convergence of the series $x(s)$ equals one.

Lemma 1. *Let $x(t)$ be a power series with c_n as the nth coefficient of t^n and*

$$\lim_{n \to \infty} \frac{c_n}{c_{n+1}} = z \in \mathbb{C} \setminus \{0\}. \tag{16}$$

Then the series $x(t = |z|s)$ has convergence radius equal to one.

Proof. Consider the effect of the substitution $t = |z|s$, respectively on the nth and the $(n+1)$th term in the series $x(t)$:

$$c_n t^n \rightarrow \underbrace{c_n |z|^n}_{=: d_n} s^n, \quad c_{n+1} t^{n+1} \rightarrow \underbrace{c_{n+1} |z|^{n+1}}_{=: d_{n+1}} s^{n+1}. \tag{17}$$

Then d_n is the coefficient of s^n in the series $x(s)$ and

$$\left| \frac{d_n}{d_{n+1}} \right| = \left| \frac{c_n}{c_{n+1}} \right| \left| \frac{1}{|z|} \right|.$$

By (16), $\displaystyle \lim_{n \to \infty} \left| \frac{d_n}{d_{n+1}} \right| = 1$. Thus, $x(s)$ has a convergence radius equal to one.   □

If interested only in the magnitude of the radius, then in the natural application of Lemma (15), $|z|$ is used. Using complex arithmetic, the series $x(t = z \cdot s)$ has radius of convergence equal to one.

In practice, the transformation as defined in as defined in (15) has numerical benefits. In theory, it implies that without loss of generality, we may assume that all series we consider all have convergence radius one.

Proposition 1. *Assume $x(t)$ is a series which satisfies the conditions of Theorem 1, with a radius of convergence equal to one. Let c_n be the coefficient of t^n in the series. Then $|1 - c_n/c_{n+1}|$ is $O(1/n)$ for sufficiently large n.*

Proof. Expressing the Taylor series of $x(t)$ as

$$x(t) = x(0) + x'(0)t + \frac{x''(0)}{2!}t^2 + \frac{x'''(0)}{3!}t^3 + \cdots + \frac{x^{(n)}(0)}{n!}t^n + \cdots \qquad (18)$$

leads to a formula for the coefficient of t^n as

$$c_n = \frac{x^{(n)}(0)}{n!} \quad \text{and} \quad c_{n+1} = \frac{x^{(n+1)}(0)}{(n+1)!}. \qquad (19)$$

Then the error is

$$\left| 1 - \frac{c_n}{c_{n+1}} \right| = \left| 1 - \left(\frac{x^{(n)}(0)}{x^{(n+1)}(0)} \right)(n+1) \right| \approx 0, \quad \text{for large } n. \qquad (20)$$

Under the assumption that the radius of convergence is equal to one, without loss of generality we may assume that the singularity occurs at $t = 1$. Otherwise, if $t = z$ for some complex number z, with $|z| = 1$, we can rotate the coordinate system so $z = 1$ in the rotated coordinate system. Therefore, we may assume there is a power series $p(t)$, so

$$x(t) = \frac{p(t)}{1 - t} = u(t)p(t), \quad u(t) = \frac{1}{1 - t} = 1 + t + t^2 + t^3 + \cdots . \qquad (21)$$

The $p(t)/(1 - t)$ can be viewed as the limit of the Padé approximant of degree $[n/1]$, for $n \to \infty$. This Padé approximant is well defined under the assumption of Theorem 1. In the limit reasoning for $n \to \infty$, we work with sufficiently large n, but never take ∞ for n.

Applying Leibniz rule to the nth derivative of $x(t)$ leads to

$$x^{(n)}(t) = \sum_{k=0}^{n} \left(\frac{n!}{k!(n-k)!} \right) u^{(n-k)}(t) p^{(k)}(t). \qquad (22)$$

At $t = 0$, we have $u^{(n-k)}(0) = (n-k)!$ and we obtain

$$x^{(n)}(0) = \sum_{k=0}^{n} \left(\frac{n!}{k!} \right) p^{(k)}(0). \qquad (23)$$

We rewrite the expression for $x^{(n+1)}(0)$ as

$$x^{(n+1)}(0) = \sum_{k=0}^{n+1} \left(\frac{(n+1)!}{k!} \right) p^{(k)}(0) \qquad (24)$$

$$= \sum_{k=0}^{n} (n+1) \left(\frac{n!}{k!} \right) p^{(k)}(0) + p^{(n+1)}(0) \qquad (25)$$

$$= (n+1)x^{(n)}(0) + p^{(n+1)}(0). \qquad (26)$$

Then we can write (20) as

$$\left| 1 - \frac{c_n}{c_{n+1}} \right| = \left| 1 - \left(\frac{x^{(n)}(0)}{(n+1)x^{(n)}(0) + p^{(n+1)}(0)} \right)(n+1) \right| \tag{27}$$

$$= \left| 1 - \frac{1}{1 + \frac{1}{n+1}\left(\frac{p^{(n+1)}(0)}{x^{(n)}(0)} \right)} \right|. \tag{28}$$

Note that we may divide by $x^{(n)}(0)$, because $x^{(n)}(0) \neq 0$ by the assumption that c_n/c_{n+1} is well defined for all values of n, otherwise the limit would not exist. Denote

$$C = \frac{p^{(n+1)}(0)}{x^{(n)}(0)}. \tag{29}$$

Then the result follows from another series expansion:

$$\frac{1}{1 + \frac{C}{n+1}} = 1 - \left(\frac{C}{n+1} \right) + \left(\frac{C}{n+1} \right)^2 - \cdots. \tag{30}$$

Substituting the right hand side of (30) into (28) gives

$$\left| 1 - \frac{c_n}{c_{n+1}} \right| = \left| \left(\frac{C}{n+1} \right) - \left(\frac{C}{n+1} \right)^2 + \cdots \right| \tag{31}$$

What remains to prove is that C does not depend on n. Dividing (26) by $x^{(n)}(0)$ leads to

$$\frac{x^{(n+1)}(0)}{x^{(n)}(0)} = n + 1 + \frac{p^{(n+1)}(0)}{x^{(n)}(0)}. \tag{32}$$

The assumption that $x(t)$ has a radius of convergence equal to one implies $c_{n+1} \approx c_n$ and that

$$\frac{x^{(n+1)}(0)}{x^{(n)}(0)} = n + O(1), \tag{33}$$

and thus we have

$$n + O(1) = n + 1 + \frac{p^{(n+1)}(0)}{x^{(n)}(0)} \quad \text{or equivalently} \quad \frac{p^{(n+1)}(0)}{x^{(n)}(0)} \text{ is } O(1). \tag{34}$$

Therefore C is a constant, independently of n. This shows that the error is $O(1/(n+1))$. For large n, $O(1/(n+1))$ is $O(1/n)$. □

Observe that the above proof does not make any assumptions on the type of homotopy used, other than the existence of a limit as in the theorem of Fabry. Then the main result of this section can be stated as below.

Corollary 1. *Assuming the convergence radius equals one, applying Richardson extrapolation N times on a Taylor series truncated after n terms, results in an $O(1/n^{N+1})$ error on the radius of convergence.*

Proof. By Proposition 1, and in particular the expansion in (31), we have

$$1 + \gamma_1 \left(\frac{1}{n}\right) + \gamma_2 \left(\frac{1}{n}\right)^2 + \gamma_3 \left(\frac{1}{n}\right)^3 + \cdots \tag{35}$$

as the expansion for the error to the limit 1.

For $N = 1$, the first extrapolated values have error $O(1/n^2)$, because the leading terms of the errors are $O(1/n)$ and running Richardson extrapolation once (for $j = 2$ and $i = 2, 3, \ldots, N$ in (10)) eliminates this leading term.

Using the formulas in (10) to compute the next columns in the triangular table eliminates the next terms in the error expansion in (35). After extrapolating $N - 1$ more times, we then obtain an $O(1/n^{N+1})$ error term.   □

The assumption that the radius of convergence equals one makes the Richardson extrapolation superfluous, as the outcome of the extrapolation is already known. We can remove this assumption. Consider for example the homotopy $h(x,t) = x^2 - 2 + t = 0$ and $x(t) = \sqrt{2 - t}$ as the positive solution branch. If c_n is the nth coefficient of the Taylor series, then

$$\frac{c_n}{c_{n+1}} = 2\left(\frac{2(n+1)}{2n-1}\right) = 2f(n), \tag{36}$$

where $f(n)$ is the formula from (4). Similarly, for the homotopy $h(x,t) = x^2 - 1/2 + t = 0$ and $x(t) = \sqrt{1/2 - t}$ as the positive solution branch, with c_n as the nth coefficient of the Taylor series, we have

$$\frac{c_n}{c_{n+1}} = \frac{1}{2}\left(\frac{2(n+1)}{2n-1}\right) = \frac{1}{2}f(n). \tag{37}$$

This implies that for those two examples, the series development of $f(n)$ in $1/n$ is multiplied respectively with 2 or $1/2$, and that therefore Richardson extrapolation applies.

Theorem 2. *Let c_n be the coefficient with t^n in $x(t)$ and denote $f(n) = c_n/c_{n+1}$. If*

$$\lim_{n \to \infty} \frac{c_n}{c_{n+1}} = z \in \mathbb{C} \setminus \{0\}, \tag{38}$$

then

$$f(n) = z + \gamma_1 z \left(\frac{1}{n}\right) + \gamma_2 z \left(\frac{1}{n}\right)^2 + \gamma_3 z \left(\frac{1}{n}\right)^3 + \cdots . \tag{39}$$

Proof. By Lemma 1, we transform $x(t)$ into $x(s) = x(t = z \cdot s)$, which has convergence radius one. Let d_n be the coefficient of s^n in $x(s)$ and denote $g(n) = d_n/d_{n+1}$. For $g(n)$, we have the expansion (35):

$$g(n) = 1 + \gamma_1 \left(\frac{1}{n}\right) + \gamma_2 \left(\frac{1}{n}\right)^2 + \gamma_3 \left(\frac{1}{n}\right)^3 + \cdots . \tag{40}$$

The above series development is unique. Therefore, transforming $s = t/z$, gives the series (39).   □

Theorem 2 provides the justification for the application of Richardson extrapolation and the statement of Corollary 1 holds in theory for any series, not only for those with radius of convergence equal to one. However, in practice, series with a radius of convergence smaller than one will have very large coefficients which cause numerical instabilities and unavoidably arithmetical overflow.

If the convergence radius of a power series equals one, then it is safe to calculate the coefficients of the power series from sample points at nearby locations.

4   Fourier Series

In computational complex analysis [18], the discrete Fourier transform is applied to compute the coefficients of the Taylor series. For general references on the application of Fourier transforms in computer algebra and numerical analysis, we refer to [14] and [6].

As described in [29], many derivatives are computed simultaneously with an accuracy close to machine precision, for a suitable step size, using complex arithmetic, extending the complex-step differentiation method [37] to higher order derivatives. Figure 3 illustrates the problem: the step size must be smaller than the radius of convergence. This problem is addressed in Sect. 5.2.

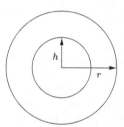

Fig. 3. The radius of convergence r and step size h. We want $h \ll r$.

To introduce the application of the discrete Fourier transform to compute the Taylor series, consider the development of f at z, using step size $h\omega$:

$$f(z + h\omega) = f(z) + h\omega f'(z) + \frac{h^2}{2}\omega^2 f''(z) + \frac{h^3}{3!}\omega^3 f'''(z)$$
$$+ \frac{h^4}{4!}\omega^4 f^{(iv)}(z) + \frac{h^5}{5!}\omega^5 f^{(v)}(z) + \frac{h^6}{6!}\omega^6 f^{(vi)}(z)$$
$$+ \frac{h^7}{7!}\omega^7 f^{(vii)}(z) + \frac{h^8}{8!}\omega^8 f^{(viii)}(z) + \cdots,$$

where ω is the eight complex root of unity: $\omega^8 = 1$. Regrouping in powers of ω then gives

$$f(z + h\omega) = f(z) + \frac{h^8}{8!} f^{(\text{viii})}(z) + \cdots$$

$$+ \omega \left(h f'(z) + \frac{h^9}{9!} f^{(\text{ix})}(z) + \cdots \right)$$

$$+ \omega^2 \left(\frac{h^2}{2!} f''(z) + \frac{h^{10}}{10!} f^{(\text{x})}(z) + \cdots \right)$$

$$+ \omega^3 \left(\frac{h^3}{3!} f'''(z) + \frac{h^{11}}{11!} f^{(\text{xi})}(z) + \cdots \right)$$

$$+ \omega^4 \left(\frac{h^4}{4!} f^{(\text{iv})}(z) + \frac{h^{12}}{12!} f^{(\text{xii})}(z) + \cdots \right)$$

$$+ \omega^5 \left(\frac{h^5}{5!} f^{(\text{v})}(z) + \frac{h^{13}}{13!} f^{(\text{xiii})}(z) + \cdots \right)$$

$$+ \omega^6 \left(\frac{h^6}{6!} f^{(\text{vi})}(z) + \frac{h^{14}}{14!} f^{(\text{xiv})}(z) + \cdots \right)$$

$$+ \omega^7 \left(\frac{h^7}{7!} f^{(\text{vii})}(z) + \frac{h^{15}}{15!} f^{(\text{xv})}(z) + \cdots \right).$$

For k from 1 to 7, the coefficients of ω^k allow the extraction of the kth derivative of f at z, at a precision of $O(h^8)$.

The Discrete Fourier Transform

$$\text{DFT}_\omega : \quad \begin{matrix} \mathbb{C}^n & \to & \mathbb{C}^n \\ (f_0, f_1, \ldots, f_{n-1}) & \mapsto & (F(\omega^0), F(\omega^1), \ldots, F(\omega^{n-1})) \end{matrix} \tag{41}$$

takes the coefficients of the polynomial F with coefficients $f_0, f_1, \ldots, f_{n-1}$, where $\omega^n = 1$ and returns the values of F at the powers of ω. The inverse of DFT_ω returns the coefficients of ω^k needed in the Taylor series of $f(z + h\omega)$.

As illustrated by Table 3, the derivatives grow as fast as $n!$ and therefore, except for small n, we may not expect to obtain highly accurate values.

The step size of $h = 0.5$ used in Table 3 is a compromise value. Values of h smaller than 0.5 give more accurate results for the lower order derivatives but give then too inaccurate values for the higher order derivatives. The opposite happens for values of h larger than 0.5.

Fortunately, we do not need the derivatives $x^{(n)}(0)$, but the coefficients of the Taylor series, $c_n = x^{(n)}(0)/n!$. Table 4 shows the application of the DFT to compute the series coefficients. Compared to the derivatives in Table 3, the computations in double precision arithmetic give six decimal places of accuracy for $n = 64$. The step size $h = 0.85$ gave the most accurate results.

In machine double precision, the results in Table 3 and Table 4 are close to optimal, with the step sizes respectively equal to 0.5 and 0.85. Using those large step sizes in multiprecision will not give more accurate results, but multiprecision will allow to select smaller step sizes. In particular with 33 decimal places (using

Table 3. Derivatives of $x(t) = \sqrt{1-t}$ at $t = 0$. The approximate values are computed with step size $h = 0.5$. The last column is the relative error.

n	exact $x^{(n)}(0)$	approximation $x^{(n)}(0)$	error
0	1.000000000000	0.999999968596	3.14E−08
1	−0.500000000000	−0.500000028787	5.76E−08
2	−0.250000000000	−0.250000053029	2.12E−07
3	−0.375000000000	−0.375000147155	3.92E−07
4	−0.937500000000	−0.937500546575	5.83E−07
5	−3.281250000000	−3.281252546540	7.76E−07
6	−14.765625000000	−14.765639282757	9.67E−07
7	−81.210937500000	−81.211031230822	1.15E−06
8	−527.871093750000	−527.871798600561	1.34E−06
9	−3959.033203125000	−3959.039180858922	1.51E−06
10	−33651.782226562500	−33651.838679975779	1.68E−06
11	−319691.931152343750	−319692.518875707698	1.84E−06
12	−3356765.277099609375	−3356771.966430745088	1.99E−06
13	−38602800.686645507812	−38602883.297614447773	2.14E−06
14	−482535008.583068847656	−482536106.155545711517	2.27E−06
15	−6514222615.871429443359	−6514238371.741491317749	2.42E−06
16	−94456227930.135726928711	−94456466497.677398681641	2.53E−06

mpmath 1.1.0 [21] with SymPy 1.4 [23] in Python 3.7.3), the 16th derivative is computed with an accuracy of 15 decimal places, with step size 0.1 and the error on the 64th coefficient coefficient on the series drops to 10^{-11}, with step size 0.5.

Instead of working with the same step size for all series coefficients, alternatively, one could explore using different step sizes. In this context, one classical and very common application of Richardson extrapolation is to improve the accuracy of numerical differentiation.

When z is a complex number, the complex step derivative is generalized in [24] and [33] with quaternion arithmetic. Using the quaternion Fourier transform [10,34], the coefficients of the Taylor series can be computed.

Table 4. Coefficients c_n of the Taylor series of $x(t) = \sqrt{1-t}$ at $t = 0$. The approximate values are computed with step size $h = 0.85$. The last column is the relative error.

n	exact c_n	approximation	error
0	1.000000000000	0.999999986011	1.40E−08
1	−0.500000000000	−0.500000013671	2.73E−08
2	−0.125000000000	−0.125000013365	1.07E−07
4	−0.039062500000	−0.039062512786	3.27E−07
8	−0.013092041016	−0.013092052762	8.97E−07
32	−0.001576932599	−0.001576940258	4.86E−06
64	−0.000554221198	−0.000554226120	8.88E−06

5   Polynomial Homotopies

The homotopies in this section have multiple singularities in the complex plane, for complex values of t, with real part < 1, but only one singularity at $t = 1$. Knowing the location of the last pole leads to the reconditioning of the homotopy and to series with convergence radius equal to one.

5.1   The Last Pole

Let $\mathbf{f}(\mathbf{x}) = \mathbf{0}$ be the system we want to solve and assume we have at least one solution of $\mathbf{g}(\mathbf{x}) = \mathbf{0}$. Then the homotopy

$$\mathbf{h}(\mathbf{x}, t) = \gamma(1 - t)\mathbf{g}(\mathbf{x}) + t\mathbf{f}(\mathbf{x}) = \mathbf{0}, \quad t \in [0, 1], \quad \gamma \in \mathbb{C}, |\gamma| = 1, \quad (42)$$

defines a path starting at $t = 0$, at a solution of $\mathbf{g}(\mathbf{x}) = \mathbf{0}$ and ending at $t = 1$, at a solution of $\mathbf{f}(\mathbf{x}) = \mathbf{0}$. The constant γ is a random complex number. If $\mathbf{g}(\mathbf{x}) = \mathbf{0}$ has no singular solutions, then it follows from the main theorem of elimination theory that all paths defined by $\mathbf{h}(\mathbf{x}, t) = \mathbf{0}$ are regular and bounded for $t \in [0, 1)$, except for finitely many *complex* values for t. In [36], this constructive argument is illustrated by examples of homotopies of small degrees and dimension.

The key point is the existence of a polynomial $H(t)$ of finite degree, with $H(0) \neq 0$, as $\mathbf{g}(\mathbf{x}) = \mathbf{0}$ has no singular solutions. Moreover, by the random *complex* choice of γ, all roots of H are in the complex plane, except for $t = 1$, if the system $\mathbf{f}(\mathbf{x}) = \mathbf{0}$ has a singular solution. By construction of $\mathbf{h}(\mathbf{x}, t) = \mathbf{0}$, we can introduce the notion of *the last pole*, as the complex number ρ, for which $H(\rho) = 0$ and of all roots of H, ρ has the largest real part less than one.[1]

Figure 4 illustrates that ρ is the last complex singular value detected by the radar of a path tracker which applies the theorem of Fabry to set its step size.

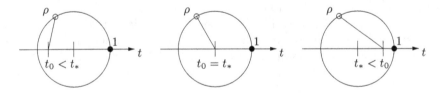

Fig. 4. Schematic of the last pole ρ marked by the hollow circle. At the center, at $t = t_0 = t_*$, ρ and 1 are at the same distance. At $t_0 < t_*$, the proximity of ρ determines the step size, while for $t_* < t_0$, the singularity at one will be detected.

By construction of the homotopy $\mathbf{h}(\mathbf{x}, t) = \mathbf{0}$, and in particular by the random choice of the complex constant γ, the solution at $t = t_*$ is regular, and well conditioned. This implies that Newton's method for the series coefficients converges quadratically. One could then already discover the singular solution for

---

[1] If all real parts of the roots of H are larger than one, then we are in the case similar to a monomial homotopy, a case that is then already solved.

$t = 1$, via the computation of a Padé approximant with quadratic denominator. Via a perturbation argument, for $t = t_* + \delta$, for suitable $\delta > 0$, the application of the theorem of Fabry will detect $t = 1$ as a singular solution, *without computing* $\mathbf{x}(t)$ *for* $t \approx 1$.

5.2 Homotopy Reconditioning

Once the path tracker reaches a value for the continuation parameter t, that is past the last pole towards an isolated singularity at $t = 1$, at the end of a path, the coefficients of the power series will grow very fast, which is already an indication for the trouble to come.

For the reliable numerical computation of the power series, consider the transformation in the homotopy $\mathbf{h}(\mathbf{x}, t) = \mathbf{0}$:

$$t = rs + t_0, \quad r = 1 - t_0, \quad t_0 = t_* + \delta, \tag{43}$$

where t_* is value as in Fig. 4, at the same distance from the last pole ρ and the end point $t = 1$, and δ is a suitable positive value so at $t_0 = t_* + \delta$, the application of the theorem of Fabry will detect $t = 1$ as the location for the closest singular solution.

After applying (43), the series development of the path $\mathbf{x}(s)$ defined by the homotopy $\mathbf{h}(\mathbf{x}(s), s) = \mathbf{0}$ will have convergence radius equal to one. The term *reconditioning* is justified as the coefficients of the Taylor series in the reconditioned homotopy do not grow exponentially fast.

6 Computational Experiments

The new methods are illustrated with computational experiments on two well known examples in the literature, with ad hoc tools, using test procedures in version 2.4.85 PHCpack [42] (with QDlib [19] and CAMPARY [22] for multiple double arithmetic), version 1.1.1 of phcpy [31], version 1.4 of sympy [23], and version 1.1.0 of mpmath [21], in Python 3.7.3. The computations were done on a CentOS 6.10 Linux computer with 23.4 GB of memory and a 12-core Intel Xeon X5690 at 3.47 GHz.

6.1 Ojika's First Example

One example in [30] (known in benchmarks as ojika1, used in [16, 25–27]) is

$$\mathbf{f}(x, y) = \begin{cases} x^2 + y - 3 = 0, \\ x + 0.125y^2 - 1.5 = 0. \end{cases} \tag{44}$$

This system has one regular solution at $(-3, 6)$ and a triple root at $(1, 2)$. Using $\gamma = -0.917153159675641 - 0.398534919043474I$, $I = \sqrt{-1}$, in the homotopy (42) with start system

$$\mathbf{g}(x, y) = \begin{cases} x^2 - 1 = 0, \\ y^2 - 1 = 0 \end{cases} \tag{45}$$

makes that the path starting at $(1, 1)$ converges to the triple root.

The value t_0 after t_* (the location of the last pole) that was used is $t_0 = 0.955647336181678$. At this value for t, the coordinates of the corresponding solution are

$$x \approx 1.17998166418735 + 0.0181391513338172\,I,$$
$$y \approx 1.60871001974391 - 0.0423866308603763\,I,$$

with the inverse of the condition number estimated at $8.9E{-}03$. In double precision, a condition number of about 10^3 is within the range of what is considered well conditioned. Observe that the coordinates of the solution corresponding to t_0 are far from the location of the triple root $(1, 2)$.

The value for $r = 1 - t_0$ is 0.044352663818322036, which implies that, without reconditioning, the magnitude of the Taylor series coefficients will increase with about two decimal places. At that pace, as $2^{64} \approx 10^{+19}$, numerical difficulties arise without reconditioning.

After reconditioning, with $n = 64$, the ratio, based on the power series for the first coordinate $x(s)$, is estimated at

$$1.0265192231142901 + 2.9197227799819557E{-}05\,I$$

and the magnitude of the imaginary part corresponds to the magnitude of the coefficients c_n in the series of $x(s)$. This mild decline of the exponents corresponds to the over estimation of the radius at about 1.0265. Applying Richardson extrapolation yields

$$0.9999729580138075 + 8.484367218447337E{-}06\,I,$$

which thus locates the singularity with an error of 10^{-6}.

The above computations were done in double precision. In double precision (≈ 32 decimal places), with $n = 512$, the ratio is first estimated at 1.00326 and Richardson extrapolation then improves the accuracy, to obtain an error of 10^{-6} on the value $t = 1$, the location of the singularity, confirming the result obtained in double precision.

6.2 One Fourfold Root of Cyclic 9-Roots

The cyclic n-roots problem

$$\mathbf{f}(\mathbf{x}) = \begin{cases} x_0 + x_1 + \cdots + x_{n-1} = 0 \\ i = 2, 3, \ldots, n-1 : \displaystyle\sum_{j=0}^{n-1} \prod_{k=j}^{j+i-1} x_{k \bmod n} = 0 \\ x_0 x_1 x_2 \cdots x_{n-1} - 1 = 0 \end{cases} \tag{46}$$

is a well known benchmark problem in polynomial system solving, which arose in the study of biunimodular vectors [13]. The cyclic 9-roots problem was solved in [12], and its roots of multiplicity four were used in the development of deflation in [25]. This system was used to illustrate the computation of the multiplicity structure in [8].

The start system $\mathbf{g}(\mathbf{x}) = \mathbf{0}$ in a homotopy to solve $\mathbf{f}(\mathbf{x}) = \mathbf{0}$ was obtained by running the plain blackbox solver (the extended version is described in [43]) on 12 cores tracking 11,016 is less than two minutes. For reproducibility, the seed in the random number generators was 7131. That $\mathbf{g}(\mathbf{x})$ was then used in the homotopy (42) with $\gamma = -0.917153159675641 - 0.398534919043474 I$, $I = \sqrt{-1}$. One path was selected that ended at one of the fourfold roots.

The value for t after t_*, the location of the last pole is $t_0 = 0.998315512784621$, with coordinates of the corresponding solution

$$x_0 \approx +1.00000126517819 \quad + 2.90396442439194\mathrm{E}-07$$
$$x_1 \approx -2.61867609654276 \quad - 2.06312686218454\mathrm{E}-03$$
$$x_2 \approx -0.381725080860952 + 6.25420054941098\mathrm{E}-05$$
$$x_3 \approx +1.00151501674915 \quad + 1.11189386260303\mathrm{E}-03$$
$$x_4 \approx +0.381629266681896 - 3.62839287359460\mathrm{E}-04$$
$$x_5 \approx +2.62034316800711 \quad + 2.49236777820171\mathrm{E}-03$$
$$x_6 \approx +0.998483898493147 - 1.11096857563447\mathrm{E}-03$$
$$x_7 \approx -2.61970187995193 \quad - 4.30339092688366\mathrm{E}-04$$
$$x_8 \approx -0.381870388949536 + 3.01610075581641\mathrm{E}-04$$

with inverse condition number estimated at $5.3\mathrm{E}\text{-}5$. Although the homotopy does not respect the permutation symmetry, the orbit structure of the solution can already be observed, at the limited accuracy of about three decimal places.

The value for $r = 1 - t_0$ is 0.0016844872153789492 and without reconditioning the homotopy, the coefficients in the power series expansions of the solution increase at a very high pace. After reconditioning, with $n = 32$, the convergence radius is estimated at

$$1.00000000099639 + 4.319265\mathrm{E}-09\,I$$

and confirmed in double precision. Because of the close proximity to the singularity, no extrapolation is necessary in this case.

7   Conclusions

Richardson extrapolation is effective to locate the closest singularity as shown by the asymptotic expansions on the ratio of two consecutive coefficients in the Taylor series of the solution curves, under the condition of the theorem of Fabry.

The homotopy continuation parameter can always be adjusted so the convergence radius of the power series equals one, which allows for a safe step size selection in the application of the discrete Fourier transform to compute all coefficients of the series efficiently and accurately.

Deflation restores the quadratic convergence of Newton's method on an isolated singular solution via reconditioning. The homotopy reconditioning using the location of the last pole provides an apriori justification for the application of the deflation method via the Richardson extrapolation on the ratios of the coefficients of power series.

The theorem of Fabry provides a radar to detect singularities. In this paper we have shown that this radar can accurately locate the nearest singular solution of a polynomial homotopy. We apply this radar at a safe distance from singularities, at a regular solution where the quadratic convergence of Newton's method holds.

Acknowledgments. Some of the results in this paper were presented by the first author on 27 March 2022 in a preliminary report at the special session on Optimization, Complexity, and Real Algebraic Geometry, which took place online. The authors thank the organizers, Saugata Basu and Ali Mohammad Nezhad, for their invitation. We thank the three reviewers of this paper for their useful comments which helped to improve the exposition.

References

1. Baker, G.A., Jr., Graves-Morris, P.: Padé Approximants. Cambridge University Press, Cambridge (1996)
2. Bliss, N., Verschelde, J.: The method of Gauss-Newton to compute power series solutions of polynomial homotopies. Linear Algebra Appl. **542**, 569–588 (2018)
3. Brezinski, C.: A general extrapolation algorithm. Numer. Math. **35**, 175–187 (1980)
4. Burr, M., Leykin, A.: Inflation of poorly conditioned zeros of systems of analytic functions. Arnold Math. J. **7**, 431–440 (2021)
5. Cheng, J.S., Dou, X., Wen, J.: A new deflation method for verifying the isolated singular zeros of polynomial systems. J. Comput. Appl. Math. **376**, 112825 (2020)
6. Corless, R.M., Fillion, N.: A Graduate Introduction to Numerical Methods. From the Viewpoint of Backward Error Analysis. Springer, New York (2013). https://doi.org/10.1007/978-1-4614-8453-0
7. Dayton, B.H., Li, T.Y., Zeng, Z.: Multiple zeros of nonlinear systems. Math. Comput. **80**(276), 2143–2168 (2011)
8. Dayton, B.H., Zeng, Z.: Computing the multiplicity structure in solving polynomial systems. In: Kauers, M. (ed.) Proceedings of the 2005 International Symposium on Symbolic and Algebraic Computation, pp. 116–123. ACM (2005)
9. Dienes, P.: The Taylor Series. An Introduction to the Theory of Functions of a Complex Variable. Dover, New York (1957)
10. Ell, T.A., Le Bihan, N., Sangwine, S.J.: Quaternion Fourier Transforms for Signal and Image Processing. Wiley, Hoboken (2014)
11. Fabry, E.: Sur les points singuliers d'une fonction donnée par son développement en série et l'impossibilité du prolongement analytique dans des cas très généraux. Annales scientifiques de l'École Normale Supérieure **13**, 367–399 (1896)
12. Faugère, J.C.: Finding all the solutions of Cyclic 9 using Gröbner basis techniques. In: Computer Mathematics - Proceedings of the Fifth Asian Symposium (ASCM 2001). Lecture Notes Series on Computing, vol. 9, pp. 1–12. World Scientific (2001)
13. Führ, H., Rzeszotnik, Z.: On biunimodular vectors for unitary matrices. Linear Algebra Appl. **484**, 86–129 (2015)
14. von zur Gathen, J., Gerhard, J.: Modern Computer Algebra. Cambridge University Press, Cambridge (1999)

15. Giusti, M., Lecerf, G., Salvy, B., Yakoubsohn, J.C.: On location and approximation of clusters of zeros: case of embedding dimension one. Found. Comput. Math. **17**(1), 1–58 (2007)
16. Hao, Z., Jiang, W., Li, N., Zhi, L.: On isolation of simple multiple zeros and clusters of zeros of polynomial systems. Math. Comput. **89**(322), 879–909 (2020)
17. Hauenstein, J.D., Mourrain, B., Szanto, A.: On deflation and multiplicity structure. J. Symb. Comput. **83**, 228–253 (2017)
18. Henrici, P.: Fast Fourier methods in computational complex analysis. SIAM Rev. **21**(4), 481–527 (1979)
19. Hida, Y., Li, X.S., Bailey, D.H.: Algorithms for quad-double precision floating point arithmetic. In: 15th IEEE Symposium on Computer Arithmetic (Arith-15 2001), pp. 155–162. IEEE Computer Society (2001)
20. Jeronimo, G., Matera, G., Solernó, P., Waissbein, A.: Deformation techniques for sparse systems. Found. Comput. Math. **9**, 1–50 (2009)
21. Johansson, F.: mpmath: a Python library for arbitrary-precision floating-point arithmetic. https://mpmath.org/
22. Joldes, M., Muller, J.-M., Popescu, V., Tucker, W.: CAMPARY: Cuda multiple precision arithmetic library and applications. In: Greuel, G.-M., Koch, T., Paule, P., Sommese, A. (eds.) ICMS 2016. LNCS, vol. 9725, pp. 232–240. Springer, Cham (2016). https://doi.org/10.1007/978-3-319-42432-3_29
23. Joyner, D., Čertík, O., Meurer, A., Granger, B.E.: Open source computer algebra systems: SymPy. ACM Commun. Comput. Algebra **45**(4), 225–234 (2011)
24. Kim, J.E.: Approximation of directional step derivative of complex-valued functions using a generalized quaternion system. Axioms **10**(206), 14 p (2021)
25. Leykin, A., Verschelde, J., Zhao, A.: Newton's method with deflation for isolated singularities of polynomial systems. Theoret. Comput. Sci. **359**(1–3), 111–122 (2006)
26. Li, N., Zhi, L.: Computing isolated singular solutions of polynomial systems: case of breadth one. SIAM J. Numer. Anal. **50**(1), 354–372 (2012)
27. Li, N., Zhi, L.: Verified error bounds for isolated singular solutions of polynomial systems. SIAM J. Numer. Anal. **52**(4), 1623–1640 (2014)
28. Morgan, A.P., Sommese, A.J., Wampler, C.W.: A power series method for computing singular solutions to nonlinear analytic systems. Numer. Math. **63**(3), 391–409 (1992)
29. Nasir, H.M.: Higher order approximations for derivatives using hypercomplex-steps. Int. J. Adv. Comput. Sci. Appl. **6**(1), 52–57 (2016)
30. Ojika, T.: Modified deflation algorithm for the solution of singular problems. I. A system of nonlinear algebraic equations. J. Math. Anal. Appl. **123**, 199–221 (1987)
31. Otto, J., Forbes, A., Verschelde, J.: Solving polynomial systems with phcpy. In: Proceedings of the 18th Python in Science Conference, pp. 563–582 (2019)
32. Piret, K., Verschelde, J.: Sweeping algebraic curves for singular solutions. J. Comput. Math. **234**(4), 1228–1237 (2010)
33. Roelfs, M., Dudal, D., Huybrechts, D.: Quaternionic step derivative: machine precision differentiation of holomorphic functions using complex quaternions. J. Comput. Appl. Math. **398**, 113699 (2021)
34. Said, S., Le Bihan, N., Sangwine, S.J.: Fast complexified quaternion Fourier transform. IEEE Trans. Signal Process. **56**(4), 1522–1531 (2008)
35. Schwetlick, H., Cleve, J.: Higher order predictors and adaptive steplength control in path following algorithms. SIAM J. Numer. Anal. **24**(6), 1382–1393 (1987)

36. Sommese, A.J., Verschelde, J., Wampler, C.W.: Introduction to numerical algebraic geometry. In: Dickenstein, A., Emiris, I.Z. (eds.) Solving Polynomial Equations. Foundations, Algorithms and Applications, Algorithms and Computation in Mathematics, vol. 14, pp. 301–337. Springer, Heidelberg (2005). https://doi.org/10.1007/3-540-27357-3_8

37. Squire, W., Trapp, G.: SIAM Rev. **40**(1), 110–112 (1998)

38. Suetin, S.P.: Padé approximants and efficient analytic continuation of a power series. Russ. Math. Surv. **57**, 43–141 (2002)

39. Telen, S., Van Barel, M., Verschelde, J.: A robust numerical path tracking algorithm for polynomial homotopy continuation. SIAM J. Sci. Comput. **42**(6), 3610-A3637 (2020)

40. Telen, S., Van Barel, M., Verschelde, J.: Robust Numerical Tracking of One Path of a Polynomial Homotopy on Parallel Shared Memory Computers. In: Boulier, F., England, M., Sadykov, T.M., Vorozhtsov, E.V. (eds.) CASC 2020. LNCS, vol. 12291, pp. 563–582. Springer, Cham (2020). https://doi.org/10.1007/978-3-030-60026-6_33

41. Timme, S.: Mixed precision path tracking for polynomial homotopy continuation. Adv. Comput. Math. **47**(5), Paper 75, 23 p (2021)

42. Verschelde, J.: Algorithm 795: PHCpack: a general-purpose solver for polynomial systems by homotopy continuation. ACM Trans. Math. Softw. **25**(2), 251–276 (1999)

43. Verschelde, J.: A blackbox polynomial system solver on parallel shared memory computers. In: Gerdt, V.P., Koepf, W., Seiler, W.M., Vorozhtsov, E.V. (eds.) CASC 2018. LNCS, vol. 11077, pp. 361–375. Springer, Cham (2018). https://doi.org/10.1007/978-3-319-99639-4_25

A General Method of Finding New Symplectic Schemes for Hamiltonian Mechanics

Evgenii V. Vorozhtsov[1(✉)] and Sergey P. Kiselev[1,2]

[1] Khristianovich Institute of Theoretical and Applied Mechanics of the Siberian Branch of the Russian Academy of Sciences, Novosibirsk 630090, Russia
{vorozh,kiselev}@itam.nsc.ru
[2] Novosibirsk State Technical University, Novosibirsk 630092, Russia

Abstract. The explicit symplectic difference schemes are considered for the numerical solution of molecular dynamics problems described by systems with separable Hamiltonians. A general method for finding symplectic schemes of high order of accuracy using parametric Gröbner bases, resultants, and permutations of variables is proposed. The implementation of the method is described by the example of four-stage partitioned Runge–Kutta (PRK) schemes of the Forest–Ruth family. All required symbolic calculations are performed using the computer algebra system *Mathematica*. 96 new PRK schemes of Forest–Ruth family have been obtained.

Keywords: Molecular dynamics · Symplectic difference schemes · Partitioned schemes · Gröbner bases · Resultants · Permutations

1 Introduction

The equations of molecular dynamics (MD) are ordinary Hamilton differential equations describing the interaction of material particles. MD equations have an exact analytical solution in a very limited number of cases [8]. Therefore, in the general case, these equations are solved numerically using difference schemes in which the differential operator is replaced by a difference operator.

When solving the Hamilton equations, it is natural to use difference schemes that preserve the symplectic properties of these equations. Violation of this condition leads to non-conservation of Poincaré invariants and the appearance of non-physical instability in numerical calculations. It follows that the difference operator of the numerical scheme must have the properties of a canonical transformation. The corresponding difference schemes are called symplectic.

This work is a continuation of the works of the authors [15,16], which provide a detailed overview of the methods for constructing symplectic schemes. In [15,16], symplectic schemes were considered in relation to the MD method, in which the Hamiltonian of interacting particles splits into the sum of kinetic

F. Boulier et al. (Eds.): CASC 2022, LNCS 13366, pp. 353–376, 2022.
https://doi.org/10.1007/978-3-031-14788-3_20

and potential energy. It was shown that the problem of constructing symplectic schemes of a given order of accuracy with a fixed number of stages is reduced to the problem of finding the roots of a polynomial system of equations that arise when the coefficients in the expression for the error of the scheme are turned to zero. It turned out that the use of the Gröbner basis technique [1] implemented in the *Mathematica* computer algebra system makes it possible to find the roots of a system of polynomial equations in the case when its Vandermonde determinant is zero. With the help of this technique, 21 new four-stage schemes of the 4th order of accuracy were obtained [16].

This paper presents a generalization of the method proposed in [15,16] to the more general case when the Vandermonde determinant of a system of polynomial equations is nonzero. The effectiveness of the proposed method is demonstrated by the example of a four-stage scheme of the 4th order of accuracy. 96 new symplectic four-stage schemes of the 4th order of accuracy of the Forest–Ruth type were obtained. The error functional for several of the best new schemes is two orders of magnitude smaller than for the Forest–Ruth scheme [6].

Verification of schemes is carried out by comparing the numerical solutions with the exact solution of the test problem. It is shown that the new Forest–Ruth four-stage schemes with the smallest norm of the leading error term provide a more accurate preservation of a balance in total energy of the particle system than the schemes of the same formal approximation order with a larger approximation error.

2 Governing Equations

For simplicity, consider the Hamilton's equations for a single particle in the one-dimensional case:

$$dx/dt = p/m, \quad dp/dt = f(x), \quad f(x) = -\partial H(x)/\partial x, \quad H = \frac{p^2}{2m} + V(x), \quad (1)$$

where t is the time, x is the coordinate, p is the momentum, m is the particle mass, f is the force acting on the particle, $V(x)$ is the potential energy, $p^2/(2m)$ is the kinetic energy, H is the Hamiltonian.

Hamilton's equations (1) generate a one-parameter group of diffeomorphisms $x^0 \to x(t, x^0)$ [5]. The smooth functions $z(x, p)$ form the algebra of the Lie group with respect to the Poisson bracket [5]

$$\frac{dz}{dt} = \{z, H\}, \quad \{z, H\} = \left(\frac{\partial H}{\partial p} \frac{\partial z}{\partial x} - \frac{\partial H}{\partial x} \frac{\partial z}{\partial p} \right). \quad (2)$$

Let us rewrite these equations in the operator form

$$\frac{dz}{dt} = \hat{L}z, \quad \hat{L} = \left(\frac{\partial H}{\partial p} \frac{\partial}{\partial x} - \frac{\partial H}{\partial x} \frac{\partial}{\partial p} \right), \quad (3)$$

where $\hat{L}$ is the Liouville operator. Write the solution of Eq. (3) in the form

$$z(t) = \hat{U}z(0), \quad \hat{U} = \exp(\hat{L}t), \quad \hat{L} = \hat{L}_1 + \hat{L}_2, \quad \hat{L}_1 = \frac{\partial H}{\partial p} \frac{\partial}{\partial x}, \quad \hat{L}_2 = -\frac{\partial H}{\partial x} \frac{\partial}{\partial p}. \quad (4)$$

Propagators $\hat{U}$ form a one-parameter Lie group and satisfy the group relations

$$\hat{U}(t_1)\hat{U}(t_2) = \hat{U}(t_1+t_2), \quad \hat{U}^{-1} = \exp(-\hat{L}t), \quad \hat{U}(0) = I, \quad \hat{U}\hat{U}^{-1} = I. \quad (5)$$

Equations (4) underlie the propagator method of constructing symplectic schemes [3,7,9,18].

3 Symplectic Partitioned Runge–Kutta Schemes

PRK schemes are based on the propagator method, which is determined by relations (4). In the case of a separable Hamiltonian, they have the following form:

$$z(t) = \hat{U}z(0), \quad \hat{U} = \exp[(\hat{L}_1 + \hat{L}_2)t],$$
$$\hat{L}_1 = \frac{p}{m}\frac{\partial}{\partial x}, \hat{L}_2 = f(x)\frac{\partial}{\partial p}, \quad f(x) = -\frac{\partial V}{\partial x}. \quad (6)$$

If the values $z^n = (x^n, p^n)$ are known at the nth time layer t^n, then we obtain at the $(n+1)$th time layer $t^{n+1} = t^n + h$

$$z^{n+1} = \exp((\hat{L}_1 + \hat{L}_2)h)z^n. \quad (7)$$

Let us subdivide the passage from t^n to t^{n+1} into K stages. Let us approximate the propagator (7) by a product of propagators with the accuracy $O(h^{\lambda+1})$:

$$\exp[(\hat{L}_1+\hat{L}_2)h] = \prod_{s=1}^{K} \exp(d_s h\hat{L}_1)\exp(c_s h\hat{L}_2) + O(h^{\lambda+1}), \sum_{s=1}^{K} c_s = 1, \sum_{s=1}^{K} d_s = 1. \quad (8)$$

Thus, the solution at the $(n+1)$th time layer is given with the accuracy of $O(h^\lambda)$ by the equation

$$z^{n+1} = \prod_{s=1}^{K} \exp(d_s h\hat{L}_1)\exp(c_s h\hat{L}_2)z^n, \quad (9)$$
$$z^K = z^{n+1} = (x^{n+1}, p^{n+1}), \quad z^{(0)} = z^n = (x^n, p^n).$$

We obtain from Eqs. (7) and (9) that at the passage from the $(i-1)$th stage to the ith stage, the following relations are valid [6]:

$$p^{(i)} = p^{(i-1)} + c_i h f(x^{(i-1)}), \quad x^{(i)} = x^{(i-1)} + d_i\frac{h}{m}p^{(i)}, \quad i = 1, \ldots, K, \quad (10)$$

which satisfy the condition $J = \frac{\partial(p^i, x^i)}{\partial(p^{i-1}, x^{i-1})} = 1$. Note that one can use instead of (9) the expansion (in the notation of the work [18]) of the form

$$z^{n+1} = \prod_{s=1}^{K} \exp(d_s h\hat{L}_2)\exp(c_s h\hat{L}_1)z^n, \quad (11)$$

which leads to the relations [18]

$$x^{(i)} = x^{(i-1)} + c_i \frac{h}{m} p^{(i-1)}, \quad p^{(i)} = p^{(i-1)} + d_i h f(x^{(i)}), \quad i = 1, \dots, K, \quad (12)$$

which also satisfy the condition $J = 1$. Equations (10) and (12) define the class of explicit symplectic schemes the accuracy of which is determined by the number of stages K and coefficients c_i, d_i.

Below we show how new sets of parameters of a four-stage multiparameter family of Forest–Ruth symplectic schemes (10) can be obtained analytically using Gröbner bases. Some of these schemes have a much smaller norm of the leading error term than the two four-stage schemes obtained in [6].

4   Forest–Ruth Scheme

Consider scheme (10) with $K = 4$. Using symbolic calculations, it is not difficult to obtain an expression for the error δp_n in the form [16]

$$\begin{aligned}
\delta p_n &= h P_1 f(x) + (h^2/2) P_2 u(t) f'(x) + h^3 [P_{31} f(x) f'(x)/m + P_{32} u^2 f''(x)]/6 \\
&+ (h^4 u)/(24m) \{ P_{41} \cdot [f'(x)]^2 + 3 P_{42} f(x) f''(x) + P_{43} m u^2 f^{(3)}(x) \} \\
&- [h^5/(120 m^2)](3 P_{51} f^2(x) f''(x) + f(x)(P_{52} \cdot [f'(x)]^2 - 6 P_{53} m u^2 f^{(3)}(x) \\
&- m u^2 (5 P_{54} f'(x) f''(x) + P_{55} m u^2 f^{(4)}(x))),
\end{aligned} \quad (13)$$

where $u(t)$ is the particle velocity. The expressions for $P_1, P_2, P_{31}, P_{32}, P_{41}, P_{42}, P_{43}, P_{51}, P_{52}, P_{53}, P_{54}$, and P_{55} are available in [16], and we give here only the expression for P_1 for the sake of brevity: $P_1 = 1 - c_1 - c_2 - c_3 - c_4$.

Let us calculate the weighted root-mean-square value of five polynomials P_{5j}, $j = 1, \dots, 5$:

$$P_{5,rms}^{(l)} = \left[(1/5) \sum_{j=1}^{5} (\sigma_j P_{5j})^2 \right]^{\frac{1}{2}}. \quad (14)$$

Here $\sigma_1, \dots, \sigma_5$ are problem-independent factors affecting the polynomials P_{5j} in (13), $\sigma_1 = -3$, $\sigma_2 = -1$, $\sigma_3 = 6$, $\sigma_4 = 5$, $\sigma_5 = 1$. The error δx_n obtained when calculating the coordinate x^{n+1} according to the scheme under consideration is given by the formula

$$\begin{aligned}
\delta x_n &= \frac{hp}{m} R_1 + \frac{h^2}{2m} R_2 f(x) + \frac{h^3}{6m} R_3 u(t) f'(x) + \frac{h^4}{24m^2} [R_{41} f(x) f'(x) \\
&+ R_{42} m u^2 f''(x)] \frac{h^5 u(t)}{120\,m^2} \{ R_{51} [f'(x)]^2 + 3 R_{52} f(x) f''(x) + R_{53} m u^2 f^{(3)}(x) \}.
\end{aligned} \quad (15)$$

The expressions for $R_1, R_2, R_3, R_{41}, R_{42}, R_{51}, R_{52}$, and R_{53} are available in [16], and we give here only the expression for R_1 for the sake of brevity: $R_1 = 1 - d_1 - d_2 - d_3 - d_4$.

The system of equations $P_1 = 0$, $P_2 = 0$, $P_{32} = 0$, $P_{43} = 0$ is linear in c_i, $i = 1, \dots, 4$. It is easy to rewrite it in the form $\mathcal{P} \mathbf{c} = \mathbf{b}$, where $\mathbf{c} = (c_1, c_2, c_3, c_4)^T$,

Table 1. The values of the parameters of the Forest–Ruth schemes FRl ($l = 1, \ldots, 6$) obtained in the case of the zero determinant (17)

FRl		c_i	d_i
FR1	$d_2 = -d_1$	1.681562217889354644	1.689032314592042095
		0.014940193405374902	−1.689032314592042095
		−1.471849481432093948	0.612326464931317799
FR2	$d2 = -d1$	0.462663328009943900	0.754092747422918106
		0.582858838825948413	−0.754092747422918106
		−1.983103761290466971	0.031209202772712671
FR3	$d_3 = -d_2$	0	φ_1^+
		φ_2^-	$-z/3$
		1/2	$z/3$
FR4	$d_3 = -d_2$	0	φ_1^-
		φ_2^+	$z/3$
		1/2	$-z/3$
FR5	$d_4 = 1$	−0.298786423182561709	0.092128124809906823
		0.781829095984937063	0.671234501323664155
		0.560639906662391246	−0.763362626133570977
FR6	$d_4 = 1$	0.447302683845444312	0.747066325751166375
		0.599527283811444124	−0.771392086019630084
		−2.142311455850704292	0.024325760268463710

Table 2. The values of the parameters of the Forest–Ruth schemes FRl, $l = 7, \ldots, 10$ in the particular cases of $c_1 = 0$ and $d_4 = 0$

FRl		c_1	c_2	c_3	c_4	d_1	d_2	d_3	d_4
FR7	$c_1 = 0$	0	φ_2^+	$\frac{1}{2}$	φ_2^-	φ_1^-	$\frac{z}{3}$	$-\frac{z}{3}$	φ_1^+
FR8	$c_1 = 0$	0	φ_2^-	$\frac{1}{2}$	φ_2^+	φ_1^+	$-\frac{z}{3}$	$\frac{z}{3}$	φ_1^-
FR9*	$c_1 = 0$	0	$2\varphi_3$	φ_4	$2\varphi_3$	φ_3	$\frac{1}{2} - \varphi_3$	$\frac{1}{2} - \varphi_3$	φ_3
FR10*	$d_4 = 0$	φ_3	$\frac{1}{2} - \varphi_3$	$\frac{1}{2} - \varphi_3$	φ_3	$2\varphi_3$	φ_4	$2\varphi_3$	0

$\mathbf{b} = (1, \frac{1}{2}, \frac{1}{3}, \frac{1}{4})^T$, and

$$\mathcal{P} = \begin{pmatrix} 1 & 1 & 1 & 1 \\ 0 & d_1 & d_1 + d_2 & d_1 + d_2 + d_3 \\ 0 & d_1^2 & (d_1 + d_2)^2 & (d_1 + d_2 + d_3)^2 \\ 0 & d_1^3 & (d_1 + d_2)^3 & (d_1 + d_2 + d_3)^3 \end{pmatrix}. \tag{16}$$

The determinant of this matrix is as follows:

$$\det \mathcal{P} = d_1 d_2 d_3 (d_1 + d_2)(d_2 + d_3)(1 - d_4). \tag{17}$$

The values c_i obtained from the system $P_1 = 0$, $P_2 = 0$, $P_{32} = 0$, $P_{43} = 0$ have the form of fractional rational functions whose numerators are polynomials in $d_1, \ldots, d_4$, and denominator is the determinant (17). If $\det \mathcal{P} = 0$, then it will be problematic to find expressions for c_i. In this connection, all six particular cases of the vanishing determinant (17) were considered in detail in [16]. The following polynomial system was handled therein:

$$R_1 = 0, \ P_1 = 0, \ P_2 = 0, \ P_{31} = 0, \ P_{32} = 0, \ P_{41} = 0, \ P_{42} = 0, \ P_{43} = 0. \quad (18)$$

The number of equations in (18) is equal to the number of unknowns c_1, c_2, c_3, c_4, d_1, d_2, d_3, d_4. We found with the aid of Gröbner bases six new analytical solutions of the above system, which are summarized in Table 1. The values $z, \zeta, \varphi_1, \varphi_2^{\pm}, \varphi_3$ were calculated as follows:

$$z = \sqrt{3}, \ \zeta = 2^{1/3}, \ \varphi_1^{\pm} = \tfrac{1}{6}(3 \pm z), \ \varphi_2^{\pm} = \tfrac{1}{12}(3 \pm 2z),$$
$$\varphi_3 = \tfrac{1}{12}(4 + 2\zeta + \zeta^2), \ \varphi_4 = -\tfrac{1}{3}(1 + \zeta)^2. \quad (19)$$

In Table 1 and in further similar tables, we give for brevity only the first three values of the parameters c_i, d_i, $i = 1, 2, 3$. The numerical values of the parameters c_4 and d_4 can be calculated using the formulas: $c_4 = 1 - c_1 - c_2 - c_3$, $d_4 = 1 - d_1 - d_2 - d_3$.

Earlier in the work [6], cases were considered when either $c_1 = 0$ or $d_4 = 0$; at the same time, additional symmetry conditions of the scheme were imposed: $d_2 = d_3$, $d_1 = d_4$ (at $c_1 = 0$) and $c_2 = c_3$, $c_1 = c_4$ (at $d_4 = 0$). We also considered in [16] special cases when either $c_1 = 0$ or $d_4 = 0$, but did not impose symmetry conditions. At $c_1 = 0$, two solutions coincide with those obtained for $d_3 = -d_2$, and the third solution coincides with the one given in [6]. For $d_4 = 0$, a unique real solution was obtained that coincides with the solution (4.9) from [6]. A more detailed description of these solutions is given in [16], see also Table 2, where symmetric schemes are marked with an asterisk.

Along with (14), we also introduce the functional $X_{5,rms}^{(l)}$, which represents the weighted root-mean square value of three polynomials R_{5j}, $j = 1, 2, 3$:

$$X_{5,rms}^{(l)} = \left\{ \left[R_{51}^2 + (3R_{52})^2 + R_{53}^2 \right] / 3 \right\}^{1/2}. \quad (20)$$

This formula accounts for multiplier "3" before R_{52} in (15). In the following, we will not give the magnitudes of the functionals $P_{5,rms}^{(l)}$ and $X_{5,rms}^{(l)}$ separately for each scheme, but we will give them in a summary Table 19.

One can see from Table 19 that in the cases of symmetric schemes FR9 and FR10, the value $P_{5,rms}^{(l)}$ obtained from the leading term of the scheme error is greater than in the cases of nonsymmetric schemes. Thus, although symmetric schemes allow calculations in the direction of decreasing time ($h < 0$), they are less accurate than nonsymmetric schemes (according to the latter schemes, problems can be solved only at $h > 0$, i.e., moving in the direction of increasing time).

4.1   The General Case

In general, when the determinant (17) is nonzero, expressions for polynomials (18) become more complicated. The *Mathematica* command

$$\texttt{GroebnerBasis[\{P1, R1, P2, P31, P32, P41, P42, P43\},} \atop \texttt{\{c1, c2, c3, c4, d1, d2, d3, d4\}]} \tag{21}$$

has enabled the obtaining of the Gröbner basis $\{G_1, \ldots, G_{19}\}$, which consists of 19 polynomials. The polynomial system is termed zero-dimensional if it has a finite number of complex solutions. A zero-dimensional system with as many equations as variables is sometimes said to be well-behaved [10]. The Bézout theorem asserts that a well-behaved system of n equations, which have degrees $d_1, \ldots, d_n$, has at most $d_1 \cdots d_n$ solutions. In our case, the polynomials $P_1, R_1, P_2,$ $P_{31}, P_{32}, P_{41}, P_{42},$ and P_{43} have according to [16] the degrees 1, 1, 2, 3, 3, 4, 4, 4, respectively. Thus, system (18) has at most 1152 complex and real roots counted with their multiplicities.

Unfortunately, in the Gröbner basis obtained from (21), there is no univariate polynomial. The shortest polynomials in the resulting basis are the polynomials G_1 and G_2, moreover, $G_1 = \mathcal{G}_1(d_3, d_4)$, $G_2 = \mathcal{G}_2(d_2, d_3, d_4)$. In order for the system $G_1 = 0$, $G_2 = 0$ to have a common root, it is necessary that its resultant, based on the Sylvester matrix, be zero [2,4]. The polynomials G_1 and G_2 depend collectively on three variables $d_2, d_3,$ and d_4. It follows that one needs to consider three options for excluding variables: excluding the variable d_2, excluding the variable d_3, and excluding the variable d_4. Therefore, we have considered three resultants: $\mathrm{Res}(G_1, G_2, d_4)$, $\mathrm{Res}(G_1, G_2, d_3)$, $\mathrm{Res}(G_1, G_2, d_2)$.

In the CAS *Mathematica*, the symbolic computation of the resultant is implemented in the function $\texttt{Resultant[\ldots]}$. The command $\texttt{Resultant[G1,G2,d4]}$ eliminates the variable d_4. As a result, it has been found that

$$\mathrm{Res}(G_1, G_2, d_4) = A_{54} d_3 \mathcal{F}_4(d_3) \mathcal{F}_8(d_3) \mathcal{F}_{13}(d_2, d_3) [\mathcal{F}_{14}(d_3)]^2 \mathcal{F}_{56}(d_3). \tag{22}$$

Here A_{54} is a constant consisting of 54 decimal digits, $\mathcal{F}_4(d_3) = -1 - 16d_3 - 96d_3^2 - 36d_3^3 + 324d_3^4$. The value $d_3 = 0$ is one of the solutions of the Eq. (22). However, it was mentioned above that with this value of the parameter d_3, the system (18) is incompatible. The solution of the equation $\mathcal{F}_4(d_3) = 0$ is in radicals, but its substitution into system (18) leads to giant symbolic expressions.

On the other hand, the roots of this equation can be found numerically with very high accuracy. But all the same, the real roots found are approximate. Substituting the root d_3, given as a floating-point machine number, into the original polynomial system (18) turns this system into a polynomial system with approximate coefficients. In the review paper [11], problems related to obtaining approximate Gröbner bases from polynomial systems with approximate coefficients were discussed. Here we note the following two problems.

- In principle, approximate solutions of the original polynomial system can be obtained from the approximate Gröbner basis. But substituting the found solution into the original system leads to the residuals that are several decimal orders of magnitude greater than machine rounding errors; this was shown in a number of examples in [11].

- There exists the problem of ill-conditioned polynomials [12,17]. The presence of errors in the coefficients of a poorly conditioned polynomial can even lead to a change in the number of real roots of the polynomial equation [17].

In cases where it is not possible to find solutions to a polynomial system in an analytical form, it is more reliable to search for these solutions by numerically solving an optimization problem formulated directly for the polynomials included in the original system. This methodology was described in [15,16].

Table 3. The values of the parameters of the Forest–Ruth schemes FRl ($l = 11, \ldots, 14$) obtained in the case of the zero resultant (23)

FRl		c_i	d_i
FR11	$d_4 = 1/3$	-3.793578045247588829	-2.015338965874309019
		-0.002556494312384837	2.010225977249539346
		4.023008448811463529	0.671779655291436340
FR12	$d_4 = 1/3$	0.215291546204744457	1.145093955234638990
		0.524182325872439832	-0.096729303489759326
		-0.717640932851958484	-0.381697985078212997
FR13	$d_4 = 1/4$	-0.456387075637887056	0.559356790721160223
		-0.065655505000061227	-0.478888139152354456
		0.903328999195627837	0.669531348431194233
FR14	$d_4 = 1/4$	0.313736873232085596	0.817539342386075480
		-0.440874365834237127	-1.966375102129151814
		-0.007941192169999430	1.898835759743076334

For the resultant $\mathrm{Res}(G_1, G_2, d_3)$, the following expression is obtained:

$$\mathrm{Res}(G_1, G_2, d_3) = 103129560704(3d_4 - 1)^9(4d_4 - 1)^5(1 - 9d_4 + 24d_4^2)^9 \times \mathcal{F}_{12}(d_2, d_4)[\mathcal{F}_{14}(d_4)]^6. \tag{23}$$

Expressions for polynomials $\mathcal{F}_{12}(d_2, d_4)$ and $\mathcal{F}_{14}(d_4)$ are not given here because of their bulky appearance.

It follows from (23) that the value $d_4 = 1/3$ is one of the roots of the equation $\mathrm{Res}(G_1, G_2, d_3) = 0$. Substituting this value into the polynomial system (18) significantly simplifies the appearance of the polynomials included in the system. This makes it easy to find the corresponding Gröbner basis consisting of seven polynomials $G_1, \ldots, G_7$. The Gröbner polynomial G_1 yields the following quartic equation for c_1: $G(1) = -25 + 340c_1 - 1920c_1^2 + 3840c_1^3 + 1152c_1^4 = 0$. Its roots are expressed in radicals, we omit them for brevity. Two of these roots are real. Since the Gröbner basis is triangular, the parameters d_3, d_2, d_1, c_4, c_3, c_2 are easily found. The corresponding schemes in Table 3 are the schemes FR11 and

FR12. The case of $d_4 = 1/4$ was handled similarly to the foregoing case. The corresponding schemes in Table 3 are the schemes FR13 and FR14. The roots of the equation $1 - 9d_4 + 24d_4^2 = 0$ are complex.

And finally, $\text{Res}(G_1, G_2, d_2) = 1$. It is a polynomial of degree zero, that is, it does not contain any of the variables d_2, d_3, d_4. Therefore, this resultant is useless when searching for solutions for these three variables.

Below we will show that with the help of parametric Gröbner bases, it is possible to obtain several dozen of further exact solutions of polynomial systems arising from the study of PRK schemes. At the same time, we can declare any of the eight variables $c_j, d_j, j = 1, \ldots, 4$ as a parameter. Following [10], we will call the quantities that are not included in the set of parameters, the indeterminates.

Our goal is to search for schemes in the Forest and Ruth family for which the functionals $P_{5,\text{rms}}$ and $X_{5,\text{rms}}$ are smaller than in the cases of symplectic schemes described in [15, 16]. Taking into account the previous considerations for working with parametric Gröbner bases, we have developed and implemented a procedure consisting of seven steps in the CAS *Mathematica* language.

Step 1. For certainty, consider the case when the variable c_4 is declared as a parameter. Then the Gröbner basis of system (18) is computed using the call

$$\texttt{GroebnerBasis}[\{\texttt{P1}, \texttt{R1}, \texttt{P2}, \texttt{P31}, \texttt{P32}, \texttt{P41}, \texttt{P42}, \texttt{P43}\},$$
$$\{\texttt{c1}, \texttt{c2}, \texttt{c3}, \texttt{d1}, \texttt{d2}, \texttt{d3}, \texttt{d4}\}] \tag{24}$$

By default, in all subsequent calls to the function $\texttt{GroebnerBasis}[\ldots]$, lexicographic ordering of monomials is used. We will show in the following that the Gröbner basis of the same polynomial system also depends on the order of the indeterminates in the call to the *Mathematica* function $\texttt{GroebnerBasis}[\ldots]$.

Step 2. The Gröbner basis $\{G_1, G_2, \ldots, G_N\}$ $(N > 2)$ obtained with the aid of the call (24) has no univariate polynomial. The first two polynomials G_1 and G_2 have the smallest length. We introduce for further convenience the notation: $\alpha_j = c_j$, $\alpha_{j+4} = d_j$, $j = 1, \ldots, 4$. As a rule, $G_1 = \mathcal{F}_1(\alpha_{i_1}, \alpha_{i_2})$, $G_2 = \mathcal{F}_2(\alpha_{i_1}, \alpha_{i_2}, \alpha_{i_3})$, where $1 \leq i_1, i_2, i_3 \leq 8$, $i_1 \neq i_2$, $i_2 \neq i_3$, $i_1 \neq i_3$. Then we calculate the following three resultants:

$$\text{Res}(G_1, G_2, \alpha_{i_k}), \quad k = 1, 2, 3. \tag{25}$$

The elimination of the variable α_{i_k} from the basis polynomials G_1 and G_2 is carried out with the aid of the call $\text{Res}[G_1, G_2, \alpha_{i_k}]$.

Step 3. All three resultants (25) are considered. First, using the *Mathematica* function $\texttt{Factor}[\ldots]$ we find out whether a particular resultant from the set (25) is factorizable. Some of the resultants (25) are irreducible over the field $\mathbb{Q}$ of rational numbers. Below we will briefly write "irreducible polynomials".

Step 4. All integer and rational roots of the equations $\text{Res}(G_1, G_2, \alpha_{i_k}) = 0$, $k = 1, 2, 3$ are found.

Step 5. Each of the found roots $\alpha_j \in \mathbb{Q}$ is substituted into system (18) and then the corresponding Gröbner basis is found. As a rule, this basis turns out to be triangular. In this case, the polynomial G_1 is univariate. If its degree is 2,

3 or 4, then the roots of the equation $G_1 = 0$ are in radicals. In order to avoid the appearance of giant symbolic expressions when substituting these closed-form roots into the remaining basic polynomials, the roots are translated into machine numbers with a mantissa length of 50 decimal digits $0, 1, \ldots, 9$. To do this, we use the *Mathematica* function `SetAccuracy[root,50]`.

If the degree of G_1 is higher than 4, then the solution of the equation $G_1 = 0$ is found with high accuracy using the command `NSolve[G[[1]]==0,c4, WorkingPrecision-> 50]`. Due to the fact that the Gröbner basis is triangular, all the required unknowns α_j, $j = 1, \ldots, 8$ are then easily found.

Step 6. Among the roots of each of the three resultants (25), there may be extraneous roots [13]. Therefore, verification of all found roots of system (18) is carried out by substituting them into this system.

Step 7. The functionals $P_{5,\mathrm{rms}}^{(l)}$ and $X_{5,\mathrm{rms}}^{(l)}$ are calculated according to (14) and (20) as criteria for evaluating the accuracy of the obtained real symplectic schemes.

The c_1 Parameter. In this case, (21) is replaced with the call

$$\text{GroebnerBasis}[\{\text{P1}, \text{R1}, \text{P2}, \text{P31}, \text{P32}, \text{P41}, \text{P42}, \text{P43}\},$$
$$\{\text{c2}, \text{c3}, \text{c4}, \text{d1}, \text{d2}, \text{d3}, \text{d4}\}] \tag{26}$$

The Gröbner basis $\{G_1, G_2, \ldots\}$ is obtained as 18 polynomials, with $G_1 = \mathcal{F}_1(c_1, d_4)$ and $G_2 = \mathcal{F}_2(c_1, d_3, d_4)$. Therefore, we have considered three resultants: $\mathrm{Res}(G_1, G_2, d_4)$, $\mathrm{Res}(G_1, G_2, c_1)$, and $\mathrm{Res}(G_1, G_2, d_3)$. The first of them is

$$\mathrm{Res}(G_1, G_2, d_4) =$$
$$A(c_1 - 1)^{10}(3c_1 - 1)^{17}(4c_1 - 1)^6(24c_1^2 - 9c_1 + 1)^{18}[\mathcal{F}_{12}(c_1)]^7 \mathcal{F}_{16}(c_1, d_3). \tag{27}$$

Here $A = 8077716527296737705984$. The system of equations $G_1 = 0$, $G_2 = 0$ has a common root if and only if the resultant $\mathrm{Res}(G_1, G_2, d_4) = 0$. It follows from (27) that this equality holds, in particular, in the following cases: $c_1 = 1$, $c_1 = 1/3$, and $c_1 = 1/4$. Below we will look at all these cases.

Substituting the value $c_1 = 1$ into the polynomial system (18) makes it easy to find the corresponding Gröbner basis $\{G_1, \ldots, G_7\}$, which proves to be triangular. It is then easy to find numerical values of all parameters from it. The equation $G_1 = 0$ has three real roots and two complex conjugate roots. The schemes FR15, FR16, and FR17 in Table 4 correspond to three real roots.

In the case when $c_1 = 1/3$, a triangular Gröbner basis is also obtained. The Gröbner polynomial G_1 is univariate and it yields the following quartic equation for d_4: $960d_4^4 - 3552d_4^3 + 1778d_4^2 - 340d_4 + 25 = 0$. Its roots are expressed in radicals, we omit them for brevity. Two of these roots are real. The corresponding schemes in Table 4 are the schemes FR18 and FR19 (see also [16]). The case of $c_1 = 1/4$ is handled similarly to the case of $c_1 = 1/3$, the corresponding schemes in Table 4 are the schemes FR20 and FR21 (see also [16]).

The equation $24c_1^2 - 9c_1 + 1 = 0$ has no real roots. The equation $\mathcal{F}_{12}(c_1) = 0$ has two real roots and five pairs of complex conjugate roots. At the same time, there

Table 4. The values of the parameters of the Forest–Ruth schemes FRl ($l = 15, \ldots, 23$) obtained in the case of the zero resultants (27) and (30)

FRl		c_i	d_i
FR15	$c_1 = 1$	1.000000000000000	-0.289496692562112655
		0.480967226556355638	0.095557008516098051
		-1.051024192239418045	0.957728476679205942
FR16	$c_1 = 1$	1.000000000000000	1.166062538245970983
		0.094779042188058442	-1.130360771387392979
		-0.969092132852183332	0.449341881021738872
FR17	$c_1 = 1$	1.000000000000000	-0.418994738647895922
		-0.517065975450449051	1.310130928129584835
		0.421590609795226645	-1.858323611577820880
FR18	$c_1 = 1/3$	1/3	0.854618769602128809
		-0.340323030871210685	-1.755412024452683413
		-0.013979395075754704	1.663063053955831823
FR19	$c_1 = 1/3$	1/3	0.717790337479662637
		0.676663198257488871	-2.867617920531312327
		2.019993063181644409	0.003543429387338222
FR20	$c_1 = 1/4$	1/4	1.053722261159673065
		-0.120984256091361380	-0.341956531133635430
		0.867350020601588225	2.076656572336811881
FR21	$c_1 = 1/4$	1/4	1.627745859492881698
		0.381426792344745990	-0.082614832312547523
		-0.458605339881521336	-0.834595847450495149
FR22	$c_1 = 1/6$	1/6	0.842603502826414075
		0.947753813816637058	-0.103841775357201984
		-1.284460514173413455	-0.182943493689021194
FR23	$c_1 = 1/6$	1/6	0.460642622151786919
		0.660127726674692834	0.491578202509501365
		0.215313914255060158	-0.735889167477648790

is no rational root among the real roots. Obtaining an approximate Gröbner basis leads to the difficulties discussed above at the beginning of this subsection.

In an effort to get several further roots of the equation $\text{Res}(G_1, G_2, d_4) = 0$, we substitute the value of $c_1 = 1$ into the polynomial $\mathcal{F}_{16}(c_1, d_3)$. As a result, we obtain the equation $\mathcal{F}_{16}(1, d_3) = -36(4418 - 34584d_3 + 93477d_3^2 - 87708d_3^3 - 11520d_3^4 + 36864d_3^5) = 0$. This equation has three real roots and two complex conjugate roots. The real roots are as follows: -1.858323611577820880, 0.4493418810217388717, and 0.9577284766792059415. It is easy to see that these values of the parameter d_3 coincide with those given in Table 4 for schemes FR17, FR16, and FR15. The substitution of the value $c_1 = 1/3$ in $\mathcal{F}_{16}(c_1, d_3)$ has enabled us to obtain an equation $\mathcal{F}_{16}(1/3, d_3) = 0$ whose roots give the schemes FR18 and FR19. Finally, the equation $\mathcal{F}_{16}(1/4, d_3) = 0$ gives as the output the familiar schemes FR20 and FR21. Thus, the equation in question did not add new roots.

The resultant $\mathrm{Res}(G_1, G_2, c_1) = 0$ is as follows:

$$\mathrm{Res}(G_1, G_2, c_1) = A_{50}(d_4 - 1)^2(3d_4 - 1)(4d_4 - 1)(24d_4^2 - 9d_4 + 1)^3 \\ \times \mathcal{F}_3(d_4)\mathcal{F}_4(d_4)\mathcal{F}_6(d_4)[\mathcal{F}_8(d_4)]^2\mathcal{F}_{10}(d_3, d_4)[\mathcal{F}_{12}(d_4)]^2\mathcal{F}_{60}(d_4), \tag{28}$$

where A_{50} is a large integer number containing 50 decimal digits. The polynomials $\mathcal{F}_3(d_4)$, $\mathcal{F}_4(d_4)$, $\mathcal{F}_6(d_4)$, $\mathcal{F}_8(d_4)$, $\mathcal{F}_{10}(d_3, d_4)$, $\mathcal{F}_{12}(d_4)$, and $\mathcal{F}_{60}(d_4)$ are not presented here for the sake of brevity. The case when $d_4 = 1$ was already considered above, see the schemes FR5 and FR6 in Table 1. The cases when $d_4 = 1/3$ and $d_4 = 1/4$ were already considered above, see Table 3.

The third resultant $\mathrm{Res}(G_1, G_2, d_3)$ is the 13th degree irreducible polynomial in two variables c_1 and d_4. It adds no new solutions to system (18).

We have used above in the call (26) to the *Mathematica* function GroebnerBasis[..] the sequence of the indeterminates in the form c2,c3,c4, d1,d2,d3, d4. Below we show that permutation of elements in this sequence can lead to new solutions of the original polynomial system. To do this, consider instead of (26) the following call to the function GroebnerBasis[...]:

$$\mathrm{GroebnerBasis}[\{\mathrm{P1, R1, P2, P31, P32, P41, P42, P43}\}, \\ \{\mathrm{d1, d2, d3, d4, c2, c3, c4}\}] \tag{29}$$

In this case, a Gröbner basis is obtained in which $G_1 = \mathcal{G}_1(c_1, c_4)$, $G_2 = \mathcal{G}_2(c_1, c_3, c_4)$. Therefore, we have considered three resultants: $\mathrm{Res}(G_1, G_2, c_4)$, $\mathrm{Res}(G_1, G_2, c_3)$, and $\mathrm{Res}(G_1, G_2, c_1)$. The first of them is as follows:

$$\mathrm{Res}(G_1, G_2, c_4) = A_{20}(4c_1 - 1)^6(6c_1 - 1)^{12}(24c_1^2 - 9c_1 + 1)^7 \\ \times (27c_1^2 - 10c_1 + 1)^5(96c_1^3 - 60c_1^2 + 12c_1 - 1)^5[\mathcal{F}_{25}(c_1)]^7. \tag{30}$$

Table 5. Successful permutations

Permutation No.	Permutation	New solutions
8	$\{d_1, d_2, d_3, c_2, d_4, c_4, c_3\}$	$c_3 = 0,\ c_3 = 1/2$
32	$\{d_1, d_2, d_4, c_2, d_3, c_4, c_3\}$	$c_3 = 0,\ c_3 = 1/2$
40	$\{d_1, d_2, d_4, c_3, c_2, c_4, d_3\}$	$d_3 = 0,\ d_3 = 3/4$
48	$\{d_1, d_2, d_4, c_4, c_3, c_2, d_3\}$	$d_3 = 0$
56	$\{d_1, d_2, c_2, d_4, d_3, c_4, c_3\}$	$c_3 = 1/2,\ c_1 = 1/6,\ c_3 = 0$
64	$\{d_1, d_2, c_2, c_3, d_4, c_4, d_3\}$	$d_3 = 0,\ d_3 = 3/4$
80	$\{d_1, d_2, c_3, d_4, d_3, c_4, c_2\}$	$c_2 = 0$
88	$\{d_1, d_2, c_3, c_2, d_4, c_4, d_3\}$	$d_3 = 0,\ d_3 = 3/4$
104	$\{d_1, d_2, c_4, d_4, d_3, c_3, c_2\}$	$c_2 = 0$
1000	$\{d_2, d_4, d_3, c_3, c_2, c_4, d_1\}$	$d_1 = 0$
3000	$\{c_2, d_1, c_4, c_3, d_4, d_3, d_2\}$	$d_2 = 0$
4000	$\{c_3, d_4, d_2, c_2, d_3, c_4, d_1\}$	$d_1 = 0$
5000	$\{c_4, c_3, d_4, d_2, d_1, c_2, d_3\}$	$c_1 = 1/6$

Here A_{20} is an integer consisting of 20 decimal digits. The equation $\mathrm{Res}(G_1, G_2, c_4) = 0$ has two rational roots $c_1 = 1/4$ and $c_1 = 1/6$. Root $c_1 = 1/6$ is a new root that is missing in (27) and (28). Its consideration leads to two new schemes FR22 and FR23, see Table 4.

The resultant $\mathrm{Res}(G_1, G_2, c_3)$ is an irreducible polynomial of the 13th degree in two variables c_1 and c_4. The third resultant has the following form: $\mathrm{Res}(G_1, G_2, c_1) = A_{72}c_4^2(8c_4^2 - 7c_4 + 2)^2(64c_4^3 - 40c_4^2 + 8c_4 + 1)[\mathcal{F}_6(c_4)]^2\mathcal{F}_8(c_4) \times \mathcal{F}_{12}(c_3, c_4)\mathcal{F}_{125}(c_4)$. The root $c_4 = 0$ of the equation $\mathrm{Res}(G_1, G_2, c_1) = 0$ has proved to be an extraneous root of the polynomial system (18).

The above two examples of using different sequences of seven indeterminates point to the following: if we want to find as many solutions of the original polynomial system as possible, then we must consider all permutations of the indeterminates. In the case of seven indeterminates, the number of permutations is equal to $7! = 5040$. For each particular permutation, one needs to get three resultants based on the first two polynomials included in the Gröbner basis.

To do this, one first needs to get expressions for G_1 and G_2 and see which specific indeterminates and parameters they depend on. As a rule, G_1 and G_2 depend on two indeterminates and one variable selected as a parameter. Thus, we need to consider $5040 \cdot 3 = 15120$ resultants. The whole procedure could be implemented in the language of the *Mathematica* system. However, this requires considerable programming effort. Instead of such a complete analysis of all 5040 permutations, we considered the permutations P_j with the following numbers j: $j = 8, 16, 24, 32, 40, 48, 56, 64, 72, 80, 88, 96, 104, 1000, 2000, 3000, 4000, 5000$. In Table 5, not all of these permutations are given, but only those that led to new solutions for the eight unknowns $c_j, d_j, j = 1, \ldots, 4$. To obtain all permutations of seven elements, the *Mathematica* function `Permutations[perm0]` was used. Here $perm0$ is the initial permutation, we took $perm0 = \{d_1, d_2, d_3, d_4, c_2, c_3, c_4\}$.

Table 5 contains a number of new solutions for the unknowns c_j and d_j. We have at first considered six cases when $c_j = 0, j = 2, 3, 4$ or $d_j = 0, j = 1, 2, 3$. In all these cases, the polynomial system (18) proved to be incompatible.

At $c_3 = 1/2$, a triangular Gröbner basis $\{G_1, \ldots, G_7\}$ is obtained, where $G_1 = (1 - 3c_4 + 6c_4^2)(-1 - 24c_4 + 48c_4^2)(1 - 32c_4 + 364c_4^2 - 4000c_4^3 + 10288c_4^4 - 8064c_4^5 - 29696c_4^6 + 73728c_4^7)$. The roots of the equation $1 - 3c_4 + 6c_4^2 = 0$ are complex. The roots of the equation $48c_4^2 - 24c_4 - 1 = 0$ are real: $(c_4)_{1,2} = \frac{1}{12}(3 \pm 2\sqrt{3})$. This case was handled above, see the schemes FR3 and FR4. The equation $1 - 32c_4 + 364c_4^2 - 4000c_4^3 + 10288c_4^4 - 8064c_4^5 - 29696c_4^6 + 73728c_4^7 = 0$ has three real roots and two pairs of complex conjugate roots. Real roots lead to three new schemes FR24, FR25, and FR26.

At $d_3 = 3/4$, a triangular Gröbner basis $\{G_1, \ldots, G_7\}$ is also obtained. The equation $G_1 = 0$ has only two real roots. The corresponding new schemes are presented in Table 6 as the schemes FR27 and FR28.

So, above we considered only 18 permutations, or 3.57 % of 5040 permutations. Therefore, it is natural that considering all 5040 permutations will allow one to obtain several further new solutions to the system (18).

The Remaining Variables as Parameters. The cases when one of seven variables $c_2, c_3, c_4, d_1, d_2, d_3, d_4$ is declared as a parameter were handled similarly to the above-considered case of the variable c_1 as a parameter, see Table 7, 8, 10, 11, 12, 13, 14, 15, and 16. Table 17 shows the number of new schemes obtained at the consideration of each of the variables $c_i, d_i, i = 1, \ldots, 4$ as a parameter.

Table 18 presents the orders of smallness of the residuals $r_{\min}$ and $r_{\max}$ for several new symplectic schemes of the Forest–Ruth family, which were obtained above. Naturally, in cases where the analytical expressions for c_i and $d_i, i = 1, \ldots, 4$ are relatively simple, it is possible to check their correctness by substituting them into the original polynomial system without switching to machine floating-point numbers. If the solution found in the analytical form is correct, then all the equations of the polynomial system are exactly satisfied. Therefore, in these cases, the equalities $r_{\min} = 0$ and $r_{\max} = 0$ take place. Formally, the schemes FR3, FR4, FR7, FR8, FR9, and FR10 are the best in terms of the smallness of the residuals $r_{\min}$ and $r_{\max}$. However, they are far from optimal in terms of the smallness of the functionals $P_{5,rms}^{(l)}$ and $X_{5,rms}^{(l)}$, see Table 19.

In other cases, numerical values of c_i and $d_i, i = 1, \ldots, 4$ were substituted into the initial system in the form of machine floating-point numbers and with a mantissa length of 18 decimal digits. As can be seen from Table 18, schemes FR27, FR50, FR51, FR52, and FR73 have the smallest residuals $r_{\min} = O(10^{-50})$ and $r_{\max} = O(10^{-49})$. The schemes FR42, FR47, FR50, FR51, FR52, FR55, and FR59 have the least approximation errors, these are the best found schemes (see Table 19). These conclusions partially correlate with the conclusions from Table 18 concerning the schemes with the least residuals $r_{\min}$ and $r_{\max}$. The magnitude of the error functional $P_{5,rms}^{(l)}$ of scheme FR10 obtained previously in [6] is 121 times larger than in the cases of new schemes FR47, FR50, FR51, and FR52.

5   Kepler's Problem

The problem of the motion of a system consisting of two interacting particles (the problem of two bodies, the Kepler's problem) admits a complete analytical solution in the general form [8]. This solution was presented in [14] for the case of equal masses of the both bodies, and in [15], it was given for the bodies with different masses.

This problem is described by a system of eight ordinary differential equations (ODEs), in which four equations describe the time evolution of the coordinates (x_k, y_k) of each of the two bodies, and the quantities (p_{kx}, p_{ky}), $k = 1, 2$ describe the evolution of the impulses of the bodies. The mentioned system of ODEs is solved at the given initial positions of the bodies and their impulses. With equal masses of bodies $m_1 = m_2 = 1$, the total initial energy E_0 of both bodies $|E_0| = |v_0^2 - 1/(2a_0)|$, where $2a_0$ is the given initial distance between the bodies $(a_0 > 0)$, v_0 is the given absolute value of the initial velocity of each body.

Table 6. The values of the parameters of the Forest–Ruth schemes FRl ($l = 24, \ldots, 28$) obtained in the cases when $c_3 = 1/2$ or $d_3 = 3/4$

FRl		c_i	d_i
FR24	$c_3 = 1/2$	0.391025572210943674	0.967398350433141497
		0.678124844465627657	2.202427111618248827
		1/2	−0.111003864212194508
FR25	$c_3 = 1/2$	0.835311238373158736	−0.350920339289419778
		−0.378958977995565850	1.171732061213387849
		1/2	−1.814913008091942417
FR26	$c_3 = 1/2$	0.094132364293527189	0.723711816433472820
		−0.046380082078234786	−0.399429463535403217
		1/2	0.497003465452825098
FR27	$d_3 = 3/4$	0.200343862556644811	0.645182676558732474
		−1.010914108245682741	−0.024497356766309670
		1.772663714786654245	3/4
FR28	$d_3 = 3/4$	−1.742065593683020234	−0.631721835302486836
		−0.296628561098109581	0.410158533684560698
		1.724192327420640529	3/4

Table 7. The values of the parameters of the Forest–Ruth schemes FRl ($l = 29, \ldots, 36$) obtained in the case when c_2 is the parameter

FRl		c_i	d_i
FR29	$c_2 = 2/3$	0.364254344284583814	0.869516211247767583
		2/3	2.792690380329816512
		0.587980436123375884	0.054227787182224768
FR30	$c_2 = 2/3$	1.878392245341283130	−0.148939912684807651
		2/3	0.072050469147999406
		−2.119991445360319150	0.835740764826119230
FR31	$c_2 = 2/3$	0.197431266262198278	0.997479482407853902
		2/3	−0.097322556987957231
		−0.933855594163194543	−0.268583735242126327
FR32	$c_2 = 2/3$	0.168148869366041101	0.464264589751261379
		2/3	0.493986641393048766
		0.2088349346809210962	−0.737725321402639611
FR33	$c_2 = 2/3$	−0.170487789371228258	0.124713923957057081
		2/3	0.644865752568700276
		0.5457035136145351390	−0.695444714544033987
FR34	$c_2 = 2/3$	0.374170872627683649	0.717310607339155385
		2/3	−1.033722587512287717
		−6.140624051165247365	0.001454385616574405
FR35	$c_2 = 1/3$	0.305530612248020483	4.937738540440601523
		1/3	−0.027239637450445254
		−0.339218512202854752	−4.168275877672941372
FR36	$c_2 = 1/3$	0.680077031402771998	0.881570343081302874
		1/3	−0.790588680725968301
		−1.186453739372546212	0.176773784101247734

Table 8. The values of the parameters of the Forest–Ruth schemes FRl ($l = 37, \ldots, 46$) obtained in the case when c_2 is the parameter

FRl		c_i	d_i
FR37	$c_2 = (1+\sqrt{5}])/3$	3.782386995165049639	−0.076051961619127312
		1.078689325833263232	0.043960674924039808
		−4.443316201850377574	0.786839835443811639
FR38	$c_2 = (1 + \sqrt{5})/3$	0.144891076593792282	0.784982662463530111
		1.078689325833263232	−0.115777773963705551
		−1.415006067291033615	−0.165458346668356931
FR39	$c_2 = (1 + \sqrt{5})/3$	0.213103271189846452	0.573068437077953834
		1.078689325833263232	1.002219472013933223
		0.021085719326262061	−1.091461204437575932
FR40	$c_2 = (1 + \sqrt{5})/3$	−0.611238566964510911	0.052429960499692211
		1.078689325833263232	0.702434707725292420
		0.580716563209192545	−0.860370750278669990
FR41	$c_2 = (1 - \sqrt{5})/3$	0.219311236012542665	0.760452067359764565
		−0.412022659166596565	−0.091641127721282675
		1.178509694121097509	1.100220634320858561
FR42	$c_2 = (1 - \sqrt{5})/3$	0.1394284522075851837	0.486323892019975997
		−0.412022659166596565	−0.042011883595088195
		0.967231066343295769	0.441924899417734389
FR43	$c_2 = (1 - \sqrt{5})/3$	0.318257154554292683	0.826446621342308992
		−0.412022659166596565	−1.905814461246662400
		−0.009200018152718083	1.832414525155164045
FR44	$c_2 = (1 - \sqrt{5})/3$	−2.671203027064244664	0.127955739058414071
		−0.412022659166596565	−0.099395301108287731
		3.473342068227190026	0.715057655495029390
FR45	$c_2 = (1 - \sqrt{5})/3$	−1.873117657996741733	−0.649164960959432341
		−0.412022659166596565	0.364748420484928815
		1.780573723084865883	0.775557276840459310
FR46	$c_2 = (1 - \sqrt{5})/3$	−1.873117657996741733	−0.649164960959432341
		−0.412022659166596565	0.364748420484928815
		1.780573723084865883	0.775557276840459310

When numerically solving the Kepler problem using symplectic difference schemes, the energy e^n at $t = t_n > 0$ is calculated with an error. Let $\delta E^n = (E^n - E_0)/E_0$, where $E^n = (1/2)[(p_{1x}^n)^2 + (p_{1y}^n)^2 + (p_{2x}^n)^2 + (p_{2y}^n)^2] - 1/r^n$, $r^n = [(x_1^n - x_2^n)^2 + (y_1^n - y_2^n)^2]^{1/2}$. It was shown in [16] that for ensuring the zero eccentricity e of the particles orbits it is sufficient to set $v_0 = 0.5/\sqrt{a_0}$. Along

Table 9. The values of the parameters of the Forest–Ruth schemes FRl ($l = 47, \ldots, 53$) obtained in the cases when $c_4 = 3/4$ or $d_1 = 1/2$

FRl		c_i	d_i
FR47	$c_4 = 3/4$	4.290878395464025169	3.434400270740872001
		0.000963763146154897	−3.435707719128205871
		−4.041842158610180066	0.656514850491044033
FR48	$c_4 = 3/4$	0.277948154623496169	2.459832693693553053
		0.343888647781502973	−0.056699486586081620
		−0.371836802404999142	−1.672913435532665059
FR49	$c_4 = 3/4$	0.520981254725582490	1.113980205862570658
		−0.161927364012237208	−1.472633501812066773
		−0.109053890713345282	1.213681750558400100
FR50	$d_1 = 1/2$	0.136825942475053071	1/2
		−0.295364245574992759	−0.062976137694193308
		0.837926908632179149	0.440948891327570560
FR51	$d_1 = 1/2$	−0.586438190415940380	1/2
		−0.075359198960140901	−0.430061540610022957
		1.038753638680670670	0.676446958625960263
FR52	$d_1 = 1/2$	0.182572224329682333	1/2
		0.744884110251947176	0.538588670363744296
		0.141395996554118555	−0.758360900822698969
FR53	$d_1 = 1/2$	−3.530966905753254684	1/2
		6.547681802370795675	−0.068515429495817908
		−7.087388112997812705	−0.375425403877225227

with the error in the energy δE^n, the values of $\delta r_{1,max}$ and $\delta y_{1,mean}$ were also calculated, where $\delta r_{1,max} = \max_j((x_{1,j}^2 + y_{1,j}^2)^{1/2} - a)/a$ for the case of the zero eccentricity. The quantity $\delta y_{1,mean}$ was computed as the arithmetic mean of the quantities $\delta y_{1j} = y_{1j} - y_{1,ex}$. Here $y_{1,ex}$ is the exact value of the y coordinate at the point of the intersection of line $x = x_{1j}$ with the ellipse of the first particle.

As one can see in Table 20, the symmetric Forest–Ruth scheme FR9 produces much larger errors $|\delta E|_{mean}$ than the schemes FR3 and FR50. This is consistent with the behavior of the norm of the leading error term of these schemes.

Table 21 shows the results obtained at the nonzero eccentricity by the time $t = 500$ (10^5 time steps). As in the case of the zero eccentricity, the schemes FR3 and FR50 demonstrate a higher accuracy than the scheme FR9. The error in energy δE_{mean} amounts in the case of the scheme FR50 to only about 1/8 of the error δE_{mean} obtained in the case of the scheme FR3. This means that the scheme FR50 produces a much smaller error in energy than the scheme FR3.

Table 10. The values of the parameters of the Forest–Ruth schemes FRl ($l = 54, \ldots, 63$) obtained in the case when c_3 is the parameter

FRl		c_i	d_i
FR54	$c_3 = 1$	0.232552203620762051	0.864933853224711811
		−0.240171870993110794	−0.173521835714879970
		1	1.450514067565065171
FR55	$c_3 = 1$	0.139963046708770169	0.484042373117847543
		−0.442222496742616455	−0.038391737134385132
		1	0.442340759342235601
FR56	$c_3 = 1$	−0.549628408322132739	0.516200660144340638
		−0.072468674149454992	−0.443663538175265205
		1	0.674727683742161504
FR57	$c_3 = 1$	2.302601066465332119	0.112742296458862205
		−1.983691962779933225	0.511421634072704643
		1	−0.936108794035271761
FR58	$c_3 = 2/3$	0.351122270415848032	0.822180431133326332
		0.663430432875993088	3.675251576576031878
		2/3	−0.029344841230348257
FR59	$c_3 = 2/3$	0.130496314992084781	0.545153172452857850
		−0.152112316952245843	−0.129275520560538766
		2/3	0.445295236274126733
FR60	$c_3 = 2/3$	−0.214714867139629529	0.684249379826096300
		−0.050505925257735998	−0.566790506272220428
		2/3	0.644799229296194684
FR61	$c_3 = 2/3$	12.985132914037646926	0.000735492767526182
		−12.559557711764700905	0.717113943685514226
		2/3	−1.050357583178864997
FR62	$c_3 = -1/3$	0.323096342154017994	13.50376205182945044
		0.332614327860069729	−0.009534841164435389
		−1/3	−12.74671531967306675
FR63	$c_3 = -1/3$	1.549722230386746638	−3.514818811551034895
		1.429077640949028648	−1.457098486909627231
		−1/3	2.622648492232692156

The absolute value of the error $\delta y_{1,mean}$ amounts in the case of the scheme FR50 to only about $1/5$ of the error $\delta y_{1,mean}$ obtained in the case of the scheme FR3. Similar errors $\delta y_{2,mean}$ obtained for the second particle have the same absolute values as in the case of the first particle, but their signs are opposite to the signs of $\delta y_{1,mean}$.

Table 11. The values of the parameters of the Forest–Ruth schemes FRl ($l = 64, \ldots, 71$) obtained in the case when c_4 is the parameter

FRl		c_i	d_i
FR64	$c_4 = -1$	0.923752664541586022	16.023236171683106527
		1.028858031528064696	10.667066118735512987
		0.047389303930349282	−9.439832216877909698
FR65	$c_4 = -1$	0.230401316001649608	0.614993313027275731
		1.767802001353397688	2.349801988037760294
		0.001796682644952704	−2.372282095213200527
FR66	$c_4 = -1$	2.693266252232378788	0.272765858595981272
		−2.687864699530211844	0.258660862508406001
		1.994598447297833056	−0.704601530894185303
FR67	$c_4 = 2/3$	0.693911807455204925	1.389601984542431907
		−0.180750033711638009	−1.751837222444102726
		−0.179828440410233583	1.391280849389145194
FR68	$c_4 = (-1 + \sqrt{5})/3$	0.114099408934045299	0.649850775109876324
		−0.066472584291344747	−0.282718024274440564
		0.540350516190702883	0.469755418612118481
FR69	$c_4 = (-1 - \sqrt{5})/3$	0.231348624383101124	0.617413961551446993
		1.845849356904791962	2.505944087496861072
		0.001491344545370146	−2.526048748369218988
FR70	$c_4 = (-1 - \sqrt{5})/3$	2.763558393992365262	0.282057553866780170
		−2.795071176619652551	0.244409846615326875
		2.110202108460550521	−0.690943228859052633
FR71	$c_4 = (-1 - \sqrt{5})/3$	0.332267944750488842	0.743399694345549828
		0.667975398001819257	−21.981819840398356626
		1.078445983080955133	0.001613923746367732

Table 12. The values of the parameters of the Forest–Ruth schemes FRl ($l = 72, \ldots, 75$) obtained in the case when $c_4 = 1/6$ or $c_4 = 16/27$

FRl		c_i	d_i
FR72	$c_4 = 1/6$	1.075394753302521569	−0.638191122739963413
		−0.562921915373837090	1.672633703407438461
		0.320860495404648855	−2.181423749195410509
FR73	$c_4 = 16/27$	0.818672678175588272	1.833608220246540205
		−0.239249580730862964	−2.402016364561991597
		−0.172015690037317901	1.987452002208430171
FR74	$c_4 = 16/27$	−0.176216464958336337	0.705396326757519899
		−0.048272736660086400	−0.577957534527034398
		0.631896609025830144	0.637881726660246176
FR75	$c_4 = 16/27$	0.494971671137937628	−0.948190157578799342
		0.369511495225417267	0.079013950429035923
		−0.457075758955947487	1.633761849965980780

Table 13. The values of the parameters of the Forest–Ruth schemes FRl ($l = 76, \ldots, 85$) obtained in the case when d_1 is the parameter

FRl		c_i	d_i
FR76	$d_1 = 2/3$	0.204275083815869887	2/3
		−0.821240613630061829	−0.034251150245736271
		1.586167530241972297	0.808482764907095971
FR77	$d_1 = 2/3$	0.110529614169578658	2/3
		−0.060536883821488756	−0.308218556075319167
		0.529316569535254555	0.475003634825955049
FR78	$d_1 = 2/3$	−0.246981683746760770	2/3
		−0.052395107294725361	−0.556352049684461481
		0.696641970456244612	0.649688395520743066
FR79	$d_1 = 2/3$	−0.029918619502392475	2/3
		1.391696101106459054	−0.179948437412172961
		−1.753316304811621968	−0.180888740802033290
FR80	$d_1 = 3/4$	0.217797377255593747	3/4
		−0.439474733154072297	−0.084286494754561427
		1.206367201598754224	1.065889648917041605
FR81	$d_1 = 3/4$	0.082220658475439553	3/4
		−0.042043274315474989	−0.446288824714853250
		0.490110712558433011	0.511001742424794030
FR82	$d_1 = 3/4$	−0.093660943038827783	3/4
		−0.043524035940067826	−0.593590536527777054
		0.562491408233619701	0.617331657398234449
FR83	$d_1 = 3/4$	0.282751320628652862	3/4
		−0.875414602261540111	−2.866262183797889275
		−0.001749298330001449	2.839324076485054340
FR84	$d_1 = 3/4$	0.122962367199072599	3/4
		1.154710480553913358	−2.866262183797889275
		−1.483263553884396613	2.839324076485054340
FR85	$d_1 = 3/4$	0.453772840879245976	3/4
		0.592568295889989923	−0.763378994607420991
		−2.069852675023231005	0.027157631288323441

Table 14. The values of the parameters of the Forest–Ruth schemes FRl ($l = 86, \ldots, 90$) obtained in the cases when $d_1 = 1/3$ or $d_2 = 1/6$

FRl		c_i	d_i
FR86	$d_1 = 1/3$	-1.061210383861491179	$1/3$
		-0.124764893770052270	-0.283051205857396058
		1.561210383861491179	0.690936406092486640
FR87	$d_1 = 1/3$	0.105100678126442684	$1/3$
		0.529988715863071753	0.502653125702352770
		0.394899321873557316	-0.609454883379762706
FR88	$d_1 = 1/3$	3.383237864740253990	$1/3$
		-3.727448818616042444	$1/6$
		3.082935363183809675	-0.615610764345786215
FR89	$d_2 = 1/6$	-4.276220596064195618	-0.859926774390249674
		-1.952856967612454919	$1/6$
		3.168223353300535282	0.943721911449059608
FR90	$d_2 = 1/6$	3.383237864740253990	$1/3$
		-3.727448818616042444	$1/6$
		3.082935363183809675	-0.615610764345786215

Table 15. The values of the parameters of the Forest–Ruth schemes FRl ($l = 91, \ldots, 94$) obtained in the case when d_2 is the parameter

FRl		c_i	d_i
FR91	$d_2 = 1/2$	-1.656461349913587036	-0.636061115269224549
		-0.161698671975507331	$1/2$
		1.728130415131394716	0.716119776115948032
FR92	$d_2 = 1/2$	-0.927682804781144334	0.821495974995660855
		0.567079873156935371	$1/2$
		0.999351869998952014	-0.741437314148937372
FR93	$d_2 = 1/2$	0.108591275285244788	0.339219840428187569
		0.532722086799126630	$1/2$
		0.388349357158891234	-0.616031036746662437
FR94	$d_2 = 1/2$	2.293935572964878270	0.119753909827438190
		-1.982466778384829536	$1/2$
		1.023930342194913976	-0.926323914261688727

Table 16. The values of the parameters of the Forest–Ruth schemes FRl ($l = 95, \ldots, 98$) obtained in the case when d_3 is the parameter

FRl		c_i	d_i
FR95	$d_3 = 2/3$	0.193635546310194233	0.610720278963038252
		-1.514131240893582026	-0.012201087465723859
		2.266721524338190638	$2/3$
FR96	$d_3 = 2/3$	-0.414351004195240352	0.579788510045530680
		-0.062798023340654284	-0.494940129503867069
		0.860535223163841919	$2/3$
FR97	$d_3 = 1/2$	0.091689601485633373	0.729954875905782486
		-0.045246631531655207	-0.410133860075899430
		0.497436187438423795	$1/2$
FR98	$d_3 = 1/2$	1.095187697120513955	1.250031937811706533
		0.066378351994972941	-1.227061470621614945
		-1.003038424429919768	$1/2$

Table 17. The number of new schemes found in the Forest–Ruth family at the consideration of each of the variables c_i, d_i, $i = 1, \ldots, 4$ as a parameter

Parameter	c_1	c_2	c_3	c_4	d_1	d_2	d_3	d_4
Number of new schemes	28	25	10	12	15	4	4	0

Table 18. The residuals $r_{\min}$ and $r_{\max}$ for several schemes from the Forest–Ruth family

l	$r_{\min}$	$r_{\max}$	l	$r_{\min}$	$r_{\max}$	l	$r_{\min}$	$r_{\max}$
9	0	0	50	$O(10^{-50})$	$O(10^{-49})$	73	$O(10^{-50})$	$O(10^{-49})$
10	0	0	51	$O(10^{-50})$	$O(10^{-49})$			
27	$O(10^{-50})$	$O(10^{-49})$	52	$O(10^{-50})$	$O(10^{-49})$			

Table 19. The values of the error functionals $P_{5,rms}^{(l)}$ and $X_{5,rms}^{(l)}$ for several schemes from the Forest–Ruth family

l	$P_{5,rms}^{(l)}$	$X_{5,rms}^{(l)}$	l	$P_{5,rms}^{(l)}$	$X_{5,rms}^{(l)}$	l	$P_{5,rms}^{(l)}$	$X_{5,rms}^{(l)}$
9	6.3431	2.5624	47	0.0386	0.0708	52	0.0386	0.0708
10	4.6743	8.3036	50	0.0386	0.0708	55	0.0498	0.0915
42	0.0471	0.0876	51	0.0386	0.0708	59	0.0559	0.0550

Table 20. Errors δE_{mean}, $|\delta E|_{mean}$, and $\delta r_{1,\max}$ at $e = 0$ and $t = 7140h$, $h = 0.005$ for the fourth-order Forest–Ruth methods from Tables 1, 2, and 9

Forest–Ruth scheme	δE_{mean}	$|\delta E|_{mean}$	$\delta r_{1,\max}$
FR9	$-1.878e - 14$	$1.878e - 14$	$4.636e - 13$
FR3	$-5.369e - 15$	$6.693e - 15$	$5.462e - 14$
FR50	$3.533e - 15$	$7.654e - 15$	$1.954e - 14$

Table 21. Errors δE_{mean}, $|\delta E|_{mean}$, and $\delta y_{1,mean}$ at $v_0 = 0.15$ for the fourth-order Forest–Ruth methods from Tables 1, 2, and 9

Forest–Ruth scheme	δE_{mean}	$|\delta E|_{mean}$	$\delta y_{1,mean}$
FR9	$3.292e - 9$	$3.292e - 9$	$-3.542e - 6$
FR3	$9.684e - 10$	$9.684e - 10$	$-8.455e - 7$
FR50	$1.226e - 10$	$1.284e - 10$	$1.714e - 7$

6   Conclusions

The problem of constructing higher-order symplectic integration techniques for molecular dynamics problems with separable Hamiltonians is considered. A general method for finding symplectic schemes of high order of accuracy using parametric Gröbner bases, resultants, and permutations of variables is proposed. The implementation of the method is described by the example of four-stage partitioned Runge–Kutta (PRK) schemes of the Forest–Ruth family. This method has enabled us to find 96 new symplectic four-stage schemes in the Forest–Ruth family. Among these schemes, several schemes have been found that are the best in terms of the smallness of the leading term of the approximation error. It turned out that the value of the error functional of the best of the two schemes obtained earlier in [6] is 121 times greater than in the case of the new best PRK schemes found in this paper.

All required symbolic calculations are performed using the computer algebra system *Mathematica*. When searching for new schemes, it turned out to be effective to combine the technique of Gröbner bases with Sylvester's resultants and with permutations in the order of variables in the call of *Mathematica* function that calculates the Gröbner basis.

We emphasize that the real solutions of polynomial systems found using the technique described above do not exhaust the entire variety of solutions of these systems, since in this paper we focused mainly on finding solutions to the original polynomial system obtained using integer and rational roots of resultants, as well as roots of quadratic multipliers of resultants. Further solutions can be found with sufficient accuracy for applications by numerically solving the problems of minimizing the functional, which takes into account the value of one of the desired parameters, found using the resultants. This work is quite feasible and can lead to even more accurate PRK schemes compared to those schemes that were discovered in the framework of the study described above.

The presented study shows that before increasing the number of stages, it is advisable to conduct a detailed search for optimal parameters of the scheme at a fixed number of stages.

References

1. Adams, A.L., Loustaunau, P.: An Introduction to Gröbner Bases. Graduate Studies in Mathematics, vol. 3. American Mathematical Society, Providence (1996)
2. Akritas, A.G.: Elements of Computer Algebra with Applications. Wiley-Interscience, New York (1989)
3. Blanes, S., Casas, F.: A Concise Introduction to Geometric Numerical Integration. CRC Press, Boca Raton (2016)
4. Cox, D., Little, J., O'shea, D.: Ideals, Varieties, and Algorithms, 2nd edn. Springer, New York (1997). https://doi.org/10.1007/978-3-662-41154-4
5. Dubrovin, B.A., Novikov, S.P., Fomenko, A.T.: Modern Geometry – Methods and Applications. Springer, New York (1992). https://doi.org/10.1007/978-1-4612-1100-6

6. Forest, E., Ruth, R.D.: Fourth-order symplectic integration. Physica D **43**, 105–117 (1990)
7. Hairer, E., Lubich, C., Wanner, G.: Geometric Numerical Integration. Structure-Preserving Algorithms for Ordinary Differential Equations, 2nd edn. Springer, Heidelberg (2006). https://doi.org/10.1007/3-540-30666-8
8. Landau, L.D., Lifshitz, E.M.: Mechanics. Course of Theoretical Physics, vol. 1, 3rd edn. Elsevier, Amsterdam (1976)
9. Leimkuhler, B., Reich, S.: Simulating Hamiltonian Dynamics. Cambridge University Press, Cambridge (2004)
10. Liang, J., Gerhard, J., Jeffrey, D.J., Moroz, G.: A new Maple package for solving parametric polynomial systems. Commun. Comput. Algebra **43**(3), 61–72 (2009)
11. Lichtblau, D.: Solving polynomial systems using numeric Gröbner bases. In: Davenport, J.H., Kauers, M., Labahn, G., Urban, J. (eds.) ICMS 2018. LNCS, vol. 10931, pp. 335–342. Springer, Cham (2018). https://doi.org/10.1007/978-3-319-96418-8_40
12. Sasaki, T., Kako, F.: Floating-point Gröbner basis computation with ill-conditionedness estimation. In: Kapur, D. (ed.) ASCM 2007. LNCS (LNAI), vol. 5081, pp. 278–292. Springer, Heidelberg (2008). https://doi.org/10.1007/978-3-540-87827-8_23
13. Soylu, R., Akbulut, M.B.: Extraneous roots and kinematic analysis of spatial mechanisms and robots. Mech. Mach. Theory **32**(7), 775–788 (1997)
14. Vorozhtsov, E.V., Kiselev, S.P.: Comparative study of the accuracy of higher-order difference schemes for molecular dynamics problems using the computer algebra means. In: Boulier, F., England, M., Sadykov, T.M., Vorozhtsov, E.V. (eds.) CASC 2020. LNCS, vol. 12291, pp. 600–620. Springer, Cham (2020). https://doi.org/10.1007/978-3-030-60026-6_35
15. Vorozhtsov, E.V., Kiselev, S.P.: Optimal four-stage symplectic integrators for molecular dynamics problems. In: Boulier, F., England, M., Sadykov, T.M., Vorozhtsov, E.V. (eds.) CASC 2021. LNCS, vol. 12865, pp. 420–441. Springer, Cham (2021). https://doi.org/10.1007/978-3-030-85165-1_24
16. Vorozhtsov, E.V., Kiselev, S.P.: Higher-order symplectic integration techniques for molecular dynamics problems. J. Comput. Phys. **452**, 110905 (2022). https://doi.org/10.1016/j.jcp.2021.110905
17. Wilkinson, J.H.: The evaluation of the zeros of ill-conditioned polynomials. Part 1. Numerische Math. **1**, 150–166 (1959)
18. Yoshida, H.: Construction of higher order symplectic integrators. Phys. Lett. A **43**(5–7), 262–268 (1990)

A Mechanical Method for Isolating Locally Optimal Points of Certain Radical Functions

Zhenbing Zeng[1]([⊠])[iD], Yaochen Xu[1,2][iD], Yu Chen[1][iD], and Zhengfeng Yang[3]

[1] Department of Mathematics, Shanghai University, Shanghai 200444, China
zbzeng@shu.edu.cn

[2] Center for Excellence in Molecular Cell Science, Shanghai Institute of Biochemistry and Cell Biology, Chinese Academy of Science, Shanghai 200031, China
xuyaochen@sibcb.ac.cn

[3] School of Software Engineering, East China Normal University, Shanghai 200062, China
zfyang@sei.ecnu.edu.cn

Abstract. In this paper, we present a symbolic computation method for constructing a small neighborhood U around a known local optimal maximal or minimal point x_0 of a given smooth function $f : \mathbb{R}^n \to \mathbb{R}$ that contains radical or rational expressions of several variables, so that x_0 is also the global optimal point of $f(x)$ restricted to the small neighborhood U. The constructed small neighborhood can be used to prove that $f(x_0)$ is the global optimum of f in a rather large region M with $U \subset M$ via exact numeric computation like interval evaluation and branch-and-bound technology.

Keywords: Locally optimal points · Isolating algorithm · Radical function · Symbolic computation

1 Introduction

In some geometric optimization problems, we want to calculate the maximal value of a multivariate function $f : \mathbb{R}^n \to \mathbb{R}$ over some domain $M \subset \mathbb{R}^n$ which contains radical (or rational, trigonometrical) expressions. Usually, the objective function f is smooth, i.e., it has continuous derivatives up to any desired order over M. Therefore, applying numerical experiments the *de facto* optimal point of f can be observed with very big confidence, and it is also relatively easy to verify that the optimal point x_0 obtained from numerical searching is actually a local optimal point, namely, the partial derivatives of f with respect to each variable is zero at x_0, and the Hessian matrix of f at x_0 is positive-(semi-)definite or negative-(semi-)definite. In many cases, the numerical computation also shows that x_0 is the unique local optimal point of the objective function,

---

Supported by National Natural Science Foundation of China (12171159, 12071282).

but a strict mathematical proof is hard due to intermediate expression swell in symbolic computation.

For example, let $P_i = (x_i, y_i, z_i)$ $(i = 1, 2, \ldots, 6)$ be six points on the unit sphere S^2 and suppose we want to find the maximum of the sum of their pairwise Euclidean distances, $d = \sum_{1 \le i < j \le 6} \|P_i - P_j\|_2$, where

$$\|P_i - P_j\|_2 = \sqrt{(x_i - x_j)^2 + (y_i - y_j)^2 + (z_i - z_j)^2}.$$

To avoid the manifold solution of this optimal problem generated by the rigid movements on S^2, we may assume that one point has been fixed at the North Pole, and another point has been fixed on the prime meridian. Then, both of Monte Carlo search and the grid search (see, e.g., [1,9,10]) show that the maximum of d is $22.9705\ldots \approx 6 + 12\sqrt{2}$, and the local optimal points are the following unique ones:

$$(0,0,1), \ (0,0,-1), \ (1,0,0), \ (-1,0,0), \ (0,1,0), \ (0,-1,0).$$

To the best of our knowledge, no mathematical proof has been given to this conjecture yet.

Generally, if $x_0 \in M$ is the unique local maximal point of a continuous function $f(x)$ formed by finitely many steps of the four basic arithmetic operations, the radical, the exponential, and the trigonometrical functions of n variables $x_1, x_2, \ldots, x_n$ on a compact domain $M \subset \mathbb{R}^n$, then, by interval evaluation of $f(x)$, we can construct a neighborhood

$$U(x) = [x_1 - \varepsilon, x_1 + \varepsilon] \times [x_2 - \varepsilon, x_2 + \varepsilon] \times \cdots \times [x_n - \varepsilon, x_n + \varepsilon] \subset M,$$

for any point $x = (x_1, x_2, \ldots, x_n) \in M \setminus \{x_0\}$, where $\varepsilon = \varepsilon(x) > 0$ is dependent on x, so that the upper bound of $f(x)$ on $U(x)$ is less than $f(x_0)$. If we can also find a neighborhood $U_0 = U(x_0)$ of x_0 so that restricted on $U(x_0)$, $f(x) \le f(x_0)$, then we will get a family of neighborhood $\{U(x) | x \in M\}$ that covers the set M. According to the compactness of M, we would find a finite subset $\{U_1, U_2, \ldots, U_N\}$ of the family that satisfies

$$M \subset U(x_0) \cup U_1 \cup U_2 \cup \cdots \cup U_N,$$

and on each U_i, $f(x) \le f(x_0)$. Clearly, if we could generate all neighborhoods $U(x)$ for every point $x \in M \setminus \{x_0\}$ in advance, then we would be able to produce a proof to the original optimization problem. To utilize this idea on computer for a machine proof, we may implement this through the following two procedures:

Procedure 1: isolate the local optimal point. Construct a function $g(x)$ which has x_0 as the unique maximal point with $g(x_0) = f(x_0)$, and a neighborhood U_0 of x_0 that satisfies $f(x) \le g(x)$ for $x \in U_0$, and, therefore, $f(x) \le f(x_0)$ on U_0.

Procedure 2: ``divide-and-conquer'' outside the isolated regions. Partition $M \setminus U_0$ into a sequence of cubes $D_1, D_2, \ldots, D_m$ in $\mathbb{R}^n$ where D_i, D_j have no common interior for $1 \le i < j \le m$, and apply the interval evaluation (or grid interpolation) to estimate the upper bound $u(D_i)$ of f on each cube $D_i (1 \leq i \leq m)$. If $u(D_i) \geq f(x_0)$ for some $i (1 \leq i \leq m)$, then divide D_i into 2^n smaller cubes $D_{i,j}$ $(j = 1, 2, \ldots, 2^n)$ in $\mathbb{R}^n$ whose edge length is one half of that of D_i, and estimate the upper bounds $u(D_{i,j})$ of $f(x)$ on the newly obtained cubes $D_{i,j}$. Recursively do this until the upper bound of every cube D_I produced in this process satisfies $u(D_I) < f(x_0)$. This process will be terminated after finitely many steps of subdivision provided $\sup_{x \in M \setminus U_0} f(x) < f(x_0)$, since according to Taylor's theorem, we have

$$f(x) = f(x_{D_I}^c) + (x - x_{D_I}^c) \nabla f(t x_{D_I}^c + (1 - t)x)$$

$$\leq f(x_{D_I}^c) + \frac{\sqrt{n}}{2} \text{edge}(D_I) \cdot B_0, \tag{1}$$

for all $x \in D_I$. Here $x_{D_I}^c$ is the barycenter of D_I, $t = t(x) \in [0, 1]$, $\text{edge}(D_I)$ is the edge length of D_I, and B_0 is the following constant:

$$B_0 = \sup_{x \in M \setminus U_0} \sqrt{\left(\frac{\partial f}{\partial x_1}\right)^2 + \left(\frac{\partial f}{\partial x_2}\right)^2 + \cdots + \left(\frac{\partial f}{\partial x_n}\right)^2} < +\infty,$$

and we may assume that the estimated upper bound $u(D_I)$ of $f(x)$ on every cube D_I satisfies the following inequality

$$u(D_I) \leq f(x_{D_I}^c) + \frac{\sqrt{n}}{2} \text{edge}(D_I) \cdot B_0. \tag{2}$$

Therefore, if the subdivision cannot be completed in finite steps, we would get a sequence $D_i, D_{i,j_1}, D_{i,j_1,j_2}, \ldots, D_{i,j_1,j_2,\ldots,j_k}$ $(1 \leq i \leq m, 1 \leq j_k \leq 2^n, k = 1, 2, \ldots)$ that satisfies $u(D_{I_k}) \geq f(x_0)$ for $D_{I_k} = D_{i,j_1,j_2,\cdots,j_k}$ $(k = 1, 2, \cdots)$, which leads to

$$\lim_{k \to \infty} f(x_{D_k}^c) \geq f(x_0),$$

and contradicts the assumption $\sup_{x \in M \setminus U_0} f(x) < f(x_0)$.

To our knowledge, this approach to automated proof of inequalities was suggested by Jingzhong Zhang in the late 1980s for proving an inequality of Zirakzadeh (see [14] and [2]). A detailed description of Zhang's method can be found in [13] in Chinese. Later the method was used in [5] and [12] for proving two other geometric inequalities related to optimal distribution of points on sphere and hemisphere. However, the technique of **Procedure 1** is not described in a general term in these case studies, so it is still difficult to apply the new method to process other unsolved or complicated problems directly.

This paper is aiming to give a general symbolic algorithm of **Procedure 1** for a class of smooth functions formed by a sum of several radical expressions. Namely, assume that $f : \mathbb{R}^n \to \mathbb{R}$ has the following form:

$$f = c_1 \sqrt{g_1(x_1, x_2, \ldots, x_n)} + \cdots + c_k \sqrt{g_k(x_1, x_2, \ldots, x_n)},$$

where $c_1, \ldots, c_k$ are real numbers and $g_j(x_1, x_2, \ldots, x_n)$ $(j = 1, \ldots, k)$ are polynomials or rational functions of polynomials, and the point $x_0 \in \mathbb{R}^n$ satisfies the conditions

$$\frac{\partial f}{\partial x_i}(x_0) = 0, \ i = 1, 2, \ldots, n;$$

and

$$H_0 := \begin{pmatrix} \frac{\partial^2 f}{\partial x_1^2} & \frac{\partial^2 f}{\partial x_1 \partial x_2} & \cdots & \frac{\partial^2 f}{\partial x_1 \partial x_n} \\ \frac{\partial^2 f}{\partial x_2 \partial x_1} & \frac{\partial^2 f}{\partial x_2^2} & \cdots & \frac{\partial^2 f}{\partial x_2 \partial x_n} \\ \vdots & \vdots & \ddots & \vdots \\ \frac{\partial^2 f}{\partial x_n \partial x_1} & \frac{\partial^2 f}{\partial x_n \partial x_2} & \cdots & \frac{\partial^2 f}{\partial x_n \partial x_n} \end{pmatrix}(x_0)$$

is negative-semi-definite. We explain how to construct a quadratic form $q(x) = q(x_1, x_2, \ldots, x_n)$ and a neighborhood $U_0 \subset \mathbb{R}^n$ of x_0 so that

(1) $q(x_0) = f(x_0)$, x_0 is the unique maximal point of $q(x)$, and
(2) $f(x) \leq q(x)$ for $x \in U_0$.

Note that the methods of local analysis in [5,12,13] are implemented for triangular functions and some special radical functions. We shall present our algorithm in a more general form. Actually, our algorithm gives a constructive approach to a special case (for $k = 0$ or n) of the Morse Lemma (see, e.g., [3,7]), which asserts that if $f : \mathbb{R}^n \to \mathbb{R}$ is a function of class C^∞ for which $x_0 = 0$ is a non-degenerate critical point, namely $\nabla f(0) = 0$ and the Hessian at x_0 has trivial kernel, then in *some neighbourhood* U of x_0 there is a local C^∞ coordinate system, namely a C^∞ diffeomorphism $\varphi : U \to V \subset \mathbb{R}^n$ with $\varphi(0) = 0$ and such that $\tilde{f} = f \circ \varphi^{-1}$ takes the form

$$\tilde{f}(x) = f(0) - x_1^2 - \cdots - x_k^2 + x_{k+1}^2 + \cdots + x_n^2.$$

Several quantitative forms of the Morse Lemma can be found in [4,6,8,11], yet a symbolic computation method cannot be directly derived from the literature.

The paper is organized as follows. In Sect. 2, we show how to find quadratic bounds of an algebraic surface in the neighborhoods of a critical point; in Sect. 3, we extend the method to rational and certain radical functions. In Sect. 4, we shall apply the method to do local critical analysis for the spherical six-point problem. The Maple computation in this paper is implemented on Maple version 18.00.

2   Quadratic Local Upper Bound of Polynomials

The following analytic definition of local optimal (extremum, maximal or minimal) of a real-valued function can be found in any calculus text book.

Definition 1. *A real-valued function f defined on a real-line is said to have a local (or relative) maximum point at the point x_0, if there exists some $\varepsilon > 0$ such that $f(x) \leq f(x_0)$ when $|x - x_0| < \varepsilon$. The value of the function at this point is called maximum of the function. Similarly, a function has a local minimum point at x_0, if $f(x) \geq f(x_0)$ when $|x - x_0| < \varepsilon$. The value of the function at this point is called minimum of the function.*

For functions of several variables, a neighborhood $U(x_0, \varepsilon)$ of the point x_0 is used to substitute the interval $|x - x_0| < \varepsilon$. It is well known that the local extrema can be found by Fermat's theorem, which states that they must occur at critical points (also called stationary points).

Theorem 1 (Fermat's theorem). *Let* $f : (a, b) \rightarrow R$ *be a function and suppose that* $x_0 \in (a, b)$ *is a local maximum of* f. *If* f *is differentiable, then* $f'(x_0) = 0$. *And exactly the same statement is true in higher dimensions.*

One can distinguish whether a critical point is a local maximum or local minimum by using the second derivative test. In calculus, the *second derivative test* is a criterion for determining whether a given critical point of a function is a local maximum or a local minimum using the value of the second derivative at the point. The test states: if the function f is twice differentiable at a stationary point x_0, then

- If $f''(x_0) < 0$ then f has a local maximum at x_0.
- If $f''(x_0) > 0$ then f has a local minimum at x_0.
- If $f''(x_0) = 0$, the second derivative test says nothing about the point x_0.

For a function of more than one variable, the second derivative test generalizes to a test based on the eigenvalues of the function's Hessian matrix at the stationary point. In particular, assuming that all second order partial derivatives of f are continuous in a neighbourhood of a stationary point x_0, and the eigenvalues of the Hessian at x_0 are all positive, then x_0 is a local minimum. If the eigenvalues are all negative, then x_0 is a local maximum, and if some are positive and others are negative, then the point x_0 is a saddle point. If the Hessian matrix is singular, then the second derivative test is inconclusive. Note that the second derivative test concludes only the existence of a neighbourhood U_0 of x_0, where the function f satisfies $f(x) \geq f(x_0)$, or $f(x) \leq f(x_0)$, for all points $x \in U_0$.

It is easy to see that for a quadratic polynomial $p(x)$ with n-variables, if $x_0 = (0, 0, \ldots, 0) \in \mathbb{R}^n$ is a local maximum point, then

$$p(x) = p_0 + \frac{1}{2}(x_1, x_2, \ldots, x_n)H_0(x_1, x_2, \ldots, x_n)^T,$$

where $p_0 = p(0, 0, \ldots, 0)$ and H_0 is a negative-semi-definite symmetric matrix, so under certain orthogonal transform of Cartesian coordinates

$$(x_1, x_2, \ldots, x_n) = (y_1, y_2, \ldots, y_n) \cdot P.$$

We may express the polynomial p using the new coordinates as

$$p(y_1, y_2, \ldots, y_n) = p_0 + \frac{1}{2}(\lambda_1 y_1^2 + \lambda_2 y_2^2 + \cdots + \lambda_n y_n^2), \tag{3}$$

where P is an orthogonal matrix and $\lambda_i \leq 0$ ($i = 1, 2, \ldots, n$) are the eigenvalues of H_0, which also shows that $x = 0$ is the global optimal of $p(x)$. In geometry, this shows that in $\mathbb{R}^{n+1}$, the algebraic surface

$$F := \{(x_1, x_2, \ldots, x_n, z) | z - p(x_1, x_2, \ldots, x_n)\} = 0$$

lies at one side of the tangent space $T_0F : z = 0$ of F at 0.

For polynomial $p(x)$ of degree $d \geq 3$, if $x = 0$ is a local maximum of p, then there is a neighborhood U_0 of $0 \in \mathbb{R}^n$ so that in the local region $U_0 \times \mathbb{R} \subset \mathbb{R}^{n+1}$, the surface $F : z - p(x) = 0$ and the tangent space T_0F can be separated by a quadratic surface $F_1 : z - q(x) = 0$, where $q(x)$ is a quadratic polynomial which has $x = 0$ as its maximal point, and therefore, the algebraic surface $z - p(x) = 0$ lies under its tangent space at 0. We will show that the quadratic polynomial can be constructed using symbolic computation. Namely, we have the following result.

Theorem 2. *Assume that* $p(x) = p(x_1, x_2, \ldots, x_n)$ *is a polynomial of degree* $d \geq 3$ *and* $x = (0, 0, \ldots, 0)$ *is a local maximum of* $p(x)$ *satisfying the condition that the eigenvalues* $\lambda_1, \lambda_2, \ldots, \lambda_n$ *of the Hessian matrix of* $p(x)$ *at* $x = 0$ *are all negative. Then we can construct a neighborhood* U_0 *of* $x = 0$ *and a quadratic polynomial* $q(x_1, x_2, \ldots, x_n)$ *satisfies*

(i) $q(0) = p(0)$,
(ii) $\frac{\partial q}{\partial x_i} = 0$ *for* $i = 1, 2, \ldots, n$,
(iii) *the Hessian matrix* $H_0(q)$ *is negative-definite, and*
(iv) $p(x) \leq q(x)$ *for all* $x \in U_0$.

We may call $q(x)$ in Theorem 2 a quadratic local upper-bound of polynomial $p(x)$. In order to prove this theorem, we need to consider the degree-j homogeneous part of polynomial $p(x)$ for each degree $j \geq 3$. We have the following lemma.

Lemma 1. *For any integer* $j \geq 3$ *and homogeneous polynomial*

$$h_j(x_1, x_2, \ldots, x_n) = \sum_{\substack{d_1, d_2, \ldots, d_n \geq 0 \\ d_1 + d_2 + \cdots + d_n = j}} c_{d_1, d_2, \ldots, d_n} x_1^{d_1} x_2^{d_2} \cdots x_n^{d_n},$$

with real coefficients, then there exists constant numbers $k_1, k_2, \ldots, k_n \geq 0$, *such that for any positive number* N *and for real numbers* $x_1, x_2, \ldots, x_n \in (-1/N, 1/N)$, *the inequality*

$$|h_j(x_1, x_2, \ldots, x_n)| \leq \frac{1}{jN^{j-2}} (k_1 x_1^2 + k_2 x_2^2 + \cdots + k_n x_n^2)$$

holds.

Proof. For any j, real numbers $z_1, z_2, \ldots, z_j \in (-1/N, 1/N)$ and any combination (k, l) of $1, 2, \ldots, j$, we have

$$z_1 z_2 \cdots z_j \leq \frac{1}{2N^{j-2}} (z_k^2 + z_l^2). \tag{4}$$

Construct this inequality for all $\binom{j}{2} = j(j-1)/2$ two-member combinations of $1, 2, \ldots, j$, and sum up them to obtain

$$\binom{j}{2} z_1 z_2 \cdots z_j \leq \frac{1}{2N^{j-2}} (j-1)(z_1^2 + z_2^2 + \cdots + z_j^2).$$

Therefore

$$z_1 z_2 \cdots z_j \le \frac{1}{jN^{j-2}}(z_1^2 + z_2^2 + \cdots + z_j^2),$$

and

$$x_1^{d_1} x_2^{d_2} \cdots x_n^{d_n} = \overbrace{x_1 \cdots x_1}^{d_1} \times \overbrace{x_2 \cdots x_2}^{d_2} \times \cdots \times \overbrace{x_n \cdots x_n}^{d_n}$$

$$\le \frac{1}{jN^{j-2}}(d_1 x_1^2 + d_2 x_2^2 + \cdots + d_n x_n^2), \qquad (5)$$

for any monomial $x_1^{d_1} x_2^{d_2} \cdots x_n^{d_n}$ of degree j. Applying inequality (5) to each monomial of the homogeneous polynomial $h_j(x_1, x_2, \ldots, x_n)$, we have

$$|h_j| \le \sum_{d_1 + d_2 + \cdots + d_n = j} |c_{d_1, d_2, \ldots, d_n} x_1^{d_1} x_2^{d_2} \cdots x_n^{d_n}|$$

$$\le \sum \frac{1}{jN^{j-2}} |c_{d_1, d_2, \ldots, d_n}| (d_1 x_1^2 + d_2 x_2^2 + \cdots + d_n x_n^2)$$

$$= \frac{1}{jN^{j-2}} \left(k_1 x_1^2 + k_2 x_2^2 + \cdots + k_n x_n^2 \right). \qquad (6)$$

Here $k_1, k_2, \ldots, k_n$ are positive real numbers defined by

$$k_i = \sum_{d_1 + \cdots + d_n = j} d_i |c_{d_1, d_2, \ldots, d_j}|, \quad i = 1, 2, \cdots, n.$$

This completes the proof of Lemma 1. $\qquad \Box$

Remark 1. Taking $C_j = \max\{k_1, k_2, \ldots, k_n\}$, then inequality (6) can be written in the following simple form:

$$|h_j(x_1, x_2, \ldots, x_n)| \le \frac{C_j}{jN^{j-2}}(x_1^2 + x_2^2 + \cdots + x_n^2). \qquad (7)$$

Now we give a proof of Theorem 2.

Proof. Let $p_0 = p(0)$, H_0 be the Hessian matrix of $p(x)$ at $x = 0$, and $\lambda_1, \lambda_2, \ldots, \lambda_n$ the eigenvalues of H_0. Then $\lambda_i < 0 \, (i = 1, 2, \ldots, n)$ according to the assumption. We can express p as follows:

$$p(x) = p_0 + \frac{1}{2}(x_1, x_2, \ldots, x_n) H_0 (x_1, x_2, \ldots, x_n)^T$$

$$+ H_3(x_1, x_2, \ldots, x_n) + \cdots + H_d(x_1, x_2, \ldots, x_n), \qquad (8)$$

where $H_j \, (j = 3, \ldots, d)$ are homogeneous polynomials of degree j, respectively.

Applying Lemma 1, for each $j \, (j = 3, \ldots, d)$ we compute a sequence of constants $k_1^{(j)}, k_2^{(j)}, \ldots, k_n^{(j)}$ that satisfy the following inequality

$$|H_j(x_1, x_2, \ldots, x_n)| \le \frac{1}{jN^{j-2}}(k_1^{(j)} x_1^2 + k_2^{(j)} x_2^2 + \cdots + k_n^{(j)} x_n^2).$$

For each $N > 0$, define a quadratic polynomial $q_N(x)$ as follows:

$$q_N(x_1, x_2, \ldots, x_n) = p_0 + \frac{1}{2}(x_1, x_2, \ldots, x_n) H_0 (x_1, x_2, \ldots, x_n)^T$$

$$+ \sum_{j=3}^{d} \frac{1}{jN^{j-2}} \left(k_1^{(j)} x_1^2 + k_2^{(j)} x_2^2 + \cdots + k_n^{(j)} x_n^2 \right). \tag{9}$$

It is clear that the requirements (i) and (ii) of Theorem 2 are satisfied, and the requirement (iv), i.e., the inequality

$$p(x) \le p_0 + \frac{1}{2} x \cdot H_0 \cdot x^T + \sum_{j=3}^{d} |H_j(x)| \le q_N(x)$$

is also true for any $x_1, x_2, \ldots, x_n \in (-1/N, 1/N)$ according to Lemma 1. To see that the requirement (iii) is satisfied for sufficient large N, observe that the Hessian matrix $H_{q_N}(0)$ of $q_N(x)$ at $x = 0$ can be written as $H_0 + 2G_N$, where G_N is the diagonal matrix

$$\begin{pmatrix} g_1(1/N) & & & \\ & g_2(1/N) & & \\ & & \ddots & \\ & & & g_n(1/N) \end{pmatrix},$$

where

$$g_i(y) = \sum_{j=3}^{d} \frac{k_i^{(j)}}{j} \cdot y^{j-2}, \quad i = 1, 2, \cdots, n.$$

Notice that $\lambda_i < 0$, $k_i^{(j)} > 0$ for all i, j ($i = 1, 2, \ldots, n; j = 3, \cdots, d$), so for each i, the equation

$$\frac{1}{2}\lambda_i + g_i(y) = \frac{1}{2}\lambda_i + \frac{k_i^{(3)}}{3} \cdot y + \cdots + \frac{k_i^{(d)}}{d} \cdot y^{d-2} = 0$$

has a unique positive real root y_i^*. Thus, if the number N satisfies

$$\frac{1}{N} < \min\{y_1^*, y_2^*, \ldots, y_n^*\},$$

then the eigenvalues of $H_0 + 2G_N$, i.e., $\lambda_i + 2g_i(1/N)$ ($i = 1, 2, \ldots, n$) are all negative, and, therefore, the Hessian matrix of quadratic polynomials $q_N(x)$ is negative-definite, as claimed in (iii).

Theorem 2 is proved.   □

Remark 2. Let $\lambda_0 = \max\{\lambda_1, \lambda_2, \ldots, \lambda_n\} < 0$ be the largest eigenvalue of H_0, $C_j = \max\{k_1^{(j)}, k_2^{(j)}, \ldots, k_n^{(j)}\}$ for $j = 3, \cdots, d$, and $1/N$ the smallest positive real root of the following equation:

$$\frac{1}{2}\lambda_0 + \frac{C_3}{3} \cdot \left(\frac{1}{N}\right) + \cdots + \frac{C_d}{d} \cdot \left(\frac{1}{N}\right)^{d-2} = 0.$$

Then, for any $x_2, x_2, \ldots, x_n \in (-1/N, 1/N)$, we have

$$p(x) \leq q_N(x) \leq p(0).$$

3 Local Critical Analysis of Rational and Radical Functions

In this section, we explain how to extend the local critical analysis method to rational functions and certain radical functions of several variables.

3.1 Rational Functions

The method we have described in Theorem 2 can be easily generalized to functions $f(x) = p(x)/q(x)$ where $p(x)$ and $q(x)$ are polynomials of $x \in \mathbb{R}^n$. Let x_0 be a local maximal (or minimal) point such that the Hessian matrix of f at the point x_0 is negative-definite (or positive-definite, respectively). Without loss of generality, we may assume that x_0 is a local minimal point of $f(x)$ and $q(x_0) > 0$. Clearly, if $q(x)$ is positive-definite, then the task of finding a neighborhood $U_0 \subset \mathbb{R}^n$ of x_0 where

$$\frac{p(x)}{q(x)} \geq f(x_0),$$

for all $x \in U_0$ can be simply transformed to finding the neighborhood U_0 where

$$p(x) - f(x_0) \cdot q(x) \geq 0,$$

for $x \in U_0$, which is same as we have done in the previous section. If $q(x)$ is neither positive-definite nor negative-definite in certain known region, we need first to construct such a neighborhood V_0 of x_0 so that $q(x_0) \cdot q(x) > 0$ for all points $x \in V_0$. To implement this work, we have the following theorem.

Theorem 3. *Let $q(x)$ be a polynomial in n variables of degree s, x_0 a point in $\mathbb{R}^n$ with $x_0 = (x_1^*, x_2^*, \ldots, x_n^*)$, and $q(x_0) > 0$,*

$$K_1 = \max\{|\frac{\partial q}{\partial x_i}(x_0)|, \ i = 1, 2, \ldots, n\} > 0,$$

$$K_2 = \max\{|\frac{\partial^2 q}{\partial x_i \partial x_j}(x_0)|, 1 \leq i, j \leq n\} > 0,$$

and

$$K_j = \max\{|\frac{\partial^j q}{\partial x_{i_1} \ldots \partial x_{i_j}}(x_0)|, 1 \leq i_1, \ldots, i_j \leq n\} > 0,$$

for $j = 3, \ldots, s$. Let δ_0 be the unique solution of the equation

$$q(x_0) = K_1 u + \frac{1}{2!}K_2 u^2 + \frac{1}{3!}K_3 u^3 + \cdots + \frac{1}{s!}K_s u^s. \tag{10}$$

Then, the inequality $q(x) > 0$ is valid for any $x = (x_1, x_2, \ldots, x_n)$ with

$$||x - x_0||_1 = |x_1 - x_1^*| + |x_2 - x_2^*| + \cdots + |x_n - x_n^*| < \delta_0.$$

Proof. Let

$$u_1 = x_1 - x_1^*, u_2 = x_2 - x_2^*, \ldots, u_n = x_n - x_n^*,$$

and h_j $(j = 3, \ldots, s)$ the homogeneous polynomials defined by

$$h_j(u_1, \ldots, u_n) = \frac{1}{j!} \left[\sum_{i=1}^{n} u_i \frac{\partial}{\partial x_i'} \right]^j q(x_1', \ldots, x_n') |_{x_1' = x_1^*, \ldots, x_n' = x_n^*}$$

$$= \frac{1}{j!} \sum_{d_1 + \cdots + d_n = j} \binom{j}{d_1, d_2, \ldots, d_n} \prod_{i=1}^{n} \left(u_i \frac{\partial}{\partial x_i'} \right)^{d_i} q(x') |_{x' = x_0}.$$

Then, we may expand $q(x)$ in a Taylor series at the point x_0 as follows:

$$q(x_1, x_2, \ldots, x_n) = q(x_1^*, x_2^*, \ldots, x_n^*)$$

$$+ \left[u_1 \frac{\partial q}{\partial x_1} + u_2 \frac{\partial q}{\partial x_2} + \cdots + u_n \frac{\partial q}{\partial x_n} \right]_{x_1 = x_1^*, \ldots, x_n = x_n^*}$$

$$+ \frac{1}{2!} \left[u_1^2 \frac{\partial^2 q}{\partial x_1^2} + 2 u_1 u_2 \frac{\partial^2 q}{\partial x_1 \partial x_2} + \cdots + u_n^2 \frac{\partial^2 q}{\partial x_n^2} \right]_{x_1 = x_1^*, \ldots, x_n = x_n^*}$$

$$+ h_3(u_1, u_2, \ldots, u_n) + \cdots + h_s(u_1, u_2, \ldots, u_n).$$

It is obvious that

$$\mathbf{abs} \left(u_1 \frac{\partial q}{\partial x_1} + u_2 \frac{\partial q}{\partial x_2} + \cdots + u_n \frac{\partial q}{\partial x_n} \right)_{x_1 = x_1^*, \ldots, x_n = x_n^*}$$

$$\leq K_1(|u_1| + |u_2| + \cdots + |u_n|), \tag{11}$$

$$\mathbf{abs} \left[u_1^2 \frac{\partial^2 q}{\partial x_1^2} + 2 u_1 u_2 \frac{\partial^2 q}{\partial x_1 \partial x_2} + \cdots + u_n^2 \frac{\partial^2 q}{\partial x_n^2} \right]_{x_1 = x_1^*, \ldots, x_n = x_n^*}$$

$$\leq K_2(|u_1|^2 + 2|u_1||u_2| + \cdots + |u_n|^2)$$

$$= K_2(|u_1| + |u_2| + \cdots + |u_n|)^2. \tag{12}$$

For $h_j(u_1, u_2, \ldots, u_n)$ $(j = 3, \ldots, s)$, we have

$$\mathbf{abs} \left(h_1(u_1, u_2, \ldots, u_n) \right)$$

$$\leq \frac{1}{j!} \sum_{d_1 + \cdots + d_n = j} \binom{j}{d_1, d_2, \ldots, d_n} \left(K_j \cdot \prod_{i=1}^{n} (|u_i|)^{d_i} \right)$$

$$= \frac{1}{j!} \cdot K_j \cdot [|u_1| + |u_2| + \cdots + |u_n|]^j. \tag{13}$$

Therefore,

$$q(x) \geq q(x_0) - K_1 ||x - x_0||_1 - \frac{1}{2} K_2 ||x - x_0||_1$$

$$- \frac{1}{3!} K_3 ||x - x_0||_1^2 \cdots - \frac{1}{s!} K_s ||x - x_0||^s. \tag{14}$$

which immediately implies that $q(x) > 0$ if $||x - x_0||_1 < \delta_0$ and δ_0 is the (unique) real root of the equation (10).

Theorem 3 is proved. $\qquad\qquad\qquad\qquad\qquad\qquad\qquad\qquad\qquad\qquad\quad$ □

3.2 Sum of Radicals

Now we consider the radical functions of the following form:

$$f(x) = c_1 \sqrt{1 + \frac{p_1(x)}{q_1(x)}} + c_2 \sqrt{1 + \frac{p_2(x)}{q_2(x)}} + \cdots + c_k \sqrt{1 + \frac{p_k(x)}{q_k(x)}}, \qquad (15)$$

where $p_j(x)$ and $q_j(x)$ are the polynomials in n variables. We can prove the following result.

Theorem 4. *Assume that $f(x)$ is function defined in (15), $x_0 = 0$, and*

$$p_j(x_0) = 0, \quad q_j(x_0) > 0$$

for $j = 1, 2, \ldots, k$. Then using symbolic computation we can construct a neighborhood U_0 of x_0 and rational functions

$$h(x) = \sum_{j=1}^{k} c_j + \frac{P_1(x)}{Q_1(x)}, \quad g(x) = \sum_{j=1}^{k} c_j + \frac{P(x)}{Q(x)}, \qquad (16)$$

where $P_1(x), Q_1(x), P(x), Q(x)$ are polynomials such that

$$P_1(0) = 0, \quad P(0) = 0,$$

and

$$Q_1(x) > 0, \quad Q(x) > 0, \quad h(x) \leq f(x) \leq g(x), \qquad (17)$$

for all $x \in U_0$.

To prove this theorem, we need the following Lemma 2 and Lemma 3.

Lemma 2. *For any real number x with $-0.3777 < x < 0.7145$, the following inequality is true:*

$$1 + \frac{1}{2}x - \frac{5}{32}x^2 \leq \sqrt{1 + x} \leq 1 + \frac{1}{2}x - \frac{3}{32}x^2. \qquad (18)$$

$\qquad\qquad\qquad\qquad\qquad\qquad\qquad\qquad\qquad\qquad\qquad\qquad\qquad\qquad\quad$ □

Lemma 3. *Assume that $p(x)$ and $q(x)$ are the polynomials in n variables $x_1, x_2, \ldots, x_n$, $x_0 = (0, 0, \ldots, 0) \in \mathbb{R}^n$, and*

$$p(x_0) = 0, \quad q(x_0) > 0.$$

Then for any $\varepsilon > 0$, we can find a constant $\delta = \delta(\varepsilon) > 0$ such that

$$\sqrt{x_1^2 + x_2^2 + \cdots + x_n^2} < \delta \implies -\varepsilon < \frac{p(x_1, x_2, \ldots, x_n)}{q(x_1, x_2, \ldots, x_n)} < \varepsilon$$

by symbolic computation.

Proof. Indeed, the existence of the $\delta(\varepsilon)$ for each $\varepsilon > 0$ is guaranteed by the continuity of $p(x)/q(x)$ at the point $x_0 = 0$. Here we show that $\delta(\varepsilon)$ can be obtained by symbolic computation. For this purpose, we may assume that in Theorem 3 we have a neighborhood

$$U_0 := \{(x_1, x_2, \ldots, x_n), |x_1| + |x_2| + \cdots + |x_n| < \delta_0\}$$

that satisfies $q(x) > q(0)/2 > 0$ for all $x \in U_0$. Assume $\deg(p) = r$ and

$$p(x) = x \cdot \nabla_{x=0} p(x) + \frac{1}{2} x \cdot H_0 x^T + h_3(x) + \cdots + h_r(x),$$

here h_j are homogeneous polynomials in $x_1, x_2, \ldots, x_n$ for $j = 3, \ldots, r$. Then applying the method described in Lemma 1 and Remark 1 given in the previous section, we can obtain constants $C_j > 0 \, (j = 3, \ldots, s)$ so that

$$|h_3(x) + \cdots + h_r(x)| \leq \sum_{j=3}^{r} \frac{C_j}{jN^{j-2}}(x_1{}^2 + x_2{}^2 + \cdots + x_n{}^2) \qquad (19)$$

for all $x \in \mathbb{R}^n$ with $x_1, x_2, \ldots, x_n \in (-1/N, 1/N)$ for any $N > 0$. Thus, for $(x_1, x_2, \ldots, x_n) \in U_0$, we have $x_1, x_2, \ldots, x_n \in (-\delta_0, \delta_0)$, and inequality (19) implies that

$$|h_3(x) + \cdots + h_r(x)| \leq C \cdot (x_1{}^2 + x_2{}^2 + \cdots + x_n{}^2). \qquad (20)$$

Here

$$C = \sum_{j=3}^{r} \frac{C_j}{j} \delta_0^{j-2}.$$

Let $\lambda_1, \lambda_2, \ldots, \lambda_n$ be the eigenvalues of H_0, P the orthogonal matrix, (i.e., $P^T P = I$) satisfying

$$H_0 = P^T \cdot \Lambda \cdot P = P^T \begin{pmatrix} \lambda_1 & & & \\ & \lambda_2 & & \\ & & \ddots & \\ & & & \lambda_n \end{pmatrix} P,$$

and

$$(x_1', x_2', \ldots, x_n') = (x_1, x_2, \ldots, x_n) P^T.$$

Then we have

$$
\begin{aligned}
|x \cdot H_0 \cdot x^T| &= |x P^T \cdot \Lambda \cdot P x^T| = |x' \Lambda (x')^T| \\
&= |\lambda_1| x_1'^2 + |\lambda_2| x_2'^2 + \cdots + |\lambda_n| x_n'^2 \\
&\leq \sqrt{\lambda_1^2 + \lambda_2^2 + \cdots + \lambda_n^2} \left(x_1'^2 + x_2'^2 + \cdots + x_n'^2 \right),
\end{aligned}
$$

Note that

$$\lambda_1^2 + \lambda_1^2 + \cdots + \lambda_n^2 = \mathrm{tr}(H_0 H_0^T) = \sum_{i=1}^{n} \left(\sum_{j=1}^{n} \frac{\partial^2 p}{\partial x_i \partial x_j} \frac{\partial^2 p}{\partial x_j \partial x_i} \right)$$

$$= \sum_{i,j=1}^{n} \left(\frac{\partial^2 p}{\partial x_i \partial x_j} \right)^2 = ||H_0||_F^2,$$

and

$$x_1'^2 + x_2'^2 + \cdots + x_n'^2 = x' x'^T = x P^T P x^T = x_1^2 + x_2^2 + \cdots + x_n^2,$$

hence, we get

$$\frac{1}{2} |x \cdot H_0 \cdot x^T| \le \frac{1}{2} ||H_0||_F \cdot (x_1^2 + x_2^2 + \cdots + x_n^2). \tag{21}$$

In view of the Cauchy–Schwarz inequality, we have

$$|x \cdot \nabla_{x=0} p(x)| = |x_1 \frac{\partial p}{\partial x_1}(0) + x_2 \frac{\partial p}{\partial x_2}(0) + \cdots + x_n \frac{\partial p}{\partial x_n}(0)|$$

$$\le ||\nabla_0 p||_2 \sqrt{x_1^2 + x_2^2 + \cdots + x_n^2}. \tag{22}$$

Here

$$||\nabla_0 p||_2 = \sqrt{\sum_{i=1}^{n} \left(\frac{\partial p}{\partial x_i}(0) \right)^2}.$$

Combining (20), (21), and (22), we obtain the following inequality

$$|p(x)| \le ||\nabla_0 p||_2 \sqrt{x_1^2 + x_2^2 + \cdots + x_n^2}$$

$$+ \left(\frac{1}{2} ||H_0||_F + C \right) (x_1^2 + x_2^2 + \cdots + x_n^2). \tag{23}$$

Therefore, if we take $\delta(\varepsilon) < \min\{\delta_0/\sqrt{n}, \delta_1\}$, where δ_1 is the unique real root of

$$\frac{1}{2} q(0) \varepsilon = ||\nabla_0 p||_2 \, u + \left(\frac{1}{2} ||H_0||_F + C \right) u^2,$$

then, from

$$\sqrt{x_1^2 + x_2^2 + \cdots + x_n^2} < \delta(\varepsilon),$$

we have

$$|p(x)| < \frac{1}{2} q(0) \varepsilon$$

and $|x_1| + |x_2| + \cdots + |x_n| < \sqrt{n} \delta(\varepsilon) \le \delta_0$, which implies that $q(x) > q(0)/2 > 0$, and, therefore,

$$\left| \frac{p(x)}{q(x)} \right| < \frac{|p(x)|}{\frac{1}{2} q(0)} < \varepsilon,$$

as claimed by Lemma 3. $\qquad\qquad\square$

Proof (Proof of Theorem 4). Without loss of generality, we may assume that $c_j > 0$ for $j = 1, \ldots, l$ and $c_j < 0$ for $j = l+1, \ldots, k$. Then, applying Lemma 3 we can construct a neighborhood U_j so that

$$\left| \frac{p_j(x)}{q_j(x)} \right| < 0.3777, \quad q_j(x) > 0$$

for each j $(1 \le j \le k)$. Therefore, for point $x \in U_0 := U_1 \cap U_2 \cap \cdots \cap U_k$, we have

$$f(x) \le \sum_{1 \le j \le l} c_j \left(1 + \frac{p_j(x)}{2q_j(x)} - \frac{3p_j(x)^2}{8q_j(x)^2} \right)$$

$$+ \sum_{l+1 \le j \le k} c_j \left(1 + \frac{p_j(x)}{2q_j(x)} - \frac{5p_j(x)^2}{8q_j(x)^2} \right) =: g(x).$$

Let

$$P(x) := (g(x) - c_1 - c_2 - \cdots - c_k) \cdot Q(x),$$
$$Q(x) := (\mathrm{lcm}(q_1(x)q_2(x) \cdots q_k(x)))^2.$$

Then $Q(x) > 0$ for $x \in U_0$ obviously, $f(x) \le g(x)$ for $x \in U_0$ as defined, and

$$g(0) = \sum_{1 \le j \le l} c_j \left(1 + \frac{p_j(0)}{2q_j(0)} - \frac{3p_j(0)^2}{8q_j(0)^2} \right)$$

$$+ \sum_{l+1 \le j \le k} c_j \left(1 + \frac{p_j(0)}{2q_j(0)} - \frac{5p_j(0)^2}{8q_j(0)^2} \right)$$

$$= c_1 + c_2 + \cdots + c_k,$$

therefore, $P(0) = 0$.

The rational function $h(x)$ and the polynomials $P_1(x)$ and $Q_1(x)$ can be constructed by a similar computation. Theorem 4 is proved.   □

Our goal is to process the situation when $x_0 = 0$ is a local maximal or minimal point of f. Namely, we wish that the upper-bound rational function $g(x)$ (and the lower-bound rational function $h(x)$, resp.) constructed by Theorem 4 has also taken the point x_0 as the local maximal (minimal, resp.) point if it is a local maximal (minimal, resp.) point of the original radical function $f(x)$, which means, $g(x)$ satisfies the following properties:

– $g'(0) = 0$, and, at best,
– the Hessian matrix $H_g(0)$ is negative-definite,

if x_0 is, for example, a maximal point of $f(x)$. To see this, we have

$$g'(0) = \sum_{j=1}^{k} \left(\frac{1}{2} c_j - \frac{c_j'}{8} \cdot \frac{2p_j(0)}{q_j(0)} \right) \left(\frac{q_j(0)p_j'(0) - q_j'(0)p_j(0)}{q_j(0)^2} \right) = \sum_{j=1}^{k} c_j \cdot \frac{p_j'(0)}{2q_j(0)},$$

here $c'_j = 3c_j$ for $1 \leq j \leq l$ and $c'_j = 5c_j$ for $l+1 \leq j \leq k$. Meanwhile, we have

$$f'(0) = \sum_{j=1}^{k} c_j \frac{[q_j(0)p'_j(0) - q'_j(0)p_j(0)] / [q_j(0)^2]}{2[1 + (p(0)/q(0))^2]} = \sum_{j=1}^{k} c_j \cdot \frac{p'_j(0)}{2q_j(0)},$$

which means that if $x_0 = 0$ is a local optimal point of the radical function defined by (15), then it is also a critical point of the upper-bound (or lower bound) rational function $g(x)$ (or $h(x)$, resp.) obtained by Theorem 4.

Remark 3. Notice that $x_0 = 0$ might not be a local maximal point of the upper-bound rational function $g(x)$ even if it is a local maximal point of $f(x)$. To ensure that

$$H_0(f) \text{ is negative-definite} \Longrightarrow H_0(g) \text{ is negative-definite},$$

we may need to refine inequalities (18) of Lemma 2. For example, we may use the following inequality

$$\sqrt{1+x} \leq 1 + \frac{1}{2}x - \frac{1}{8}x^2 + \frac{1}{16}x^3 \ (-1 < x < +\infty), \tag{24}$$

for c_j $(1 \leq j \leq l)$, and the inequality

$$\sqrt{1+x} \geq 1 + \frac{1}{2}x - \frac{1}{8}x^2 + \frac{1}{16}x^3 - \frac{1}{16}x^4 \ (-0.5161 < x < 3), \tag{25}$$

for c_j $(l+1 \leq j \leq k)$. The upper-bound rational function $g(x)$ generated by (24) and (25) satisfies $H_0(g) = H_0(f)$ since

$$\frac{\partial^2 f}{\partial x_i \partial x_j} = \frac{\partial^2 g}{\partial x_i \partial x_j} \ (1 \leq i, j \leq n).$$

We omit its proof here.

4 Local Critical Analysis of the Spherical Six-Point Problem

In this section, we discuss the optimization spherical point problem we have mentioned in Sect. 1. Recall that the numerical result says that best arrangement is

$$\Gamma_6 := \{(0,0,1), (0,-1,0), (1,0,0), (0,1,0), (-1,0,0), (0,0,-1)\}, \tag{26}$$

up to certain rotation of the sphere. We will prove the following theorem.

Theorem 5. *Assume that the six points $P_1, P_2, \ldots, P_6$ are placed on the unit sphere S^2 as follows:*

$$P_1 = (0,0,1), \quad P_2 = (0, -\sqrt{1-z_2^2}, z_2),$$

$$P_3 = (\sqrt{1-y_3^2-z_3^2}, y_3, z_3), \quad P_4 = (x_4, \sqrt{1-x_4^2-z_4^2}, z_4),$$

$$P_5 = (-\sqrt{1-y_5^2-z_5^2}, y_5, z_5), \quad P_6 = (x_6, y_6, -\sqrt{1-x_6^2-y_6^2}),$$

so that

$$-\frac{1}{22.9} \le z_2, y_3, z_3, x_4, z_4, y_5, z_5, x_6, y_6 \le \frac{1}{22.9}, \quad z_2 \ge 0, \tag{27}$$

then

$$\sum_{1 \le i < j \le 6} d(P_i, P_j) \le 6 + 12\sqrt{2},$$

and the equality holds if and only if $P_1, P_2, \ldots, P_6$ *are congruent to* Γ_6.

Proof. Without loss of generality, we may assume that

$$z_2 = \frac{2p}{1+p^2}, \quad y_3 = \frac{2q}{1+q^2+r^2}, \quad z_3 = \frac{2r}{1+q^2+r^2},$$

$$x_4 = \frac{2u}{1+u^2+v^2}, \quad z_4 = \frac{2v}{1+u^2+v^2},$$

$$y_5 = \frac{2s}{1+s^2+t^2}, \quad z_5 = \frac{2t}{1+s^2+t^2},$$

$$x_6 = \frac{2x}{1+x^2+y^2}, \quad y_6 = \frac{2y}{1+x^2+y^2},$$

where

$$-1/45.7 \le p, q, r, s, t, u, v, x, y \le 1/45.7, \quad p \ge 0.$$

Then, we have

$$d(P_i, P_j) = \begin{cases} 2 \cdot \sqrt{1+w_{ij}}, & (i,j) \in \{(1,6),(2,4),(3,5)\}, \\ \sqrt{2} \cdot \sqrt{1+w_{ij}}, & \text{otherwise}, \end{cases}$$

here w_{ij} are rational functions of $p, q, r, s, t, u, v, x, y$, for example,

$$w_{23} = \frac{-2(p^2q + 2\,pr - q)}{(p^2+1)\,(q^2+r^2+1)}, \quad w_{24} = \frac{-(p^2 + 2\,pv + u^2 + v^2)}{(p^2+1)\,(u^2+v^2+1)}.$$

Applying inequality (24) we have

$$\sum_{1 \le i < j \le 6} d(P_i, P_j) \le 6 + 12\sqrt{2} + G(p, q, r, s, t, u, v, x, y),$$

here

$$G = \sum_{1 \le i < j \le 6} c_{ij}\Big(\frac{1}{2}w_{ij} - \frac{1}{8}w_{ij}^2 + \frac{1}{16}w_{ij}^3\Big)$$

is a rational function and

$$c_{ij} = \begin{cases} 2, & \text{for } (i,j) \in \{(1,6),(2,4),(3,5)\}, \\ \sqrt{2}, & \text{otherwise}. \end{cases}$$

Using `Maple` we obtain $G = P/Q$, where

$$Q = 8 \left(p^2+1\right)^3 \left(q^2+r^2+1\right)^3 \left(u^2+v^2+1\right)^3 \cdot \left(s^2+t^2+1\right)^3 \left(x^2+y^2+1\right)^3,$$

and the $P = \text{numer}(G)$ is polynomial of degree 30 with $543, 609$ monomials, of which the least degree is 2. Therefore, we can write P as a sum of 29 homogeneous polynomials as follows:

$$P = H_2 + H_3 + \cdots + H_{30},$$

where H_{30} can be factorized into

$$H_{30} = -11\, p^6 \left(x^2 + y^2\right)^3 \left(u^2 + v^2\right)^3 \left(s^2 + t^2\right)^3 \left(q^2 + r^2\right)^3 \le 0.$$

The number of monomial in $H_j (2 \le j \le 29)$ are:

$$34, 37, 217, 279, 947, 1221, 3165, 3885, 8142, 9559, 17033,$$
$$18977, 29766, 30993, 43117, 41763, 51880, 46416, 52178,$$
$$42108, 42910, 30102, 27244, 16388, 13536, 6080, 4544, 832.$$

Using **Maple** we can check that the quadratic form H_2 is negative-definite. For simplicity, we show this later.

Assume that $p, q, r, \ldots, x, y \in (-1/N, 1/N)$. Then, applying Lemma 1, we can obtain the following inequalities:

$$|H_3| \le J_3 = 4\sqrt{2}(26\, p^2 + 35\, q^2 + 31\, r^2 + 35\, s^2 + 31\, t^2$$
$$+\, 38\, u^2 + 43\, v^2 + 38\, x^2 + 44\, y^2)/3N,$$

$$|H_4| \le J_4 = \frac{466\sqrt{2} + 170}{N^2} \cdot S_9,$$

$$|H_5| \le J_5 = \frac{2177}{N^3} \cdot S_9, \ \ldots, \ |H_{29}| \le J_{29} = \frac{44743}{N^{29}} \cdot S_9,$$

here

$$S_9 := p^2 + q^2 + r^2 + s^2 + t^2 + u^2 + v^2 + x^2 + y^2,$$

so $J_3, J_4, \ldots, J_{29}$ can be considered as quadratic forms with a parameter N. We will show more information of J_k at the end of this section. Let $J_{30} = 0$. Then $H_{30} \le J_{30}$, and we can check that if $N > 45.6866$, $H_2 + (J_3 + J_4 + \cdots + J_{30})$ is also a negative-definite quadratic form. Therefore,

$$P = \sum_{k=2}^{30} H_k \le H_2 + \sum_{k=3}^{30} J_k \le 0,$$

and

$$\sum_{1 \le i < j \le 6} d(P_i, P_j) \le 6 + 12\sqrt{2},$$

for $p, q, \ldots, x, y \in (-1/45.7, 1/45.7)$, also for $P_1, P_2, \ldots, P_6$ that satisfy (27). This proves Theorem 5. $\qquad\qquad\Box$

Now we show that H_2 is negative-definite. We can write H_2 as follows.

$$H_2 = 4(1 + \sqrt{2})(p, q, r, s, t, u, v, x, y)A(p, q, r, s, t, u, v, x, y)^T,$$

where

$$A = \begin{bmatrix}
-1 & 0 & a & 0 & a & 0 & b & 0 & c \\
0 & -1 & 0 & b & 0 & c & 0 & 0 & a \\
a & 0 & -1 & 0 & b & 0 & a & -c & 0 \\
0 & b & 0 & -1 & 0 & -c & 0 & 0 & a \\
a & 0 & b & 0 & -1 & 0 & a & c & 0 \\
0 & c & 0 & -c & 0 & -1 & 0 & a & 0 \\
b & 0 & a & 0 & a & 0 & -1 & 0 & -c \\
0 & 0 & -c & 0 & c & a & 0 & -1 & 0 \\
c & a & 0 & a & 0 & 0 & -c & 0 & -1
\end{bmatrix},$$

and

$$a = \sqrt{2} - 2, \quad b = 1 - \sqrt{2}, \quad c = \sqrt{2}/2 - 1.$$

The characteristic polynomial of A is

$$f(\lambda) = \left(\lambda^3 + 12\lambda\sqrt{2} + 3\lambda^2 + 27\sqrt{2} - 15\lambda - 38\right)$$
$$\times \left(-\lambda^2 + 2\sqrt{2}\lambda - 5\lambda + 3\sqrt{2} - 5\right) \times \left(\lambda^2 + \lambda + 5\sqrt{2} - 7\right)$$
$$\times \left(\lambda - 4 + 3\sqrt{2}\right) \times \left(-\lambda - 4 + \sqrt{2}\right) = 0.$$

Using `Maple` it is easy to see that $f(\lambda) = 0$ has 9 zeros and all of them are are negative numbers. The largest one is

$$-1/2 + 1/2\sqrt{29 - 20\sqrt{2}} \approx -0.07699\cdots < 0.$$

Therefore, H_2 is a negative-definite quadratic form.

To conclude the paper we show more details about H_k and J_k for $k \geq 3$. As for $4 \leq k \leq 29$, the degree-k homogeneous polynomial H_k has more than 200 monomials, here we only show the cubic homogeneous polynomial H_3 and the construction of J_3 and J_4.

The cubic homogeneous polynomial H_3 has 37 monomials, and all coefficients have a common factor $4\sqrt{2}$.

$$H_3 = -4\sqrt{2}(\underline{4px^2 + py^2} + 4p^2q + 4p^2s + p^2y - 4u^2y)$$
$$-q^2u - 4q^2x - qu^2 - 4qv^2 + 4vx^2 + vy^2 + 4ty^2 - v^2y - 4r^2u - r^2x$$
$$+rx^2 + 4ry^2 + s^2u + tx^2 + 4s^2x - su^2 - 4sv^2 + 4t^2u + t^2x$$
$$\underwave{-4uvx + 4uxy + 4qrv - 4qry + 4qxy - 4sxy}$$
$$\underwave{-4tuv - 4pqr - 4pst + 4ruv + 4stv - 4sty.}$$

We observe that there are two types of monomials in H_4: those monomials of the form $c \cdot w_1^2 w_2$ in the first three lines, and those monomials of the form $c \cdot w_1 w_2 w_3$ in the last two lines (printed with underwave), where $w_1, w_2, w_3 \in \{p, q, r, s, t, u, v, x, y\}$ and $w_i \neq w_j$ for $i \neq j$. Notice also that

$$-4\sqrt{2}(\underline{4px^2 + \sqrt{2}py^2}) \leq 0,$$

for $p > 0$. Applying the above inequality to the first two monomials (underlined) of H_3 and the following inequalities

$$cw_1^2 w_2 \leq \frac{|c|}{3N}(2w_1^2 + w_2^2), \quad cw_1 w_2 w_3 \leq \frac{|c|}{3N}(w_1^2 + w_2^2 + w_3^2)$$

to the remaining 35 monomials of H_2 of corresponding types, we obtain the upper bound quadratic form of H_3.

$$J_3 = \frac{4\sqrt{2}}{3N}(31p^2 + 35q^2 + 31r^2 + \cdots + 46x^2 + 46y^2).$$

For H_4, the monomials can be classified into five types and for each type we have its corresponding upper bound form as follows:

(1) monomials in the form $c \cdot w_i^4$, which upper bounds are $c' w_i^2 / N^2$, with $c' = \max\{0, c\}$;
(2) monomials in the form $c \cdot w_i^2 w_j^2$, which upper bounds are $c'(w_i^2 + w_j^2)/(2N^2)$, with $c' = \max\{0, c\}$;
(3) monomials in the form $c \cdot w_i^3 w_j$, the corresponding upper bounds are $|c|(3w_i^2 + w_j^2)/(4N^2)$;
(4) monomials in the form $c \cdot w_i^2 w_j w_k$, their upper bounds are $|c|(2w_i^2 + w_j^2 + w_k^2)/(4N^2)$;
(5) monomials in the form $c \cdot w_i w_j w_k w_l$, their upper bounds are $|c|(w_i^2 + w_j^2 + w_k^2 + w_l^2)/(4N^2)$.

where $c \in R$ and $w_i, w_j, w_k, w_l \in \{p, q, r, \ldots, x, y\}$. For obtaining tighter upper bound, we have taken

$$c' = \begin{cases} 0, & \text{if } c < 0, \\ c, & \text{otherwise}, \end{cases}$$

in the first two cases. Therefore, we obtain the following result:

$$J_4 = \frac{452\sqrt{2} + 170}{N^2}p^2 + \frac{374\sqrt{2} + 168}{N^2}q^2 + \cdots + \frac{490\sqrt{2} + 72}{N^2}y^2.$$

The largest coefficient of J_4 is $(466\sqrt{2} + 170)/N^2$, thus we have

$$J_4 \leq \frac{466\sqrt{2} + 170}{N^2}(p^2 + q^2 + r^2 + s^2 + t^2 + u^2 + v^2 + x^2 + y^2).$$

Similarly, we have

$$J_k \leq \frac{c_k}{N^{k-2}}(p^2 + q^2 + r^2 + s^2 + t^2 + u^2 + v^2 + x^2 + y^2)$$

for $k = 5, 6, \ldots, 29$, where we can take integer c_k as follows:

$2177, 9031, 21156, 61636, 121551, 284559, 476083, 938831, 1425542, 2280819,$

$3167178, 4135346, 5315958, 5594346, 6708463, 5568363, 6210033, 3953535,$

$4035347, 1890122, 1715600, 543585, 421574, 70800, 44743.$

Clearly,

$$P = H_2 + H_3 + H_4 + \cdots + H_{29} + H_{30}$$
$$\leq H_2 + (J_3 + J_4 + \cdots + J_{29}) =: P'(N, p, q, r, \ldots, x, y).$$

It is easy now to use `Maple` to verify that $P'(N, \cdot)$ is negative-definite.

Acknowledgments. We are grateful to the anonymous reviewers of the ISSAC 2022 and the CASC 2022 for their insightful comments and helpful suggestions, both on mathematics and language aspects, to our manuscripts.

References

1. Berman, J., Hanes, K.: Optimizing the arrangement of points on the unit sphere. Math. Comput. **31**(140), 1006–1008 (1977)
2. Bollobas, B.: An extremal problem for polygons inscribed in a convex curve. Can. J. Math. **19**, 523–528 (1967)
3. Bott, R.: Lectures on morse theory, old and new. Bull. Amer. Math. Soc. (N. S.) **7**, 331–358 (1982)
4. Chattopadhyay, A., Vegter, G., Yap, C.K.: Certified computation of planar Morse-Smale complexes. J. Symb. Comput. **78**, 3–40 (2017)
5. Hou, X.: Spherical Distribution of 5 points with maximal distance sum. Discrete Comput. Geom. **46**, 156–174 (2011)
6. Loi, T.L., Phien, P.: The quantitative Morse theorem. Int. J. Math. Anal. **6**(10), 481–491 (2012)
7. Morse, M.: The Calculus of Variations in the Large, vol. 18. American Mathematical Society Colloquium Publications, New York (1934)
8. Schur, V.: A quantitative version of the Morse lemma and ideal boundary fixing quasiisometries. J. Funct. Anal. **264**(3), 815–836 (2013)
9. Stolarsky, K.B.: Sums of distances between points on a sphere. Proc. AMS **35**(2), 547–549 (1972)
10. Stolarsky, K.B.: Sums of distances between points on a sphere II. Proc. AMS **41**(2), 575–582 (1973)
11. Yomdin, Y.: The geometry of critical and near-critical values of differentiable mappings. Math. Annal. **264**, 495–515 (1983)
12. Zeng, Z., Lu, J., Xu, Y., Wang, Y.: Maximizing the sum of the distances between four points on the unit hemisphere. In: Janičić, P., Kovács, Z. (eds.) Electronic Proceedings in Theoretical Computer Science, vol. 352, pp. 27–40 (2021)
13. Zeng, Z., Zhang, J.: A mechanical proof to a geometric inequality of Zirakzadeh through rectangular partition of polyhedra. J. Syst. Sci. Math. Sci. **30**(11), 1430–1458 (2010)
14. Zirakzadeh, A.: A property of a triangle inscribed in a convex curve. Can. J. Math. **16**, 778–786 (1964)

Author Index

Printed in the United States
by Baker & Taylor Publisher Services